Astronomy & Astrophysics

Astronomy & Astrophysics

Edited by Morton S. Roberts

The American Association for the Advancement of Science

Library of Congress Cataloging in Publication Data
Main entry under title:

Astronomy and astrophysics.

Includes index.
1. Astronomy—addresses, essays, lectures. 2. Astrophysics—addresses, essays, lectures.
I. Roberts, Morton S.
QB51.A855 1985 520 85-13380
ISBN 0-87168-311-3
ISBN 0-87168-275-3 (pbk.)

The material in chapters 1 through 24 has been updated and corrected from the original articles that appeared in ***Science*** issues published between 1982 and 1984. ***Science*** is the official journal of the American Association for the Advancement of Science.

AAAS Publication No. 84-5

Printed in the United States of America

1333 H Street, NW, Washington, DC 20005

Contents

Preface vii

Introduction

Astronomical Frontiers 1
Morton S. Roberts

I. Solar System

1. The Sun's Influence on the Earth's Atmosphere and Interplanetary Space 7
J.V. Evans
2. Solar Flares, Proton Showers, and the Space Shuttle 25
David M. Rust
3. Cosmic-Ray Record in Solar System Matter 41
Robert C. Reedy, James R. Arnold, Devendra Lal
4. Ultraviolet Spectroscopy and the Composition of Cometary Ice 59
Paul D. Feldman

II. Structure and Content of the Galaxy

5. The New Milky Way 77
Leo Blitz, Michel Fich, Shrinivas Kulkarni
6. The Most Luminous Stars 92
Roberta M. Humphreys and Kris Davidson
7. Chromospheres, Transition Regions, and Coronas 105
Erika Böhm-Vitense
8. Interstellar Matter and Chemical Evolution 120
M. Peimbert, A. Serrano, S. Torres-Peimbert
9. The Formation of Stellar Systems from Interstellar Molecular Clouds 131
Robert D. Gehrz, David C. Black, Philip M. Solomon
10. Binary Stars 143
Bohdan Paczyński
11. Dynamics of Globular Clusters 154
Lyman Spitzer, Jr.

12. The Magnetic Activity of Sunlike Stars — 171
Arthur H. Vaughan

13. On Stars, Their Evolution and Their Stability — 188
S. Chandrasekhar

III. Galaxies and Cosmology

14. The Most Distant Known Galaxies — 209
Richard G. Kron

15. Galactic Evolution: A Survey of Recent Progress — 219
K.M. Strom and S.E. Strom

16. The Rotation of Spiral Galaxies — 241
Vera C. Rubin

17. Quasars and Gravitational Lenses — 254
Edwin L. Turner

18. Windows on a New Cosmology — 263
George Lake

19. The Origin of Galaxies and Clusters of Galaxies — 275
P.J.E. Peebles

20. Jets in Extragalactic Radio Sources — 290
David S. De Young

21. The Quest for the Origin of the Elements — 301
William A. Fowler

22. The Dark Night-Sky Riddle: A "Paradox" That Resisted Solution — 332
E.R. Harrison

IV. Instrumentation

23. Radio Astronomy with the Very Large Array — 345
R.M. Hjellming and R.C. Bignell

24. Space Research in the Era of the Space Station — 359
Kenneth J. Frost and Frank B. McDonald

About the Authors — 371

Index — 373

☼ Color Plates — following page 240

Preface

Each week the pages of *Science* carry reports on current research, news from the scientific and political scenes, information on current publications and, under the heading "Articles," a series of minireviews of active areas of science research. Prepared by leaders in their fields, these reviews are like having permanent records of colloquia presentations: slides, viewgraphs, and all.

I have selected two dozen such articles, including two Nobel Prize Lectures, for this volume. All are astronomically oriented, and in their subject matter they cover extremes in distance from the solar system to quasars at the very edge of the observable universe. Research techniques and instruments range over such diverse topics as searching for proton decay, the Very Large Array (VLA) of radio telescopes, and the planned Space Station as a platform for future experiments.

The articles were selected from issues of *Science* published between 1982 and 1984. Within the limitations of a reasonable number of such articles, the primary criteria were astronomical orientation and breadth in coverage. Just as the pages of *Science* are intended for the entire scientific community, so it is hoped that this compilation will appear to a community wider than that of astronomers and astrophysicists.

Morton S. Roberts
Charlottesville, Virginia
29 April 1985

Astronomy & Astrophysics

Introduction

Astronomical Frontiers

Morton S. Roberts

The imagination of nature is far, far greater than the imagination of man. For instance, how much more remarkable it is for us all to be stuck—half of us upside down—by a mysterious attraction to a spinning ball that has been swinging in space for billions of years, than to be carried on the back of an elephant supported on a tortoise swimming in a bottomless sea.—Richard P. Feynman

The articles in this volume separate into four major categories: (1) solar system, (2) the structure and content of the Galaxy, (3) extragalactic, and (4) instrumentation. Each article is self-contained, yet a broad, coherent and contemporary picture of our astronomical Universe emerges.

From solar system to local stellar environment to our Milky Way galaxy populated by a remarkable collection of variable and exploding stars, dwarf, normal, giant and even supergiant stars, rapidly spinning neutron stars and clusters of a hundred thousand stars are bound together in a beautifully spheroidal arrangement. All of this is a galaxy which itself is of several components: a disk with a central bulge surrounded by a tenuous halo, and within this disk is a prominent spiral pattern of interstellar dust and gas defining the locus of ongoing star formation.

Many other galaxies lie beyond our own. Their morphology is varied, though most can be ordered into a simple sequence. Their environment ranges from the relatively rare isolated system to groupings of ever increasing number: pairs, small groups—our Galaxy is in such a group—to clusters of large population. The morphology of the constituent galaxies is related to the richness of the environment. Clusters themselves appear to cluster into entities of over a hundred million light years in size, aptly invoking the title of superclusters.

In this extragalactic world are interacting galaxies, the result of tidal encounters. There is evidence of mergers, cannibalism on an heroic scale. Other systems are seen to spew out gigantic structures visible as radio jets that dwarf the dimensions of the parent optical galaxy. Some argue that quasars are born in such ejections. Another view, that quasars are in the central part of galaxies lit up by a surge of energy, seems more likely as detailed imaging of quasars finds them surrounded by luminous fuzz that appears to have the properties of a faint background galaxy. The "quasi-stellar" aspect of quasars is giving way to modern imaging techniques.

A common theme in many of these articles is an attempt to describe the life cycle of stars, interstellar matter, galaxies, and even the Universe as a whole. Such a goal is not a display of hubris, it is achievable. We see stages of such evolutionary history throughout the sky. The death throes of a supernova, clusters of stars in various stages of evolution, dust shrouded infrared sources, and white dwarfs—beginnings and ends, bright superluminous stars with such prodigious energy consumption that they can only have been born recently. The very expansion of the Universe itself clearly marks a beginning, descriptively called the "Big Bang," with the remnant background radiation of that event surrounding us.

A basic topic is only touched upon in this compilation. It is the number (or number density) of objects which populate the universe. I will discuss it briefly here together with two related areas of contemporary astronomical research: the determination of the rate of expansion of the Universe, measured by the Hubble Constant, and the problem of the "missing" mass.

There are several relatively high precision determinations of the Hubble Constant, H_o, which relates expansion velocity, V, and distance, D, such that $V = H_oD$, that is, the expansion velocity increases linearly with distance. There are two controversies in this area: The first relates to the value of H_o. The errors in its determination are generally quoted at 10 to 20 percent, but there are two sets of well-supported and well-documented determinations that differ by a factor of two. It is likely that $50 < H_o < 100$ km s^{-1} Mpc^{-1}. This range changes distances by a factor of two, luminosities by four, and volumes by eight. The inverse of the Hubble Constant has the dimensions of time. It measures the expansion age of the Universe. Ages derived by other techniques must obviously be no larger than H_o^{-1}. Such ages present an important constraint on the value of H_o or, conversely, the correct value of H_o constrains other techniques of determining the age of the Universe.

Another controversy concerns the interpretation of the redshift of galaxian spectral lines from their laboratory wavelengths, $\lambda - \lambda_o = \Delta\lambda$. It is widely held that this displacement, $\Delta\lambda$, is Doppler in origin due to the expansion of the Universe. Hubble himself, however, was very cautious about this interpretation. In his Rhodes Memorial Lectures delivered at Oxford in 1936, Hubble (*1*) states "Thus the familiar interpretation of redshifts as velocity-shifts leads to strange and dubious conclusions; while the unknown, alternative interpretation leads to conclusions that seem plausible and even familiar."

The largest measured redshift, $\Delta\lambda/\lambda_o \equiv z$, at the time of the Hubble's remarks was 0.13, a value derived for a galaxy in the Boötes cluster by Hubble's collaborator, Milton Humason. Some twenty years later the largest redshift, z, had been extended to 0.20 (*2*). The discovery and study of quasars with redshifts of 0.1 to close to 4, imply large distances and remarkably high luminosities. It also reopened the question as to the Doppler interpretation of these redshifts. If not all, was at least some fraction of z due to some other mechanism; in Hubble's words, an "unknown, alternative interpretation"? It is not uncommon at scientific meetings and discussions to hear the cautionary remark, "If these redshifts measure expansion then. . . ." We keep questioning and testing the paradigms of cosmology, but are not at the stage of needing a Bohr or a Planck to rescue the situation.

The same spectroscopic studies that measure the expansion redshift also measure motions within individual galaxies and the motion, within a cluster, of one galaxy with respect to another. The analysis of motions within a galaxy will yield the mass of that system which, together with its intrinsic luminosity, yields a useful and informative parameter, the mass-to-luminosity ratio. If galaxies are similar, or nearly so, then this ratio will allow the mass of any system to be determined given its luminosity. Thus, the "luminous" mass of a cluster of galaxies can be determined by just adding up the luminosities of the constituent galaxies.

Another method of measuring the mass of this cluster is via the virial theorem using the motions of the member galaxies to evaluate the kinetic energy and their separations to derive the potential energy. This will yield a "dynamical" mass of the cluster. This latter approach, the dynamical, always, or nearly always, yields a higher mass than the luminosity method. This is the case if the material is in our galaxy (using stars rather than galaxies as the test particles) or if it be the masses of pairs of galaxies, groups of galaxies, or giant, rich clusters of galaxies. In fact, the discrepancy gets larger as the system gets larger.

"Missing mass" is a catchy title for this discrepancy, but is misleading. A more appropriate, more descriptive title is "nonluminous matter," for the comparison is between a dynamical determination and a luminosity-derived mass. The difference is such that of order 90 percent of the material in the Universe is nonluminous. This surprising result—that we "see" so little of the Universe—is referenced in several of the papers in this volume. There are many proposed explanations which generally invoke various forms of dark matter, for example, neutrinos, planet-sized bodies, black holes. The current state of "our ignorance is encapsulated by the statement that there are more than 70 powers of ten uncertainty in the masses of the entities that constitute more than 90 percent of the content of the Universe." (*3*)

We have counted objects (stars, galaxies) in the above discussion. What do we know of their number in general? The quantity we seek is the number of objects within luminosity bins and within a specified space volume. Such a description of the number density is called the "luminosity function." In concept, simply described and derived, in practice, be it for stars or galaxies, difficult and poorly known, especially at fainter luminosities. Luminosity functions have been derived for stars, clusters of stars, galaxies and clusters of galaxies. Each is smoothly varying and carries with it the imprint of its origin and subsequent evolution. Each has a constraint that the integral luminosity density be less than infinite as the luminosity tends to zero. The high slope of the number count of bright galaxies and galaxy clusters changes to avoid such a catastrophe.

The bright end of the globular cluster luminosity function matches that of the faint end of the galaxy luminosity function. This faint end is well described by an exponential, a form suggested on theoretical grounds by Fritz Zwicky in 1957 (*4*). He argued that the galaxian luminosity function does not have a maximum, that it should be extended to "individual intergalactic stars." We are hard pressed to identify such stars because of their extreme faintness, but one can smoothly pass from the combined luminosity function described above to that for the stars in our galaxy after allowance for the evolution of the very brightest stars and for a density normalization.

Over a range of 10^{19} in luminosity the slope is always positive, gradually lessening at the fainter stellar magnitudes. A vast number of intrinsically faint galaxies, though the most common variety, are yet to be recognized. It is an observationally difficult task, and thus far, knowledge of their high frequency results from searches in groups and within the Virgo cluster.

There is no theoretical reason to expect a sudden decrease in the number of faint galaxies, and none is observed. There is however a natural limit to the faint end of the stellar luminosity function. At very low masses, less than a few hundredths of that of the sun, nuclear burning ceases. Here is a limit to the source of star light as well as to the luminosity function. A "mass function" would more aptly describe the distribution of objects in space. Perhaps the nonluminous mass (*5*) resides in such cold stars as the continuing extension of the luminosity function to (close to) zero luminosity.

References and Notes

1. E. Hubble, *The Observational Approach to Cosmology* (Oxford Univ. Press, Oxford, 1937), especially p. 43.
2. M. Humanson, in *Vistas in Astronomy*, A. Beer, Ed. (Pergamon Press, London, 1956), vol. 2, p. 1620.
3. M. J. Rees, in *Early Evolution of the Universe and its Present Structure*, G. O. Abell and G. Chincarini, Eds. (Reidel, Dordrecht, Netherlands 1983), p. 299.
4. F. Zwicky, *Morphological Astronomy* (Springer-Verlag, Berlin, 1957), especially p. 222. See also *Physical Review* **61**, 489 (1942).
5. S. S. Kumar, *Astrophysics and Space Science* **17**, 219 (1972).

Part I
Solar System

1. The Sun's Influence on the Earth's Atmosphere and Interplanetary Space

J. V. Evans

The sun's energy comes from thermonuclear reactions, such as the synthesis of helium from hydrogen, which proceed with a loss of mass and liberation of energy. This accounts for the vast release of energy that must have occurred for many millennia and which might be expected to continue well into the future. There is, however, no certainty that the rate of energy release will remain constant; any variations would be of considerable scientific interest and profoundly important for man's well-being.

Solar-terrestrial research deals with the response of the earth's upper atmosphere, ionosphere, magnetosphere, and intervening space to the emission of energy from the sun and to any fluctuations therein. Excluded by tradition has been the troposphere, the lowest 10 kilometers of the earth's atmosphere, where the earth's weather occurs (*1*). The earth's ionosphere is formed at altitudes above about 70 km (*2*). Here the gas becomes partially ionized as a result of the absorption of solar ultraviolet and extreme ultraviolet (EUV) radiation, forming a region that affects the propagation of radio waves. The magnetosphere is that region of space surrounding the earth in which the earth's magnetic field dominates solar or other fields. It extends great distances in the antisunward direction but only 10 to 12 earth radii (R_E) on the sunward side.

As early as the 18th century it was noted that changes on the earth (such as magnetic field fluctuations and the appearance of visible auroras) occur that seem linked to changes in the sun (such as the appearance of dark sunspots). Only since the advent of space research has it been possible to establish the coupling mechanism and fully study the earth's upper atmosphere, the magnetosphere, and interplanetary space.

The National Academy of Sciences has recently released a report on the field of solar-terrestrial research and its development during the 1980's (*3*). The report argues (i) that understanding this vast system entails more than the study of its component parts because of the existence of complex linkages and feedback mechanisms and (ii) that there should be appropriate recognition of the unity of the subject by the scientific community, the funding agencies, and teachers at universities. In this chapter some of these connections are discussed and an attempt is made to sketch what is known about the solar-terrestrial system.

Solar Processes

The energy produced in the sun's core by thermonuclear processes is transported outward by radiation. It undergoes successive transitions to longer wavelengths as high-energy gamma rays are absorbed and reradiated in the form of x-rays and as these in turn are absorbed and reemitted in the form of EUV radiation. At about five-sixths of the distance to the visible surface, or photosphere, the ionization of the gas has decreased to the point where it is convectively unstable. The transport of energy to the surface through much of this outer zone is by means of turbulent convection, not radiation. The convection zone organizes the solar magnetic field that reaches the surface and is a source of mechanical energy; both phenomena have important consequences for the sun's atmosphere (*4*).

In the convection zone the temperature is low enough that some of the free electrons can be recaptured by protons or other nuclei to form atoms. These are more opaque to radiation than the highly ionized gas below, and this leads to an increased temperature gradient. An upward-moving parcel of gas thus finds itself in a region of decreasing temperature where it is heated by energy released by further electron recombination and its upward motion is accelerated; the reverse holds for a parcel moving downward. The whole process is analogous to the convective processes in thunderstorms. The transport of energy again becomes radiative below the visible surface of the sun, but some of the convective motions penetrate to the photosphere, where they are visible as rising and falling granular structures.

In the chromosphere, a region extending only a few thousand kilometers above the surface, the temperature rises rapidly from about 5,000 to 500,000 K. There is a further increase to 1 million to 2 million K over a distance of about half a solar radius in the corona which constitutes the outermost part of the sun's atmosphere. The rise in temperature in the chromosphere and corona cannot be produced by radiative transfer of energy from the lower cooler layers; the energy is thought to be supplied by mechanical power in the motions of the convection zone. The chromosphere and corona are the source of the ultraviolet and x-ray radiation emitted by the sun. The visible light scattered by the corona can readily be photographed during an eclipse (or from a satellite above the earth's atmosphere). Near sunspot minimum the corona often exhibits considerable gross structure, with irregular equatorial extensions, and may be nearly absent above the poles. Near sunspot maximum the corona appears to be much more structured and jagged, but overall exhibits greater spherical symmetry.

In the convection zone the energy density of the gas motion greatly exceeds that of the solar magnetic field, and thus the motions tend to organize the fields. (Whenever the energy density of the ionized particles in space greatly exceeds that of the magnetic field, the field must follow the particle motions as if frozen into the medium.) The reverse is true in the chromosphere and corona, where the motion of the completely ionized gas is confined and controlled by the magnetic field. Sunspots are thought to be cool regions where the magnetic fields at the photosphere are strong enough to influence the convective motions and upward flow of heat. The 11-year cycle in their number and latitude on the disk may be a result of an observed differential rotation of the convection zone with latitude; the equatorial regions rotate faster than the poles. This differential motion is thought to twist the sun's dipole field until loops or knots of field lines break through the photosphere. Very large loops are sometimes seen on the limb of the sun. These loops represent magnetic "bottles" capable of containing high-density plasma. Solar flares are thought to result when the plasma containment collapses as a result of the merging of magnetic fields in regions where they are strong, complex, and highly stressed. The magnetic energy released is transformed into radiation through heating and acceleration of the trapped ionized gas. In the very largest flares, it is estimated that more than 10^{32} ergs can be released in a period of minutes, but this output remains small compared to the sun's overall luminosity ($\sim 4 \times 10^{33}$ ergs per second).

The Solar Constant

The solar constant is the total electromagnetic energy radiated by the sun at all wavelengths per unit time through a given area at the mean distance of the earth; the contemporary value is $S = 1.37 \pm 0.02$ kilowatts per square meter (*5*). The term solar constant is somewhat misleading since the total energy fluctuates by a few tenths of 1 percent. About 99 percent of the sun's energy is emitted at wavelengths between 0.3 and 10 micrometers from the photosphere (which can be characterized as a blackbody radiator with a temperature of roughly 5800 K).

Stellar structure theory suggests that a G-type star such as the sun can be expected to exhibit a change in luminosity of approximately 30 percent during its main-sequence lifetime ($\sim$ 10 billion years) (*6*). Evidence that life has existed on the earth without interruption for 2.5 billion years implies that much of the planet has had temperatures between 0° and 100°C throughout this period and places constraints on the rate and extent of secular changes in the solar constant (*7*). There is, however, evidence for climatic changes, such as glaciations, that were striking departures from the norm. Since the cause of these is not well understood, small fluctuations in solar output cannot be ruled out solely on the basis of the climatic record. Computer model studies suggest that fluctuations in the solar constant exceeding a few tenths of 1 percent for a long period would have significant climatic effects (*8*).

Ground-based measurements of the solar constant are limited by the highly varying and uncertain effects of the atmosphere, which influences the transmission of visible light and blocks out most of the energy below 0.3 μm and above 2.5 μm. Despite these difficulties, C. G. Abbot and co-workers at the Smithsonian Astrophysical Observatory made daily measurements of the solar constant from several stations between 1923 and 1952 and obtained a nearly continuous set of data (*9*). Analysis of the data secured at the two best stations rules out any sustained variation of the flux reaching the ground (about 80 percent of the total) exceeding 0.17 percent over 30 years (*10*). However, the removal of slow trends that may be of instrumental origin reveals a significant fluctuation ($\sim$ 0.1 percent) with a 28-day period (*11*). The solar constant tends to be low when parts of the disk are covered with dark sunspots and high when the disk is covered by bright features (faculae), which generally are also associated with a locally strong magnetic field.

The latter finding has been confirmed

by recent spacecraft observations of the sun. High-precision measurements of total solar irradiance made with the Solar Maximum Mission satellite for 150 days showed two decreases of > 0.1 percent below the mean, each lasting about 1 week (*12*). These same two decreases were seen independently by the Nimbus 7 satellite (*13*) and are shown as points a and b in Fig. 1A. The Nimbus 7 data appear noisier than those reported by Willson *et al.* (*12*) because of limitations in the data sampling system, but extend earlier in time and show a decrease exceeding 0.3 percent (point c in Fig. 1A). Figure 1B shows the variation that would be expected at the earth if the radiation blocked by sunspots is not emitted simultaneously elsewhere from the disk and if allowance is made for the position of the spot relative to the central meridian (*14*). This function is capable of predicting all the major dips, but not their relative amplitudes.

It seems extremely unlikely that there is any variation in the rate of energy production from nuclear fusion in the sun's core over these time scales. Thus, what is being seen must be a temporary reduction in the rate at which heat is convected through the outer layers of the sun. The "missing" energy presumably is stored as increased thermal and potential energy and released slowly after a spot disappears (*14*).

Ultraviolet and X-ray Emissions

Approximately 1 percent of the total solar irradiance lies in the ultraviolet at wavelengths between 3000 and 1200 angstroms. Over much of this wavelength interval the spectrum appears to be that of a continuum with superimposed absorption lines. This energy does not reach the earth's surface but is absorbed by ozone between 20 and 50 km and by molecular oxygen at still higher altitudes. Ultraviolet at wavelengths shorter than 2424 Å can photodissociate O_2 into its constituent oxygen atoms; these in turn oxidize other O_2 molecules to produce ozone. Thus the energy in this wavelength region is important for controlling the oxygen chemistry in the earth's atmosphere.

Figure 2 shows, as a function of wavelength, the level at which the ultraviolet penetrating the earth's atmosphere from an overhead sun has been attenuated by a factor of $1/e$. (This is also the level at which the absorption per unit height interval is greatest.) At wavelengths shorter than about 1500 Å, emission lines of gas in the chromosphere and corona dominate the spectrum. The EUV radiation at wavelengths between 300 and 1200 Å is absorbed by O_2, O, N_2, and N in the earth's atmosphere, chiefly above 90 km (Fig. 2). Radiation at wavelengths below 1027, 911, and 796 Å can photoionize O_2, O, and N_2, respectively. Thus, EUV radiation and x-rays govern the production of free electrons in the earth's ionosphere. In all, only about 1 to 3 milliwatts per square meter, or one-millionth of the total solar flux, is absorbed above 120 km, but because the atmospheric density is so low, this energy is capable of raising the temperature of the uppermost levels of the earth's neutral atmosphere (above 300 km) to 700 to 1500 K. The diurnal cycle of heating and cooling creates pressure variations that set up winds at this level with speeds on the order of 100 meters per second.

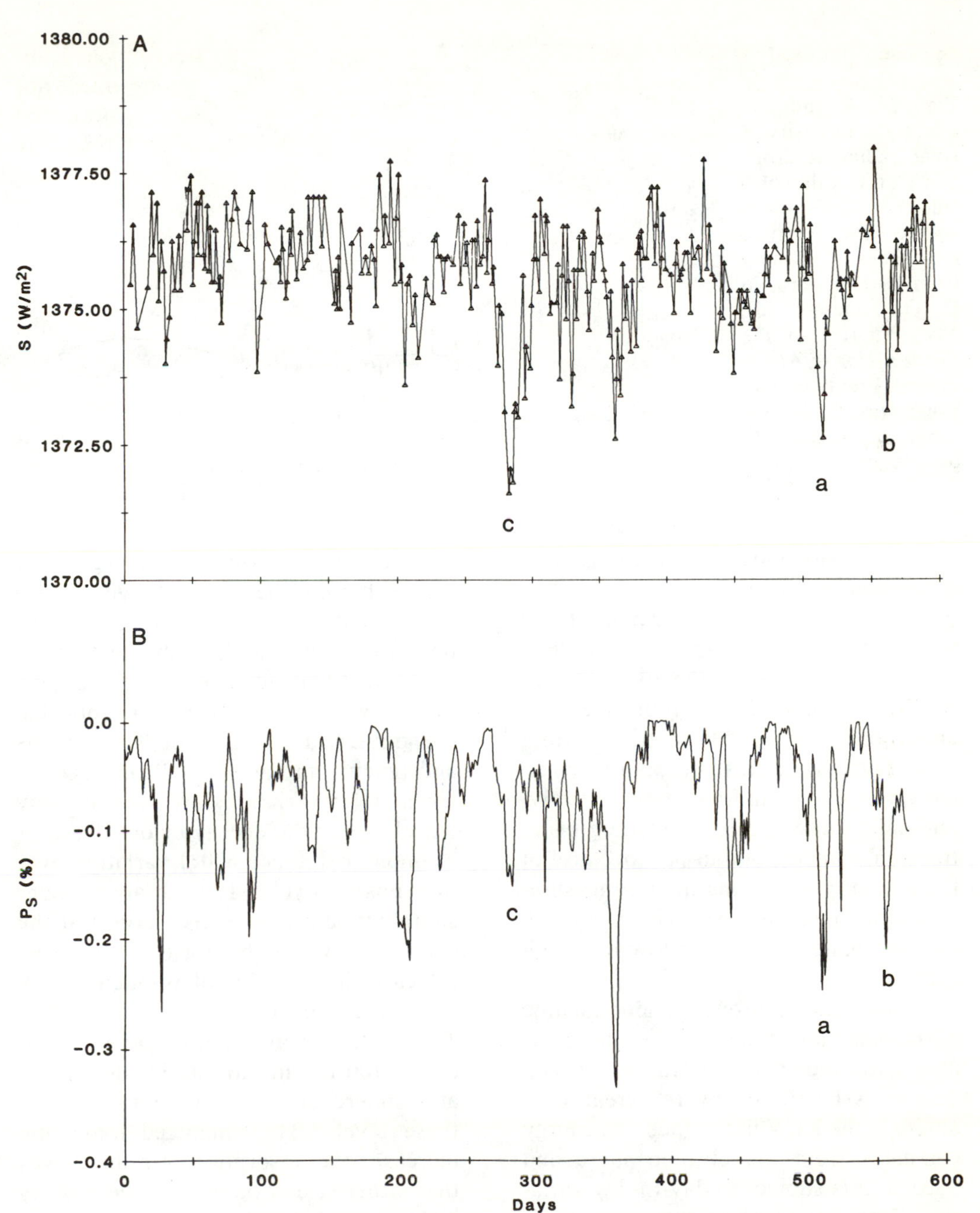

Fig. 1. (A) Nimbus 7 measurements of total solar irradiance from 15 November 1978 through June 1980 (*13*). The dips a and b were detected in observations made with the Solar Maximum Mission satellite (*12*). Point c, the deepest depression seen to date, occurred in August 1979. (B) Predicted variation (P_s) when allowance is made for the fraction of the sun's disk obscured by sunspots. [Reprinted with permission of P. Foukal (*14*)]

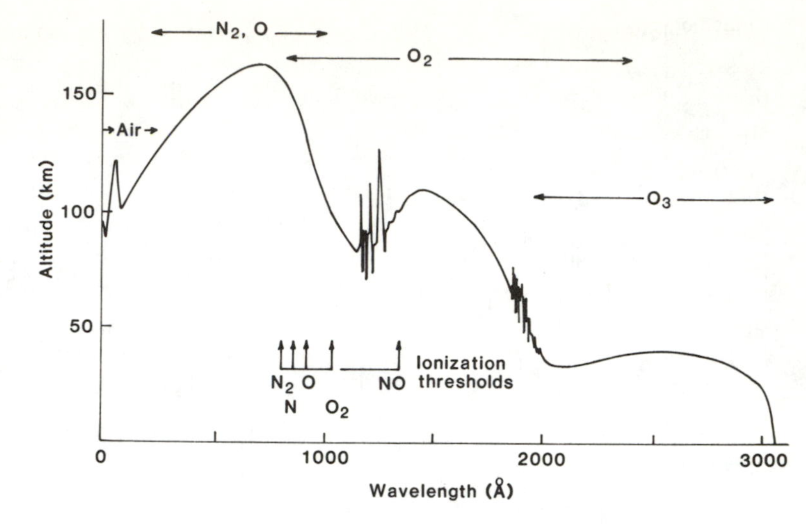

Fig. 2. Altitude at which the intensity of solar radiation drops to $1/e$ of its value outside the earth's atmosphere for vertical incidence. [Adapted from L. Herzberg, in C. O. Hines et al., eds., *Physics of the Earth's Upper Atmosphere* (Prentice-Hall, Englewood Cliffs, NJ, 1965), reprinted with permission]

Recent satellite investigations and ground-based radar studies have led to better understanding of the neutral and ion photochemistry (*15*) and dynamic behavior (*16*) of the region above 90 km, the thermosphere. Less has been learned about the regions below the thermosphere, in part because the photochemistry appears to be more complex (involving many more species, including many trace substances) and in part because of the difficulty in making in situ measurements at these altitudes, which are too high for balloons and too low for satellites.

Despite the improved understanding of the behavior of the thermosphere, our ability to predict the neutral or electron density is hampered by the great variability in the EUV flux. Since this energy emanates from the chromosphere and corona, it is subject to day-to-day variations associated with solar activity, such as the formation of bright plages, and with cyclical variations caused by the sun's rotation (period of 27 days) as active centers grow and decay. The amount of activity is itself subject to the 11-year sunspot cycle, which introduces about a 2:1 variation in the energy input to the thermosphere over a cycle and a corresponding variation in mean temperature. Direct satellite observations of solar spectral irradiance (*17*) have shown that the variations over the sunspot cycle are greatest at the shorter wavelengths and reach a factor of 10:1 near 300 Å. Even larger variations are seen at x-ray wavelengths; the energy below 10 Å, for example, exhibits a 500:1 variation over the sunspot cycle. The variability over shorter time scales is also largest at the shortest wavelengths; for example, brief increases in x-ray flux of as much as 100-fold have been observed during flares. These short wavelengths penetrate to below 100 km in altitude in the earth's atmosphere and contribute to ionizing these levels. The enhanced ionization increases the absorption of radio waves that otherwise would be reflected by higher levels of the ionosphere, creating a shortwave blackout.

The Solar Wind

The earth's upper atmosphere derives a further input of energy from the solar wind, a general outflow of ionized particles from the sun (*18*). The first indications of such a particle flux came from studies of comet tail behavior (*19*); later it was shown, on theoretical grounds, that the solar wind is a natural consequence of the conditions in the corona (*20*). The continued increase in coronal temperature out to large distances from the sun causes the gas pressure to fall off less rapidly than the sun's gravitational attraction. This leads to a situation in which there can be no statically stable atmosphere that is gravitationally bound; instead, there must be a continuous outward expansion of the gas. The gas speed increases and becomes supersonic at a distance of a few solar radii. Thereafter the speed and direction remain constant, as there are no forces large enough to change them. Eventually the gas must undergo a shock transition at a distance where the dynamic pressure of the wind drops to the level of the pressure of the gas in interstellar space. This transition is thought to lie beyond the solar system; Pioneer 10, now midway between the orbits of Uranus and Neptune (about 25 times the distance of the earth from the sun), has not encountered any evidence of it.

At the orbit of the earth, the solar wind plasma consists principally of protons and electrons, with a small percentage of helium and other ions, all fully ionized. The number density varies from 2 to 100 ions per cubic centimeter, with a mean of about 10. Likewise, the bulk velocity is found to be variable over the range 200 to 800 km/sec, with a mean of about 450. The plasma carries with it the disordered solar magnetic field in the corona. At the orbit of the earth, the intensity is typically about 5 gammas (1 gamma = 10^{-5} gauss or 10^{-9} tesla) (*21*). This weak, disordered solar magnetic field influences the intensity of galactic cosmic rays reaching the earth. The gas temperature typically is about 10^5 K, but varies from 10^4 K at quiet times when the bulk flow is low to 10^6 K when conditions are disturbed and the bulk speed is high. On average, the total thermal energy is only about 1 percent of the energy of the bulk motion.

Increasingly sophisticated in situ measurements of the solar wind have been made since the early 1960's, when the first deep-space satellites were launched. The early observations were, for the most part, concerned with the microstructure of the local plasma properties that describe the overall flow (*22*). Yet, even before the first satellite observations, the existence of large structures, usually termed streams, had been surmised on the basis of observed variations in the earth's magnetic field (*23*). Recent satellite observations have shown that the solar wind is considerably more complex than might be supposed from these ground-based studies. One does not see discrete streams embedded in steady, low-speed flows. Rather, the latter are few and rarely persist for more than 2 days, while there are several kinds of high-speed flows that might be called streams. Many of these seem to be contiguous or even superimposed on one another (*24*). In the simplest cases, the solar wind velocity increases by several hundred kilometers per second over a period of 1 to 2 days and then decreases monotonically over 2 to 7 days. Compound streams show more complex variations of velocity with time and may be the result of the interaction of two or more simple streams. It is now believed that these recurrent streams originate from a relatively small number of localized regions of the sun known as coronal holes. These regions of relatively low

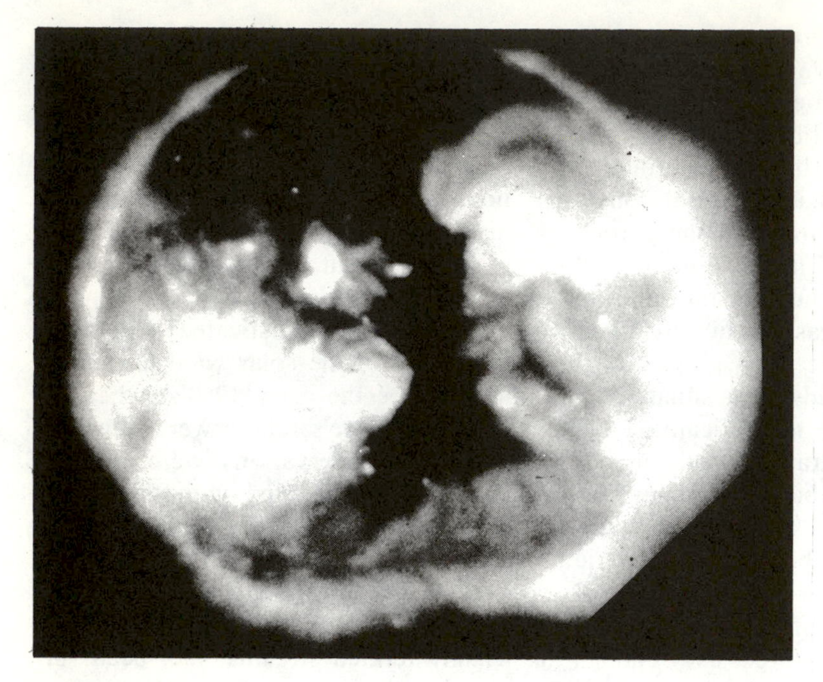

Fig. 3. Image of the sun taken in soft x-rays on 1 June 1973. A large coronal hole is visible as a dark area extending from the north pole down the center of the disk. [Courtesy of the Solar Physics Group, American Science and Engineering, Inc.]

density and temperature are best seen in images of the sun made at x-ray wavelengths (Fig. 3) (*25*). Coronal holes were first observed at a time near the sunspot minimum. The holes over the poles seemed to be semipermanent; those at lower latitudes had lifetimes of several solar rotations (*26*). In recent years, during which sunspots have been approaching their maximum, the polar holes have become less prominent and the mean lifetime of the equatorial holes has decreased to only one or two rotations.

No direct measurements have been made of the magnetic field in coronal holes, but observations of emission features in their vicinity suggest that the field is open and diverging (*27*). Timothy *et al.* (*28*) developed a plausible model for the existence of coronal holes. According to this model, holes form whenever the magnetic fluxes from bipolar magnetic regions interact to produce large regions of locally unbalanced flux, that is, regions that are dominated by a single polarity (this automatically includes the polar regions). In such regions, the field lines tend to extend far into the corona and do not impede the outward flow of plasma, which eventually succeeds in convecting the field away.

Efforts have been made to model the flow through coronal holes (*29*). Thus far it has been necessary to adopt some model for the streamlines a priori. It is generally assumed that the particles and the field lines diverge rapidly above the hole (*30*). Although the solar wind plasma moves radially outward from the sun, a stream appears to occupy a spiral path because the plasma in it moves with constant radial velocity while its source rotates with the sun. As may be deduced from Fig. 3 and from photographs taken during eclipses, semipermanent holes are thought to exist at the north and south poles near sunspot minimum when the sun's polar fields are strong. These polar holes are important in controlling the properties of the solar magnetic field at the orbit of the earth.

As observed at the earth, the solar magnetic field carried outward by the solar wind may be directed away from the sun for periods of many days. Comparable intervals then follow, corresponding to many degrees of solar longitude, during which the opposite polarity is observed (*31*). These intervals of similarly directed field have been termed sectors. Except for occasional short intervals in which the polarity is not well-defined, two, four, and occasionally six major sectors are observed per rotation. The transition from one sector to the next, the sector boundary, often precedes the arrival of a fast stream. While satellite measurements have not been made at very high heliocentric latitudes, there are indications that this sector structure disappears some distance above or below the solar equator; the dominant polarity at high latitudes corresponds to the polar magnetic field of the sun in the hemisphere in which the observations are made (*32*). According to this model (Fig. 4), the magnetic fields from the polar coronal holes are kept apart by the outward flow of ionized gas, which forms a thin current sheet. At the earth one dominant polarity or the other is seen depending on whether the observer is above or below the current sheet. It also appears that the current sheet is inclined with respect to the solar equator, so that at the earth two sectors are seen as the sun rotates. Four or six sectors can appear when the current sheet itself develops large-scale warps or undulations (a condition likened to a ballerina's skirt). The warps in the current sheet appear to be caused by the fast streams, which carry out fields from coronal holes located at more central parts of the disk.

Increased disturbance of the earth's magnetic field and increased energy input into the earth's atmosphere are associated with sector boundary crossings. These changes appear to be a result of the fast streams overtaking the slower solar wind ahead and compressing it, creating interaction regions in which the plasma density increases. The magnetic field carried by the plasma is also compressed in the interaction regions and has a tendency to turn southward and then northward. This southward turning of the field is an important parameter governing the interaction of the solar wind with the earth.

Somewhat less is known about flare-produced streams. It is believed that the flare initiates a large, explosive release of coronal material which arrives at the earth after 2 to 3 days as a shock front that has swept up the plasma and magnetic field ahead of it. Behind the shock is a shell of anomalously high He^{++} abundance. The highest flow speed occurs behind the He^{++} shell. It has been suggested that magnetic fields are looped back to the site of the flare, forming an ever-expanding magnetic bottle.

The Magnetosphere

The flow of solar wind around the earth has been a subject of considerable study since the space program began. This research culminated in the International Magnetosphere Study (1976 to 1979), a program involving spacecraft and ground-based observations by many nations (*33*).

The earth and its associated magnetic field represents an obstacle in the path of the solar wind. The result of the interaction of the two magnetized plasmas is to confine the extent of the earth's magnet-

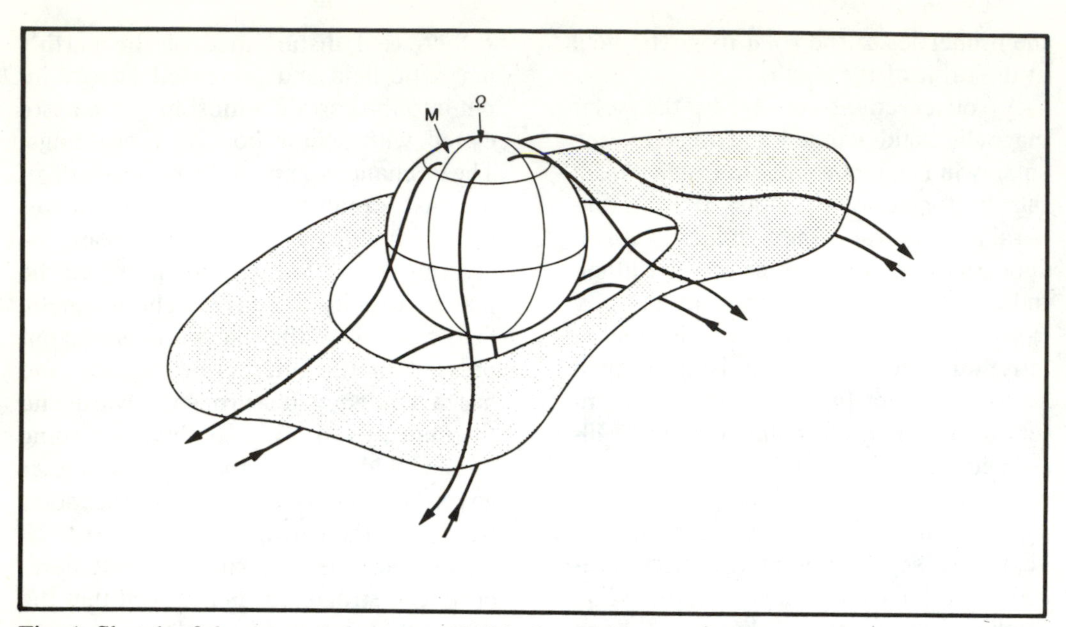

Fig. 4. Sketch of the current sheet responsible for the sector structure. The axis of the current sheet, M, appears to be tilted with respect to the solar rotation axis, Ω (*32*).

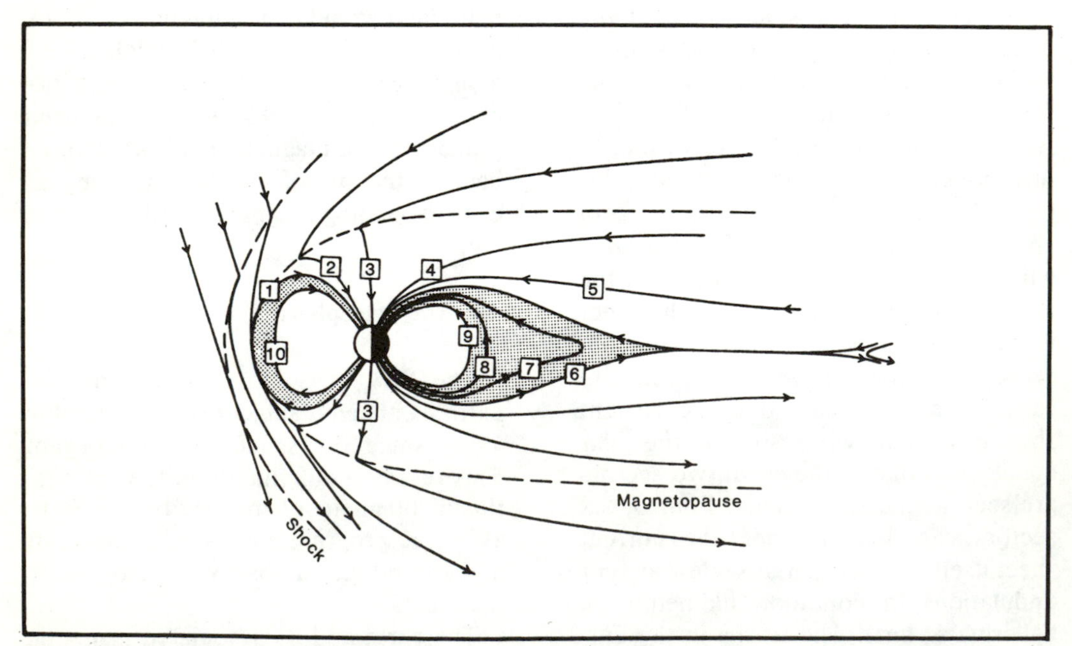

Fig. 5. Sketch illustrating the distorted configuration of the outer parts of the earth's magnetic field and how this may be maintained by merging of field lines of solar and terrestrial origin. The numbers indicate in sequence the position of the field line after it has merged on the dayside (*35*).

ic field in space. The volume occupied by the earth's field is known as the magnetosphere; its outer boundary is the magnetopause (*34*).

As a result of compression of the earth's magnetic field by the solar wind, the magnetopause is closest to the earth on the sunward side, where it typically extends about 10 R_E (Fig. 5). On the night side the magnetosphere stretches out like a comet's tail for at least 100 R_E, or well past the orbit of the moon.

There has been considerable controversy over the mechanism responsible for the distorted configuration of the earth's magnetic field. One view is that it results from some form of "viscous" drag applied by the solar wind to the magnetopause boundary, presumably through wave-particle interactions. The other view is that the distorted configuration results from the conversion of magnetic energy to particle kinetic energy through merging on the dayside between field lines carried by the solar wind (the interplanetary magnetic field, or IMF) with those of the earth. The merging should be especially effective when the IMF is directed south (Fig. 5) (*35*). In this case, the field lines of the IMF have the correct orientation to sever and connect with the lines of the earth's magnetic field. The effect of the merging is to connect the terrestrial field lines to the solar wind flowing past the earth, which carries them in an antisunward direction across the poles and into the tail, where they become very extended. Eventually, the terrestrial field line reconnects with itself, leaving the IMF field line to be carried off beyond the orbit of the earth. This picture has received considerable support from observations showing that (i) the amount of energy coupled from the solar wind into the earth's atmosphere increases when the IMF turns southward (*36*) and (ii) the pattern of current flow produced in the upper atmosphere at high latitudes depends on whether the IMF is directed toward or away from the sun (*37*). However, the distorted field configuration and weak auroral activity persist even when the IMF is directed northward; therefore, it may be that the viscous mechanism is also operative (*38*).

There cannot be a continuous accumulation of magnetic field lines on the nightside; once the field lines have reconnected, they circle back (through either the dawn- or the duskside of the earth) to the dayside, where they may undergo the whole process again. This picture of moving field lines is a convenience in describing the motion of magnetospheric plasma in a process known as convection. As seen from above one of the earth's poles, the feet of the field lines move in two vortices (Fig. 6). At ionospheric heights the velocity of motion of the field lines is typically a few hundred meters per second, but can reach 1 to 2 km/sec during disturbed times (*39*).

At altitudes above about 120 km the ions (and electrons) in the earth's ionosphere are constrained to gyrate around the field lines; thus the motion of the field lines induces a corresponding motion of the ionospheric plasma. Frictional heating of the thermosphere is then produced by the convective motion of the ionized particles through the neutral air whenever the ion speed exceeds a few hundred meters per second. This heating, usually termed joule heating, is greatest along two arcs through dawn and dusk, where the feet of the field lines return to the dayside along a narrow belt of latitudes between 65° and 75° magnetic latitude (Fig. 6) (*40*).

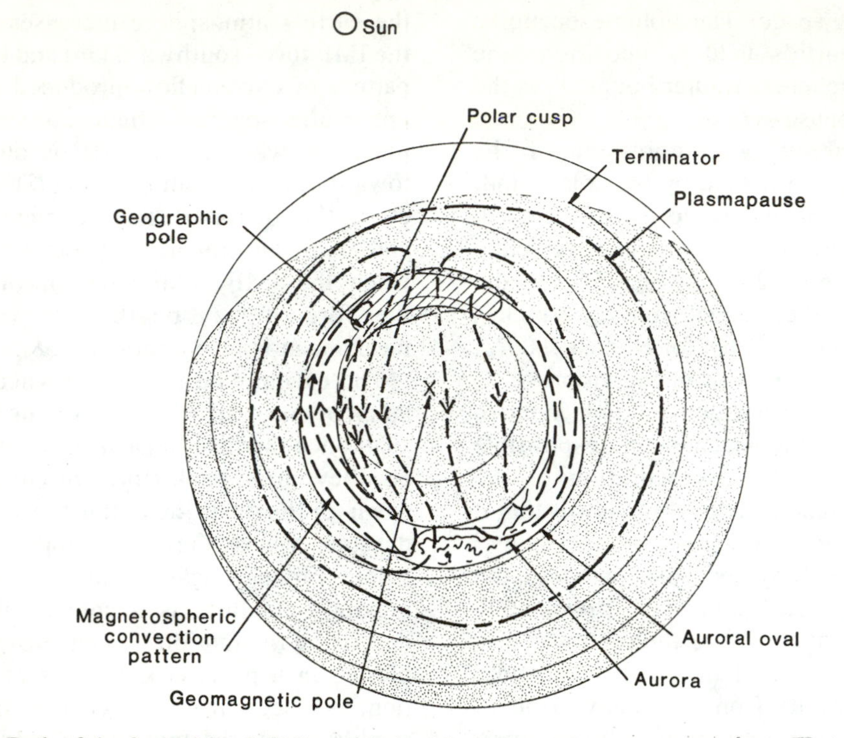

Fig. 6. Path of the feet of the field lines (dashed lines) in the ionosphere above the north pole as a result of the convection process. The actual pattern and speed depend on the orientation of the interplanetary magnetic field. Also indicated is the location of the oval where visible auroras are most often seen.

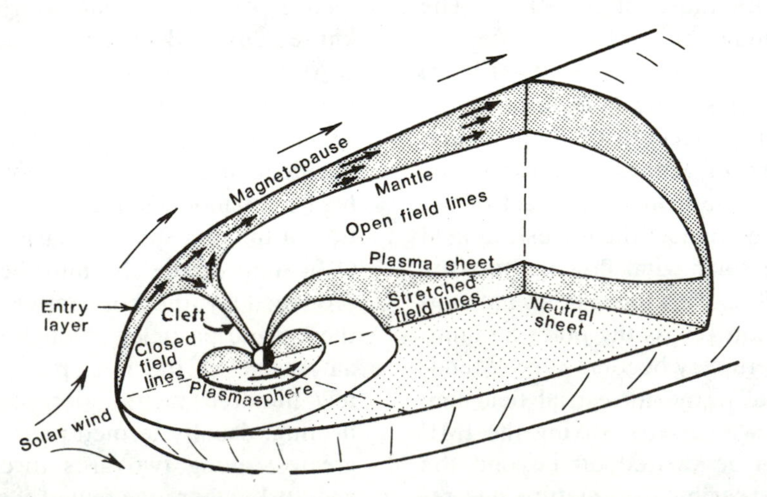

Fig. 7. Topology of the magnetosphere in terms of its particle populations. The plasmasphere is a region of closed field lines populated by H^+ ions. These are produced in the ionosphere and diffuse to fill the flux tubes. The other populations are discussed in text (*33*). [Courtesy of J. Roederer]

In addition to the heating produced by convection, there is heating associated with the bombardment of the atmosphere by energetic particles that produce visible auroras. These occur along an oval that surrounds the magnetic pole (Fig. 6).

The origin of auroral particles is not clear, since they appear to be considerably more energetic than those in the solar wind. The solar wind particles approaching the earth are slowed as they pass through a shock (the bow shock) about 15 R_E sunward of the earth. Much of the bulk motion of the particles carried through the shock is converted to random, disordered motion. While most of the solar wind flows around the flanks of the magnetosphere, a portion appears to penetrate the magnetopause in a thin region on the dayside (the entry layer) and to flow along the inside of the magnetosphere, forming what is called the plasma mantle (Fig. 7). The plasma in the mantle may also include ions of ionospheric origin released when the terrestrial field lines on which they were convecting became broken and connected to the solar wind. We do not yet have a complete picture of the motion of the plasma in the mantle, but it seems likely that this plasma becomes distributed throughout the magnetosphere. Current thinking is that some of the mantle plasma becomes captured by terrestrial field lines after they reconnect and is then carried toward the equatorial plane, where it forms what is termed the plasma sheet. The subsequent shortening of the terrestrial field lines as they return to a more dipolar configuration serves to energize the plasma sheet particles through the betatron effect. The accelerated ions precipitate into the atmosphere, giving rise to the diffuse or mantle aurora, although some are scattered into trapped orbits on permanently closed field lines; these became Van Allen radiation particles. However, this is not the whole story, since the most intense auroral arcs appear to be formed by the precipitation of electrons that have been accelerated by electric fields (of several kilovolts) oriented parallel to the earth's magnetic field lines and a few thousand kilometers in altitude. These parallel electric fields presumably are set up by the need to close the electric circuit linking the ionosphere and the magnetosphere, and they are the subject of considerable research (*41*).

The magnetospheric convection process is not smooth, but undergoes repeated intensifications (substorms) at intervals of 2 to 3 hours. Following an increase in the velocity of the solar wind or a southward turning of the IMF, the rate of transfer of magnetic flux from the dayside to the nightside increases without a corresponding immediate increase in the return flux. As a result, magnetic flux accumulates on the nightside until some form of instability triggers its release. In the process, some of the magnetic energy is converted to the energy of the plasma sheet particles in what may be the terrestrial analog of a solar flare. There is then a short-lived intensification of the aurora and an increase in the speed with which the field lines return to the dayside.

The Upper Atmosphere and Ionosphere

Although the properties of the upper atmosphere are largely controlled by the sun's ionizing ultraviolet radiation, below about 150 km there are pressure variations and winds caused by propagating atmospheric tides. These tides are the result of the absorption of ultraviolet by ozone and water vapor in the atmosphere below 100 km (*42*). Above about 200 km, the diurnal cycle of heating and cooling gives rise to a temperature variation of several hundred degrees (*43*). The resulting pressure variations establish winds that follow roughly great circle paths from the hottest region on the dayside to the coolest on the nightside. That is, unlike winds at the earth's surface, they blow directly from regions of high to low pressure, rather than circling them. This results from the presence of the ionospheric ions, which are constrained to follow the earth's magnetic field lines. Ion drag thus serves to balance the pressure force and limit the wind speed attained (*44*). The speeds increase on the nightside of the earth, where ion density decays.

Above about 100 km, the constituents of the earth's atmosphere are not mixed and each establishes its own density distribution, with the abundance of lighter constituents decreasing less rapidly with altitude than that of heavier constituents. To satisfy flow continuity and conservation of mass, the horizontal wind velocity increases with height so as to compensate for the altitude variation of the principal species (N_2) in the region where the winds are established. Since lighter constituents (such as O and He) decrease in abundance less rapidly than N_2, there is an overall flux of these lighter constituents both from day to night and (at solstice) from the summer to winter hemisphere. This transport creates diurnal and seasonal variations in the neutral composition at a given location.

The wind pattern established by absorption of ultraviolet radiation is modified by the effects of auroral heating and ion motion. The total heat input into the auroral regions as a result of joule heating and particle precipitation is thought to be only 1 to 10 percent of the global heating caused by ultraviolet and EUV (about 10^{12} W). Its importance lies in the fact that it is concentrated over a narrow interval of latitude and hence can introduce large latitudinal changes in pressure, unlike solar heating. Figure 8 illustrates the change in the mean meridional motion of the upper atmosphere which occurs at equinox as a result of increased auroral heating (*45*). Associated with this change is a redistribution of the light constituents of the upper atmosphere. In this case, high latitudes are depleted of atomic oxygen and equatorial regions are enriched (*16*). There is a corresponding change in the ionosphere, which is formed principally through ionization of this species.

It is believed that the change in mean meridional motion is brought about chiefly by an increase in the speed of the equatorward nighttime winds. Winds at 300 km as large as 500 m/sec have been observed at night blowing southward over North America during magnetically disturbed conditions (*46*). The rapid heating of the air at auroral latitudes during substorms is also observed to launch gravity waves in the atmosphere which propagate to lower latitudes and perturb the ionosphere.

The Lower Atmosphere and Weather

The idea that the earth's weather may in some way depend on solar events goes back for a century, yet there remains no conclusive evidence for this. Correlations that suggest a connection have been noted (*47, 48*), but until a mechanism can be identified these must be treated with considerable caution.

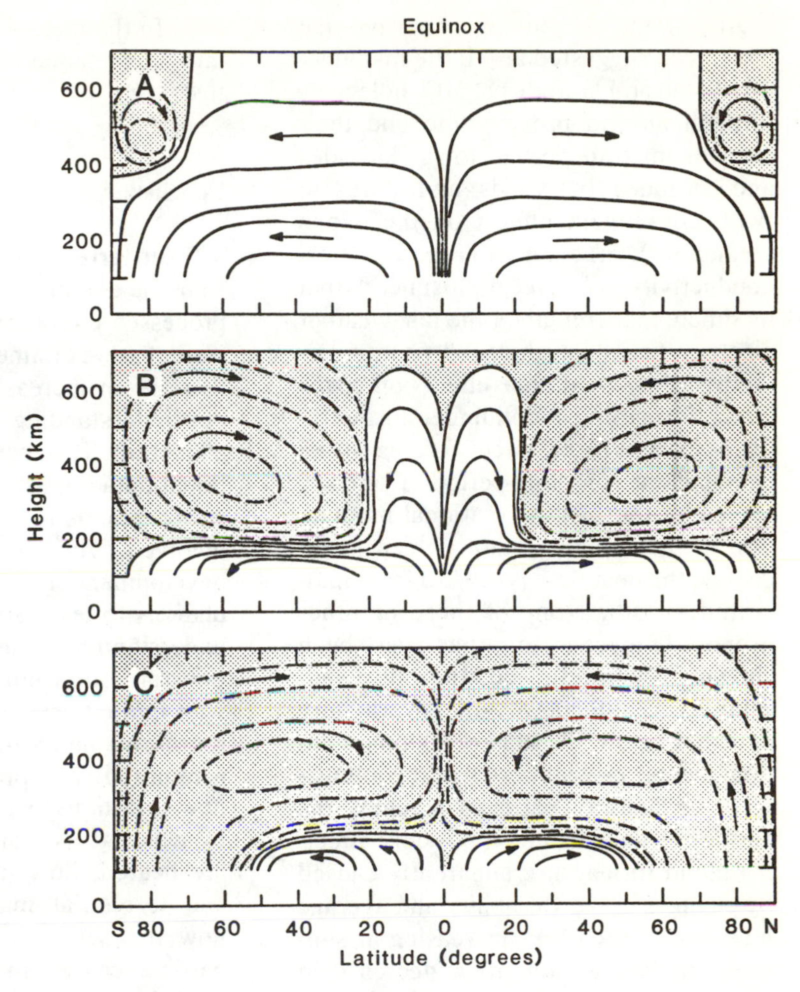

Fig. 8. Mass flow lines indicating the net transport of air (mean meridional flow) at equinox brought about by winds above 100 km during (A) quiet times with no significant auroral heating, (B) moderately disturbed periods, and (C) times of intense magnetic disturbances. [Courtesy of R. G. Roble]

The shortest period variations in atmospheric properties which appear to be correlated with a solar-dependent quantity are tied to the passage of a sector boundary in the solar wind; both the storminess at the level of the jet stream (defined by a vorticity-area index) and the vertical fair-weather electric field appear to decrease at such times. On a longer time scale, there is evidence for 11- and 22-year variations in rainfall rate that are correlated with the sunspot cycle. There have also been studies suggesting climatic variations. For example, it has been suggested that cold periods in Europe during the Middle Ages (such as the so-called ''little ice age'') were related to the absence of spots on the sun for several cycles (such as the Maunder minimum).

Most meteorologists are quite skeptical about such connections, since the variable fraction of the sun's energy appears to be quite small. In addition, they have questioned the validity of many of the correlations that have been found (*49*). It appears that if such a coupling does exist, then there must be a trigger mechanism whereby a small amount of energy organizes a much larger amount.

Among the coupling mechanisms that have been suggested are (i) the manufacture at auroral latitudes of NO molecules through auroral precipitation and their subsequent transport to lower latitudes and altitudes by winds and diffusion (with consequent effects on the ozone abundance); (ii) changes in the electrical conductivity of the atmosphere that modulate the strength of the fair-weather electric fields produced by thunderstorms (this could have effects on cirrus cloud formation and thunderstorm frequency); and (iii) a direct forcing in the troposphere of long-period planetary waves by the 27-day rotational modulation of the solar constant (which might change the heat flow poleward from midlatitudes) (*50*). None of these or other proposed mechanisms has yet been shown to work. Indeed, the only well-established effects of solar variability detected below 100 km are (i) a modulation over an 11-year period of the intensity of cosmic-ray particles reaching the surface of the earth (and possibly affecting cloud formation), apparently caused by changes in the variability and average intensity of the IMF (increasing at sunspot maximum) and (ii) a decrease in ozone abundance at a level of 60 to 70 km, induced by solar flare–produced particles penetrating to this level of the atmosphere (*51*).

In the mistaken belief that the only way in which the sun could cause terrestrial magnetic storms was through the emission of magnetic waves, Lord Kelvin in 1892 concluded that the sun was not the cause of these disturbances (*52*). His mistake lay in assuming that he knew all the facts. Lest we fall into the same trap, we must be wary before reaching conclusions on a sun-weather connection. In the meantime, the topic remains intriguing, tantalizing, and far from settled.

Conclusion

Solar-terrestrial research is moving from the exploratory phase, in which the processes operating in different regions were first examined, to one in which there is an increasing effort to quantify our understanding and develop an overall picture. The accumulation of observational data has spurred attempts to construct models for the various processes involved. This, in turn, has led to the development of numerical computer simulations to test particular aspects of our understanding. These numerical simulations are still rudimentary, dealing only with limited aspects of the models they represent; much more theoretical work is required. This progress has also raised new questions, such as the mechanism by which the chromosphere and corona are heated, how (and where) the solar and terrestrial magnetic fields merge, how magnetic energy is converted to particle energy in the sun and in the earth's magnetosphere, and how intense electric fields are established above the earth's poles parallel to the terrestrial magnetic field. There is a complex and imperfectly understood set of linkages between the source of our energy—the sun—and the earth's atmosphere, and the study of these phenomena remains an important intellectual challenge.

References and Notes

1. The troposphere is beginning to be included in definitions of solar-terrestrial research as evi-

dence is gathered suggesting a connection between sunspots and other forms of solar variability and the weather or climate.

2. Solar-terrestrial researchers distinguish between the middle atmosphere [which includes the stratosphere (10 to 50 km) and the mesosphere (50 to 90 km)] and a region above 90 km, the upper atmosphere [which includes the thermosphere (90 to 400 km) and the exosphere (> 400 km)].
3. Committee on Solar-Terrestrial Research, *Solar-Terrestrial Research for the 1980's* (National Academy of Sciences, Washington, D.C., 1981).
4. E. G. Gibson, Ed., *The Quiet Sun* (NASA SP-303, Government Printing Office, Washington, D.C., 1973).
5. C. Frolich, in *The Solar Output and Its Variation*, O. R. White, Ed. (Colorado Associated Univ. Press, Boulder, 1977), pp. 93–109.
6. G. Newkirk, in *Proceedings of the Conference on the Ancient Sun*, R. O. Pepin and J. A. Eddy, Eds. (Merrill, Boulder, Colo., 1979), p. 293.
7. It seems unlikely that the earth was ever completely ice-covered, since it would not recover from this situation; this probably rules out fluctuations in the sun's energy exceeding about 10 to 15 percent, provided that the earth's cloud cover and atmospheric constituents did not also change. See the recent review by G. R. North, R. F. Cahalan, and J. A. Coakley [*Rev. Geophys. Space Phys.* **19**, 91 (1981)].
8. M. Budyko, *Tellus* **21**, 611 (1969); *J. Appl. Meteorol.* **8**, 392 (1969). Such studies are, however, model-dependent. The sensitivity of the terrestrial atmosphere to changes in the input of solar energy can be demonstrated independent of climate models by the presence of variations in ocean cores that correspond to periods of the orbital variation of the earth [J. D. Hays, J. Imbrie, N. J. Shackleton, *Science* **194**, 1121 (1976)].
9. C. G. Abbot and L. B. Aldrich, *Ann. Smithson. Astrophys. Obs.* **6**, 85 (1942); L. B. Aldrich and W. Hoover, *ibid.* **7**, 26 (1954).
10. T. Sterne and N. Dieter, *Smithson. Contrib. Astrophys.* **3** (No. 3), 9 (1958).
11. P. Foukal, P. Mack, J. Vernazza, *Astrophys. J.* **215**, 952 (1977); P. Foukal and J. Vernazza, *ibid.* **234**, 707 (1979).
12. R. C. Willson, S. Gulkis, M. Janssen, H. S. Hudson, G. A. Chapman, *Science* **211**, 700 (1981).
13. J. R. Hickey, F. Griffin, H. Jacobowitz, L. Stowe, P. Pellegrino, R. M. Haschhoff, *Eos* **61**, 365 (1980).
14. P. Foukal, paper presented at the Workshop on Sunspots, Sacramento Peak Observatory, New Mexico, 14 to 17 July 1981.
15. D. J. Torr and M. A. Torr, *J. Atmos. Terr. Phys.* **41**, 797 (1979).
16. H. G. Mayr and I. Harris, *Rev. Geophys. Space Phys.* **17**, 492 (1979); J. V. Evans, *ibid.* **16**, 195 (1978).
17. H. E. Hinteregger, in *Advances in Space Research* (Pergamon, Oxford, England, 1981), pp. 39–52.
18. C. F. Kennel, L. J. Lanzerotti, E. N. Parker, Eds., *Solar System Plasma Physics* (North-Holland, New York, 1979), vol. 1.
19. L. Biermann, *Z. Astrophys.* **29**, 274 (1951); J. C. Brandt and D. A. Mendis, in *Solar System Plasma Physics*, C. F. Kennel, L. J. Lanzerotti, E. N. Parker, Eds. (North-Holland, New York, 1979), vol. 2, pp. 255–291.
20. E. N. Parker, *Astrophys. J.* **128**, 664 (1958); in *Solar Terrestrial Physics*, J. W. King and W. S. Newman, Eds. (Academic Press, London, 1967), chap. 2.
21. W. C. Feldman, in *The Solar Output and Its Variability*, O. R. White, Ed. (Colorado Associated Univ. Press, Boulder, 1977).
22. H. J. Völk, *Space Sci. Rev.* **17**, 255 (1975).
23. Investigations of the fluctuations of the earth's magnetic field revealed two general types of disturbance. One is a sudden decrease in the horizontal component of the earth's magnetic field which lasts one or more days and tends to follow large flares on the sun. The other disturbance also lasts for one or more days but does not follow a flare, is less intense, and recurs one or more times at 27-day intervals. These events are called sudden commencement storms and recurrent storms, respectively. See S. I. Akasofu and S. Chapman, *Solar-Terrestrial Physics* (Oxford Univ. Press, Oxford, England, 1972).
24. L. F. Burlaga and K. W. Ogilvie, *J. Geophys. Res.* **78**, 2028 (1973); L. F. Burlaga, *Space Sci. Rev.* **17**, 327 (1975).
25. J. W. Harvey and M. R. Sheeley, *Space Sci. Rev.* **23**, 139 (1979).
26. Many of the best observations were made by Skylab astronauts over the period May 1973 to February 1974. See *Coronal Holes and High Speed Wind Streams*, J. B. Zirker, Ed. (Colorado Associated Univ. Press, Boulder, 1977).
27. A. J. Hundhausen, in *ibid.*, p. 226.
28. A. F. Timothy, A. S. Krieger, G. S. Vaiana, *Sol. Phys.* **42**, 135 (1975).
29. S. T. Suess, *Space Sci. Rev.* **23**, 159 (1979).
30. R. H. Levine, M. D. Altschuler, J. W. Harvey, *J. Geophys. Res.* **82**, 1061 (1977).
31. J. M. Wilcox and N. F. Ness, *ibid.* **70**, 5793 (1965).
32. E. J. Smith, B. T. Tsurutani, R. L. Rosenberg, *ibid.* **83**, 717 (1978).
33. J. G. Roederer, *Science* **183**, 37 (1974).
34. ———, in *Solar System Plasma Physics*, C. F. Kennel, L. J. Lanzerotti, E. N. Parker, Eds. (North-Holland, New York, 1979), vol. 2, chap. 1.
35. I. Axford, *Rev. Geophys.* **7**, 421 (1969).
36. S. I. Akasofu, *Planet. Space Sci.* **27**, 425 and 1039 (1979).
37. J. Heppner, *J. Geophys. Res.* **82**, 1115 (1977).
38. P. H. Reiff, R. W. Spiro, T. W. Hill, *ibid.* **86**, 7639 (1981).
39. Movement of the field lines cannot be detected at ground level and occurs fully only above about 120 km. It is the layer of neutral air between the earth and the ionosphere which permits the field lines embedded in these two conducting regions to move with respect to one another.
40. There are several magnetic coordinate systems in use. That employed in Fig. 6 is defined by tracing a field line out to the equator, where its distance L from the earth's center (in earth radii) fixes the latitude Λ (= arc sec $\sqrt{L}$).

41. S. I. Akasofu, *Am. Sci.* **69**, 492 (1981).
42. J. M. Forbes and H. B. Garrett, *Rev. Geophys. Space Phys.* **17**, 1951 (1979).
43. A. E. Hedin *et al.*, *J. Geophys. Res.* **82**, 2193 (1977).
44. In the lower atmosphere the pressure force is balanced by Coriolis force, causing the winds to circle regions of high and low pressure.
45. R. G. Roble, in *The Upper Atmosphere and Magnetosphere* (National Academy of Sciences, Washington, D.C., 1977), chap. 3.
46. D. W. Sipler and M. A. Biondi, *J. Geophys. Res.* **84**, 37 (1979); G. Hernandez and R. G. Roble, *ibid.* **81**, 5173 (1976).
47. J. Eddy, Ed., *Solar Variability, Weather and Climate* (National Academy of Sciences, Washington, D.C., in press).
48. A recent summary of papers published in support of possible connections between solar variability and weather or climate may be found in J. R. Herman and R. A. Goldberg, *Sun, Weather and Climate* (NASA SP-426, Government Printing Office, Washington, D.C., 1978).
49. The climatic connection has been questioned by M. Stuiver, *Nature (London)* **286**, 868 (1980).
50. H. Volland, in *Solar-Terrestrial Influences on Weather and Climate*, B. M. McCormac and T. A. Seliga, Eds. (Reidel, Dordrecht, 1979), p. 263.
51. P. J. Crutzen, I. S. A. Isaksen, G. C. Reid, *Science* **189**, 457 (1975); D. F. Heath, A. J. Krueger, P. J. Crutzen, *ibid.* **197**, 886 (1977).
52. J. K. Hargreaves, *The Upper Atmosphere and Solar-Terrestrial Relations* (Van Nostrand Reinhold, New York, 1979), p. 267.
53. C. O. Hines *et al.*, Eds., *Physics of the Earth's Upper Atmosphere* (Prentice-Hall, Englewood Cliffs, N.J., 1965).
54. I am grateful to R. Carovillano, P. Foukal, A. F. Nagy, G. C. Reid, and P. H. Reiff for helpful comments on the manuscript. This work was supported by grant ATM 7909189 from the National Science Foundation.

This material originally appeared in *Science* **216**, 30 April 1982.

2. Solar Flares, Proton Showers, and the Space Shuttle

David M. Rust

Flares, which are sudden brightenings on the solar surface, affect our environment in many ways. Some effects are harmless, even beautiful, but others may be costly and have important consequences for the space shuttle program. The 10 April 1981 flare heated the earth's upper atmosphere and increased the drag on the space shuttle Columbia during its maiden flight of 12 to 14 April. The flare caused a great red aurora on the night of 12 April and an ionospheric storm that interfered with police communications, but not with shuttle communications. In fact, the flare posed no serious problems for the Columbia. If the shuttle mission had included an extravehicular activity in a polar orbit, however, the mission probably would have been put off to avoid exposing astronauts John Young and Robert Crippen to a proton shower. Solar flare protons with an energy of more than 10 million electron volts can penetrate the aluminum layers of most space suits (*1*).

Solar flares were the subject of an intensive study in the international Solar Maximum Year, which ended in February 1981. The National Science Foundation and the National Aeronautics and Space Administration (NASA) led U.S. support of the effort. In all, more than 400 solar physicists from 17 countries worked together to document and model every stage in the buildup, release, and propagation of flare energy. Before discussing the models and the first results of the Solar Maximum Year, I will review the most important consequence of flares for space missions—the hazard from a proton shower.

Radiation Hazard

An ambitious shuttle program will be confronted with threats posed by various sorts of space radiation (*2*). Charged particles trapped in the Van Allen belts, for example, may cause gradual damage to materials on missions of long duration and may damage large structures in space through static discharges. But the greatest challenge will be to ensure that astronauts and others who work in space will not be exposed to lethal or disabling solar radiation. During the Apollo program the radiation hazard was qualitatively the same, and NASA created a Solar Proton Alert Network to evaluate and warn of proton shower risks. Studies of the radiation hazards made in the 1960's for the Apollo program are still the most comprehensive available (*2*). No way has yet been found to avoid proton showers, but so far astronauts have been at risk for only a few days each year. The risk from proton showers was small compared to other risks that the astronauts faced. In the shuttle era, however, the relative importance of the hazard from proton showers will require reconsideration since there may be people in space almost continuously.

NASA hopes to have a permanent manned space station in orbit by 1990, and this and other projects may rely on extensive extravehicular activity. One study (*3*) estimated that NASA could save about $3 billion by using extravehicular activity to deploy, maintain, and repair satellites.

The nearly complete shuttle launch facility at Vandenberg Air Force Base will be used primarily to launch missions into polar orbits, where the radiation hazard is substantially higher than that in the low-inclination orbits used by shuttles launched from the Kennedy Space Center.

The risks to those who work in space for 3 months or more are potentially great. Careers in space may not be feasible if a worker can absorb a lifetime's radiation allowance in a few missions. Both long- and short-term exposure levels will have to be considered since background cosmic radiation, particles trapped in the earth's magnetic fields, and six to ten proton-producing flares a year will deliver a continuous dose of radiation. The long-term effects of such radiation will not be discussed here. Radiological studies do indicate, however, that the effects of exposure are cumulative (*4*).

In this chapter, I will describe the risks posed by major solar flares, which start suddenly and can deliver in a few hours a disabling or even lethal dose of radiation. The largest doses can be two to three times the level lethal to a man. A lethal dose could be delivered over a period of several days, but at peak flux rates, a 1-hour exposure would cause nausea and possibly vomiting (*5*, *6*).

Space radiation risks for short and long missions currently can be evaluated only in a statistical way. Stassinopoulos and King (*7*) at the Goddard Space Flight Center used modified Poisson statistics to estimate radiation doses at geosynchronous altitude (35,000 kilometers) during a 90-day mission. They found that if a major proton shower occurred, astronauts in a shielded working area would receive a dose of 155 rads. Hospital x-rays, by comparison, give less than 0.01 rad. An astronaut outside the shielded working area during a major proton shower could receive about 1000 rads. Over 90 days the total dose due to ordinary proton showers would be about 10 rads. But this estimate is strongly dependent on statistical assumptions, and it does not account for trapped radiation and cosmic rays. Another estimate (Table 1), prepared at the Marshall Space Flight Center (*8*), of exposure for 1 year to all penetrating radiation inside a shielded work space, when compared to the suggested exposure limits for Apollo astronauts (Table 2), indicates the level of risk for lengthy manned space activities in geosynchronous and polar orbits is unacceptable because of flare protons. Standards in industry are much more stringent than those proposed for the astronauts (*9*).

It may be possible to design around hazards from trapped radiation, but better short-term forecasting of solar flares will be required to avoid aborting missions unnecessarily or exposing astronauts to a hazardous stream of protons.

Table 1. Predicted radiation dose in shielded work space for 1-year exposure.

Orbit		Dose (rads) from		
Altitude	Inclination	Trapped and cosmic rays	Solar flares*	Total
400 km	30°	32		32
400 km	90°	28	80	108
Geosynchronous	Geosynchronous	300	250	550

*For a 99 percent confidence level at maximum, that is, one chance in 100 that a 1-year exposure at solar maximum will exceed these doses.

Table 2. Suggested proton exposure limits in rads for Apollo crew members.

Constraint	Bone marrow	Skin	Ocular lens	Testes
Average daily rate	0.1	0.3	0.15	0.05
30 days	12	40	20	6
Yearly	40	110	55	20
Career	200	600	300	100

Insight into the physical processes that cause proton showers will be essential because short-term forecasting is reaching the limit that can be achieved with purely statistical methods (*10*). At present, flare forecasts are based on statistical correlations between flares and sunspots, solar radio emission level, earlier flares, and other solar parameters. Many of the characteristics of solar activity used in forecasts are measured only once each day. More frequent measurements usually reveal no changes above the background noise level, which is determined by atmospheric conditions at the solar observatories. Thus, there is a fundamental limit to the accuracy of the forecasts. To understand this, suppose that the average rate of flares that disturb the space environment is 0.25 per day. Even if this average rate is known to a high degree of precision, Poisson statistics tell us that a forecast of "flare" will be wrong 80 percent of the time.

A Flare Chronology

The physical processes operating in flares are not well understood, although the effects are well documented. To trace these phenomena in detail, I will describe the flare phenomena of 10 to 13 April 1981 and their effects on the earth. When first observed on 2 April, the sunspot group that would give rise to the flare of 10 April was a single spot. Spots themselves have no effect on our planet, but their magnetic fields twist and stretch until some instability releases pent-up energy and causes a flare. There was little change in the spot group until 7 April when a moderate flare occurred. On 8 April, 16 new, small spots burst into the region, and five more flares occurred. On 9 April, the number of sunspots grew to 29, and there were eight more small flares. Still, there was no reason to suspect that a major flare was only 48 hours away.

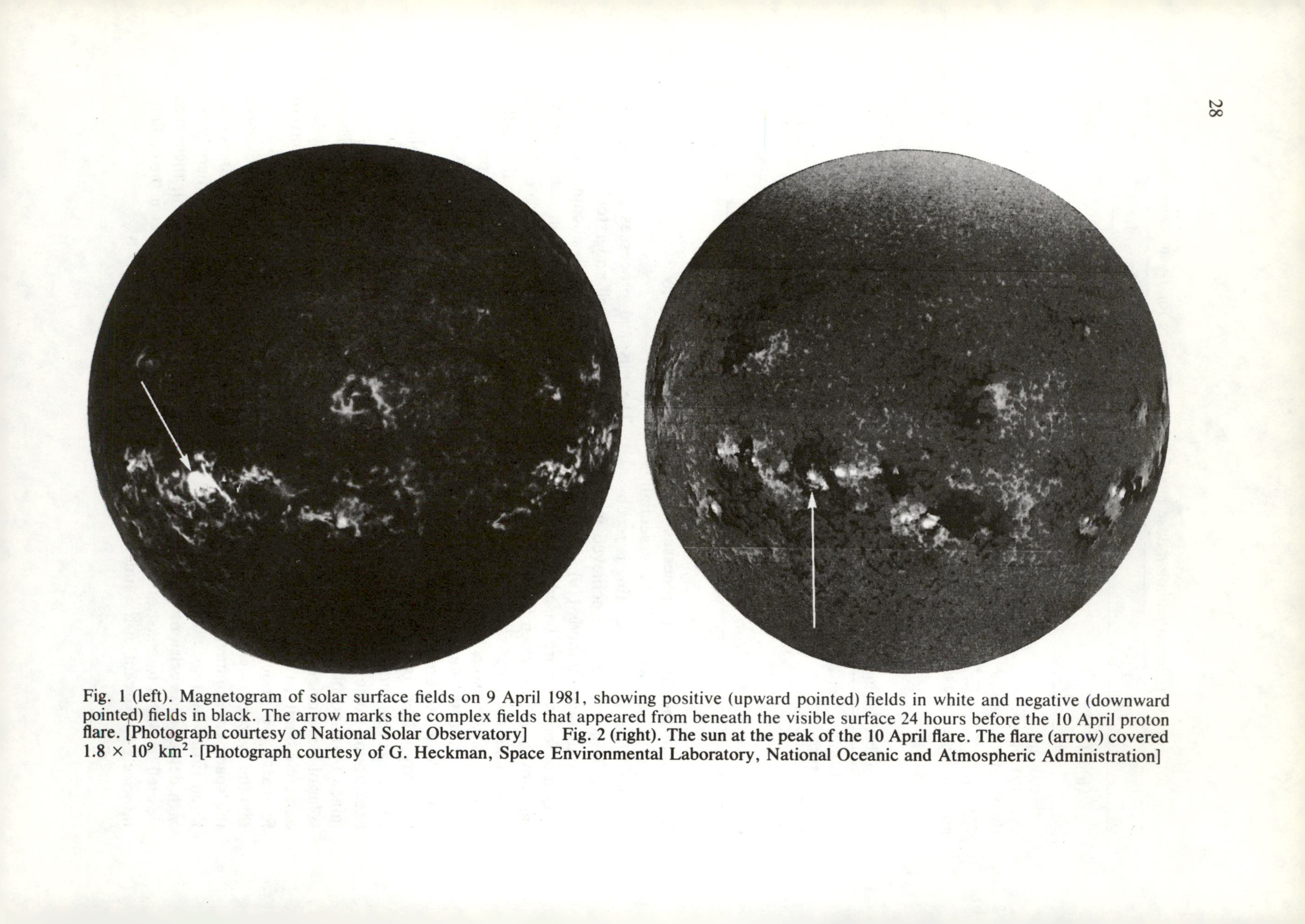

Fig. 1 (left). Magnetogram of solar surface fields on 9 April 1981, showing positive (upward pointed) fields in white and negative (downward pointed) fields in black. The arrow marks the complex fields that appeared from beneath the visible surface 24 hours before the 10 April proton flare. [Photograph courtesy of National Solar Observatory] **Fig. 2 (right). The sun at the peak of the 10 April flare. The flare (arrow) covered** 1.8×10^9 km^2. [Photograph courtesy of G. Heckman, Space Environmental Laboratory, National Oceanic and Atmospheric Administration]

On 9 April, the sunspot group was classified as a complex magnetic region. The magnetic fields in the group formed an unusual and complicated mosaic of upward and downward pointed fields (Fig. 1). A chain of tiny positive and negative spots encircled the large positive spot that had been observed since 2 April. Only about 10 percent of all spot groups attain such complexity, and almost all large flares come from such complex groups.

By the morning of 10 April the number of spots had increased by only one, but flare activity increased dramatically. Seven subflares preceded the giant flare, which engulfed the sunspot group at 1645 Greenwich mean time, 5 hours after the originally scheduled launch of the Columbia and 43 hours before the actual launch. In only a few minutes, electrons and protons began streaming outward toward the earth and inward to denser layers of the solar atmosphere. Intense emission appeared throughout the electromagnetic spectrum (*11*).

The flare lasted for 3½ hours. At its peak, it covered an area of 2×10^9 square kilometers on the solar disk (Fig. 2). The flare released over 10^{25} joules or about 10^{19} kilowatt-hours of energy or about 400,000 times the total yearly energy consumption of the United States. This amount of energy could easily have been stored in the magnetic fields of the sunspot group even though the sunspots showed no changes as a result of the flare. Gradually, solar physicists have been able to rule out many other invisible storehouses once proposed as flare energy sources. The invisible energy stored in twisted magnetic fields is the only plausible energy source for flares (*12*).

In the first 10 minutes after flare onset on 10 April a shock wave started out through the corona. After 58 hours the shock hit the earth's magnetosphere and triggered a magnetic storm. The actual decrease in the horizontal component of the magnetic field at the earth's surface was barely 1 percent of the quiet time value, but the effect was sufficient to accelerate billions of electrons in the magnetosphere and produce a power outage in Canada. The 12 April magnetic storm, which began only 15 hours after the launch of the Columbia, was the largest of the current sunspot cycle (*13*).

The earth's upper atmosphere was swamped by electrons that rushed in from the storm. The electrons raised the temperature at 260 km over Boulder, Colorado, from a normal temperature of 1200 K to about 2200 K. This temperature increase, the greatest recorded by the National Oceanic and Atmospheric Administration (NOAA) in 15 years of measurements, swiftly changed the orbit of the Columbia. The heating expanded the atmosphere and dragged the shuttle (and other satellites) down into lower orbits 60 percent faster than expected (*14*). The early demise of Skylab in 1978 was due to such an unexpected increase in atmospheric drag. In general the present sunspot cycle is much more active than predicted, and our upper atmosphere is much more extended because of this. An extended atmosphere exerts more drag on all satellites and pulls them into lower orbits. The drag effect increases exponentially with density since, by Kepler's law of planetary motion, a satellite moves faster in a lower orbit, thereby losing more energy to atmospheric drag; it drops into still denser layers of the atmosphere, moves faster still, suffers more drag, and quickly burns up. On the Columbia, Young and Crippen simply fired the retrorockets early to correct for the overshoot on

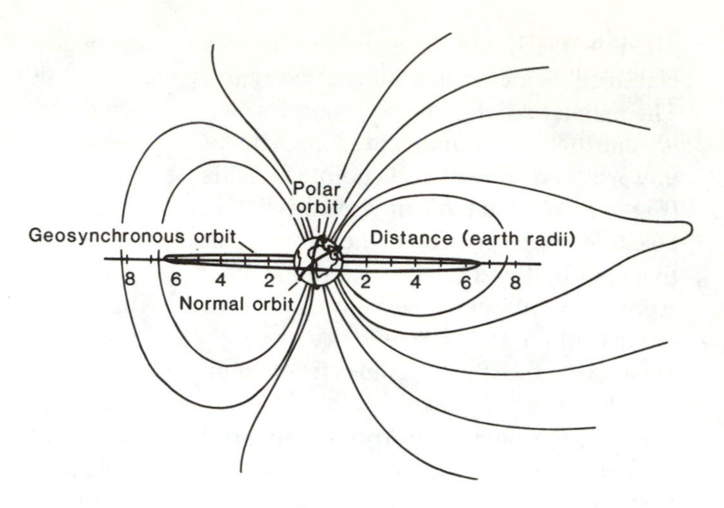

Fig. 3. Sketch showing the most frequently used satellite orbits and the magnetic fields that deflect protons. Geosynchronous orbits are fully exposed to proton showers. A satellite in polar orbit is exposed about 30 percent of the time. Satellites in 30° orbits are well protected from solar protons.

reentry. But even a little extra drag on satellites built to operate for years could be costly if they are lost prematurely or required to carry expensive on-board maneuvering capability.

Solar Protons

The most energetic protons from the 10 April flare reached the earth's atmosphere an hour after flare onset, and they would have posed a potentially lethal threat to astronauts on an extravehicular activity in a polar or geosynchronous orbit (Fig. 3). The peak flux of protons with energies of 50 to 500 megaelectron volts had passed by the Columbia's launch time, but 10-MeV protons continued to stream into the magnetosphere until early on 14 April, the day of reentry (*13*). Young and Crippen were never in any danger, however, because the Columbia flew in a low-latitude, low-altitude orbit, where the earth's magnetic field provides adequate shielding against everything except very energetic cosmic rays. The average proton flux on 10 April was about 300 particles per square centimeter per second. The event was an ordinary proton shower. About ten such showers can be expected each year near the sunspot cycle maximum. The highest fluxes recorded in the past quarter-century were more than 100 times higher than those on 10 April (*7*), and they occurred in anomalously large events that happen perhaps one to six times each solar cycle during the rise and fall from solar maximum. The next solar maximum will occur in 1990.

How are solar protons accelerated to high energies? Direct production of a beam of 10-MeV protons requires a drop in electric potential of 10 million volts. It is unlikely that such a potential could be established on the sun since the solar atmosphere is composed of highly ionized gas, or plasma, that conducts electricity with higher efficiency than pure copper. Electrons in the plasma will move almost instantaneously to short-circuit any electrical differences that might accelerate protons.

The search for another way to explain how these solar cosmic rays are accelerated began shortly after announcement of their discovery in 1946 (*15*), but solar flare data were intermittent before the International Geophysical Year in 1958.

Then, a worldwide network of observatories was established to monitor the sun 24 hours a day. At about the same time the riometer, which measures the relative ionospheric opacity to cosmic radio noise, was invented. Solar protons with energies of 10 to 100 MeV spiral into the polar cap and increase the ionization there. The added ionization absorbs cosmic radio waves and can be directly related to the proton flux (*16, 17*). Before development of the riometer, only proton showers at energies above 500 MeV could be recorded. Satellite-borne detectors have since lowered the threshold for detection of flare protons to about 0.5 MeV.

From the start it was clear that only some large flares cause proton showers. Also, proton showers generally miss the earth when emitted from a flare on the sun's eastern hemisphere. The preference for the western hemisphere (as seen from the earth) was explained in 1962 as a result of channeling by the solar wind (*18*). Parker (*19*) had proposed the existence of a solar wind to explain why ionized comet tails point away from the sun. The solar wind stretches the sun's magnetic fields in an Archimedes' spiral out to well beyond the earth's orbit, and protons released at the base of a spiral follow it outward (Fig. 4). Flares that occur in the sun's western hemisphere are well connected to the earth, and the most energetic protons can reach the earth in as little as 20 minutes (*20*).

Even after the channeling effects of the solar wind were accounted for, solar physicists were not able to distinguish proton-producing flares from other flares. They tried to find clues in pictures of the optical emissions (Fig. 2), but the photographs show plasma with an average particle energy of only 1 electron volt. Solar magnetograms and pictures of

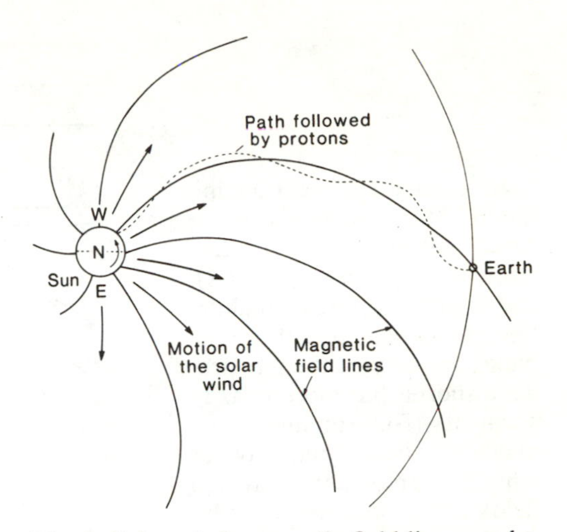

Fig. 4. Solar wind, magnetic field lines, and a proton trajectory. The solar wind pulls magnetic fields radially away from the sun at about 400 km/sec, but the solar rotation of 2 km/sec twists the fields into a gradual spiral. Protons released at the base of magnetic fields 20° to 80° west of the central meridian are said to be well connected because they do not have far to drift across the spiral pattern to reach the earth.

sunspots (Fig. 1) do show what the surface magnetic fields are doing and, since Fermi (*21*) had shown that protons could be accelerated in colliding magnetic fields, it seemed reasonable to search for colliding magnetic fields in flares.

Magnetic Fields and Neutral Sheets

Giovanelli (*22*) showed that flare occurrence is statistically associated with complex and changing sunspots. Martres and her colleagues (*23*) and Severny (*24*) established that flares start near growing or decaying sunspots. I found that the changing magnetic features were actually fresh magnetic fields emerging from below the solar surface. Flares frequently start precisely at these emerging fields. On this basis, Heyvaerts, Priest,

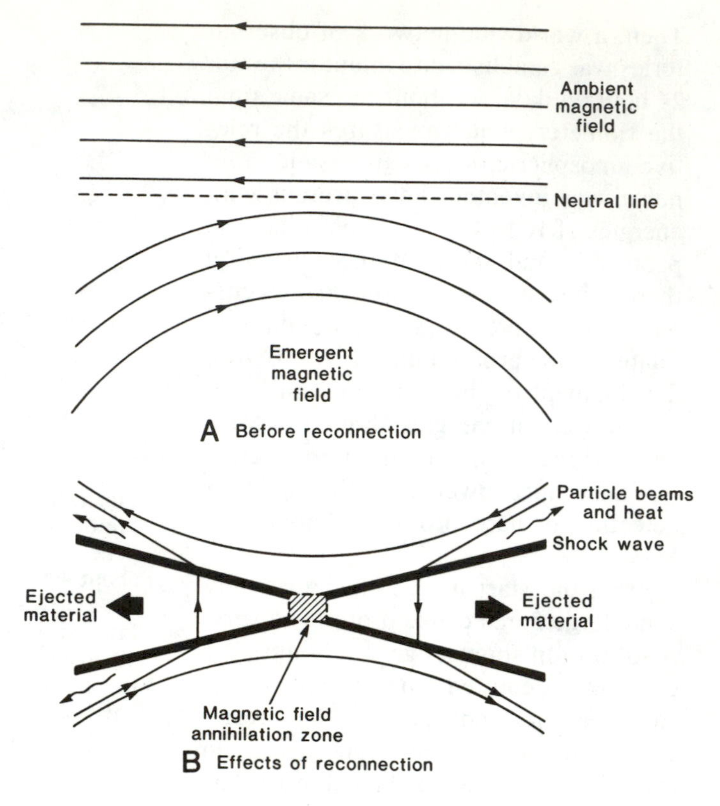

Fig. 5. Magnetic field annihilation. Magnetic fields in the solar plasma behave like stretched rubber bands (A), and the effect of field annihilation is like that of cutting the bands and retying them to produce shorter but more relaxed segments (B). Annihilation of magnetic fields may occur where oppositely directed fields are pushed together by outside forces, such as convection or a shock wave. Energy is released in heat, shocks, and energetic particles near the neutral sheet between the fields.

and I proposed (*25*) that the magnetic energy was released principally as heat and in beams of particles accelerated in collisions between the old and new fields.

Our model invoked two earlier ideas about magnetic fields in flares. One, proposed by Sweet (*26*) held that energy can be drained out of the solar magnetic fields where two oppositely directed fields come into contact (Fig. 5). Along what Sweet termed a "neutral sheet," strong electric currents will flow and rapidly heat the plasma. The mechanism, however, could not explain the rapidity of the heating. Then Petschek (*27*) showed that magnetic fields, if pushed together with sufficient speed, would generate shocks, which would greatly increase the rate of energy conversion. Petschek called this process "magnetic field annihilation." Solar physicists had realized that magnetic fields were the only plausible energy source for flares, and now Petschek's theory could explain how magnetic energy could be converted to intense heat in 1 minute, or less, as observed in flares. Additionally, the neutral sheets might provide escape routes for accelerated protons into interplanetary space.

Skylab. In 1973, pictures of flares obtained by American Science and Engineering and the Naval Research Laboratory strongly challenged the emerging flux model. These x-ray and ultraviolet pictures from Skylab showed nothing that looked like the current sheets that the model requires. Every flare consisted of a single bright loop (Fig. 6) or, sometimes, an arcade of bright loops. Proponents of current sheets argued that

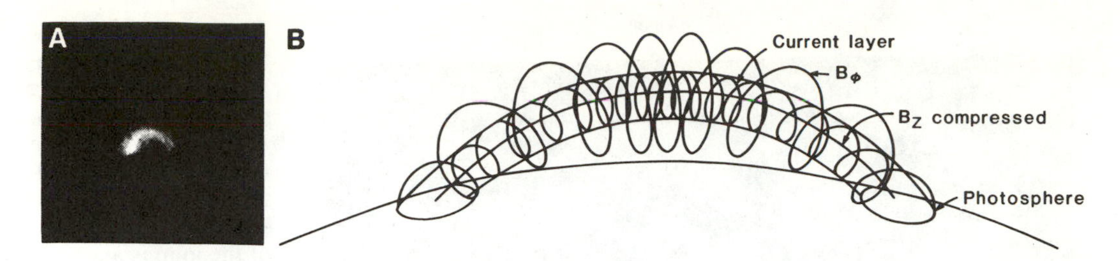

Fig. 6. The 10^7 K plasma kernel in flares, as shown by x-ray telescopes aboard the Skylab. (A) In most cases the plasma resembles half a torus. (B) Plasmas of similar temperature are confined in tokamaks by sheared magnetic fields. The laboratory experience of plasma physicists suggests a number of ways to convert the energy of the confining fields into the intense heat of the plasma. The loop's length in (A) is 1/13th of the sun's diameter.

the sheets were too thin to photograph and that the loops simply showed trapped hot plasma. Hot plasma and accelerated particles should quickly exit from any current sheet. Another possible explanation was that the Skylab telescopes showed only the aftermath of the energy conversion process (*12*, p. 212; *28*). Then Spicer (*29*) proposed a model based on experience in laboratories with tokamaks in which toroidal fields confine hot plasma for fusion research. Such fields are subject to internal instabilities that are well known to laboratory researchers, and Spicer suggested that solar magnetic fields are annihilated not at neutral sheets, but within the elemental magnetic loops themselves, as in tokamaks.

Spicer's model seemed to describe the Skylab flare pictures. The model required no more than what was seen—no hard-to-detect colliding fields or impossible-to-detect neutral lines—but it did not provide a route by which protons could enter interplanetary space.

Solar Maximum Year: First Results

Until recently, efforts to understand flares and proton showers were hampered by a lack of comprehensive observations of all the phenomena involved. During the Solar Maximum Year (*30*) solar physicists assembled the ground-based and satellite instruments needed to record everything, from the first tentative increase in x-rays to the proton shower and shock wave in space. NASA launched the International Sun-Earth Explorer to monitor the particle fluxes and solar wind near the earth as well as a $100-million observatory, called the Solar Maximum Mission. Most instruments for the Solar Maximum Mission (*31*) were specially designed to probe flare energy release processes with x-ray and ultraviolet images and spectra. There was also a coronagraph to photograph the large magnetic loops (Fig. 7) that burst into interplanetary space during some flares. Thousands of flares were recorded before the Solar Maximum Mission lost solar pointing control; the experiments on the Sun-Earth Explorer and a cluster of solar telescopes aboard Air Force satellite P78-1 are still functioning. Although enough flares have now been studied to provide new insights, the conclusions drawn must be treated with caution because flares vary widely. Nevertheless, the results seem to point the way to understanding the origins of proton showers.

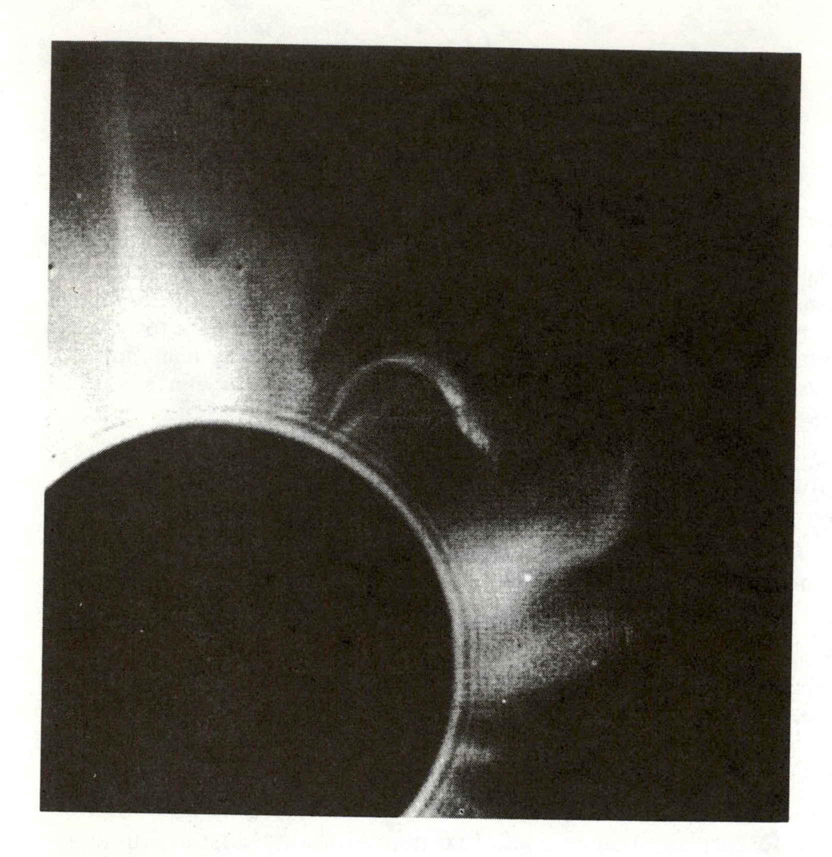

Fig. 7. Solar Maximum Mission coronagraph image showing expanding loop on 14 April 1980 pushing material out of the solar atmosphere at 700 km/sec. The diameter of the loop is approximately the same as that of the sun. Most of the energy in a large solar flare is carried away by such coronal mass ejection. In front of the leading edge of the bright material is a bow shock, detectable only by the radio emission it excites. This photograph was made in filtered sunlight reflected from electrons trapped in the loop. The sun is blocked by a disk. [Photograph courtesy of L. House, National Center for Atmospheric Research]

The hot flare loops discovered with the Skylab telescopes are about 6000 km wide and about 30,000 km long—just within the resolving power of the hard x-ray imaging spectrometer (HXIS) on the Solar Maximum Mission (*29*). Using the HXIS, Hoyng and other Solar Maximum Mission investigators compared maps of electron emission with magnetic field maps and optical images (*32*). The optical images (Fig. 2) obtained by "flare patrol" telescopes operated by NOAA and the U.S. Air Force are especially useful for clarifying magnetic field structure because they show features as small as 500 km in diameter. We were able to show that a flare on 21 May 1980 started with two bursts of electrons in adjacent magnetic loops. The electrons, identified by characteristic x-ray bremsstrahlung as they penetrated the chromosphere, must have been accelerated somewhere in the magnetic loops or at a neutral sheet between loops.

Within a minute of the x-ray bursts, the loops filled with 70×10^6 K plasma. Spectra from another Solar Maximum Mission experiment, the x-ray polychromator, showed that the plasma boiled up from the chromosphere. As it cooled, it began to emit ultraviolet and optical radiation that illuminated the magnetic loops only a few minutes after the first release of energy.

From the 21 May observations electron beams were located unambiguously in the heart of a flare. Zirin (*33*) and others had guessed that electron beams caused the intense optical emission at the bases of flare loops, but without HXIS images, it was not possible to distinguish between the action of electron beams and conduction from the flare plasma. The HXIS observations clearly separated the x-ray emission of the electron beams from the equally intense x-ray emission of the 70×10^6 K plasma.

Besides clarifying the acceleration and heating sequence in flare loops, the 21 May observations provide a clue to what triggered the flare. Maps of the underlying magnetic fields made at Kitt Peak National Observatory showed that tiny strands of magnetic field were emerging from beneath the sun's surface an hour before the flare (*32*). The location of the emergent fields indicated that they probably collided with the loops that flared. Colliding magnetic fields also appeared to be responsible for flares on 30 April and 5 November 1980. Thus, emerging magnetic fields do seem to trigger the process that converts magnetic energy into heat and beams of accelerated particles.

Investigators using the Very Large Array of the National Radio Astronomy Observatory found more evidence for electron beams in the loops (*34*, *35*). During a half-dozen flares, they mapped emission from electrons with energies of 100 to 500 kiloelectron volts. Such electrons spiral in magnetic loops and emit synchrotron radiation, whose polarization and location reveal the nature of the electron population and outline magnetic fields. The radio wave–emitting loops pictured by the Very Large Array stretch over the magnetic fields in exactly the same way as do the x-ray–emitting loops shown by the HXIS, except that at radio wavelengths the electron beams emit mostly at the loop tops (Fig. 8). Some of

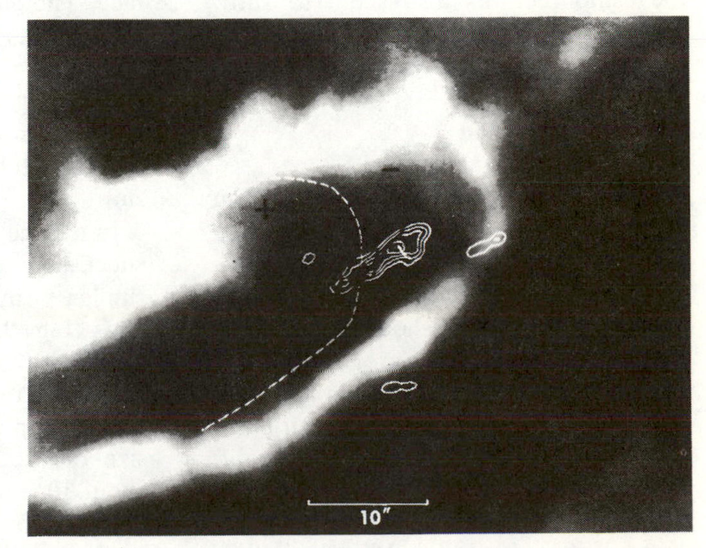

Fig. 8. Contours (solid lines) of circularly polarized radio emission level from electrons trapped in a magnetic loop in the corona. The bipolar nature of the radio emission reflects the upward- and downward-facing fields, respectively, on the positive (+) and negative (−) sides of the magnetic field boundary (marked by a dashed line). The radio emission contours are superimposed on a photograph of the chromosphere, where the flare appears as several bright ribbons. Radio observations were made with the Very Large Array; the chromospheric flare was photographed at the Big Bear Solar Observatory in California. [Photograph courtesy of G. Hurford, California Institute of Technology]

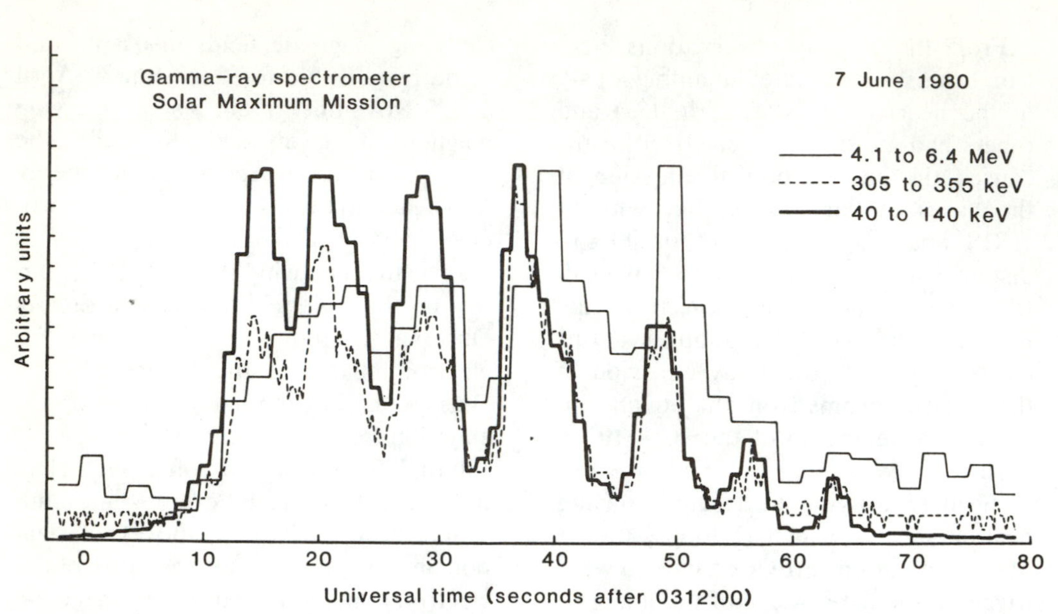

Fig. 9. X-rays and gamma rays from the flare of 7 June 1980. The bursts of x-rays lead the gamma rays by less than 2 seconds. The x-rays are emitted exclusively by electrons. The gamma rays can be excited only by protons. Both emissions result from beams penetrating the solar surface. The electrons and protons must have been accelerated at about the same time in a loop no more than 10^5 km long. Otherwise, time-of-flight differences between the electrons, whose velocity is near the speed of light, and the protons, at one-fourth the light speed, would have been more apparent. [Reprinted with permission of E. L. Chupp, *Solar Physics* 86, 386 (1983), © D. Reidel Publishing Company]

the radio pictures also seem to show colliding loops (*35*).

Proton Acceleration

Electrons in flares stream away from points, perhaps neutral sheets where magnetic fields collide. All the theoretical mechanisms for particle energization indicate that protons should be accelerated whenever electrons are, but protons do not betray their presence in flares by direct radiation; we must infer their presence from nuclear reaction products. The affected nuclei, mostly of hydrogen, carbon, nitrogen, and oxygen, emit gamma rays when they decay after excitation by protons that penetrate the solar surface. The gamma rays were detected by instruments on the Solar Maximum Mission and other spacecraft. Although flares that produce detectable gamma rays are rare, they show when protons are definitely being accelerated at the sun.

In a large flare on 7 June 1980 several successive gamma-ray bursts followed similar x-ray bursts by less than 2 seconds (Fig. 9) (*36*). The inescapable conclusion is that the proton beam was being created in the same small (by solar standards) loop as the electron beam. If the size of the loop had been more than about 10^5 km, the fluctuations observed in the gamma rays would have been washed out by the spread in flight time of the protons. The fastest protons would arrive first and excite some nuclei; slower protons would arrive up to several seconds later and excite nuclei too, but the gamma rays emitted would be indistinguishable from those produced first. Thus, a big acceleration region would produce more gradual variations in gamma-ray emission than was observed. Evidently the acceleration region was small, and the protons were accelerated in the first seconds of the flare. Optical observations of the 7 June flare indicated that the proton acceleration region was probably a magnetic loop only a few thousand kilometers long (*37*). Yet the magnetic fields nearby were quite complex. Neutral sheets there could guide protons from the accelerator into space.

It was widely expected that gamma-ray emitting flares well connected to the earth would produce the strongest proton showers, and Pesses *et al.* (*38*) examined records of the International Sun-Earth Explorer for proton showers following three of the largest gamma-ray bursts. Two of the flares produced minor enhancements and one produced no shower at all. The study was then extended to all reported solar gamma-ray bursts, and still no correlation was found. If no correlation turns up in more detailed studies, then proton beams in gamma-ray flares are probably directed toward the sun's surface. Relatively few protons would escape in that case, and proton showers could not be predicted from gamma-ray flares.

Actually, the acceleration of a proton beam in a few seconds in a loop about the size of a sunspot came as a surprise. Most acceleration models are based on the Fermi mechanism and require 2 to 20 minutes to produce protons with energies greater than 10 MeV. The models divide proton flares into two stages. In the first stage, particles are accelerated to 100 keV in a turbulent neutral sheet. Whenever the impacts of the particles heat the chromosphere sufficiently, the dense plasma there expands explosively. A shock should streak through the corona at about 1500 km per second. Some of the protons from the first stage, then, might be accelerated as the shock sweeps through the overlying magnetic fields. The effect on particles would be the same as in Fermi's (*21*) original description of cosmic-ray acceleration in turbulent magnetic clouds; that is, protons already accelerated in the first stage could be accelerated to energies as high as 10^9 eV.

There were problems with the two-stage scenario even before the data from the Solar Maximum Mission showed that 10-MeV protons can be created in the first stage. Forecasts of proton showers based on the first-stage emissions are notably inaccurate. There is a poor correlation between first-stage emission level and proton shower intensity. Also, Mathews and Lanzerotti (*20*) had shown that the protons that struck the earth after one particularly large flare must have been accelerated into interplanetary space at the beginning of the flare. Delay times between 32 giant flares and their proton showers also indicate that proton acceleration occurs during flare onset (*39*). Further, Gloeckler *et al.* (*40*) showed that heavy nuclei in proton showers originate in the ambient solar atmosphere. If the nuclei are injected into the accelerator from the first-stage particle population, they would have ionization states characteristic of the hotter flare plasma. We needed a fast ion acceleration mechanism that does not require injection of hot flare particles. The accelerator would have to be triggered by some flare phenomenon but not fed by the flare particles themselves. It should be able to produce fast protons whether or not fast protons are produced in the flare itself.

Unnoticed by most solar physicists, Armstrong and his colleagues (*41*) were studying just such a mechanism in order to explain how ions are accelerated in shocks in interplanetary space (*42*). They succeeded in explaining the time profiles, anisotropies, and energy spectra of proton streams in shocks, and their acceleration model does not require first-stage injection from flares (*41*). Protons can be accelerated as they spiral in and out of the shock wherever it is almost parallel to the magnetic field. There is ample evidence for shocks and complex magnetic fields near large flares, and shocks and fields could meet at just the right angle to produce bursts of protons.

Pictures of the corona (Fig. 7) may show what the solar atmosphere looks like when a proton shower is developing. Munro, MacQueen, and their colleagues (*43*) found that every fast coronal eruption pushes a shock through the atmosphere ahead of it. Kahler, Hildner, and van Hollebeke (*44*) showed that every proton shower they studied was preceded by such a coronal mass ejection and shock wave. But sometimes there are shocks without coronal mass ejections. These seem to be blast waves triggered by the explosion of the chromosphere reacting under the intense heat generated by the flare. There is some evidence that the blast wave shocks decay quickly. Shocks driven by coronal mass ejections, on the other hand, can probably propagate to the earth (*45*). These shocks accelerate protons not only at the sun, but all the way along the route. The protons escape ahead of the shock face, and that is why they started arriving at earth an hour after the 10 April flare and kept coming until after the shock hit 58 hours later.

Conclusions

It now appears that proton-producing flares signal their onset from the sun not by their x-ray spectra, nor by covering an unusually large surface area, nor by copious emission of gamma rays. Spacecraft observations indicate that protons are energized in shocks, many of which are driven by coronal mass ejections. The faster an ejection moves, the steeper is the shock and the more energetic the protons.

Proton showers require two stages of particle acceleration as well as a favorable magnetic field geometry. The first stage of acceleration usually involves electrons and, sometimes, protons in small magnetic loops near sunspots. Most particles do not escape from the sun (*46*) but lose all their energy in penetrating the solar surface. The heat generated there and in the flaring fields themselves causes the atmosphere near the flare to explode. Shock waves from the explosion as well as shocks from the mass ejections that accompany many flares probably accelerate protons where they strike the ambient magnetic fields of the corona at an angle of about 3° (*41*). The origins of coronal mass ejections and of pre-flare magnetic energy are still not known, however. Although solar physicists believe that they have discovered where particles are accelerated in flares, they cannot distinguish among the proposed mechanisms.

As a result of the new insights gained from the Solar Maximum Year programs, we can mount a new attack on the problems of forecasting proton showers, but many uncertainties will have to be resolved before forecasts turn into confident predictions. The acceleration process requires an angle of about 3° between shock face and magnetic field. The magnetic fields in the solar atmosphere are complicated. How can we predict whether a shock will hit them at the required angle?

A cluster of small telescopes could be launched to provide real-time pictures of the magnetic fields in the corona as well as of the solar surface. The fields show up as giant loops when photographed with an x-ray telescope. Such observations cannot be obtained from the ground. Coronal mass ejections passing through the loops should show where protons are being accelerated. Surface magnetic fields should be monitored from space where the measurements will not be degraded by weather conditions. Subtle shifts in the fields could be tracked by observations at 30-minute intervals and the measurements used to continuously correct magnetic field models. Magnetic field growth and collision courses could be recognized quickly. Even though the connection between emerging magnetic fields and flares is not clearly understood, flare forecasts based on known statistical associations could be freed from the fundamental precision limit imposed now by the diurnal pace of ground-based observations. Many other statistical parameters used in flare forecasting could be made more reliable by obtaining more data at shorter intervals.

From a practical point of view, improved solar data are needed to predict the weather that men will encounter as they work more regularly in space. Had the Columbia been launched on 10 April as scheduled and had the flight plan called for a polar orbit with an extravehicular activity, Young and Crippen would have encountered the proton shower. Polar missions are planned for the next decade, but without improved forecasting many may be short and costly.

References and Notes

1. The range [*AIP Handbook* (McGraw-Hill, New York, ed. 3, 1972), pp. 8–161] of 11 MeV protons in aluminum exceeds 0.2 g cm^{-2}, which is the surface density of most spacesuits.
2. R. F. Donnelly, Ed., *Solar-Terrestrial Predictions Proceedings* (Government Printing Office, Washington, D.C., 1979); A. Reetz, Jr., Ed., *NASA Spec. Publ. SP-71* (1964); E. A. Warman, Ed., *NASA Tech. Memo. TMX-2440* (1972).
3. *Shuttle EVA Description and Design Criteria* (JSC-10615, NASA Johnson Space Flight Center, Houston, Texas, 1976); *Study to Evaluate the Effect of EVA on Payload Systems* (Rockwell International Space Division, Pittsburgh, Pa., 1975).
4. A. C. Upton, *Sci. Am.* **246**, 41 (February 1982).
5. C. C. Lushbaugh, *NASA Tech. Memo. TM X-2440* (1972), p. 398.
6. C. C. Lushbaugh, F. Comas, R. Hofstra, *Radiat. Res. Suppl.* **7**, 398 (1967).
7. E. G. Stassinopoulos, *J. Spacecr. Rockets* **17**, 145 (1980). Stassinopoulos has also shown that trapped electrons at geosynchronous altitude will deliver a prohibitively large dose of radiation within a few hours to a worker in a conventional spacesuit [E. G. Stassinopoulos and J. H. King, *IEEE Trans. A^rosp. Electron. Syst.* **10**, 442 (1974)].
8. J. W. Watts, Jr., and J. J. Wright, *NASA Tech. Memo. TM X-73358* (1976).
9. Radiation levels [*Natl. Res. Counc. Publ. 1487, File Memo FA 2-10-67* (Washington, D.C., 1967), table 2] were given in rems. A quality factor of 2 rems/rad has been applied to adapt the values to solar flare proton spectra [R. Madey and T. E. Stephenson, *NASA Spec. Publ. SP-71* (1964), p. 229]. Application of this quality factor to the annual maximum permissible doses recommended by the National Council on Radiation Protection yields the following yearly limits for occupational exposure: bone marrow, testes, or ocular lens, 2.5 rads; skin, 15 rads.
10. P. Simon *et al.*, in *Solar-Terrestrial Predictions Proceedings*, R. F. Donnelly, Ed. (Government Printing Office, Washington, D.C., 1979), vol. 2, p. 287.

11. H. E. Coffey, Ed., "Solar-geophysical data, prompt reports" (NOAA Environmental Data and Information Service, Boulder, Colo., 1981), Nos. 441 and 442.
12. Z. Svestka, *Solar Flares* (Reidel, Dordrecht, Netherlands, 1976).
13. B. Springer, "Space environmental support to the Space Transportation System orbital test flight one" (Air Force Global Weather Center Report AFGWC/OL-B, Offutt Air Force Base, Omaha, Neb., 1981).
14. *STS-1 Low Speed Ground Navigation Console Mission Report JSC-17495* (NASA Johnson Space Flight Center, Houston, Tex., 1981).
15. S. E. Forbush, *Phys. Rev.* **70**, 771 (1946).
16. D. K. Bailey, *J. Geophys. Res.* **62**, 431 (1957); W. I. Axford and G. C. Reid, *ibid.* **68**, 1743 (1963).
17. G. C. Reid, in *Intercorrelated Satellite Observations Related to Solar Events*, V. Manno and D. F. Page, Eds. (Reidel, Dordrecht, Netherlands, 1970), p. 319.
18. K. G. McCracken, *J. Geophys. Res.* **67**, 447 (1962).
19. E. N. Parker, *Astrophys. J.* **128**, 664 (1958); *ibid.* **132**, 175 (1960); *ibid.*, p. 821.
20. T. Mathews and L. J. Lanzerotti, *Nature (London)* **241**, 335 (1973).
21. E. Fermi, *Phys. Rev.* **75**, 1169 (1949); *Astrophys. J.* **119**, 1 (1954).
22. R. Giovanelli, *Astrophys. J.* **89**, 555 (1939).
23. M. J. Martres, R. Michard, I. Soru-Iscovici, T. T. Tsap, *Sol. Phys.* **5**, 187 (1968).
24. A. B. Severny, *Annu. Rev. Astron. Astrophys.* **2**, 363 (1964).
25. D. M. Rust, *Sol. Phys.* **47**, 21 (1976); J. Heyvaerts, E. R. Priest, D. M. Rust, *Astrophys. J.* **216**, 123 (1977).
26. P. Sweet, *Annu. Rev. Astron. Astrophys.* **7**, 149 (1969).
27. H. Petschek, *NASA Spec. Publ. SP-50* (1964), p. 425.
28. P. A. Sturrock, Ed., *Solar Flares* (Colorado Associated Univ. Press, Boulder, 1980).
29. D. S. Spicer, *Sol. Phys.* **53**, 305 (1977).
30. Z. Svestka, G. Van Hoven, P. Hoyng, M. Kuperus, *ibid.* **67**, 377 (1980); C. de Jager and Z. Svestka, *ibid.* **62**, 9 (1979).
31. *Solar Physics* **65** (1980), entire issue.
32. P. Hoyng *et al.*, *Astrophys. J. Lett.* **246**, L155 (1981).
33. H. Zirin, *Sol. Phys.* **58**, 95 (1978).
34. K. A. Marsh, G. J. Hurford, H. Zirin, R. M. Hjellming, *Astrophys. J.* **242**, 352 (1980); K. A. Marsh and G. J. Hurford, *Astrophys. J. Lett.* **240**, L111 (1980).
35. M. R. Kundu, E. J. Schmahl, T. Velusamy, *Astrophys. J.*, in press;———, L. Vlahos, *Nature (London)*, in press.
36. E. Chupp, in *Proceedings of the La Jolla Institute Workshop on Gamma Ray Transients and Related Astrophysical Phenomena* (American Institute of Physics, New York, in press).
37. D. M. Rust, A. Benz, G. J. Hurford, G. Nelson, M. Pick, V. Ruzdjak, *Astrophys. J. Lett.* **244**, L179 (1981).
38. M. E. Pesses, B. Klecker, G. Gloeckler, D. Hovestadt, *Proc. Intl. Conf. Cosmic Rays 17th*, **3**, 36 (1981).
39. E. W. Cliver, S. W. Kahler, M. A. Shea, D. F. Smart, *Astrophys. J.*, in press.
40. G. Gloeckler, R. K. Sciambi, C. Y. Farr, D. Hovestadt, *Astrophys. J. Lett.* **209**, 93 (1976).
41. G. Chen and T. P. Armstrong, *Proc. Intl. Conf. Cosmic Rays 14th* **5**, 1814 (1975); T. P. Armstrong and R. B. Decker, in *Proceedings of the Workshop on Particle Acceleration Mechanisms in Astrophysics*, J. Arons, C. Max, C. McKee, Eds. (American Institute of Physics, New York, 1979), p. 101.
42. E. T. Sarris and J. A. Van Allan, *J. Geophys. Res.* **79**, 4157 (1974).
43. R. H. Munro, J. T. Gosling, E. Hildner, R. M. MacQueen, A. I. Poland, C. L. Ross, *Sol. Phys.* **61**, 201 (1979).
44. S. W. Kahler, E. Hildner, M. A. I. van Hollebeke, *ibid.* **57**, 429 (1978).
45. H. V. Cane, R. G. Stone, J. Fainberg, J. L. Steinberg, S. Hoang, *NASA Tech. Memo. TM-82049* (1980).
46. T. Bai and H. S. Hudson, *Bull. Am. Astron. Soc.* **13**, 912A (1981).
47. I am grateful to S. Kahler, D. Neidig, and L. Acton for helpful comments on the manuscript. Supported by NSF contract ATM 7918412 and by the National Aeronautics and Space Administration contract NAS5-25545.

This material originally appeared in *Science* **216**, 28 May 1982.

3. Cosmic-Ray Record in Solar System Matter

Robert C. Reedy, James R. Arnold, Devendra Lal

The earth moves in a region of space once thought of as empty. The density of ambient gas is low, though measurable, as is the density of solar thermal photons, 3 K universal black body photons, and other low-energy particles. In this chapter we discuss the fluxes of high-energy nuclear particles in the earth's neighborhood, the inner solar system.

We define high-energy particles as those with energies on the order of 1 million electron volts or more per nucleon and consider only two types of energetic particles in the earth's environment: the galactic cosmic rays (GCR), which come from outside the solar system, and the solar cosmic rays (SCR), emitted irregularly by major flares on the sun. These particles have sufficient energy to penetrate deeply into solids and to interact with them, and they produce appreciable chemical effects, atomic displacements, and ionization in suitable media. Particles with higher energies, above ~ 10 megaelectron volts, also produce nuclear reactions. Because these various effects leave persistent records, the study of our high-energy particle environment is rich in historical possibilities.

It is easy to study solar and galactic cosmic rays separately. The GCR particles have high energies and are continuously incident on the solar system, but their flux and energy spectrum is modulated by solar activity. They produce effects to considerable depths, and are therefore important in larger bodies. The sporadic SCR particles, on the other hand, have much lower energies but higher average fluxes; the distinctive effects due to solar particles can be seen in the surface layers of extraterrestrial matter.

The records of both types of particles have been studied in terrestrial as well as extraterrestrial samples, such as meteorites and lunar rocks and soils. To study cosmic rays, the irradiation history of extraterrestrial samples and the transport processes on the earth must be understood. Although it is possible to study the history of asteroidal and planetary surface processes by studying the effects of charged-particle irradiation, this is a bootstrap process, involving studying one with assumptions about the other. The big differences in the nature and the interactions of the two types of cosmic-ray particles, the availability of a suite of terrestrial and extraterrestrial materials, and significant technological developments in studying the effects of charged-particle irradiation, have made it possible in the last decade to make significant progress in both fields.

Table 1. Energies, mean fluxes, and interaction depths of various types of cosmic-ray particles.

Radiation	Energies (MeV nucleon^{-1})	Mean flux (particles cm^{-2} sec^{-1})	Effective depth (cm)
	Solar cosmic radiation		
Protons and helium nuclei	5 to 100	~ 100	0 to 2
Iron group and heavier nuclei	1 to 50	~ 1	0 to 0.1
	Galactic cosmic radiation		
Protons and helium nuclei	100 to 3000	3	0 to 100
Iron group and heavier nuclei	~ 100	0.03	0 to 10

Nature of the Cosmic Rays

The energetic nuclei in the solar system have a vast variety of energies and compositions (Table 1). The nuclei in both the GCR and SCR are mainly protons and alpha particles (ratio ~ 10 to 20), with about 1 percent heavier nuclei (lithium to atomic number $Z \sim 90$ or more). The cosmic-ray intensities vary with geomagnetic latitude and with solar phenomena (*1*). Alterations in the mean flux of solar particles, and in the GCR flux, are both tied to the 11-year sunspot cycle (*1, 2*). Near sunspot maximum, the GCR flux decreases whereas the mean intensity of particles emitted by solar flares increases. Although these and other properties of cosmic rays were established by earth-based observations, measurements from spacecraft, especially beyond 1 astronomical unit from the sun, have refined and extended our knowledge of the origin, nature, and distribution of these energetic particles (*2*).

The initial sources of the GCR particles and the mechanisms for their acceleration are not known well but probably involve supernovae (discrete sources), the interstellar medium (diffuse sources), or both (*3*). As the particles diffuse or are transported to the solar system, various interactions, including acceleration, may occur. Finally, the solar magnetic fields modulate the spectrum of GCR particles as the particles enter the heliosphere zone, which extends out to ~ 50 AU. The modulation is due to scattering of particles on irregularities in the interplanetary magnetic fields, which are convected outward by the highly conductive solar wind plasma (*4*). Although changes in the sources, acceleration, or interstellar propagation of the particles can change their fluxes in the solar system, solar modulation is the dominant source of the observed GCR variability (Fig. 1).

The fluxes of 200 to 500 MeV particles are modulated by an order of magnitude during a solar cycle. At $E > 5$ to 10 gigaelectron volts per nucleon, the spectrum of GCR particles is not influenced much by solar activity, and its shape can be described roughly by a power law in energy, $E^{-2.5}$. Most nuclear interactions that produce records are induced by particles with $E > 1$ GeV per nucleon, which are affected only slightly by solar modulation.

An important question in examining the cosmic-ray record is how the GCR spectra varied during sustained periods of low or high solar activity. During long periods of essentially no solar activity, such as the Maunder minimum from 1645 to 1715 (*5*), the GCR particles would not

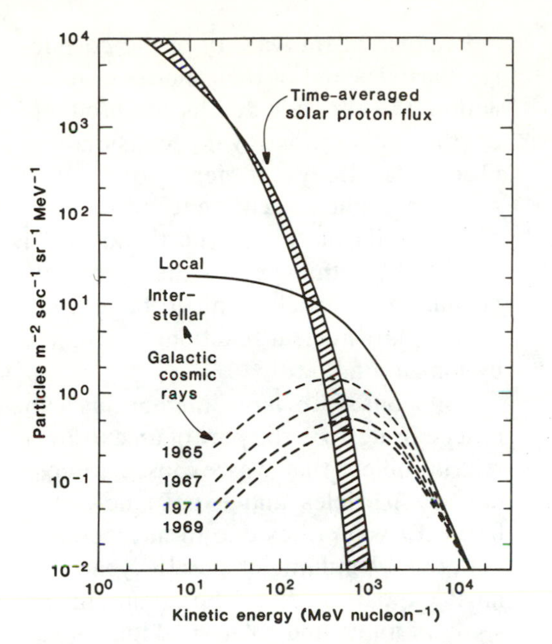

Fig. 1. The long-term average fluxes of solar protons determined from lunar data and GCR-proton fluxes for different modulation levels. The fluxes of GCR protons near the earth for 1965, 1967, 1969, and 1971 are calculated fits to satellite data and the curve for local interstellar space was estimated from modulation theory (*75*). [GCR curves and calculations courtesy of M. Garcia-Munoz and J. A. Simpson]

be hindered from reaching the inner solar system. The local interstellar spectrum (*6*) is estimated (*7*) to be similar to the one that is in the inner solar system during such periods of low solar activity.

At a distance of 1 AU from the sun, solar flare particles constitute an important source of medium energy (< 100 MeV per nucleon) corpuscular radiation. More than 100 solar flare cosmic-ray events have been observed near the earth since 1950. Only a few large flares produce most of the SCR particles emitted during an 11-year sunspot cycle (Fig. 2). When solar activity is low (sunspot numbers below 50), few energetic particles are emitted. The fluence of protons with $E > 10$ MeV in individual events has ranged from below 10^5 to more than 10^{10} protons per square centimeter (*8, 9*). Individual events last for a couple of days and have time-averaged fluxes that are several orders of magnitude higher than the total GCR flux.

The energy spectrum of solar particles is soft, with many particles of $E > 10$ MeV but few with $E > 100$ MeV. In

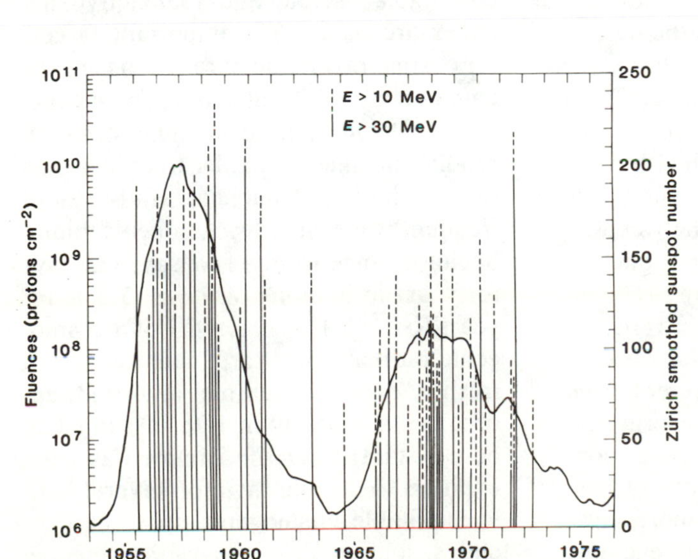

Fig. 2. Zürich smoothed sunspot number (continuous solid curve) and the omnidirectional integral fluxes of protons above 10 and 30 MeV emitted by solar flares from 1954 to 1976. Fluences for several flares that occurred close to each other have been combined (*8*).

general, the spectrum of SCR particles can be represented fairly well by $E^{-\gamma}$, where E is the kinetic energy per nucleon. For proton energies between 20 and 80 MeV γ typically ranges from 2 to 4, with an average of 2.9 at the time of maximum proton intensity (*10*). For $E < 20$ MeV, γ is generally lower and for $E > 100$ MeV the energy spectrum is usually steeper. Several other spectral shapes are also used for solar protons (*9*).

Cosmic-Ray Particle Interactions with Matter

The energy, charge, and mass of an energetic particle and the mineralogy or chemistry of the target mainly determine which interaction processes are important and which cosmogenic (cosmic ray–produced) products are formed. Energetic nuclear particles mainly interact with matter in two ways (*11*). (i) All charged nuclei continuously lose energy by ionizing atoms as they pass through matter, and the damage produced by radiation can accumulate in matter and be detected as thermoluminescence. The paths traveled by individual nuclei with $Z > \sim 20$ and with energies from 0.1 to 1 MeV per nucleon can be etched by certain chemicals and made visible as tracks (*12*). (ii) A nuclear reaction between an incident particle and a target nucleus involves the formation of new, secondary particles (for example, neutrons, pions, and gamma rays) and of a residual nucleus which is usually different from the initial one. Nuclei of low energy and high Z lose energy rapidly through ionization and come to rest. Those of high energy and low Z lose energy more slowly and usually induce nuclear reactions before stopping.

Because of the variety of the cosmic-ray particles and of their modes of interaction, the effective depths and products of the interactions vary considerably (Table 1). Below a depth of ~ 1000 grams per square centimeter (a few meters in solid matter or the thickness of the earth's atmosphere) there are few cosmic-ray particles; most have been removed by nuclear reactions or stopped by ionization.

Collisions with large meteoroids can reduce or destroy the part of an extraterrestrial object that was exposed to cosmic-ray particles and expose new surfaces. Erosion rates due to microcratering are ~ 1 millimeter per 10^6 years for most exposed rocks. The simultaneous study of implanted solar wind ions, solar and galactic heavy nuclei tracks, and nuclides produced by GCR and SCR nuclei can reveal changes in the irradiation geometry because each of these effects has a characteristic depth dependence.

When extraterrestrial matter is exposed to cosmic rays, a wide variety of cosmogenic stable and radioactive nuclides are made. The important targets for cosmic ray–induced reactions are the common elements like oxygen, magnesium, silicon, and iron. The mass of product nuclides is similar to or less than that of the target nucleus. Half-lives of frequently measured cosmogenic radionuclides range from a few days (35 days for ^{37}Ar) to millions of years (3.7×10^6 years for ^{53}Mn). The activity of a cosmogenic radionuclide starts near zero for a freshly exposed sample and will approach its production rate (assumed to be constant) after the sample has been exposed to cosmic rays for several half-lives. Stable cosmogenic noble gas nuclides, such as ^{21}Ne, are readily detected by mass spectrometry and are often used to determine a sample's integral exposure to cosmic rays.

The relatively low-energy solar protons and alpha particles are usually stopped by ionization energy losses near the surface of extraterrestrial matter. The SCR particles that induce nuclear reactions produce few secondary particles (and the product nucleus is close in mass to the target nucleus). Nuclear reactions induced by SCR-produced secondary neutrons are unimportant (*13*). The fluxes of SCR particles as a function of depth can be calculated accurately from ionization–energy loss relations; thus production rates of a nuclide as a function of depth can be calculated well if the cross sections for its formation are known. Activities of SCR-produced nuclides decrease rapidly with increasing depth (^{56}Co in Fig. 3).

Because high-energy GCR particles have ranges in matter that are much longer than their interaction lengths, most react before they are stopped, and each reaction with a nucleus usually produces many secondary particles, especially neutrons and pi mesons. The cascade of particles that develops from these interactions produces a population with many low- and medium-energy particles. The fluxes of the incident GCR particles decrease roughly exponentially with depth. The fluxes of secondary neutrons increase with depth near the surface, but then decrease exponentially with depth (*11*). In large objects, many secondary neutrons are slowed by scattering reactions to low energies and can produce certain nuclides, like ^{60}Co, by neutron capture reactions (*14*). The spectrum of the particles in an extraterrestrial object varies with depth and with the size of the object. Although the flux of GCR particles varies with solar activity, the shapes of their production rate-

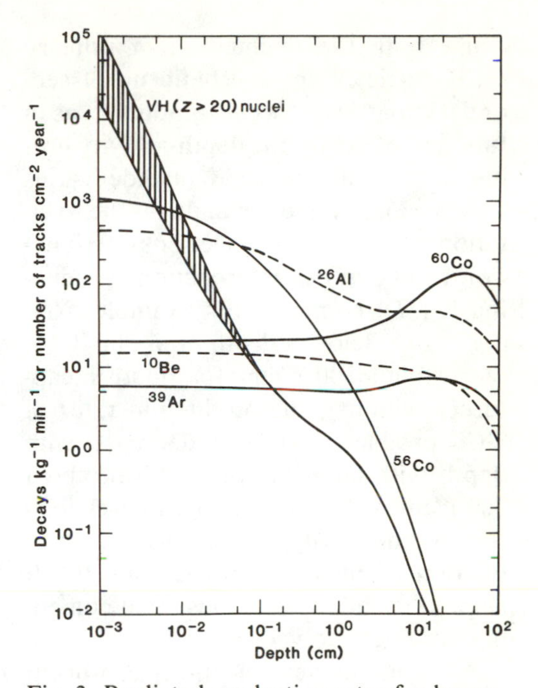

Fig. 3. Predicted production rates for heavy-nuclei tracks and various radionuclides as a function of depth in a lunar rock (density of 3.4 g cm^{-3}) directly exposed to cosmic-ray particles. The shaded region reflects uncertainties in the fluxes of low-energy VH nuclei in the SCR (*16*). The units for the production of ^{60}Co by the ^{59}Co(n,γ) reaction (*77*) are disintegrations per minute per gram of cobalt. The other curves were calculated by the Reedy-Arnold model (*11*) and the chemical composition of lunar rock 12002. The curve for ^{56}Co represents only production by solar protons by the ^{56}Fe(p,n)^{56}Co reaction. For ^{26}Al, ^{39}Ar, and ^{10}Be, production by both solar protons and GCR particles were included, although solar proton production is important only for ^{26}Al.

depth curves do not change much during a solar cycle (*13*).

The nature of the cascade caused by GCR particles and the resulting spectrum is well known from theoretical calculations (*13, 14*), bombardments of thick targets with accelerated particles (*15*), and studies of cosmogenic nuclides in extraterrestrial objects (*9, 11*). Because both the attenuation of primary GCR particles and the build up of secondary particles occur gradually as a function of depth, the depth-activity profile of a GCR-produced nuclide varies slowly (Fig. 3) and depends on the excitation functions (cross sections as a function of energy) of the reactions producing it (*11*); compare, for example, ^{39}Ar and ^{10}Be. Below a depth of about 100 g/cm^2 (about 300 g/cm^2 for neutron-capture reactions), the production rates of GCR-produced nuclides decrease with depth with an *e*-folding length of about 200 g/cm^2 (*11, 14*). The big difference in the depth-activity profiles for the production of nuclides by SCR and GCR (Fig. 3) usually allows these two components to be resolved.

Although nuclear reactions can occur in all the constituent phases of meteorites and lunar samples, observations of solid-state damage due to charged particle irradiation are usually limited to the crystalline dielectric phases of the minerals present, usually olivines, pyroxenes, and feldspars. Various techniques are used to observe solid-state damage due to solar wind , heavy cosmic-ray nuclei (usually $Z > 20$), and fission fragments (*9, 12*). We are primarily concerned here with tracks due to cosmic-ray nuclei of $Z \geq 20$. From their observed abundances, these nuclei are conveniently divided into two groups, those with $20 < Z \leq 28$, the iron or VH group, and the $Z \geq 30$ nuclei, the VVH group. The VVH nuclei (mainly, $30 \leq Z \leq 40$) are less abundant than the VH nuclei by about a factor of 500 to 700 for energies above ~ 500 MeV per nucleon (*16*). The determination of Z for the nucleus forming a track can now usually be made within ±2 charge units (*12, 16*). The identification of charge group (VH or VVH) can be made with certainty.

Profiles of track density as a function of depth have been determined in many meteorites and lunar samples with simple exposure histories and in a number of man-made materials exposed in space, including a glass filter from the Surveyor III camera returned by the Apollo 12 astronauts. In a track production profile for the moon (Fig. 3), the tracks in the top ~ 1 mm are made mainly by heavy SCR nuclei ($E <\sim 10$ MeV per nucleon); the deeper tracks are made by GCR nuclei ($E \geq 100$ MeV per nucleon) after they have slowed to $E \sim 2$ MeV per nucleon.

History of the Targets

The record of the effects of cosmic-ray bombardment on the earth and in lunar rocks and soil, meteorites of various classes, interplanetary dust collected in the stratosphere, and cosmic spherules found in deep-sea sediments is discussed in the following paragraphs. For these materials, the study of cosmogenic nuclides and nuclear tracks gives significant historical information.

Earth. The most famous cosmogenic radionuclide is ^{14}C (*17*), whose half-life is 5730 years. On the shortest time scales accessible to ^{14}C—tens to hundreds of years—correlations are observed among solar activity indices, climatic variables (especially temperature), and ^{14}C activity. When sunspots are few, ^{14}C activity is relatively high, and the climate is generally cold (*5, 18–20*). The higher ^{14}C content can be understood as a consequence of more GCR protons reaching the earth. It is less clear how solar activity affects climate.

In the 10^3- to 10^4-year range the dominant effect seems to be variation in the earth's magnetic field (Fig. 4). The main dipole field was apparently weaker one to two ^{14}C half-lives ago, so that more

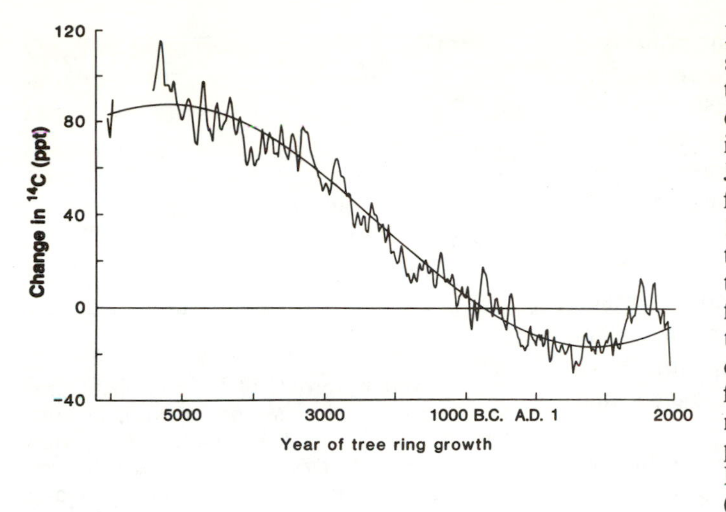

Fig. 4. Best-fit sine curve and spline functions drawn through the experimental change in ^{14}C values measured in parts per thousand at the La Jolla Radiocarbon Laboratory for dated tree rings (*19*). The 10^4-year sine curve represents the slow variation caused by the changing geomagnetic field. The high frequence fluctuations (Suess wiggles) are computer-generated spline functions through the experimental $\Delta^{14}C$ variations. [Reprinted with permission of H. E. Suess, *Endeavour* 4, 113 (1980)]

protons reached the earth, and the ^{14}C production rate was higher. The effect is modeled as a sinusoidal variation, with an amplitude in ^{14}C of about 10 percent and a period of about 10^4 years (*18*). Because only the last half-cycle is subject to detailed check, the true variation curve may not be close to the model at times further back than 10^4 years.

Longer-lived cosmogenic radionuclides enter the terrestrial environment in at least three ways. They are carried in by meteoritic material bombarded in space. This is the main source of nuclides produced from iron-group targets in meteorites, the most important of which is ^{53}Mn, with a half-life of 3.7×10^6 years. They are also produced by spallation reactions in the atmosphere; ^{10}Be (half-life, 1.6×10^6 years) is made from oxygen and nitrogen in this way (*21*), and bombardment of atmospheric argon is the main source of ^{26}Al (half-life, 7×10^5 years) (*22*). Finally, a small amount of production takes place in surface rocks and soil because of reactions of neutrons and muons.

Sensitive methods of detection, such as activation analysis for ^{53}Mn (*23*) and the use of accelerators for high-energy ion counting (*24*), have increased interest in these nuclides. For example, interest in ^{53}Mn is related to understanding the special processes that lead to the deposition of manganese-rich nodules and crusts in the deep sea. Raisbeck and coworkers (*25*) have measured the ^{10}Be content of seawater, ice cores, and sediments. We expect this work to lead to the development of models of the distribution of ^{10}Be in natural waters and sediments, which will permit this nuclide to play a useful role in geochronology and geochemistry. The chemical similarity of beryllium and aluminum suggests that atmospheric ^{10}Be and ^{26}Al should be distributed in the same way. Measurement of the $^{26}Al/^{10}Be$ ratio can provide another parameter and reduce the need for a well-supported geochemical model.

Moon. Because the moon has no atmosphere and relatively weak magnetic fields, cosmic rays produce all their effects in solid matter. Transport and mixing are slow by terrestrial standards. The transport process of "gardening" (regolith turnover) by meteoritic impact seems to predominate at the scales of distance and time accessible to our study. The radionuclides ^{53}Mn and ^{26}Al, produced abundantly by solar protons at depths less than a few centimeters, are ideally suited for gardening studies on a time scale of 10^6 to 10^7 years.

Our collections contain basically two kinds of samples. Rocks collected on the surface have been there for at least 10^6 years, with interesting exceptions. Most rocks seem to have had complex surface histories (*26*), in which tumbling, fragmentation, burial, and reexposure have all played a role. It also seems clear that on time scales as long as 10^7 years, a simple one-stage bombardment history on the moon's surface is improbable for rocks small enough to be collected. One well-characterized event, formation of the South Ray crater at the Apollo 16 site 2.0×10^6 years ago (*27*), ejected pristine material from below the zone penetrated by cosmic rays, including some rocks with simple bombardment histories. The dominant alteration effect in such rocks is erosion, produced mainly by sandblasting by micrometeorites, at a rate of ~ 1 mm per 10^6 years (*9*). The rate is dependent on the hardness of the rock.

The most instructive samples of the lunar regolith or soil are cores, ranging in length from about 20 cm to 3 m. Their history also appears to be complex, involving usually one or a few large cratering or depositional events, superimposed on a quasi-continuum of smaller disturbances. The nuclear track data (*9, 12, 16*) make clear the local heterogeneity of each layer, except for rare materials like the Apollo 17 orange glass. The lunar core track results also suggest several episodic enhancements in meteorite bombardment rates (*28*) during the last ~ 10^9 years. The data on SCR-produced ^{53}Mn (Fig. 5) and ^{26}Al show that disturbance has occurred, on a time scale of 10^6 to 10^7 years, to a depth that varies from a few to more then 10 cm. The deeper layers are undisturbed on this time scale. Monte Carlo models give a fairly satisfactory match with such observations (*29*).

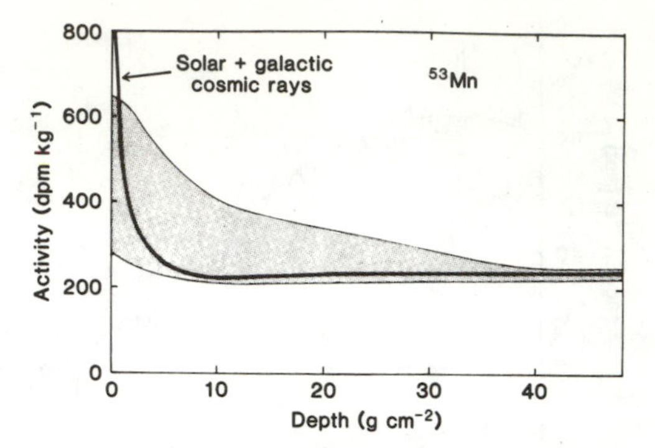

Fig. 5. Activity profile of ^{53}Mn in disintegrations per minute of ^{53}Mn per kilogram of iron. The heavy line is a calculated production profile for ^{53}Mn as a function of depth from the model of Reedy and Arnold (*11*) and parameters determined from lunar rock data. The hatched region covers the experimental ^{53}Mn concentrations determined in seven lunar cores (*78*). The broad spread results from the varying degrees of disturbance (gardening) which have occurred at the different places sampled, on a time scale of ~ 10^7 years. To convert the depth scale to centimeters, divide by ~ 1.8.

Meteorites. Cosmogenic noble gases, radionuclides (especially ^{53}Mn and ^{26}Al), and tracks measured in meteorites are the main source of evidence that most or all classes of meteorites originate in the asteroid belt, between Mars and Jupiter. The production rates for cosmogenic nuclides or tracks as a function of composition and shielding conditions (preatmospheric size and sample location) are known fairly well and are often used to determine the duration of exposure of a meteorite to cosmic-ray particles.

The exposure (or bombardment) age of a meteorite is most precisely measured with a radionuclide–stable product pair. The activity of the radionuclide is used to determine the production rate of

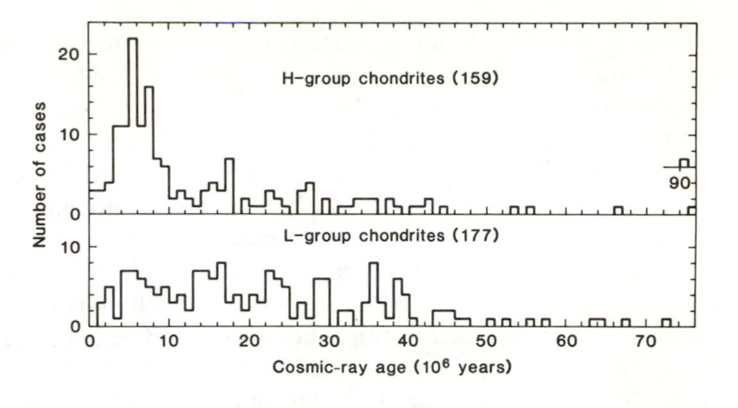

Fig. 6. Cosmic-ray exposure ages of the major types of stone meteorites calculated from measured concentrations of cosmogenic ^{21}Ne and $^{22}Ne/^{21}Ne$ ratios (for shielding corrections) taken from various sources. [Courtesy J. Smith and K. Marti. Updated version is in preparation]

the stable nuclide. The best pair is ^{81}Kr-^{83}Kr because both nuclides are produced by closely similar reactions, and both are measured by mass spectrometry at the same time. For long-lived iron meteorites, the measurement of ^{40}K (1.26×10^9 years) along with stable cosmogenic potassium isotopes is of great value (*30*). More commonly, a production rate for ^{21}Ne or some other nuclide is used. Recently, $^{22}Ne/^{21}Ne$ isotopic ratios in meteoritic samples have been used to correct ^{21}Ne production rates for the shielding conditions during exposure of the samples (*31*). Nuclear tracks and activities of long-lived radionuclides also provide information on the exposure history of meteorites. There have been uncertainties about the actual exposure ages (*32, 33*), but the ^{21}Ne data yield a good set of relative ages.

Bombardment ages of chondritic meteorites range up to a few tens of millions of years (*34*) (Fig. 6); those of iron meteorites are longer (*30*). The longest bombardment age so far, about 2×10^9 years on the ^{40}K-^{41}K scale, is that of the iron meteorite Deep Springs (*30*). There are statistically significant groups of H-group chondrites at 4×10^6 to 8×10^6 years (Fig. 6), and of coarse octahedrites (irons) at 5×10^8 to 6×10^8 years. These groupings, associated with specific meteorite types, appear to record individual events. Our picture is that meteorites spent their earlier histories contained in bodies that were large in comparison to the characteristic length of GCR penetration, a few meters. The start of the cosmic-ray exposure was a collision between two objects, which produced disruption or at least fragments of meteoritic size. These were brought into earth-crossing orbits by planetary perturbations (*35*). The observed time scales and age groupings have been essential data for the development of planetary perturbation models.

Some meteorites have been small objects for a relatively short time. The extreme example, so far, is the chondrite Farmington (*36*), with a bombardment age of 5×10^4 years. There is little time for perturbation of the orbit of such a short-lived object; thus if the orbit could be determined, it would be of major importance.

Multistage bombardments in space were noted by Chang and Wänke (*37*) for large iron meteorites. For such objects, 1 m or larger with bombardment ages $\sim 5 \times 10^8$ years, two or more stages of collisional breakup seem to be the rule. Nishiizumi (*38*) reported two cases, one very striking, of two-stage bombardment of chondrites in space.

There is evidence for precompaction irradiation—that is, the exposure to high-energy radiation preceding the assembly of the meteorite as a solid rock. Certain gas rich meteorites seem to have been exposed to the solar wind before assembly, and there are nuclear track records of ≥ 1-MeV particles in some of these (*39*). The best case for cosmogenic nuclide production before compaction in a meteorite is provided by inclusions (xenoliths) in the chondrite St. Mesmin (*40*).

The number of known meteorites is rapidly multiplying, because of the large numbers of objects being found in "blue ice" regions of the Antarctic ice sheet. We are particularly interested in them because they are "old." From the evidence of cosmogenic nuclides, it appears that nearly all of them fell to earth more than 3×10^4 years ago (*41*) and generally between 1×10^5 and 7×10^5 years ago. Some iron meteorites have been on the earth longer; Tamarugal, for example, has a terrestrial age of $\sim 2 \times 10^6$ years (*37*) and is the oldest we know.

Cosmic spherules and cosmic dust. The cosmic spherules found in deep-sea sediments have long been known, from their composition, to be of extraterrestrial origin. They are comparatively abundant (1 part in 10^7) in deep-sea sediments because of the slow rate of sedimentation of terrestrial materials there. These spherules must include materials from the meteoroids that never reach the earth's land surface (*42, 43*). The main uncertainty about their makeup is whether they are mostly ablation droplets from large meteoroids or small meteoroids that have undergone partial or total fusion. Deep-sea sediment cores can provide a continuous record of the intensity and nature of incoming extraterrestrial material for hundreds of millions of years (*43, 44*).

Brownlee and his co-workers (*45*) have collected small (~ 10 micrometers), fluffy, unaltered particles in the stratosphere that seem to be of cometary origin. These particles, whose survival was predicted by Öpik (*46*), are often chondritic in composition, but much less dense and crystalline than the carbonaceous chondrites collected on the earth's surface. Solar wind noble gases have been detected in them (*47*), verifying their extraterrestrial origin, but the small amount of material collected so far precludes measuring any cosmic-ray effects.

History of the Cosmic Rays

Many targets bombarded by the cosmic rays have relatively simple exposure histories that can be used to study the irradiating particles. Tree rings, meteorites, lunar rocks, and parts of Surveyor III returned by the Apollo 12 astronauts are examples. Concentrations of radioactive and stable cosmogenic nuclides provide data for determining fluxes of SCR and GCR particles, and track densities provide information on the heavy nuclei.

Solar cosmic rays. Indirect measurements of SCR particle flux by ionization chambers or radio-wave absorption over the polar caps extend back to 1936, but direct measurement by particle detectors in satellites began only in 1960 (*1*). The SCR record in meteorites usually is removed by ablation during passage through the earth's atmosphere, but the record in the lunar samples is well preserved.

The depth-activity profiles of SCR-produced radionuclides are clearly evi-

Table 2. Average solar-proton fluxes over various time periods as determined from lunar radioactivity measurements. Abbreviation: SPME, solar proton monitor experiment (*76*).

Period	Data source	Reference	Fluxes (protons cm^{-2} sec^{-1})			
			$E >$ 10 MeV	$E >$ 30 MeV	$E >$ 60 MeV	$E >$ 100 MeV
1965 to 1976	SPME	(*8*)	90	30	8	
1954 to 1964	^{22}Na, ^{55}Fe	(*8*)	380	140	60	26
10^4 years	^{14}C	(*52*)		70	26	9
3×10^5 years	^{81}Kr	(*53, 54*)			18	9
10^6 years	^{26}Al	(*50*)	70	25	9	3
5×10^6 years	^{53}Mn	(*50*)	70	25	9	3

dent in the surface layers of the returned lunar rocks. Generally the chemistries and cross sections used to derive mean fluxes and spectra of SCR particles, and the corrections for GCR production, are reasonably well known. The activities of short-lived radionuclides (for example, 78-day ^{56}Co and 35-day ^{37}Ar) produced in lunar rocks by solar protons were found to be in good agreement with those from satellite-measured proton fluxes for the solar flares (*8, 48*). The activities measured for ^{22}Na (2.6-years), ^{55}Fe (2.7-years), and ^{3}H (12.3-years) were made mainly by protons emitted before direct satellite measurements and, along with the relative intensities inferred from indirect measurements, were used to determine solar proton fluxes for the 1954 to 1964 solar cycle (*8*). The fluxes for the two most recent solar cycles were adopted from satellite measurements or determined from lunar radioactivities (*8*) (Table 2).

Several long-lived radionuclides are used to study past SCR fluxes: ^{14}C (5730 years), ^{59}Ni (8×10^4 years), ^{81}Kr (2.1×10^5 years), ^{26}Al (7.3×10^5 years) and ^{53}Mn (3.7×10^6 years). In lunar rocks with long exposure ages, the radionuclides can tell us about fluxes one to two half-lives before the present. The nuclides are produced mainly by solar protons, except ^{59}Ni, which is made mainly by solar alpha particles. On the basis of lunar ^{59}Ni activities and satellite measurements of solar alpha particles, Lanzerotti *et al.* (*49*) concluded that long-term and current solar alpha particle fluxes are comparable to within a factor of 4.

For the last 0.5×10^6 to 10×10^6 years, ^{26}Al and ^{53}Mn activities indicate relatively little change in the average fluxes of solar protons. The solar proton fluxes given in Table 2 are the values we prefer and are based on measurements from lunar rocks with a variety of exposure ages (*50*). Bhandari *et al.* (*51*) measured ^{26}Al activities in four rocks with exposure ages ranging from 0.5 to 3.7 million years and concluded that the average solar proton fluxes during the last 0.5 to 1.5 million years have varied less than ± 25 percent; their inferred fluxes are about three times those reported by Kohl *et al.* (*50*), however. The fluxes measured during the last few decades and the last few million years are similar, indicating that current solar activity is not atypical of what it has been in the past.

The activities of ^{14}C and ^{81}Kr indicate that average SCR fluxes for the last 10^4 and 10^5 years were considerably higher than they were during the last 10^6 years (Table 2). The ^{14}C measurements of Boeckl (*52*) for six depths in a lunar rock gave solar proton fluxes that are about three times those determined from ^{26}Al and ^{53}Mn data. Concentration-depth profiles of ^{81}Kr, measured by mass spectrometry in two lunar rocks, suggest that solar-proton fluxes above 60 MeV (the threshold energy for the main reactions producing ^{81}Kr) were considerably higher than those determined for the last 10^6 years (*53, 54*). There are, however, uncertainties in the ^{14}C- and ^{81}Kr-deduced fluxes (*54*).

Giant solar flares, much larger than any we have observed, might have occurred in the past. The tree ring record of ^{14}C seems to indicate that no flares more than ten times greater than those observed since 1956 have occurred since ~ 5000 B.C. The lunar data on ^{26}Al and ^{53}Mn set some limits on giant individual flares up to 10^7 years ago (*55*).

VH and VVH nuclei. The studies of tracks made by VVH and VH nuclei in meteorites and in lunar samples provide long-term average values of relative fluxes and energy spectra of nuclei for the Z interval 20 to 92 and in the energy region 0.5 to 2000 MeV per nucleon (*9, 56, 57*). The determination of absolute time-averaged fluxes of VH and VVH nuclei is not yet possible because of the lack of an independent measure of time; the inferences are somewhat model dependent. At low energies (< 10 MeV per nucleon), the duration of irradiation is controlled by erosion; in the high-energy region (> 100 MeV per nucleon), fragmentation, resulting in changes in the exposure geometry, becomes important. But it is possible to select rocks with a predominantly single-stage irradiation.

The flux and spectrum of iron group nuclei with energies of 100 to 200 MeV per nucleon are similar to these observed today. The tracks were formed during different intervals over the last 10^9 years. For nuclei of $Z \geq 30$, the available data primarily refer to the VVH/VH abundance ratios in the energy region 100 to 1000 MeV per nucleon. The ratio is found to be $1.5\ (\pm\ 0.2) \times 10^{-3}$, which is in good agreement with the abundance ratios in the sun or primitive chondrites (*57*). The relative abundances of heavier nuclei in the four charge groups Z = 52 to 62, 63 to 75, 76 to 83, and 90 to 96, have been found to match well with the solar and cosmic abundances.

Below 50 MeV per nucleon, the VVH/VH ratio increases as one goes to lower energies. For heavy nuclei, ~ 40 MeV per nucleon is the dividing line; above that the flux is due mainly to GCR and below to SCR particles (*57*). The ratio increases down to energies of ~ 1 MeV per nucleon, reaching about 2×10^{-2}, which is about ten times higher than the solar ratio. Preferential acceleration of heavier nuclei at lower energy also has been observed in recent SCR events (*58*).

High densities of tracks have been observed in grains inside the "gas rich" meteorites (*39*). The grains also have high concentrations of solar wind implanted noble gases (*59*). They appear to have been irradiated ~ 4×10^9 years ago while in a regolith similar to that found now on the moon's surface (*60*). Carbonaceous chondrites also have grains with tracks made during the early history of the solar system (*56*). The ratios of VVH/VH nuclei irradiating these meteoritic grains have varied but within factors of 2 or 3.

Analyses of SCR track densities and

gradients in grains from lunar cores show no evidence for periods as long as ~ 10^3 years with high flux of heavy nuclei during the last billion years or so (*61*). The shape of the energy spectrum of VH nuclei has remained remarkably similar for the epochs for which data are available.

Galactic cosmic rays. Most studies of GCR flux variations compare activities of radionuclides with different half-lives in meteoritic or lunar samples. Ratios of measured activities to those predicted by various models or to the concentrations of stable cosmogenic nuclides are used to look for variations in average fluxes over the mean lives of various radionuclides. Activities of ^{14}C and ^{10}Be in terrestrial samples have been used to study flux variations over shorter time intervals.

The orbits of only three meteorites are accurately known. They had low orbital inclinations and therefore were exposed not far from the earth's orbital plane. The activities of long-lived radionuclides in these and other meteorites agree well, indicating that almost all meteorites have been exposed to similar fluxes of GCR particles. The Malakal chondrite has an unusually high activity of ^{26}Al, which might be the result of irradiation by a high flux of cosmic-ray particles before about 2×10^6 years ago (*62*); the ^{53}Mn content, however, is normal (*38*). High activity ratios of short-lived (^{22}Na and ^{54}Mn) to long-lived (^{26}Al and ^{53}Mn) radionuclides in the Dhajala chondrite have been interpreted as being due to higher GCR fluxes at heliographic latitudes between 15° and 40°S than within ±15° of the ecliptic plane during solar minimum (*63*).

The activities of ^{22}Na (2.6 years), ^{46}Sc (84 days), and ^{54}Mn (312 days) have been measured in 24 meteorites which fell between 1967 and 1978 (*64*). The activities varied by factors of 2 or more, and the variations correlated with the sunspot cycle, with maximum activities at solar minimum. These results indicate that production rates for cosmogenic nuclides in meteorites can vary by up to factors of 3 between solar minimum and solar maximum.

Production rates for radionuclides with half-lives from 16 days (^{48}V) to 3.7×10^6 years (^{53}Mn) were calculated for iron meteorites and compared with experimental activities in the Aroos iron meteorite (*65*). The ratios of observed to calculated activities varied but did not show any systematic trend with half-life. These and other results for radioactivities in meteorites and lunar samples indicate that the fluxes of energetic GCR particles have varied less than about 25 to 50 percent during the last few million years and are similar to present fluxes.

Most studies of long-term GCR flux variations use iron meteorites or metallic (FeNi) phases of meteorites because they are chemically simple targets. Most of the radionuclides produced from iron have reaction threshold energies above 100 MeV, so that secondary particles are relatively unimportant and results from bombardments with accelerated particles can be used to predict production-rate ratios. Forman *et al.* (*66*) examined activities of ^{37}Ar (35 days) and ^{39}Ar (269 years) in metal from about 12 meteorites. They found that the ^{39}Ar activities were 10 to 18 percent higher than expected from the ^{37}Ar activities, a result consistent with long periods of reduced solar modulation during the last 500 years. Measurements of ^{39}Ar activities in meteorites that fell several centuries ago would help confirm the presence of higher GCR fluxes during the Maunder minimum. Activity ratios of ^{39}Ar to ^{36}Cl

(3×10^5 years) measured in iron meteorites are within 10 percent of the production ratios measured in iron targets bombarded by high-energy protons (*67*); thus the flux of GCR particles during the last ~ 500 years is similar to the average flux over the last $\sim 5 \times 10^5$ years.

Other long-lived radionuclides, such as ^{26}Al, ^{53}Mn, and ^{40}K, are produced by such different reactions that the calculated production ratios are somewhat model sensitive. As discussed above, a pair of radioactive and stable nuclides that are produced by similar reactions can be used to obtain exposure ages. For iron meteorites, the ages determined from $^{39}Ar/^{38}Ar$, $^{36}Cl/^{36}Ar$, and $^{26}Al/^{21}Ne$ ratios are usually similar, but those from $^{40}K/^{41}K$ ratios are usually about 50 percent higher (*68*). The differences cannot easily be explained by meteorite orbital changes $\sim 10^6$ to 10^7 years ago or space erosion (*68*). Thus, the flux of the cosmic rays to which iron meteorites were exposed during the past 10^6 years appears to have been roughly 50 percent more intense than the average for the last 10^9 years.

In stone meteorites, production rates of cosmogenic nuclides can vary considerably. Several investigators developed methods for determining exposure ages that include corrections for shielding effects due to different sizes and shapes of meteorites (*31, 69*). Studies (*33*) of radionuclides and cosmogenic neon in meteorites, including several with short exposure ages, indicate that the average GCR flux producing ^{26}Al in meteorites for the last 10^6 years could have been significantly greater (~ 40 percent) than that for ^{53}Mn over the last $\sim 5 \times 10^6$ years. The reactions that produce these two radionuclides have low threshold energies, so that this flux ratio involves lower energy particles than that for reactions in iron meteorites.

In addition to GCR flux changes, shielding changes due to multiple collisions or other causes could alter cosmogenic-nuclide production rates and apparent exposure ages, especially in meteorites with long exposure histories. Some of our data, such as half-lives, may be in error. These sources must be considered and eliminated before concluding that GCR flux variations caused production-rate changes in meteorites.

A GCR flux change could be either solar or nonsolar in origin. The movement of the solar system through the galaxy or in and out of the galactic plane could cause long-term flux changes (*70*). Rare external events (nearby supernovae) or solar variations (a Maunder minimum) can cause short-term flux changes. Such rapid fluctuations would be difficult to detect in extraterrestrial samples but are observable in terrestrial samples.

The most interesting cosmogenic radionuclides in terrestrial samples are those like ^{10}Be, ^{14}C, ^{32}Si, and ^{36}Cl, which are mainly made in the earth's atmosphere and which have half-lives that are long in comparison to the time scales for their removal from the atmosphere (*21*). These radionuclides produce "differential" records showing changes of production on a short time scale in organic matter, marine sediments, or glaciers that can be dated by independent techniques (for example, dendrochronology, geophysical events like magnetic reversals, or natural radioactivity). Most other radionuclides in the terrestrial environment yield results that are long-term averages only, like those in meteorites, because they mainly originate in interplanetary space (^{53}Mn) or reside in the atmosphere (^{39}Ar or ^{81}Kr).

The activities of ^{10}Be have been measured in deep-sea sediment cores, sections of which have been dated by paleomagnetic stratigraphy or by thorium isotope methods. Other chemical and physical properties of the cores were determined so that ^{10}Be concentrations could be converted to production rates (*71*). The ^{10}Be content varies, mainly because of changes in sedimentation rates but also possibly because of climatic changes or reversals or other variations in the earth's magnetic field. The inferred global ^{10}Be production rates for the last 2.5×10^6 years have varied by less than ± 30 percent when averaged for periods of 10^5 years and less than ± 10 percent for periods longer than 2×10^5 years. Studies of large diameter sediment cores from the equatorial Pacific Ocean indicate that, during the last 10^6 years, the global ^{10}Be production rate could have changed once by as much as 30 ± 7 percent, averaged over $\sim 10^5$ years, and had perhaps three or four smaller excursions with amplitudes of < 20 percent (*72*).

As discussed above and shown in Fig. 4, the measured ^{14}C activities in dated tree rings can be resolved into two components: one slowly varying because of geomagnetic field changes and one rapidly oscillating. The rapid variations (called Suess wiggles) have amplitudes of 1 to 2 percent and a prominent 200-year periodicity (*18*). These amplitudes agree with those predicted for changes in GCR fluxes due to extremes in solar modulation (*7*). The terrestrial ^{14}C data provide the sharpest limit available on short-term spikes, or sudden shifts, in the particle flux in the inner solar system. Because atmospheric $^{14}CO_2$ is a small part of the total ^{14}C reservoir, and because the transfer of CO_2 to other, larger parts of the reservoir is slow (*73*), a sharp spike or shift in production is well displayed in the record. Thus the production of ^{14}C by nuclear weapons testing in the early 1960's produced a quick doubling of atmospheric ^{14}C which has now decayed to a much lower level. The Suess wiggles are believed to provide evidence for long periods of unusual solar activity (*5, 18, 74*), and the magnitudes are consistent with the expected variations in production rates for a periodicity in solar activity of ~ 200 years (*7*). A larger change in the low-energy (< 1 GeV) proton flux is not ruled out, because such protons only reach the earth's atmosphere in the polar regions.

Conclusions

The most definite evidence for time variations of the cosmic rays near the earth is provided by studies of SCR products in lunar samples. The ^{14}C data seem to require a rather high proton flux on a scale of 10^4 years; limited ^{81}Kr data suggest something similar for as long as a few hundred thousand years. Over the longer periods, represented by ^{26}Al and ^{53}Mn, the flux seems to have been lower, on the average more like the present, if we knew how to define a present-day average.

We have several lines of evidence for the GCR intensity in space 1 to 3 AU from the sun. The best is ^{14}C variations in wood, which indicate a strong solar modulation effect leading to a large change in the global production of ^{14}C with an average period of about 200 years. These variations exceed in magnitude, but are similar to, solar modulation effects observed during the last few decades. The last such period of unusual solar activity, the Maunder minimum, apparently also caused enhanced production of ^{39}Ar in meteorites.

On a longer time scale the classical result is that there has been no change in the time-averaged GCR flux near the earth within some error on the order of 30 to 40 percent. Variations might well be expected either from changes in solar modulation on longer time scales, or in the sun's location in the galaxy (in or out of spiral arms). On a time scale of 10^5 to 10^7 years, we see effects that may be due to such changes, but other possibilities have not been eliminated.

On a 10^9-year scale, data on cosmogenic ^{40}K in iron meteorites require an increase in cosmic-ray flux toward the present; over $\sim 10^9$ years, the average is about one-third lower than that of the last 10^6 years. We cannot yet be sure whether this is a chapter in the history of meteorites or that of the cosmic rays.

Track and noble gas studies indicate that cosmic rays were present in the solar system near its beginning, with energy and charge spectra much like they are today. We may learn more from comparisons of the relative intensities of the solar wind, SCR, and GCR.

Studies of SCR products in lunar samples allow us to measure rates of meteorite impact, gardening, and rock fragmentation on the moon's surface. Gardening due to impact occurs to depths on the order of 10 cm in a few million years. Surfaces of lunar rocks are eroded at rates of $\sim$ 1 mm per 10^6 years.

Meteorites are broken out of larger bodies, again by impact processes, on time scales of $\sim 10^5$ to 10^9 years. Meteorites that have been in unusual orbits may show some differences in the GCR fluxes.

With the development of more sensitive techniques to measure cosmogenic radionuclides, we can expect to increase greatly the range and precision of the data available to us. The study of ^{10}Be in terrestrial ice cores may provide a detailed look at time variations in a range of time scales. The use of two or more isotopes (for example, ^{10}Be and ^{26}Al) can remove uncertainties in interpretation. Deep-sea sediments and the cosmic spherules they contain, and also Antarctic meteorites, will provide other important windows on the past.

References and Notes

1. M. A. Pomerantz and S. P. Duggal, *Rev. Geophys. Space Phys.* **12**, 343 (1974).
2. J. A. Simpson, *Astronaut. Aeronaut.* **16**, (No. 7/8), 96 (1978).
3. R. E. Lingenfelter, *Int. Cosmic Ray Conf.* **14**, 135 (1979).
4. H. Moraal, *Space Sci. Rev.* **19**, 845 (1976).
5. J. A. Eddy, *Science* **192**, 1189 (1976).
6. M. Garcia-Munoz and J. A. Simpson, personal communication.
7. G. Castagnoli and D. Lal, *Radiocarbon* **22**, 133 (1980).
8. R. C. Reedy, *Proc. Lunar Sci. Conf.* **8**, 825 (1977).
9. D. Lal, *Space Sci. Rev.* **14**, 3 (1972).
10. M. A. I. Van Hollebeke, L. S. Ma Sung, F. B. McDonald, *Sol. Phys.* **41**, 189 (1975).
11. R. C. Reedy and J. R. Arnold, *J. Geophys. Res.* **77**, 537 (1972).
12. R. L. Fleischer, P. B. Price, R. M. Walker, *Nuclear Tracks in Solids* (Univ. of California Press, Berkeley, 1975).
13. T. W. Armstrong and R. G. Alsmiller, Jr., *Proc. Lunar Sci. Conf.* **2**, 1729 (1971).
14. R. E. Lingenfelter, E. H. Canfield, V. E. Hampel, *Earth Planet. Sci. Lett.* **16**, 355 (1972).
15. T. P. Kohman and M. L. Bender, in *High-Energy Nuclear Reactions in Astrophysics*, B. S. P. Shen, Ed. (Benjamin, New York, 1967), p. 169; B. M. P. Trivedi and P. S. Goel, *J. Geophys. Res.* **78**, 4885 (1973); M. Honda, *ibid.* **67**, 4847 (1962).
16. D. Lal, *Philos. Trans. R. Soc. London Ser. A* **285**, 69 (1977).
17. W. F. Libby, *Radiocarbon Dating* (Univ. of Chicago Press, Chicago, ed. 2, 1955).
18. H. E. Suess, *Radiocarbon* **22**, 200 (1980).
19. ———, *Endeavour* **4**, 113 (1980).
20. J. A. Eddy, *Clim. Change* **1**, 173 (1977).

21. D. Lal and B. Peters, *Handb. Phys.* **46** (part 2), 551 (1967).
22. S. Tanaka, K. Sakamoto, J. Takagi, M. Tsuchimoto, *Science* **160**, 1348 (1968).
23. H. T. Millard, Jr., *ibid.* **147**, 503 (1965); K. Nishiizumi *et al.*, *Earth Planet. Sci. Lett.* **52**, 31 (1981).
24. R. A. Muller, *Science* **196**, 489 (1977); C. L. Bennett *et al.*, *ibid.* **198**, 508 (1977); G. M. Raisbeck, F. Yiou, M. Fruneau, J. M. Loiseaux, *ibid.* **202**, 215 (1978); A. E. Litherland, *Annu. Rev. Nucl. Part. Sci.* **30**, 437 (1980).
25. G. M. Raisbeck, F. Yiou, M. Fruneau, M. Lieuvin, J. M. Loiseaux, *Nature (London)* **275**, 731 (1978); G. M. Raisbeck *et al.*, *Geophys. Res. Lett.* **6**, 717 (1979); G. M. Raisbeck *et al.*, *Earth Planet. Sci. Lett.* **51**, 275 (1980); G. M. Raisbeck, F. Yiou, M. Fruneau, J. M. Loiseaux, M. Lieuvin, *ibid.* **43**, 237 (1979).
26. F. Hörz, R. V. Gibbons, D. E. Gault, J. B. Hartung, D. E. Brownlee, *Proc. Lunar Sci. Conf.* **6**, 3495 (1975); M. Wahlen *et al.*, *ibid.* **3**, 1719 (1972).
27. C. Behrmann *et al.*, *ibid.* **4**, 1957 (1973).
28. J. N. Goswami and D. Lal, *ibid.* **10**, 1253 (1979).
29. Y. Langevin and J. R. Arnold, *Annu. Rev. Earth Planet. Sci.* **5**, 449 (1977); ———, K. Nishiizumi, *J. Geophys. Res.* **87**, 6681 (1982).
30. H. Voshage and H. Feldmann, *Earth Planet. Sci. Lett.* **45**, 293 (1979).
31. G. F. Herzog and P. J. Cressy, Jr., *Geochim. Cosmochim. Acta* **38**, 1827 (1947); P. J. Cressy, Jr., and D. D. Bogard, *ibid.* **40**, 749 (1976).
32. R. C. Finkel, C. P. Kohl, K. Marti, B. Martinek, L. Rancitelli, *ibid.* **42**, 241 (1978).
33. K. Nishiizumi, S. Regnier, K. Marti, *Earth Planet. Sci. Lett.* **50**, 156 (1980); O. Müller, W. Hampel, T. Kirsten, G. F. Herzog, *Geochim. Cosmochim. Acta* **45**, 447 (1981).
34. J. Crabb and L. Schultz, *Geochim. Cosmochim. Acta* **45**, 2151 (1981).
35. E. Öpik, *Interplanetary Encounters* (Elsevier, Amsterdam, 1976); J. R. Arnold, *Astrophys. J.* **141**, 1536 (1965).
36. T. Kirsten, D. Krankowsky, J. Zähringer, *Geochim. Cosmochim. Acta* **27**, 13 (1963); E. Anders, *Science* **138**, 431 (1962).
37. C. Chang and H. Wänke, in *Meteorite Research*, P. M. Millman, Ed. (Reidel, Dordrecht, 1969), p. 397.
38. K. Nishiizumi, *Earth Planet. Sci. Lett.* **41**, 91 (1978).
39. D. Lal and R. S. Rajan, *Nature (London)* **223**, 269 (1969); P. Pellas, G. Poupeau, J. C. Lorin, H. Reeves, J. Audouze, *ibid.*, p. 272.
40. L. Schultz and P. Signer, *Earth Planet. Sci. Lett.* **36**, 363 (1977).
41. E. L. Fireman, *Proc. Lunar Planet. Sci. Conf.* **11**, 1215 (1980).
42. D. E. Brownlee, in *Cosmic Dust*, J. A. M. McDonnell, Ed. (Wiley, Chichester, 1978), p. 295.
43. M. T. Murrell, P. A. Davis, Jr., K. Nishiizumi, H. T. Millard, Jr., *Geochim. Cosmochim. Acta* **44**, 2067 (1980).
44. D. W. Parkin, R. A. L. Sullivan, J. N. Andrews, *Philos. Trans. R. Soc. London Ser. A* **297**, 495 (1980).
45. D. E. Brownlee, in *Protostars and Planets*, T. Gehrels, Ed. (Univ. of Arizona Press, Tucson, 1978), p. 134; ———, D. A. Tomandl, E. Olszewski, *Proc. Lunar Sci. Conf.* **8**, 149 (1977).
46. E. Öpik, *Publ. Tartu Astrofiz. Obs.* **29**, 51 (1937).
47. R. S. Rajan, D. E. Brownlee, D. Tomandl, P. W. Hodge, H. Farrar IV, R. A. Britten, *Nature (London)* **267**, 133 (1977); B. Hudson, G. J. Flynn, P. Fraundorf, C. M. Hohenberg, J. Shirck, *Science* **211**, 383 (1981).
48. R. C. Finkel *et al.*, *Proc. Lunar Sci. Conf.* **2**, 1773 (1971).
49. L. J. Lanzerotti, R. C. Reedy, J. R. Arnold, *Science* **179**, 1232 (1973).
50. C. P. Kohl, M. T. Murrell, G. P. Russ III, J. R. Arnold, *Proc. Lunar Planet. Sci. Conf.* **9**, 2299 (1978).
51. N. Bhandari, S. K. Bhattacharya, J. T. Padia, *Proc. Lunar Sci. Conf.* **7**, 513 (1976).
52. R. S. Boeckl, *Earth Planet. Sci. Lett.* **16**, 269 (1972).
53. A. Yaniv, K. Marti, R. C. Reedy, in *Lunar and Planetary Science XI* (Lunar and Planetary Institute, Houston, 1980), p. 1291.
54. R. C. Reedy, *Proc. Conf. Ancient Sun* (1980), p. 365.
55. R. E. Lingenfelter and H. S. Hudson, *ibid.*, p. 69.
56. J. N. Goswami, D. Lal, J. D. Macdougall, *ibid.*, p. 347.
57. D. Lal, *Philos. Trans. R. Soc. London Ser. A* **277**, 395 (1974); N. Bhandari, J. N. Goswami, D. Lal, A. S. Tamhane, *Astrophys. J.* **185**, 975 (1973); N. Bhandari and J. T. Padia, *Science* **185**, 1043 (1974).
58. E. K. Shirk, *Astrophys. J.* **190**, 695 (1974).
59. P. Eberhardt, J. Geiss, N. Grögler, *J. Geophys. Res.* **70**, 4375 (1965); H. Wänke, *Z. Naturforsch. Teil A* **20**, 946 (1965).
60. K. R. Housen, L. L. Wilkening, C. R. Chapman, R. Greenberg, *Icarus* **39**, 317 (1979).
61. G. Crozaz, *Proc. Conf. Ancient Sun* (1980), p. 331.
62. P. J. Cressy, Jr., and L. A. Rancitelli, *Earth Planet. Sci. Lett.* **22**, 275 (1974).
63. N. Bhandari, S. K. Bhattacharya, B. L. K. Somayajulu, *Earth Planet. Sci. Lett.* **40**, 194 (1978).
64. J. C. Evans, J. H. Reeves, L. A. Rancitelli, D. D. Bogard, *J. Geophys. Res.* **87**, 5577 (1982).
65. J. R. Arnold, M. Honda, D. Lal, *J. Geophys. Res.* **66**, 3519 (1961).
66. M. A. Forman, O. A. Schaeffer, G. A. Schaeffer, *Geophys. Res. Lett.* **5**, 219 (1978).
67. O. A. Schaeffer and D. Heymann, *J. Geophys. Res.* **70**, 215 (1965).
68. W. Hampel and O. A. Schaeffer, *Earth Planet. Sci. Lett.* **42**, 348 (1979).
69. G. F. Herzog and P. J. Cressy, Jr., *Geochim. Cosmochim. Acta* **41**, 127 (1977).
70. M. A. Forman and O. A. Schaeffer, *Rev. Geophys. Space Phys.* **17**, 552 (1979).
71. B. L. K. Somayajulu, *Geochim. Cosmochim. Acta* **41**, 909 (1977); S. Tanaka and T. Inoue, *Earth Planet. Sci. Lett.* **45**, 181 (1979).
72. P. Sharma and B. L. K. Somayajulu, *Earth Planet Sci. Lett.* **59**, 235 (1982).

73. J. R. Arnold and E. C. Anderson, *Tellus* **9**, 28 (1957).
74. M. Stuiver, *J. Geophys. Res.* **66**, 273 (1961).
75. M. Garcia-Munoz, G. M. Mason, J. A. Simpson, *Astrophys. J.* **202**, 265 (1975).
76. C. O. Bostrom *et al.*, *Solar Geophysical Data* (1967–1973), monthly reports.
77. M. Wahlen, R. C. Finkel, M. Imamura, C. P. Kohl, J. R. Arnold, *Earth Planet. Sci. Lett.* **19**, 315 (1973).
78. K. Nishiizumi, M. Imamura, C. P. Kohl, M. T. Murrell, J. R. Arnold, G. P. Russ III, *Earth Planet. Sci. Lett.* **44**, 409 (1975); K. Nishiizumi, personal communication.
79. We thank M. Honda, K. Marti, K. Nishiizumi, H. E. Suess, M. Garcia-Munoz, and J. A. Simpson for valuable discussions and for unpublished results and figures. Supported in part by NASA grant NSG-7027 and NASA work order 14,084.

This material originally appeared in *Science* 219, 14 January 1983.

4. Ultraviolet Spectroscopy and the Composition of Cometary Ice

Paul D. Feldman

The coming apparition of Halley's comet in 1986 has stimulated a degree of scientific activity unmatched in the history of this or any other comet. Current planning includes missions to the comet by instrumented spacecraft from the European Space Agency (ESA), the Soviet Union, and Japan. As this armada approaches the comet in March 1986, the comet will be scrutinized, both from the surface of the earth and from near-earth orbit, by a network of professional astronomers and by millions of curious amateurs and casual observers, many lured by the comet's fame or by the accounts of its "near miss" of the earth in 1910, the year of its last apparition.

Comets constitute a minuscule fraction of the total mass of the solar system. They are normally uninteresting objects, marking time in the cold outer reaches of the solar system except when their highly elongated orbits bring them close to the sun, typically within 1 or 2 astronomical units (AU; 1 AU is the mean earth-sun distance). At these distances, the sublimation of volatile material from the surface of the comet, induced by solar radiation, often produces one of the most spectacular celestial displays visible to the naked eye: a diffuse coma extending for tens of thousands of kilometers and a complex tail system, sometimes reaching more than 10^7 km from the head of the comet. All of this from an icy nucleus a few kilometers in diameter containing about 10^{-11} of the mass of the earth.

While some 10 to 20 comets are discovered and reported each year, including recoveries of known periodic comets, the number of bright, naked-eye objects is more like one or two per decade. Halley's comet, with an orbital period of approximately 76 years (this changes slightly due to planetary perturbations on each return), is the only reasonably short-term periodic comet to be a naked-eye object and to display all the phenomena usually associated with "new" comets. Apparitions of bright comets are sufficiently infrequent that such events were regarded by our ancestors as being inexorably linked with the important events of the day (for instance, the invasion of England by William the Conqueror in A.D. 1066), and so we have records and images of some comets (including Halley) that go back more than 2000 years.

Nowadays, the apparition of a bright, new comet is a stimulus to astronomers to apply the latest technology routinely available to observers of stars, planets, or extragalactic objects—all fixed in the heavens—to the study of these small, frozen visitors. However, the time scales for such apparitions, except in rare cases such as the widely publicized comet Kohoutek (1973 XII), are only of the order of a few months, making it impossible to send instrumented spacecraft for in situ exploration and limiting the space study of comets to observations made from earth-orbiting observatories designed for the study of objects quite different from comets. The predictability of Halley's return in 1985–1986 removes this problem, and in fact planning for the now nonexistent NASA mission to Halley was begun as early as 1970.

Even for comet Kohoutek the available time before perihelion was only 9 months, but this afforded the astronomical community the opportunity to plan and launch several far-ultraviolet sounding rockets; to place an existing far-ultraviolet camera on board the Skylab space laboratory, which was expected to be operating near the time of perihelion; and to obtain telescope time for an extensive program of ground-based visible and radio observations. An important lesson was learned from this experience: even though the comet "fizzled" as a naked-eye object to the casual observer, the scientific return was prodigious. Two years later, when comet West (1976 VI) appeared with scarcely 3 months notice and blossomed into the spectacular naked-eye object that comet Kohoutek was advertised to be, the scientific enterprise was markedly inferior, and a number of important discoveries made on Kohoutek were not followed up and remain questionable to this day.

As an example of the results from comet Kohoutek, a pair of images published 8 years ago (*1*) is shown in Fig. 1. The image on the left was taken in visible light, while the one on the right is an ultraviolet image in the light of HI Lyman-α (wavelength $\lambda = 1216$ Å) and shows the extent of the atomic hydrogen cloud produced by photodissociation of the primary water ice constituent. Both images were obtained from sounding rockets launched a few days apart approximately 1 week after perihelion. The extended hydrogen envelope was actually discovered a few years earlier from observations of comets Tago-Sato-Kosaka (1969 IX) and Bennett (1979 II) by two orbiting observatories, OGO-5 and OAO-2 (*2, 3*), but the rocket image of Kohoutek was the first direct image of this phenomenon, an atomic cloud several times larger than the sun. A similar camera on Skylab provided images of the hydrogen envelope over a period of a few weeks (*4*), and both rocket experiments provided spectroscopic evidence for a large abundance of carbon and oxygen atoms in a rather extended ($\sim 10^6$ km) coma (*1, 5*). The impression of the comet, viewed in the ultraviolet, is quite different from that in the visible, where what mainly is seen is sunlight reflected from the dust component in the coma and resonance fluorescence of sunlight by radicals such as CN, C_2, and NH—dissociation products of still uncertain polyatomic species that exist as impurities (< 1 percent) in the water ice. In other words, in the ultraviolet we see the "ice" products from the "dirty ice" nucleus, whereas in the visible we see, and have seen quite spectacularly for millennia, only the "dirt."

The other area in which comet Kohoutek yielded interesting results was radio astronomy. Whereas in the vacuum ultraviolet we observe the strong resonance transitions of the atoms that constitute the molecules which make up the cometary ice only after the molecules have been completely photodissociated, we can detect the parent molecules, evaporating directly from the cometary

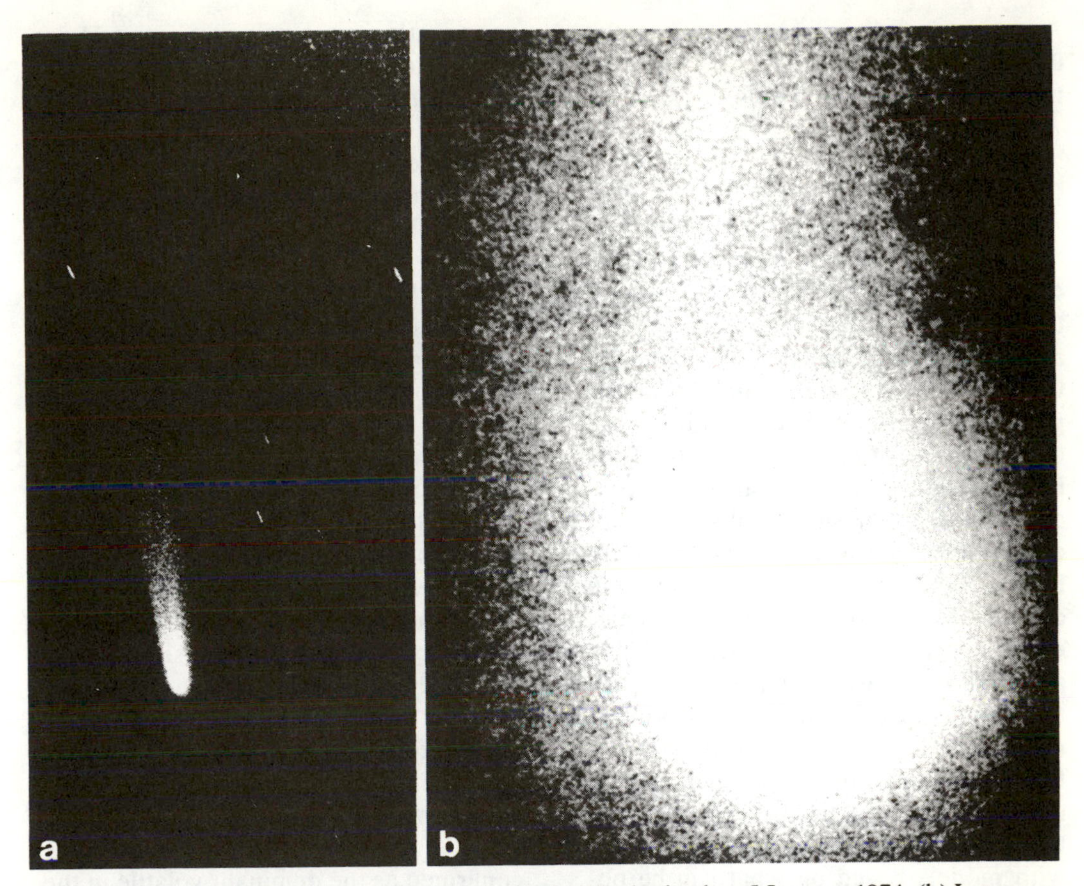

Fig. 1. (a) Visible image of comet Kohoutek (1973 XII) obtained on 5 January 1974. (b) Lyman-α image of the hydrogen envelope surrounding comet Kohoutek obtained by an ultraviolet camera on board a sounding rocket 3 days later, shown to the same scale (*1*).

nucleus, only by transitions in the infrared or radio-frequency region of the electromagnetic spectrum. At the time of the apparition of comet Kohoutek, radio astronomers were rapidly adding to the list of polyatomic molecules found in interstellar clouds, and many of these molecules, such as NH_3, HCN, HC_3N, and CH_3CN, were of interest as possible progenitors of the species then known to be present in comets, particularly C_2, C_3, CN, NH, and NH_2. Radio observations of the 18-centimeter lines of OH, a species known to be present in cometary comae from the OH ultraviolet band at 3085 Å, and detection of CH_3CN, HCN, and CH were reported, as well as upper limits for a wide range of other polyatomic species (*6*). Subsequently, OH has been observed by radio techniques in a number of comets (*7*), but none of the other species has since been detected, even though searches of a number of comets have been carried out with equipment considerably more sensitive than that used in 1974. A comprehensive survey of the radio results, both positive and negative, and a discussion of the problems inherent in these observations have recently been given by Snyder (*6*).

A continuing discrepancy between the ultraviolet and the radio OH observations will be discussed below.

Cometary Models

Comets are of particular interest because they may be frozen remnants of the primordial solar system and as such may provide clues needed to understand the evolution and differentiation of the planets and their satellites after the initial condensation of the solar nebula. Thus the basic questions to be answered concern (i) the composition and structure of comets and (ii) where and how they were formed. The answer to the first question can be obtained by direct measurement, even though the present group of Halley missions might not be equal to the task and it may require an eventual sample return mission. Also, the answer to the first question does not guarantee a unique answer to the second one. For the present, though, we must be content with models based on what can be observed by optical and radio techniques, namely the coma and tail when the comet is at small heliocentric distances and an unresolved asteroid-like inactive nucleus for comets at several astronomical units from the sun.

The generally accepted model of an icy conglomerate a few kilometers in diameter containing both volatile and refractory components was first proposed by Whipple (*8, 9*) on the basis of his analysis of the "nongravitational" force perturbation of the orbit of periodic comet Encke. These forces arise from the jet action produced by nonuniform vaporization of matter from the surface of a rotating comet nucleus. Subsequent investigations (*10*) indicated that the major volatile component controlling the vaporization must be water ice, and recent ultraviolet spectra of a number of comets provide strong confirming evidence (*11*) for this idea. However, the ultraviolet observations are of the gaseous coma, not the nucleus, and as Oppenheimer (*12*) pointed out, gas phase reactions in the inner coma (within 10^3 km of the nucleus, a distance barely resolvable with available instruments) might drastically alter the initial gaseous composition, making inferences of the nuclear constituents based on derived coma abundances incorrect. This suggestion has not been substantiated by recent detailed photochemical models, which employ many hundreds of coupled molecular reactions in attempts to reproduce the observed cometary abundances, particularly for the simple radicals (*13, 14*). Moreover, the ultraviolet observations show remarkably similar spectra for comets with widely differing gas production rates, indicating little or no dependence of the composition on gas density. As a result, not only is water ice confirmed as the dominant volatile in the nucleus, but an initial composition of volatiles similar to what is found in interstellar molecular clouds is needed to account for the observed abundances of radicals (*15*). A different approach was taken by Delsemme (*16*), who ignored the details of the chemistry and simply counted the end-product atoms, which were then assembled into a hypothetical nucleus, but again water ice emerged as the dominant constituent.

Unlike planetary atmospheres, which are gravitationally bound and exhibit only relatively mild temporal variations, the atmosphere of a comet is a highly transitory and variable phenomenon. The most rudimentary coma models (*17, 18*) assume that the parent molecules sublimated at the surface of the nucleus flow radially outward, subject only to the solar ultraviolet radiation field, which progressively decomposes them into their constituent atoms (or ions), which continue to flow radially outward. This one-dimensional picture serves as a basis for the more detailed photochemical models (*13, 14*), which must still be regarded as only approximate since they ignore the excess photochemical energy that is available to the dissociative fragments and results in velocities that differ markedly from the monokinetic, radial velocities of the models. This is a severe constraint for the lightest species, particularly hydrogen, as demonstrated by Festou (*19*). Moreover, as Mendis and Houpis point out (*20*), there are actually two classes of models: those that concentrate on the chemical processes and assume simplified geometry and dynamics, and those that minimize the chemistry and emphasize the dynamics and geometry of the outward flow. Both approaches treat the radiative transfer problem, in the ultraviolet and the infrared, in only a rudimentary fashion, and this is of some concern for the interpretation of the ultraviolet observations of H and O, the atomic products of H_2O dissociation, both of which are normally optically thick. In addition, the refractory grains dragged out of the nucleus by the vaporizing gas are assumed to decouple from the gas close to the nucleus, yielding noninteracting gas and dust comae whose subsequent expansion is usually treated separately (*21*).

Despite these difficulties, the derived abundances of H, O, and OH in the coma (*11*) appear to be consistent with an H_2O source at least an order of magnitude more abundant than any other hydrogen-bearing molecule. This interpretation can be verified by the in situ mass spectrometer measurements that will be made by the Halley fly-through missions, particularly by the ESA Giotto spacecraft, which is to fly within 10^3 km of the nucleus. In situ measurements have an additional advantage of being able to detect species not observable by remote optical means. Examples of such species are atomic nitrogen, whose principal resonance transition at 1200 Å is masked, at least to presently available instrumentation, by the nearby intense HI Lyman-α emission at 1216 Å; H_3O^+, the presumed dominant ionic species in the inner coma (*22*), whose optical spectrum is unknown; and, of course, the whole range of possible parents of the observed radicals which have not been detected by any radio search to date (*6*). Even so, a number of species may escape detection if their photochemical lifetime is so short that they have disappeared at the distance of closest encounter of the spacecraft. This is the case for CS_2, which has been suggested as the parent of the S and CS observed in the ultraviolet spectra of a number of comets, and whose calculated photochemical lifetime of $\approx 10^2$ seconds is consistent with the observed intensity and spatial distribution of these species (*23*). With an abundance $\sim 10^{-3}$ that of H_2O, CS_2 appears to be present in the ice of most comets (*24*).

It should be stressed that the in situ measurements will give only a snapshot view of the coma of Halley's comet and that complementary remote observations will be necessary to fit this snapshot into an overall description of the evolution of the comet. Moreover, unlike the similar case that arises in connection with planetary exploration, the in situ measurements provide only a partial calibration of the remote observations made on other comets, whose composition may vary considerably from that of Halley. The degree of similarity or difference of composition may reflect on the various proposals for the site of formation of the comets in the solar nebula.

Far-Ultraviolet Observations

The fundamental compositional difference among various comets, as deduced from ground-based observations, is the dust-to-gas ratio. This is related to the amount of observed continuum (sunlight reflected from solid grains) relative to gas fluorescence produced by C_2, C_3, CN, and so on, which is taken as a measure of the total gas production rate. Among the species detectable in the visible, which represent less than 1 percent of the total volatile component, there appears little variation from comet to comet, provided observations are compared at similar heliocentric distances (*25*). The abundance of these species relative to H_2O (or at least to OH presumably derived from H_2O) is also relatively constant from comet to comet. One exception does exist, and that is CO^+, whose fluorescence in the comet-tail band system ($A^2\Pi_i$–$X^2\Sigma^+$) in the blue region of the spectrum gives the ion tail its visibility. However, sometimes this species is not detectable in the coma, and at best only a weak ion tail is seen, particularly in old, short-period comets. On the grounds that spectacular ion tails are most often associated with bright, new comets (or at least those with very long periods, $> 10^6$ years) and that both CO and CO_2 are considerably more volatile than H_2O, it has been suggested that new comets contain a significant fraction of either of these species and that, for at least part of their orbit, the sublimation of gases from the nucleus is controlled by this more volatile fraction. This explanation could certainly account for the early discovery of comet Kohoutek (at ~ 5 AU from the sun) and its subsequent fizzling (*16*).

Spectroscopy in the far-ultraviolet provides a convenient means of studying compositional variations of the dominant volatile species, as H_2O, CO_2, and CO all have signatures in this spectral region. As noted above, all three dissociation products of H_2O—H (1216 Å), O (1304 Å), and OH (3085 Å)—resonantly scatter sunlight strongly, as demonstrated by the spectrum of comet Bradfield (1979 X) obtained with the International Ultraviolet Explorer (IUE) satellite observatory (*11*) shown in Fig. 2. Carbon monoxide is detectable through its fourth positive band system ($A^1\Pi$–$X^1\Sigma^+$), as seen in the rocket spectrum of comet West (1976 VI) in Fig. 3 (*26*), and its ion by its first negative system ($B^2\Sigma^+$–$X^2\Sigma^+$) between 2100 and 2500 Å, although the latter provides the same information as the blue comet-tail system. While CO_2 is not directly detectable, its dissociation often leads to the product CO molecule in the metastable $a^3\Pi$ state with subsequent emission of a photon in the forbidden Cameron band system (1800 to 2200 Å), which is one of the dominant emissions in the dayglows of the CO_2 planets (*27*). An upper limit (*26*) and a possible detection (*28*) of this emission from comet West were reported from separate rocket experiments. The $\tilde{B}^2\Sigma_u$–$\tilde{X}^2\Pi_g$ double band of CO_2^+ at 2890 Å, also a feature of the Martian dayglow, appears to be present in most comets (*29*) (see Figs. 2 and 3), and atomic carbon is detectable through several multiplets at 1561, 1657, and 1931 Å (*26*).

Since the terrestrial atmosphere is opaque to wavelengths below 3000 Å, observations in the far-ultraviolet must

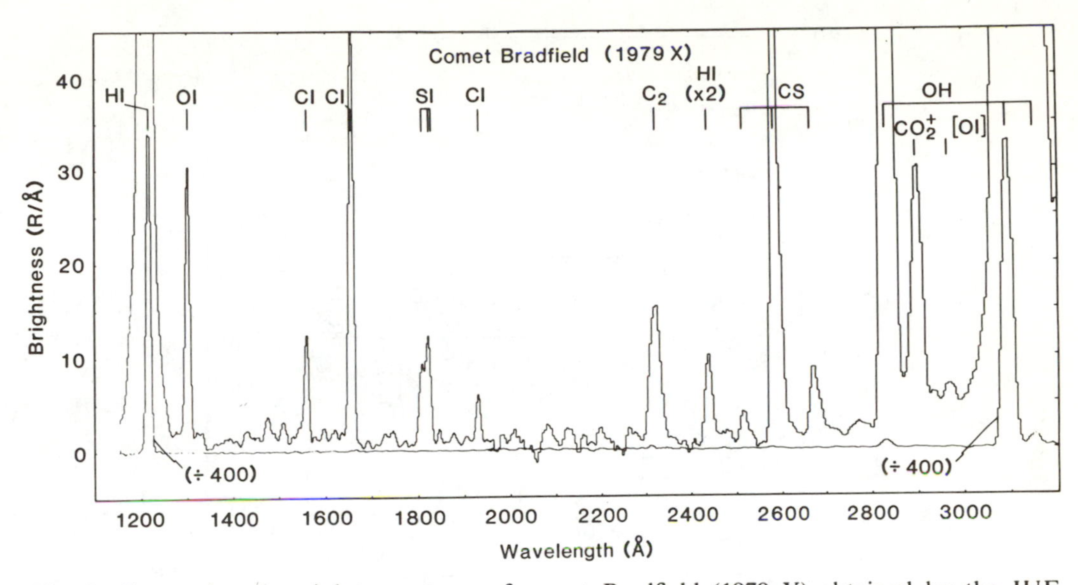

Fig. 2. Composite ultraviolet spectrum of comet Bradfield (1979 X) obtained by the IUE satellite observatory on 10 January 1980, when the comet was 0.71 AU from the sun and 0.62 AU from the earth (*11*).

be made from above the absorbing O_2 and O_3, either from earth-orbiting observatories or from sounding rocket platforms. Despite the initial successes with comet Kohoutek, comprehensive spectra spanning the entire wavelength range from 1200 to 3200 Å were not obtained until 1976, for comet West, by means of several rocket experiments (*26, 28*). Since 1978, the IUE has observed nearly a dozen comets, but most of these have been quite faint, and only comet Bradfield (1979 X), a moderately active comet similar in orbit and gas production rate to Halley (*30*), and periodic comet Encke were studied in any detail (*31*). An overview of cometary far-ultraviolet spectroscopy (*32*) and a summary of the IUE observations through 1981 (*24*) have recently been given, and I will limit the discussion here to aspects of the data relevant to the determination of compositional trends.

The interesting comparison to be made is between comet West, in Fig. 3, and comet Bradfield, in Fig. 2. However, this comparison is model-dependent as the two observations were made with widely differing viewing geometries. The IUE spectrographs have apertures of roughly 10 by 20 arc seconds, which translates to a projected size of 4500 by 8900 km^2 for the observation of comet Bradfield on 10 January 1980. The ≈ 5 arc second spatial resolution capability of IUE gives a linear resolution of 2200 km at the comet. The rocket spectrometers had much larger apertures and the linear size of the projected image on the short-wavelength slit for comet West was 3.7×10^5 by 1.3×10^6 km^2, about a factor of 10^4 in solid angle greater than that for the IUE. The higher signal-to-noise ratio of the rocket spectrum, despite the much shorter observation time, is in part due to the intrinsically brighter comet observed, but is mainly due to the increased sensitivity to extended objects

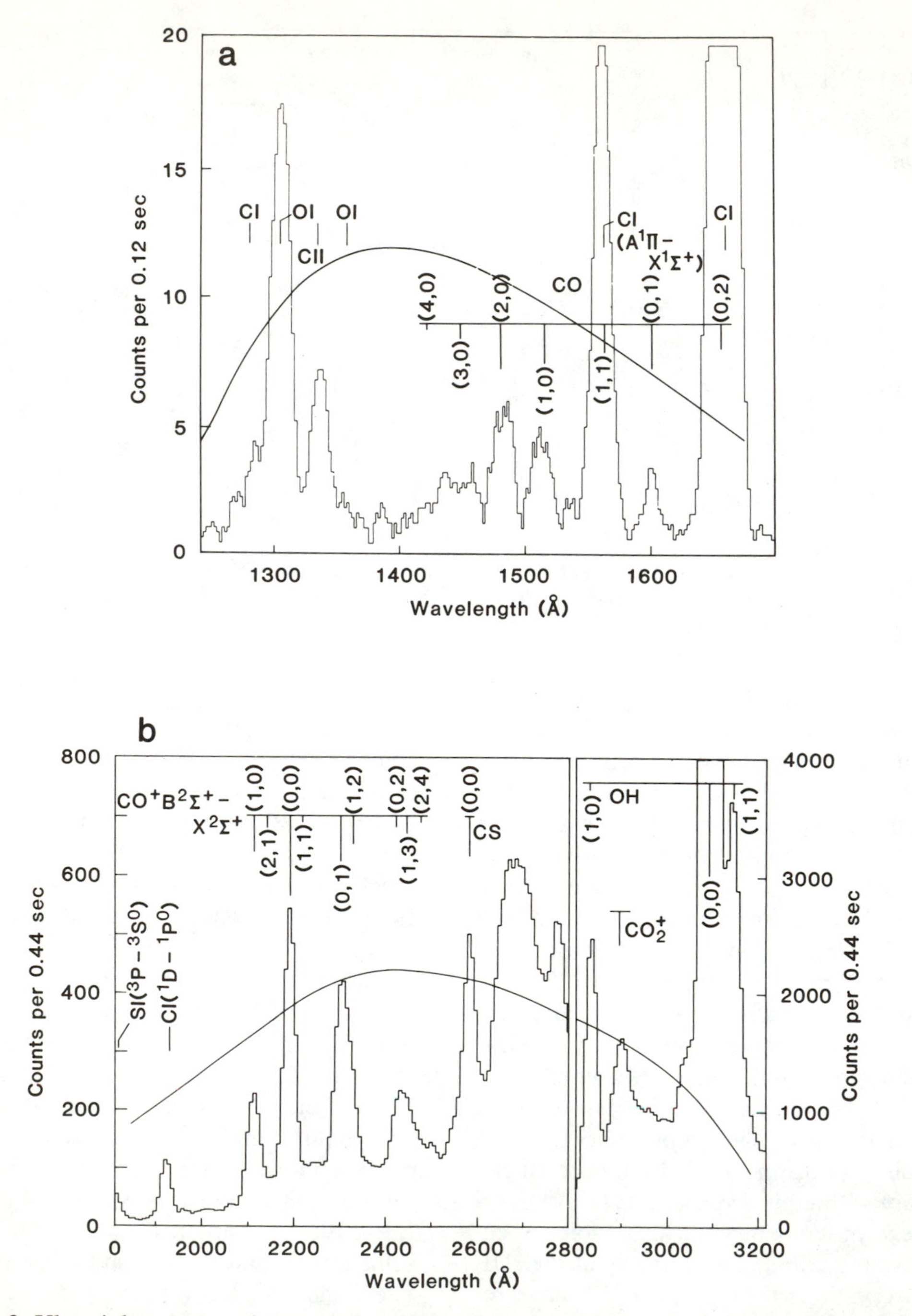

Fig. 3. Ultraviolet spectra of comet West (1976 VI) recorded by a sounding rocket payload on 5 March 1976 (*26*); (a) and (b) are short- and long-wavelength spectra, respectively. The solid line shows the relative response of the spectrometers. In (a) a CaF_2 filter was used to attenuate the transmission of HI Lyman-α to prevent grating-scattered light from masking the weaker emission features.

Table 1. Principal ultraviolet emission features.

Species	Wavelength (Å)
H	1216
O	1304
C	1561, 1657
S	1814
CO	1510
C_2	2313
CS	2576
OH	3085
CO_2^+	2890

of the rocket instruments. Furthermore, if the instrument field of view is large enough that the total photon flux in a given emission feature is collected—that is, if the projected distance of the slit on the comet is greater than the scale length for that species—then the derivation of the production rate of the species becomes model-independent. The large field of view also maximizes the sensitivity of the spectrograph to weak emission features. On the other hand, for studies of the chemistry and dynamics of the gas outflow the highest possible spatial resolution is warranted, even though it may be useful only for the strongest emissions.

A list of the spectral features usually observed in the IUE data is given in Table 1. Not all the comets observed by IUE were bright enough for all of these features to be discernible over the limiting camera noise, but in every case the strong OH emission bands were clearly detected. Table 2 summarizes briefly the comets observed to date by IUE. Since the conversion from observed surface brightness to gas production rate is model-dependent, these observations were compared (*24*) by considering groups of observations made when the comets were at roughly the same heliocentric distance, and, of course, the observations were all made with the same instruments. In the case of the water dissociation products, the different relative intensities of the observed O and OH emissions in different comets can be accounted for completely on the basis of the variation of the resonance fluorescence efficiency with heliocentric velocity, a differential Doppler shift first proposed by Swings (*33*) and more recently applied in detail to these species (*34, 35*).

Table 2. Comets observed with the IUE.

Comet	Date	Heliocentric distance (AU)
Seargent (1978 XV)	October 1978	0.92
Bradfield (1979 X)	January to March 1980	0.71 to 1.55
P/Encke (1980 XI)*	October and November 1980	0.81 to 1.02
P/Stephan-Oterma (1980 X)	December 1980	1.58
Meier (1980 XII)	December 1980	1.52
P/Tuttle (1980 XIII)	December 1980	1.02
P/Borrelley (1980i)	March 1981	1.33
Panther (1980u)	March 1981	1.73
Bowell (1980b)	April and May 1982	3.4
P/Grigg-Skjellerup (1982a)	April 1982	1.02
Austin (1982g)	July 1982	0.80 to 1.10

*P indicates that the comet is periodic.

Fig. 4. Periodic comet Encke, as observed in October 1980, showing the sunward fan-shaped coma. [Photograph courtesy of H. Spinrad and J. Stauffer, University of California, Berkeley]

For the OH bands the effect varies in an irregular way with the comet's velocity as the individual rotational lines are displaced relative to the Fraunhofer absorption lines in the solar spectrum (*35*), and this results in a fluorescent pumping of the ground state Λ sublevels that gives rise to the observed variation in the radio OH emission (or absorption) (*7*). In this regard, a comparison between derived OH parent production rates from the IUE observations and from the 18-cm observations made at Nançay shows the ultraviolet results to be consistently about a factor of 2 higher than the radio results (*36*). This discrepancy is traced to the different model parameters used to interpret the data, particularly the OH scale length, whose value derived from the radio observations is a factor of 10 larger than that derived from either ultraviolet images (*37*) or photochemical analysis (*38*). Nevertheless, these systematic radio observations of about a dozen comets, beginning with comet Kohoutek, provide an invaluable data base for relating the production rate of the most probable primary volatile constituent to the visual brightness of the comet.

Two other compositional trends were noted in the analysis of the IUE data. The abundance of both CS (*24*) and CO_2^+ (*29*) relative to OH appeared to vary by no more than a factor of 2 between comets, which, considering the numerous uncertainties in the abundance derivations, implies a nearly constant fractional abundance for the parents of these species, presumably CS_2 and CO_2, respectively, in the cometary ice. However, for comets Bradfield and Encke, the only two comets for which observations were made over a range of heliocentric distance r, the production rate of CS appears to decrease more rapidly with r than does that of H_2O (*23, 24*).

The abundance similarities are perhaps most significant for comet Encke, a periodic comet with the shortest known period (3.3 years) and smallest perihelion distance (0.34 AU), which has been observed at 52 different apparitions. The peculiar fan shape of this comet (Fig. 4) and its asymmetric light curve have been interpreted by Whipple and Sekanina (*39*) in terms of a rapidly rotating comet whose axis lies in its orbital plane and whose motion is such that opposite hemispheres, of quite different volatility, face the sun before and after perihelion. The IUE observations of this comet (*31*) show a similarly asymmetric OH cloud away from the nucleus, again supporting the idea of an irregular nuclear surface with scattered icy patches, the result of a long history within the inner solar system. The similarity of composition of the coma then implies a homogeneous ice, regardless of the details of the structure of the nucleus.

As noted above, IUE observations of comet Bradfield (1979 X) were made over a range of heliocentric distances from 0.71 to 1.55 AU over a 7-week time span in early 1980. The OH brightnesses and the derived water production rates as a function of r are shown in Fig. 5 (*11*); the surprising result is that this variation is quite steep, being proportional to $r^{-3.7}$, whereas the conventional vaporization theory, as summarized by Delsemme (*16*), predicts an r^{-2} variation for

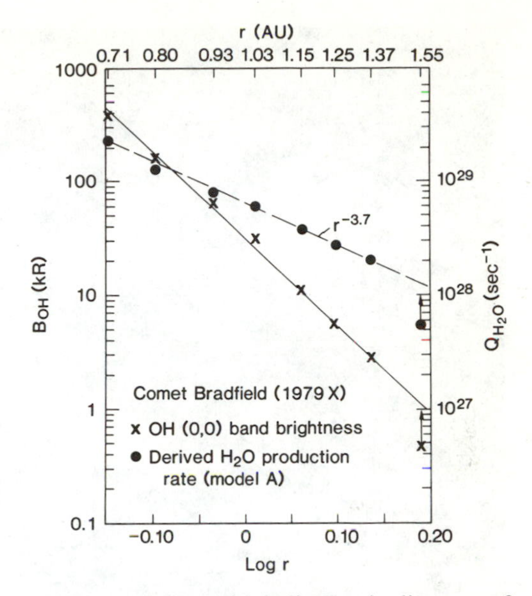

Fig. 5. Variation with heliocentric distance of the H_2O production rate in comet Bradfield (1979 X) as derived from IUE observations of the OH(0,0) band brightness (*11*).

H_2O ice at these heliocentric distances. Recent thermal modeling of the nucleus, including the effects of the infrared opacity of the dust coma (*40*), showed that the effect of the grains in absorbing additional sunlight can produce an enhancement in the vaporization rate when the comet is close to the sun. For a model of Halley at these distances, this model gives a variation proportional to $r^{-3.9}$, in good agreement with the observations of comet Bradfield. However, Bradfield was one of the least dusty comets ever recorded (*41*) (Fig. 6), and for the case of a nondusty coma the model prediction is $r^{-2.7}$ (*40*). Similar behavior was observed in July 1982 for comet Austin (1982g), an equally nondusty comet, but the observations were limited to a smaller range of heliocentric distances, 1.10 to 0.80 AU.

Ground-based observations of the OH band at 3085 Å (which is subject to severe attenuation by the atmosphere),

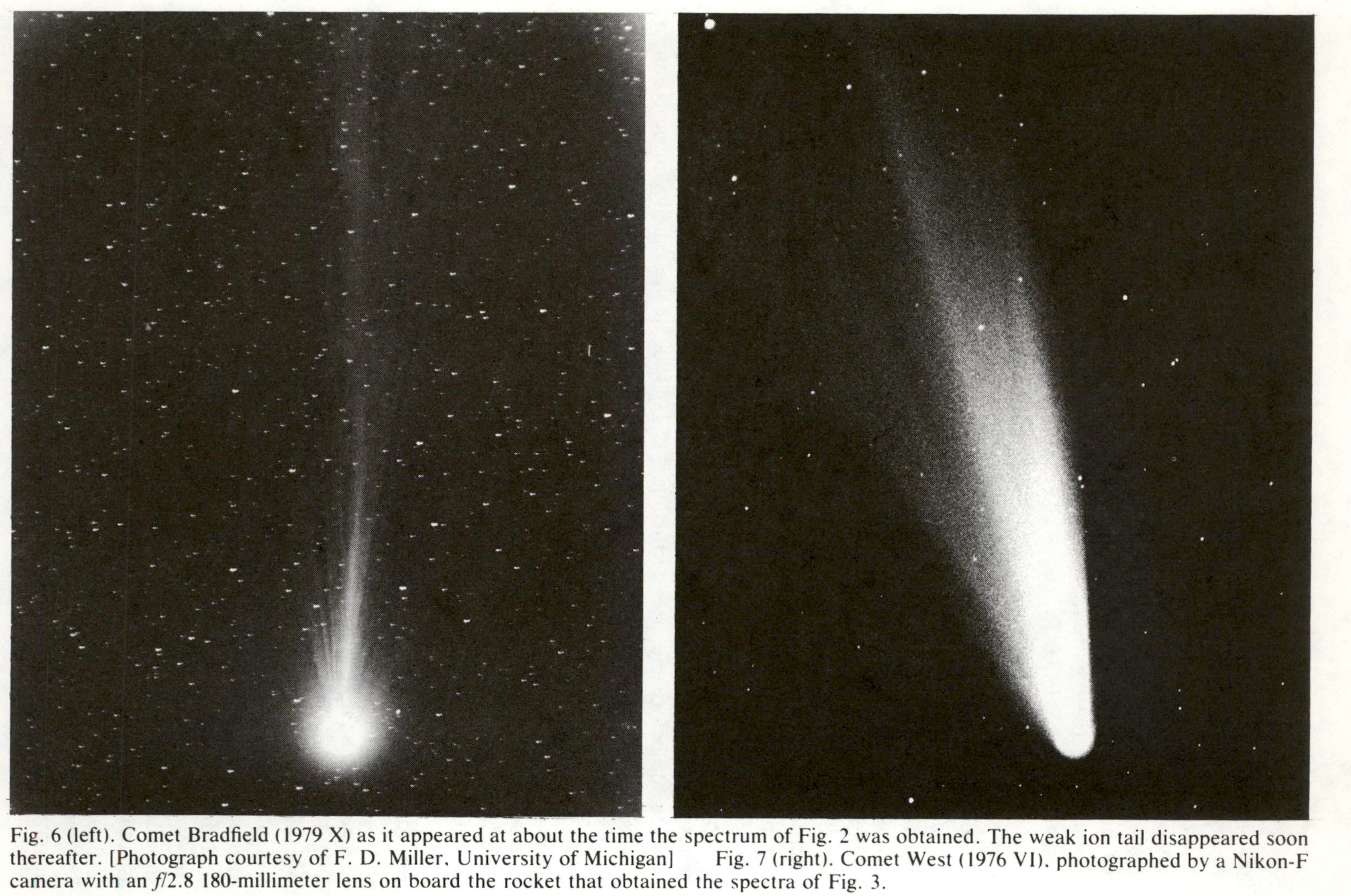

Fig. 6 (left). Comet Bradfield (1979 X) as it appeared at about the time the spectrum of Fig. 2 was obtained. The weak ion tail disappeared soon thereafter. [Photograph courtesy of F. D. Miller, University of Michigan] Fig. 7 (right). Comet West (1976 VI), photographed by a Nikon-F camera with an *f*/2.8 180-millimeter lens on board the rocket that obtained the spectra of Fig. 3.

as well as of C_2 and CN, appear to confirm the steeper variation of gas production rate for dusty comets. For periodic comet Stephan-Oterma (1980 X), Millis *et al.* (*42*) find that these three species all vary as $\sim r^{-5}$ for heliocentric distances between 1.58 and 1.95 AU. The discrepancy with the vaporization theory should be regarded not as evidence against H_2O being the dominant constituent of the cometary ice, but rather as suggestive of a rather complex surface structure in which the ice is embedded. The longer the comet has been in the inner solar system, the more complex this structure is likely to be (*43*). Over a limited range of *r*, the two new comets observed by OAO-2, Tago-Sato-Kosaka (1979 IX) and Bennett (1970 II), appeared to follow an $r^{-2.2}$ gas production law (*44, 45*)—quite unexpected for dusty comets and suggesting an almost pristine ice surface for these comets.

The same trends in gas production rate for dusty and nondusty comets have been detected from recent ground-based observations of the forbidden oxygen red lines at 6300 and 6364 Å. These lines, first identified in comet Mrkos (1957 V) by Swings and Greenstein (*46*), present two difficulties to the comet observer: they appear strongly in the night-sky spectrum, and the cometary 6300 Å line is overlapped by a broad band of NH_2. Modern techniques do allow for separation of the cometary emission, either by high-resolution interferometric spectroscopy utilizing the relative earth-comet motion (*47*) or by careful sky subtraction with high-spatial-resolution spectrophotometry (*48*). The latter has been used to observe a number of recent comets, including several also observed by IUE, and the comparison between them provides strong support for the idea that this emission is produced by direct photodissociation of H_2O sublimating from the nucleus, the branching ratio to $O(^1D)$ being $\sim$ 5 to 10 percent (*49*). The auroral green line, at 5577 Å, should also be present, about a factor of 10 weaker, but the night-sky and cometary backgrounds are even more severe than at 6300 Å. The data of Spinrad (*48*) are consistent with this red-to-green ratio. If the dominant volatile in the cometary ice were CO_2 the red and green lines would be of nearly equal strength (*49*), and this led Spinrad (*48*) to conclude that all of the comets that he had observed were water-dominated.

This brings us back to the comparison of comets West and Bradfield, the former being the most recent CO-rich comet to exhibit an extensive ion tail system in addition to its spectacular dust tail (Fig. 7). Comparing Figs. 2 and 3, and qualitatively assuming that the spatial distributions of CO^+ and CO_2^+ are similar so that effects due to the different apertures used for the two sets of data are minimal, we notice a marked difference in the CO^+/CO_2^+ ratio for the two comets, the $(B^2\Sigma^+–X^2\Sigma^+)$ first negative system of CO^+ being below the IUE detection threshold in the spectrum of comet Bradfield (*50*). Thus, the neutral CO observed in comet West must be a parent molecule contained in the cometary ice rather than a product of the dissociation of CO_2, which must also be present in the ice. Recall that the IUE data show the CO_2^+ emission to be a regular feature from comet to comet. Assuming CO to be a parent, the comet West data imply a CO production rate about 20 percent that of H_2O (*26*). Since in the solar ultraviolet radiation field CO dissociates and ionizes with roughly equal probability, the radial outflow model used for comparison with the atomic carbon emission lines must be modified to include ion-molecule reactions in the inner coma (*51*). Inclusion of dissociative recombination of CO^+ with electrons is required to match the large abundance of metastable $C(^1D)$ atoms that resonantly scatter at 1931 Å. This model was able to match the observed carbon line fluxes quite well and demonstrated that CO was most likely the dominant carbon species in the cometary ice (*51*). This was not unexpected, as CO is, after H_2, the major constituent of interstellar molecular clouds, and its abundance is significantly higher than that of the polyatomic molecules that are candidate parents of the observed cometary radicals. However, CO is considerably more volatile and might have escaped from the collapsing solar nebula as the cloud heated during its collapse. It is also possible that the high abundance of CO observed in comet West resulted from the splitting of the nucleus into four pieces only a few days before the ultraviolet observations were made.

This model was considerably less successful when applied to the comet Bradfield data (*50*). The CO fourth positive bands were marginally detected in the spectrum of Fig. 2 (great care must be taken in examining IUE spectra to avoid misidentifying noise "features" as weak emissions), and the derived CO production rate was found to be only 2 percent that of H_2O, a factor of 10 lower than in comet West. While this abundance would account for the absence of the CO^+ bands, it predicted emission rates for the atomic carbon lines that were one to two orders of magnitude smaller than the observed values. This discrepancy was much too large to be resolved even by the inclusion of a factor of 4 increase in solar extreme ultraviolet flux between 1976 and 1980 in the model (*52*). Clearly, either an additional source of atomic carbon is present in the coma, or the basic model is incorrect and the agreement with the comet West data was fortuitous. One positive result of the IUE data is that the 1931 Å carbon emission, indicative of the abundance of $C(^1D)$ atoms in the coma, decreased much faster with heliocentric distance than any other emission, implying a collisional source as presumed by the model. Further study of $C(^1D)$ is possible with ground-based telescopes, using the ^1D–^3P transitions at 9823 and 9850 Å, which are analogous to the oxygen red lines discussed above. Despite the rather limited data sample, it seems safe to conclude that it is CO, and not CO_2, that is the variable, highly volatile species in the ice of new comets.

Future Directions

Although it is regrettable that the United States, which over the past two decades has pioneered in the exploration of the solar system, will not be represented among the fleet heading for Halley's comet, NASA does have a unique earth-orbiting observatory that will be available at the time of the 1986 apparition and which contains instruments employing state-of-the-art technology to provide observational capabilities never before available to astronomers. The symbiosis with the in situ measurements of Halley will add to the enormous scientific potential of this observatory.

This orbiting observatory, known as Astro, is a group of three ultraviolet telescopes assembled on a common pointing platform; these telescopes are designed for stellar and extragalactic astronomy and are scheduled to fly aboard the space shuttle, early in 1986. The three instruments are the Hopkins Ultraviolet Telescope (HUT), the Wisconsin Ultraviolet Photo-Polarimeter Experiment (WUPPE), and the Ultraviolet Imaging Telescope (UIT). Several features of these instruments as currently designed make them very well suited to cometary observations and provide capabilities not available in the cometary ultraviolet studies made to date. For example, the monochromatic imaging capability of the UIT will produce images of the spectrally well-separated O, C, S, CS, and OH features at ≈2 arc seconds resolution, which corresponds to a spatial resolution of ≈1000 km for a comet 0.7 AU from the earth. Fine details of the chemistry and dynamics of the inner coma can be studied with the aid of these monochromatic images, particularly those of O and C, whose scale lengths are in the range 10^5 to 10^6 km.

The HUT (*53*) will provide not only one to two orders of magnitude better sensitivity and better spectral resolution than instruments used for previous observations, but an increased spectral range that extends below HI Lyman-α at 1216 to 500 Å and includes the resonance transitions of the noble gases, in particular HeI 584, as well as several lines of neon and argon. The HUT has a unique capability in this regard as both the IUE telescope currently in orbit and the Space Telescope now being built have mirror coatings and other optical elements that limit the telescope transmission to wavelengths above 1150 Å. Because of the high volatility of helium, its presence in a comet would be an important indicator of the physical conditions present at the time of comet formation in the early solar system. An estimate of the HUT sensitivity at 584 Å indicates that a helium abundance 1 percent that of water in the coma would be detectable. As noted above, atomic nitrogen has not been observed in comets, although nitrogen-containing molecules such as CN, NH, and N_2^+ are known to be present in the coma. This is due to the low efficiency for scattering of solar radiation by the NI multiplet at 1200 Å and the presence of the very intense Lyman-α feature at 1216 Å, which tends to mask any weak emission nearby. The HUT spectral range will include N at 1134 Å and N^+ at 1085 Å, providing additional means of spectroscopically detecting atomic nitrogen. Finally, the WUPPE covers the spectral range beyond the long-wavelength limit of the HUT, thus ensuring complete spectral coverage from ~ 500 to 3200 Å, and it also provides polarimetry of the solar radiation scattered by cometary dust at the shortest wavelengths practical for such observations. Both HUT and WUPPE have selectable apertures ranging from a few arc seconds to 2 arc minutes, allowing for an optimum balance between spectral resolution, spatial resolution, and sensitivity to extended objects.

The current schedule for the Astro payload includes a mission coincident in time with the European and Soviet encounters. If history is any guide, the ultraviolet perception of Halley so obtained will be as different from the visual image as the 1910 photographs were different from either the fantastic representation in the Bayeux tapestry of 1066 or the mannered drawings of Bessel's telescopic view in 1835. Whether our knowledge of the physical state of the comet's nucleus will improve accordingly remains to be seen.

References and Notes

1. C. B. Opal, G. R. Carruthers, D. K. Prinz, R. R. Meier, *Science* **185**, 702 (1974).
2. J.-L. Bertaux, J. E. Blamont, M. Festou, *Astron. Astrophys.* **25**, 415 (1973).
3. A. D. Code, T. E. Houck, C. F. Lillie, in *The Scientific Results from Orbiting Astronomical Observatory (OAO-2)*, A. D. Code, Ed. (SP-310, NASA, Washington, D.C., 1972), p. 109.
4. G. R. Carruthers, C. B. Opal, T. L. Page, R. R. Meier, D. K. Prinz, *Icarus* **23**, 526 (1974).
5. P. D. Feldman, P. Z. Takacs, W. G. Fastie, B. Donn, *Science* **185**, 705 (1974).
6. L. E. Snyder, *Icarus* **51**, 1 (1982).
7. D. Despois, E. Gerard, J. Crovisier, I. Kazès, *Astron. Astrophys.* **99**, 320 (1981).
8. F. L. Whipple, *Astrophys. J.* **111**, 375 (1950).
9. ______, *ibid.* **113**, 464 (1951).
10. A. H. Delsemme and P. Swings, *Ann. Astrophys.* **15**, 1 (1952).
11. H. A. Weaver, P. D. Feldman, M. C. Festou, M. F. A'Hearn, *Astrophys. J.* **251**, 809 (1981).
12. M. Oppenheimer, *ibid.* **196**, 251 (1975).
13. P. T. Giguere and W. F. Huebner, *ibid.* **223**, 638 (1978).
14. G. F. Mitchell, S. S. Prasad, W. Huntress, *ibid.* **244**, 1087 (1981).
15. M. B. Swift and G. F. Mitchell, *Icarus* **47**, 412 (1981).
16. A. H. Delsemme, in *Comets*, L. L. Wilkening, Ed. (Univ. of Arizona Press, Tucson, 1982), p. 85.
17. L. Haser, *Bull. Cl. Sci. Acad. R. Belg.* **43**, 740 (1957).
18. ______, *Cong. Coll. Univ. Liège* **37**, 233 (1966).
19. M. C. Festou, *Astron. Astrophys.* **95**, 69 (1981).
20. D. A. Mendis and H. L. F. Houpis, *Rev. Geophys. Space Phys.*, in press.
21. M. K. Wallis, in *Comets*, L. L. Wilkening, Ed. (Univ. of Arizona Press, Tucson, 1982), p. 357.
22. A. C. Aiken, *Astrophys. J.* **193**, 263 (1974).
23. W. M. Jackson, J. Halpern, P. D. Feldman, J. Rahe, *Astron. Astrophys.* **107**, 385 (1982).
24. H. A. Weaver, P. D. Feldman, M. C. Festou, M. F. A'Hearn, H. U. Keller, *Icarus* **47**, 449 (1981).
25. M. F. A'Hearn, in *Comets*, L. L. Wilkening, Ed. (Univ. of Arizona Press, Tucson, 1982), p. 433.
26. P. D. Feldman and W. H. Brune, *Astrophys. J. Lett.* **209**, L145 (1976).
27. R. R. Conway, *J. Geophys. Res.* **86**, 4767 (1981).
28. A. M. Smith, T. P. Stecher, L. Casswell, *Astrophys. J.* **242**, 402 (1980).
29. M. C. Festou, P. D. Feldman, H. A. Weaver, *ibid.* **256**, 331 (1982).
30. P. D. Feldman, H. A. Weaver, M. C. Festou, M. F. A'Hearn, W. M. Jackson, B. Donn, J. Rahe, A. M. Smith, P. Benvenuti, *Nature (London)* **286**, 132 (1980).
31. P. D. Feldman, H. A. Weaver, M. C. Festou, H. U. Keller, in *Advances in Ultraviolet Astronomy: Four Years of IUE Research*, Y. Kondo, J. M. Mead, R. D. Chapman, Eds. (CP-2238, NASA, Washington, D.C., 1982), p. 307.
32. P. D. Feldman, in *Comets*, L. L. Wilkening, Ed. (Univ. of Arizona Press, Tucson, 1982), p. 461.
33. P. Swings, *Astrophys. J.* **95**, 270 (1942).
34. P. D. Feldman, C. B. Opal, R. R. Meier, K. R. Nicolas, in *The Study of Comets*, B. Donn *et al.*, Eds. (SP-393, NASA, Washington, D.C., 1976), p. 773.
35. D. G. Schleicher and M. F. A'Hearn, *Astrophys. J.* **258**, 864 (1982).
36. D. Bockelée-Morvan, J. Crovisier, E. Gerard, I. Kazès, *Icarus* **47**, 464 (1981).
37. M. C. Festou, *Astron. Astrophys.* **96**, 52 (1981).
38. W. M. Jackson, *Icarus* **41**, 147 (1980).
39. F. L. Whipple and Z. Sekanina, *Astron. J.* **84**, 1894 (1979).
40. P. R. Weissman and H. H. Kieffer, *Icarus* **47**, 302 (1981).
41. M. F. A'Hearn, R. L. Millis, P. V. Birch, *Astron. J.* **86**, 1559 (1981).
42. R. L. Millis, M. F. A'Hearn, D. T. Thompson, *ibid.* **87**, 1310 (1982).
43. G. D. Brin and D. A. Mendis, *Astrophys. J.* **229**, 402 (1979).
44. H. U. Keller and C. F. Lillie, *Astron. Astrophys.* **34**, 187 (1974).
45. ______, *ibid.* **62**, 143 (1978).
46. P. Swings and J. L. Greenstein, *C. R. Acad. Sci.* **246**, 511 (1958).
47. D. Huppler, R. J. Reynolds, F. L. Roesler, F. Scherb, J. Trauger, *Astrophys. J.* **202**, 276 (1975).
48. H. Spinrad, *ibid.*, in press.
49. M. C. Festou and P. D. Feldman, *Astron. Astrophys.* **103**, 154 (1981).
50. M. F. A'Hearn and P. D. Feldman, *Astrophys. J. Lett.* **242**, L187 (1980).
51. P. D. Feldman, *Astron. Astrophys.* **70**, 547 (1978).
52. M. Oppenheimer and C. J. Downey, *Astrophys. J. Lett.* **241**, L123 (1980).
53. A. F. Davidsen *et al.*, in *Shuttle Pointing of Electro-Optical Experiments* (Society of Photo-Optical Instrumentation Engineers, Bellingham, Wash., 1981), vol. 265, p. 375.

54. Various phases of this work were supported by NASA grants NGR 21-001-001 and NSG-5393 and NASA contract NAS5-27000.

This material originally appeared in *Science* **219**, 28 January 1983.

Part II
Structure and Content of the Galaxy

5. The New Milky Way

Leo Blitz, Michel Fich, Shrinivas Kulkarni

Ten years ago, the study of the overall structure of our Galaxy, the Milky Way, seemed to be past its prime. The basic large-scale features and properties of the Galaxy were thought to have been known for some time, and a theory proposed in the middle 1960's (*1*) showed considerable promise in explaining the dynamics of the rotation of the disk. Subsequently, many of the workers most active in galactic structure research turned their attention elsewhere. There was more to be learned, it seemed, from observations of other galaxies than from observations of our own.

Over the past several years, however, it has become clear that not only have our ideas about the structure of the Milky Way needed some refinement, but the most massive component of our Galaxy had been overlooked. This component has recently been detected gravitationally, but it is not luminous enough to have been recognized by any observer at any wavelength. It may therefore be composed of a currently unknown type of astronomical object. In this chapter we describe how the large-scale picture of the Milky Way has changed over the past few years (*2*).

The Old Milky Way

The traditional picture of the Galaxy is that it is a thin round disk 30,000 parsecs (pc) in diameter with a thickness a few percent of its extent (*3*). The sun is located at a distance of about 10,000 pc (10 kpc) from the center, and has a roughly circular orbit with a velocity of about 250 km/sec (*4*). The sun thus orbits the center once every 250 million years. The center of the Milky Way cannot be seen with optical telescopes, but contains a powerful source of radio waves, much more energetic than the sun. It was, in fact, the first source detected by any radio telescope, and was discovered in 1932 by Jansky (*5*). Within a few kiloparsecs of the galactic center, the disk of stars fattens into a spheroidal bulge of old stars. Also spheroidally distributed about the center, but to distances as large as 50 kpc, are about 200 globular clusters: spherical clusters of old stars containing as many as a million members. Completing the picture is the interstellar gas, which is closely confined to the plane defined by the disk of stars.

Most of the mass of the Milky Way was thought to be in ordinary stars. Interstellar gas, almost all of which was thought (until the late 1960's) to be neutral atomic hydrogen, comprises roughly 1 to 2 percent of the total mass. The gas resides mainly in the disk and has a thickness of only 250 pc (measured between points with half of the midplane density). The hydrogen was known to have a large-scale warp with a shape like the brim of a fedora.

Observations of other disk galaxies showed that many of them have beautiful spiral structures. It was natural to wonder whether the Milky Way might also have large-scale regular spiral arms. Indeed, within 3 kpc of the sun, three segments of armlike concentrations of luminous young stars were identified (*6*) and were thought to be representative pieces of the overall large-scale spiral structure of the Galaxy. The interstellar dust (which comprises about 1 percent of the mass of interstellar gas) absorbs starlight, however, and it is difficult to see much beyond a few kiloparsecs in the plane of the Milky Way in most directions; the overall spiral structure could not be convincingly deduced from optical observations alone. It was hoped that radio observations of the atomic hydrogen gas, which is pervasive, relatively transparent, and not absorbed by the dust, would show the concentrations which could be seen in the spiral arms of other galaxies.

When the radio observations were completed, however (*7*), it turned out that the results were difficult to interpret (*8*), and the question of whether the Milky Way has large-scale spiral structure could not be answered unambiguously. Although most disk-shaped galaxies exhibit spiral structure of some sort, in many galaxies this structure is quite irregular. The length of a spiral arm segment which can be continuously identified might be quite small, and it has not been clear whether the inability to identify a system of large-scale regular spiral arms in the Milky Way on which all observers could agree results from the limitations of the observing technique or the absence of such arms in the Galaxy.

In the middle 1960's, the Lin-Shu (*1, 9*) density wave theory suggested an explanation for the major unsolved problem of spiral galaxies: Why don't the spiral arms wind themselves up and become unrecognizable? Galaxies rotate differentially, the inner portions orbiting the center more frequently than the outer portions. Over the lifetime of a spiral galaxy, there are so many orbits (about 50 in the vicinity of the sun) that any large-scale spiral pattern should be completely obliterated. The Lin-Shu theory suggested that the spiral arms were not material arms, but were manifestations of a wave phenomenon which propagated through the disk of a galaxy. The density wave theory had success in explaining a number of the detailed structural and kinematic features of the Milky Way and other spiral galaxies.

By 1970, the major stellar components of the Milky Way had been known for some time and the atomic (*7*) and ionized (*10*) gas had been surveyed. Although surveys at a number of wavelengths were to be completed later (indeed, some are not yet complete), the outlines of the large-scale distribution of matter in the Galaxy seemed to be known with some confidence, even though many of the details remained obscure.

Interstellar Molecules

In the late 1960's, our understanding of the nature of the interstellar gas in the Milky Way began to change drastically with the discovery of large quantities of molecular gas in addition to atomic gas (*11*). The early molecular observations demonstrated the presence of cold (10 K) gas in large quantities, with densities 10^2 to 10^5 times greater than were previously thought to be common in the interstellar medium [the atomic gas typically has temperatures of 30 to 5000 K (*12*) and an overall mean density of 0.3 atom per cubic centimeter (*7*)]. The discovery of interstellar formaldehyde (*13*) in 1969 indicated that organic molecules were readily manufactured in the interstellar environment. Subsequently, more than 50 molecules have been detected in the interstellar medium.

The most abundant molecule by far is molecular hydrogen. But, because it has no permanent electric dipole moment, it does not emit and absorb photons at the low temperatures typical of most of the molecular gas. Molecular hydrogen has been observed directly only when it absorbs the ultraviolet light from a hot massive background star (*14*), or under unusual high-temperature conditions such as those associated with star formation (*15*).

Nevertheless, the presence of H_2 can be inferred from the excitation of other molecules, especially carbon monoxide. Excitation of carbon monoxide to the temperatures at which it is observed in the interstellar medium requires collisions with other particles that have a density of $\sim 10^3$ cm^{-3}. It was shown (*11, 16*) that the collisions take place primarily with electrically neutral particles, and these were inferred to be molecular hydrogen (atomic hydrogen would have been detected by means of its 21-cm radio radiation). Subsequently, a few regions have been observed in which both CO and H_2 have been detected directly (*17*). These observations confirm that the H_2 is present in sufficient quantities to collisionally excite CO. Although the abundance of CO is only 10^{-4} times that of H_2, it is the next most abundant of the molecules detected to date in the interstellar medium. It is readily detected with high-frequency millimeter-wave radio telescopes by means of the $J = 1 \rightarrow 0$ rotational transition at a wavelength of 2.6 mm. The CO molecule thus acts as a tracer of the more abundant H_2. Recently, it has been shown that OH, the third most abundant molecule, can also be used as a reliable tracer of the H_2 gas (*18*).

Observations of CO in the vicinity of the sun have shown that the molecules are collected in discrete clouds with well-defined boundaries (*19*); an example is shown in Fig. 1. These clouds are almost entirely molecular, with only small quantities of atomic gas. Some of these clouds (perhaps all of the largest of them) have extended atomic envelopes. In contrast, atomic hydrogen is detected essentially everywhere in the disk of the Galaxy. The molecular gas fills only about 1 percent of the volume of the disk, and the atomic gas appears to fill most of the rest. [An alternative picture of the interstellar medium was proposed recently by McKee and Ostriker (*20*), who argued that the largest fractional volume of the disk is hot ionized hydrogen gas. The validity of their picture, which is supported by a number of observations, does not affect any of the subsequent discussion in this chapter.]

The largest of the molecular clouds, which are referred to as giant molecular clouds, are remarkable in several respects. First, they are the most massive objects in the Galaxy—typically several hundred thousand solar masses. They contain as much matter as the richest of the globular clusters, but are about 20 times as numerous (*19, 21*). Second, they are the nucleation sites for most of the present-day star formation in the Galaxy (*19, 22*). Essentially all of the stars in the Milky Way known to be young—that is, less than a few million years old—appear to have formed from the dense molecular gas (dense by interstellar standards, but still far more rarefied than the best terrestrial vacuum) that comprises the giant

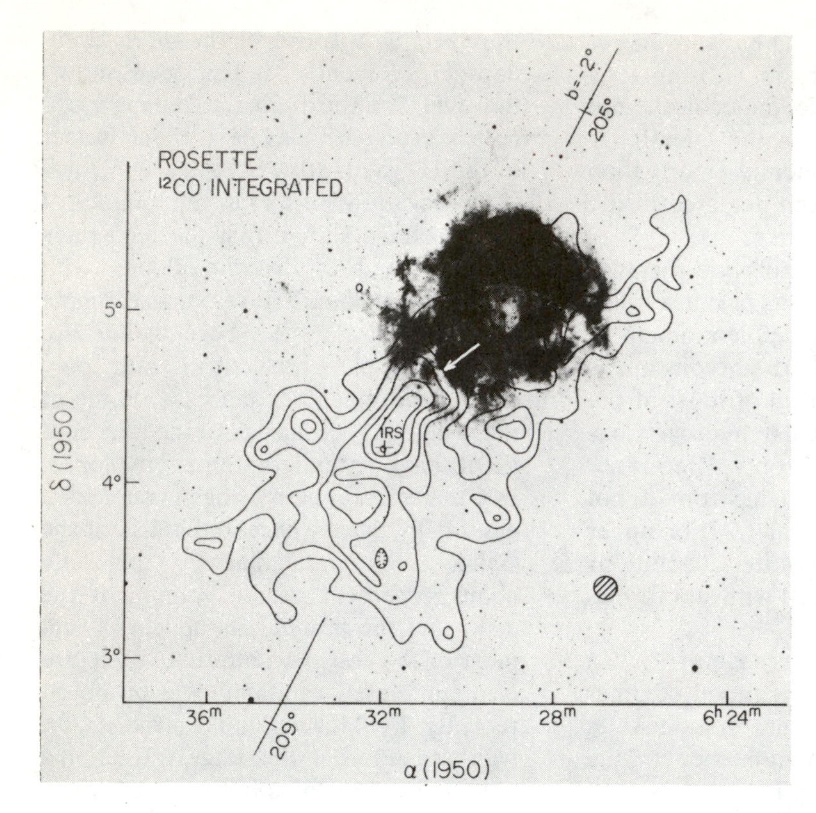

Fig. 1. A typical giant molecular cloud complex in the vicinity of the sun, associated with the Rosette Nebula (*59*). The contours are integrated CO antenna temperature and delineate the extent of the H_2 cloud associated with the nebula, a moderately bright HII region. The arrow shows the location of a clear interaction between the HII region and the molecular cloud. The extent of the cloud is 100 pc and its mass is about 1.5×10^5 solar masses.

molecular clouds. Third, the total mass of H_2 locked up in the giant molecular clouds approaches the total mass of atomic hydrogen in the Milky Way (*21, 23*). The precise ratio of atomic to molecular gas has been the subject of considerable debate (*24*), but all observers agree that the molecular gas is a substantial fraction of the total. Not only are the giant molecular clouds a new, previously unsuspected component of the Milky Way, they have played an important role in our newly emerging understanding of galactic structure.

Galactic Rotation

The dominant large-scale force on astronomical scales is gravity, and it is the mutual gravitational attraction of massive cosmic bodies that determines their state of motion. The large-scale distribution of matter in the Milky Way can therefore be revealed by the motion of its constituents—for example, by measuring the variation of the rotational velocities of objects in orbit around the galactic center as a function of distance. A plot of the velocity (V) versus distance from the galactic center (R) is called a rotation curve, and it is one of the fundamental astronomical measurements made of our own and other galaxies. Once the rotation curve is measured, the distribution of matter in the Galaxy can be inferred from the equations of motion. The solar system, for example, has a Keplerian rotation curve with $V \propto R^{-1/2}$ because almost all the mass is concentrated in the sun. If the mass in the solar system were more spatially extended, the exponent of

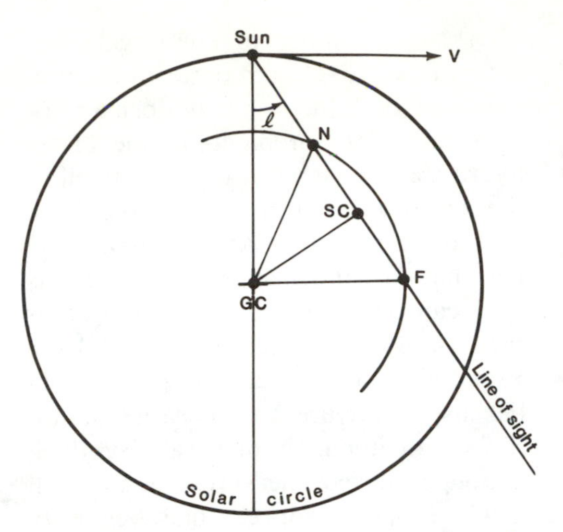

Fig. 2. Schematic diagram of the geometry related to observations of the gas in the plane of the Galaxy as viewed normal to the plane. *GC* is the galactic center and *V* is the orbital velocity of the sun, which orbits the center along the solar circle at distance *R*. The gas in the disk rotates differentially with decreasing angular velocity at increasingly large distances from the center. Along a given line of sight, the differential rotation produces the largest Doppler-shifted emission at *SC*, the subcentral point. The points *N* and *F* are equidistant from the center and therefore have the same angular velocity. Gas at these positions produces emission lines with the same radial velocity (the velocity component along the line of sight). Thus, for positions inside the solar circle, it is not possible to determine whether the emission is coming from point *N* or point *F* from measurements of the radial velocity alone. The angle ℓ is galactic longitude; it increases counterclockwise from a direction toward the galactic center (see Plates I and II).

R would be greater than $-1/2$; its precise value would depend in detail on the mass distribution. In the outer regions of a galaxy, the rotation curve is expected to approach a Keplerian form because most of the mass can be approximated to be in the central regions.

In order to measure the rotational properties of the Galaxy as a whole, it is necessary to find a pervasive component whose velocity and distance can be independently measured over a large range of galactocentric distances. Until recently, the most useful component has been the atomic hydrogen gas. Unlike starlight, the 21-cm spectral line of atomic hydrogen is unattenuated by galactic dust and can therefore be observed almost anywhere in the Milky Way. The differential rotation of the galactic disk causes the 21-cm line from gas at different distances from the center to be Doppler-shifted relative to the local gas. Along a line of sight that intersects the solar circle—the circle at the distance of the sun in the plane of the Galaxy centered on the galactic center—the biggest radial velocity comes from the point closest to the galactic center (see Fig. 2). By making observations along all such lines of sight, it is possible to measure the velocity and distance of the hydrogen gas along a particular, well-defined locus of points. This method allows the rotation curve to be measured out to the solar distance, but not beyond (*25*).

From a distance of about 4 kpc to the solar distance, *V* is not strongly dependent on *R*. There is a local maximum about 8 kpc from the galactic center, and it had been expected that *V* would continue to decrease at larger distances. Observations of other spiral galaxies showed that the disk light decreases exponentially from the center (*26*). Because the disk light is produced by ordinary stars, and because these stars were thought to be responsible for most of the mass, an exponential disk implied that there would be little mass beyond the solar distance. A mass model of the Galaxy incorporating the most reliable observations available at the time was developed by Schmidt in 1965 (*27*). The model, which became the standard for a number of years, predicted a rotation curve in which V decreased with increasing R beyond the solar circle and approached a Keplerian form 15 to 20 kpc from the center. This expectation of the Schmidt model has turned out to be grossly inconsistent with recent observations.

It has been known for some time that, in order to be dynamically stable, clusters of galaxies generally require more mass than can be accounted for in terms of their luminous mass (*28*). In the middle 1970's, evidence began to accumulate that spiral galaxies in general, and the Milky Way in particular, contain large quantities of undetected matter. Ostriker and Peebles (*29*) argued that the disks of spiral galaxies like our own should be unstable to the formation of bar-shaped structures, but could be stabilized if they possess massive, spheroidal halos. In a recent, more general analysis, Vandervoort (*30*) takes issue with the detailed results of Ostriker and Peebles, but does not dispute their fundamental qualitative conclusions. Analyses of various observational results lent support to the conjecture that galaxies have massive halos. Among these have been the comparison of the rotational velocities and light distribution of some spiral galaxies, the statistical analysis of the radial velocities of binary galaxies, and the analysis of the tidal radii of nearby dwarf spheroidal galaxies (*31, 32*). A computer simulation of the Magellanic stream, a tidally generated flow resulting from the passage of the Magellanic Clouds, also pointed to a large mass for the Milky Way (*33*).

Perhaps the most convincing evidence that spiral galaxies contain large quantities of nonluminous matter came from the measurement of the rotation curves of a large number of spiral galaxies. Rubin *et al.* (*34*), from optical measurements of the Hα line, and Bosma (*35*), from radio measurements of the 21-cm hydrogen line, showed that nearly all of the spiral galaxies they observed have rotation curves that are approximately flat (that is, V independent of R) as far as they can be measured. Thus, the outer regions of spiral galaxies do not have Keplerian rotation curves, presumably because of the presence of large quantities of nonluminous material. By 1978, although a direct measurement was still lacking, many workers had begun to suspect that the rotation curve of the Milky Way is also flat. A dynamical analysis of the distant globular clusters and dwarf spheroidal galaxies (*36*) gave results consistent with a flat rotation curve, as did a kinematic analysis of distant atomic hydrogen gas (*37*).

The New Rotation Curve

The first direct measurements of the rotation curve to large galactocentric distances beyond the sun (*38, 39*) resulted from the recognition that there are many very young star clusters in the outer parts of the Milky Way which show evidence of interaction with their placental molecular clouds. Because young clusters contain the brightest, most massive stars in the Milky Way, they can often be identified to large distances from the sun. Beyond the solar circle, the obscuration due to dust is significantly lower than in the inner parts (*40*). By comparing the brightness and color distribution of the stars in the outer clusters to those in nearby clusters, whose distances are independently known, one can obtain distances to the more outlying clusters to an accuracy of about 25 per-

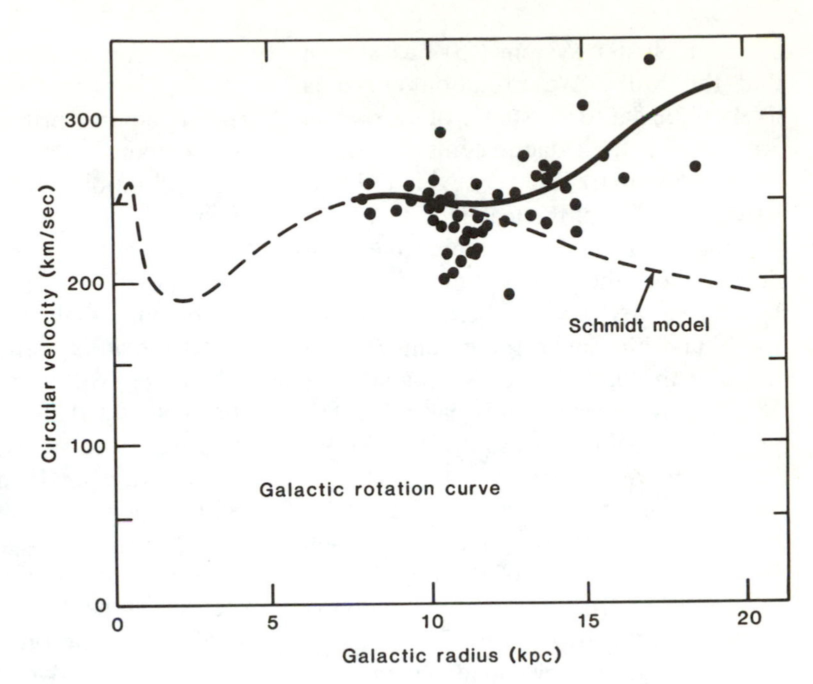

Fig. 3. Rotation curve of the Milky Way (*42*). The filled circles represent data and the solid line a fourth-order least-squares fit to the data. The dashed line is the rotation curve from the Schmidt mass model (*27*). The Schmidt model fits the rotation curve from the atomic hydrogen data inside the solar circle well, but does not fit the data beyond 12 kpc from the center.

cent. An important survey by Moffat *et al.* (*40*) obtained distances to 40 of the most distant young clusters.

The brightest cluster members often ionize the hydrogen gas in their vicinity, producing bright, fluorescent nebulae known as HII regions. These regions have been shown by many workers to interact with their placental giant molecular clouds, thus providing clear evidence for the association of the clusters with the molecular gas. A recent survey showed that about 75 percent of all HII regions have associated molecular clouds (*41*). Most of the mass of the star cluster/HII region/molecular cloud complex resides in the molecules. The molecules are observed by radio heterodyne techniques, which can provide velocities accurate to a fraction of a kilometer per second, an order of magnitude improvement over any other method. Therefore, measurement of the velocities from CO observations provides accurate radial velocities of the center of mass of a complex, whose distance can be determined from the optical observations of the young cluster. These quantities are converted to galactocentric distance and orbital velocity (assuming circular rotation), and thus a rotation curve is determined.

The rotation curve to 18 kpc from the galactic center is shown in Fig. 3 (*42*). Although the detailed shape of the curve depends somewhat on the exact value of the galactocentric distance and orbital velocity of the sun, the curve deviates markedly from the expectation of the Schmidt mass model, shown as a dashed curve in Fig. 3. The implication is that there is much more mass in the outer region of the Milky Way than was thought to exist in stars. Or, to put it another way, the dark matter that was predicted on theoretical grounds, and was found to be common in other spiral galaxies, has been detected gravitationally in the Milky Way.

Recently, a dynamical analysis of the

globular cluster system (*43*) has shown that the Milky Way rotation curve is likely to be flat to a distance of more than 50 kpc from the galactic center.

Thus the dark matter, which dominates the mass of the Galaxy 5 to 10 kpc beyond the sun, is detectable to at least 50 kpc from the center. For a flat rotation curve, $M(R) \propto R$, where $M(R)$ is the total mass interior to some point R. The mass of the dark matter is thus at least four to five times greater than all of the previously known mass in the Milky Way. But what is the nature of this material and how is it distributed? A partial answer to this question comes from an examination of the distribution of atomic hydrogen gas in the outlying portions of the Milky Way. The distribution of this gas has provided new insights into the structure of the Milky Way and is interesting in its own right.

Atomic Hydrogen Beyond the Solar Circle

In 1974 an important survey of the atomic hydrogen in the Milky Way was published by Weaver and Williams (*44*). This was a complete survey of the hydrogen near the plane of the Galaxy which is visible from the Northern Hemisphere. Although the hydrogen in the Milky Way is detected from all directions in the sky, most of it is from a region within $\pm 10°$ of the plane. Most of the rest is very near the sun and does not provide much information on the overall structure of the Galaxy. The determination of the rotation curve of the outer portions of the Milky Way has made it possible to analyze the atomic hydrogen from the Weaver and Williams survey in such a way as to derive reliable physical properties of the gas. The results have provided a much clearer picture of the distribution of the gas than has been possible in the inner regions, primarily because there is no distance ambiguity to confuse the interpretation of the results. That is, inside the solar circle the radial velocity of a parcel of gas corresponds to two different distances from the sun, whereas in the outer Galaxy the velocity uniquely determines its distance. In addition, all of the gas within $\pm 10°$ of the plane has been included in the analysis. Most of this gas is relatively transparent at the 21-cm wavelength of atomic hydrogen; that is, it is optically thin. The analysis therefore provides a clear picture of the distribution of the gas to the edge of its extent in the Milky Way. Independent analyses of the outer Galaxy data were performed by two different groups (*45*, *46*), using different techniques of analysis, and we discuss the results in some detail in order to emphasize the new features of the Galaxy discovered by these workers.

Plates I and II are plots of the mass surface density and the deviation of the gas from the midplane of the Galaxy defined by the galactic equator (specifically, the first moment of the vertical distribution of the gas). The vertical axis represents galactic longitude, ℓ: increasing azimuthal angle along the galactic plane centered on the sun (refer to Fig. 2). Zero longitude corresponds to a line toward the galactic center. The horizontal axis is radial velocity, which is converted to distance by using the outer Galaxy rotation curve. Lines of constant galactocentric distance are shown. The left edge corresponds to the solar distance.

In Plate I, detectable atomic hydrogen extends to at least 30 kpc from the center. There are several long connected features of enhanced surface density which cross lines of constant galactic radius and correspond to large-scale, coherent, trailing spiral arms. Features *B* and *C* correspond to the well-known Perseus and Orion arms. Feature *A* has been called the Cygnus arm (*46*) because it occurs chiefly in the constellation Cygnus. The lengths of the Cygnus, Perseus, and Orion arms are 25, 20, and 5 kpc, respectively. Although these features have been identified by others (*47*), the coherence and relative contrast of the arms are nowhere shown better than in Plate I. The lengths of the Cygnus and Perseus arms indicate that the Milky Way does indeed have a large-scale, coherent spiral pattern, at least in its outer parts.

The pitch angle measures the openness of the arms. The pitch angle of the Cygnus arm is 21° and that of the Perseus arm about 25°, assuming they are logarithmic spirals. For the Orion arm the pitch angle is ~ 29° but is very uncertain. The pitch angles are typical of open spirals designated Sbc or Sc. The Orion arm does not appear to be a major arm like the other two, and is probably a short spur. Figure 4 shows what the arms would look like if viewed from above the Galaxy. If the Milky Way is reasonably symmetric, one would expect two more

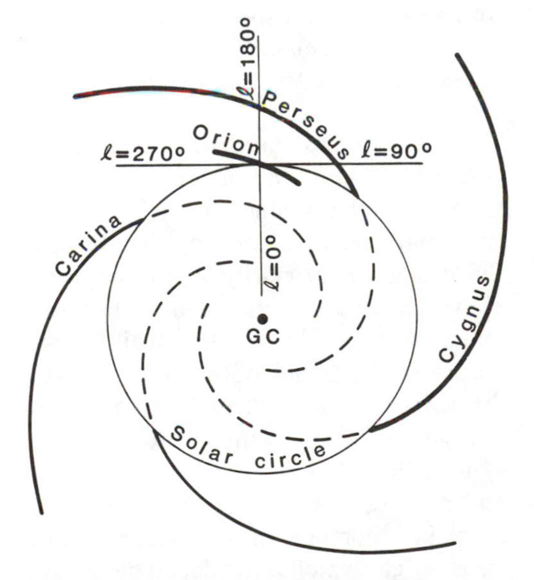

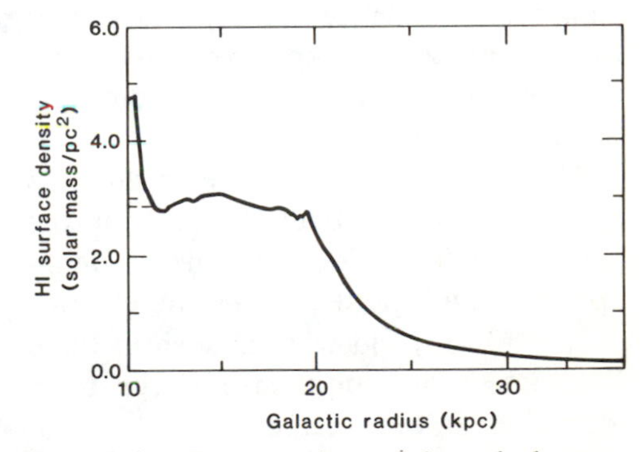

Fig. 4 (left). Face-on view of the spiral structure of the Milky Way. The dark lines marked Cygnus, Perseus, and Orion are features *A*, *B*, and *C* in Plate I. Symmetry considerations predict additional major arms beyond the solar circle as shown. The Carina arm is well known from atomic hydrogen and other studies made from the Southern Hemisphere. The fourth arm (unnamed) would be difficult to detect because emission from it would be blended with the broad, intense emission from the region of the galactic center. Nevertheless, hydrogen features have been identified that are consistent with this feature. The dashed lines show what this simple four-armed pattern would look like if it were extended into the inner regions of the Galaxy. Although this pattern can explain a number of the features seen in the inner Galaxy, the gaseous emission there is considerably more complex than is implied by this simple extrapolation. **Fig. 5 (right). Surface density of atomic hydrogen averaged over the disk.** The surface density is roughly constant at a value near 3 solar masses per square parsec to a distance of 20 kpc from the center. From about 4 to 10 kpc the hydrogen surface density is also nearly constant at a value close to this. Beyond 20 kpc, the surface density falls sharply. The peak between 10 and 11 kpc is an artifact of the reduction technique. The data between longitudes 10° and 250° were averaged to obtain this plot.

arms as shown in Fig. 4. These occur in the Southern Hemisphere and both correspond to known features in the atomic hydrogen gas—one to the well-known Carina arm, the other to features in the outer Galaxy maps of Henderson *et al.* (*45*) and Sills (*48*).

The surface density of the Orion arm is underestimated because there is a considerable amount of gas in the arm outside the latitude boundary of the Weaver and Williams survey. When the proper corrections are made, the surface densities along the arms and from arm to arm are roughly constant. Indeed, a plot of the surface density of all of the gas averaged over longitude (Fig. 5) shows that, as in the inner portions of the Galaxy, the surface density is constant to a distance of 20 kpc from the center and declines monotonically beyond that distance.

The bright portions of the spiral arms are seen only to about 20 kpc from the galactic center (from Southern Hemisphere data, the Perseus arm extends only ~10° beyond what is seen in Plate I). Some faint spiral arms have been identified at even larger distances (*48, 49*), but these would be difficult to see in Plate I because of insufficient contrast and the rapid decline in the atomic hydrogen surface density shown in Fig. 5. Nevertheless, the arms shown in Plate I are those that would be identified by extragalactic radio telescopes similar to terrestrial ones that might be observing the Milky Way.

The black dots in Plate I are the giant molecular clouds observed as part of the survey to measure the rotation curve. These are fairly well concentrated in the spiral arms. This is not surprising since the spiral arms of other galaxies are usually best identified by the bright hot stars and HII regions, which, in the Milky Way, are closely identified with giant molecular clouds. The HII region/molecular cloud complexes are not seen much beyond 20 kpc; massive star formation (and possibly all star formation) in the Milky Way apparently ceases at about that distance, although detectable atomic hydrogen extends to at least another 10 kpc.

To summarize, the outer portions of the Milky Way have large-scale, coherent spiral arms of roughly constant surface density to about 20 kpc from the center. In the outer portions, the Milky Way appears to be a fairly regular four-armed spiral with some spurs, and the giant molecular clouds are fairly well concentrated in the arms. Massive star formation appears to cease at about 20 kpc from the center, but the hydrogen extends at least 10 kpc beyond that distance.

Plate II shows the distribution of gas above and below the mean plane of the Galaxy which is defined by the gas inside the solar circle. Between $\ell = 10°$ and 180° the gas is primarily above the mean plane, and beyond $\ell = 180°$ it lies primarily below. This is the well-known fedora brim warping of the outer parts of the galactic plane. Plate II also shows two new features of the hydrogen emission. At $R = 11$ kpc, from $\ell = 30°$ to 70°, and $R = 12$ kpc, from $\ell = 95°$ to 165°, there is a corrugation—a discontinuous string of gas which is displaced below the galactic plane, although most of the gas is above. Beyond $\ell = 220°$, what appears to be a continuation of this string is again displaced in a direction opposite to the large-scale galactic warp. This string corresponds to a radial corrugation of the disk.

The large-scale warp is fairly small to a distance of ~ 18 kpc from the center and then rises very sharply. Maps of atomic hydrogen in some edge-on galaxies show a similar effect (*50*). As in the Milky Way, the warps of these galaxies become particularly pronounced toward the edge of the stellar disk.

Plate II and Fig. 6 show a wholly unexpected feature of the edge of the galactic disk—a high-frequency scalloping. If one traces the outermost edge of the gas in Plate II, starting at $\ell = 20°$, the gas is consistently above the plane of the Milky Way. The maximum value of the warp is

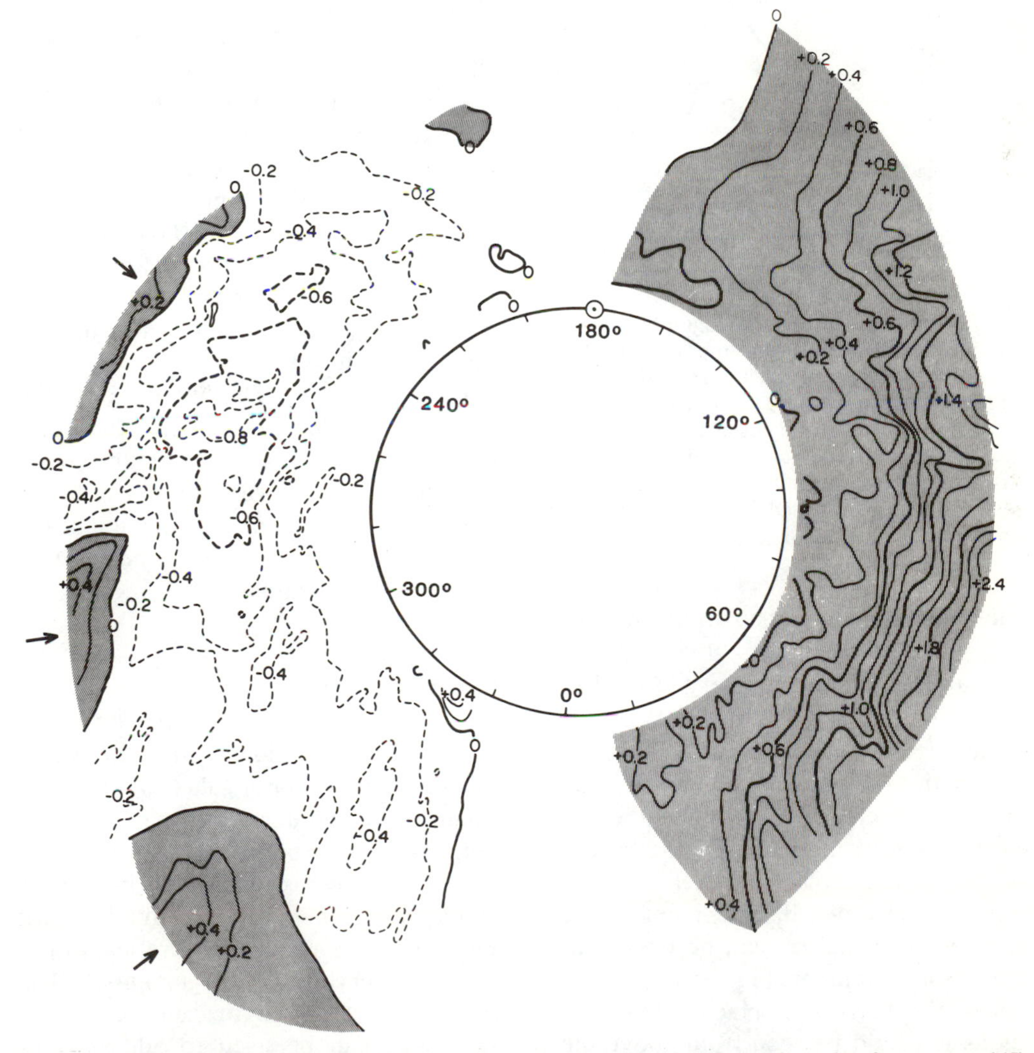

Fig. 6. Deviation of the centroid of the atomic hydrogen emission from the galactic equator (*45*). This is similar to Plate II, except that it shows the deviation as it would be seen with a face-on view of the Milky Way from the north galactic pole, and it includes data from the Southern Hemisphere. Gas that is warped above the galactic equator is indicated by shading. The scalloping in the Southern Hemisphere is indicated by arrows. (The scalloping in the Northern Hemisphere is not seen here because the sensitivity is lower than that shown in Plate II.)

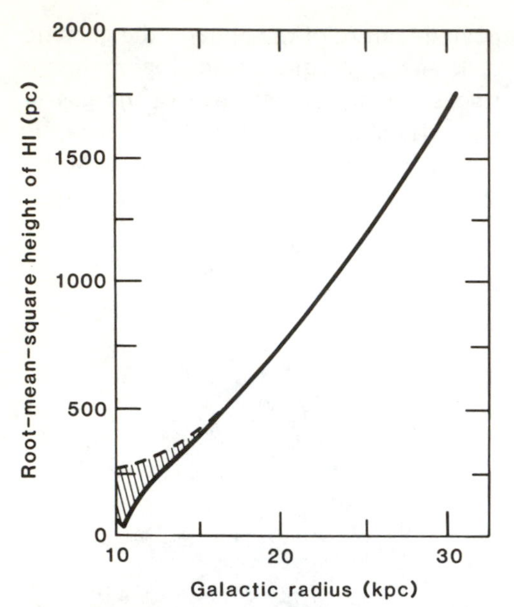

Fig. 7. Thickness of the hydrogen layer as a function of distance from the center. The shaded portion represents a correction due to the incompleteness of the latitude coverage of the data used in the analysis at low velocities. These data are used to show that the large mass implied by the flat rotation curve cannot reside in a thin disk and is thus presumably spheroidally distributed.

reached near $\ell = 45°$, slowly decreasing to $\ell = 135°$. Between $\ell = 135°$ and 150° the warp again becomes very large. But from $\ell = 145°$ to 170° the displacement of the outer lip of gas changes sign even though that of the gas at smaller radii does not. From $\ell = 190°$ to 230° there is another sign change, and from $\ell = 220°$ to the survey boundary, yet another. This up and down displacement of the outer edge of the gas can be traced for three more cycles in Fig. 6, which is a plot of the vertical displacement of the gas as it would be seen from above the Galaxy. In Plate II and Fig. 6 the lip of gas that is scalloped in a sense opposite to the large-scale warp is indicated by an arrow. This remarkable scalloping of the outer edge of the atomic hydrogen has an azimuthal wave number of ~ 10 and is remarkably regular. Its origin is unknown.

The second moment of the vertical gas distribution in the Milky Way (that is, the root-mean-square displacement of the gas from the midplane defined by Plate II) is a measure of the thickness of the gas layer. Figure 7 shows a plot of this quantity averaged over the range $\ell = 90°$ to 250°. Unlike the inner part of the Milky Way, where the thickness of the atomic gas shows little variation with distance from the center (outside the nuclear regions), the outer Galaxy shows a steeper than linear increase in the thickness as a function of distance—an increase of about a factor of 10 from the vicinity of the sun to a distance of 30 kpc from the center. The increase in the thickness of the gas beyond the solar circle has been known for many years (*51*), but it is only with the recent analyses of the outer Galaxy atomic hydrogen (*45, 46*) that the increase has been quantitatively well established.

The Massive Halo

Although the evidence that the Milky Way (like most spiral galaxies) has large quantities of nonluminous matter is reasonably well established, the evidence that this matter is distributed in a spheroidal halo rather than an extended dark disk came mainly from theoretical rather than observational arguments. The thickness of the hydrogen gas layer, however, can be used to address this question. The gas thickness is determined by the gravity of the disk, which tends to confine it, and the gas pressure, which tends to disperse it. If information on the gas pressure can be obtained as a function of distance from the center, it is possible to determine whether any of the dark matter must reside in a halo.

Under the assumption that the turbulent velocity of the gas is constant in R and that it provides the sole contribution to the gas pressure, Gunn (*52*) showed that the increasing thickness of the atomic gas is inconsistent with the dark matter being in the disk. Kulkarni *et al.* (*46*) included the effects of cosmic-ray pressure and magnetic field pressure and independently came to the same conclusion. They showed that if the ratio of the total gas pressure to the gas density was not an increasing function of R (which it appears not to be), and H_*, the thickness of the stellar layer, is larger than that of the gas (as is observed locally), then the thickness of the gas layer, T, at radii R_1 and R_2 can be written

$$\frac{T_2^2}{T_1^2} = \frac{R_2 H_{*2}}{R_1 H_{*1}}$$

if all the mass implied by the rotation curve is in a thin disk. The subscripts refer to values at each distance.

From Fig. 6, for gas at 10 and 30 kpc, the value of the left-hand side of the equation is about 100. This would require the thickness of the stellar layer at 30 kpc to be comparable to its radius, which is inconsistent with the idea of a thin disk. If the thickness of the stellar layer becomes smaller than that of the gas layer at some distance, it can be shown that the continuing increase in the thickness of the gas layer is also incompatible with the dark matter being in the disk. This heuristic argument is independent of the form of the stellar mass distribution and thus does not require any assumptions about an exponential stellar disk. A similarly motivated study of the edge-on spiral galaxy NGC 891 (*53*) concludes that the dark matter implied by the rotation curve is in a spheroidal halo.

Most of the mass of the Milky Way and of spiral galaxies in general thus appears to be contained in the dark spheroidal halos. Gunn (*52*) has argued that elliptical galaxies also have dark, massive halos. Most of the mass in the universe therefore seems to be in the form of nonluminous matter, which has so far been detected only by the gravitational effect it has on the matter we can directly observe. What can we infer about such matter?

We know from observations of our own and other galaxies that the matter is not in the form of an ordinary stellar population (*54, 55*), although very-low-mass stars are not yet entirely ruled out. The halo in the Milky Way also cannot be gas in any form. If the gas were predominantly atomic, its average density within 50 kpc would be about 0.3 atom per cubic centimeter, comparable to the mean density of atomic gas in the disk of the Galaxy. Such a large quantity of gas would be easily detectable from observations of the 21-cm line in our own and other galaxies. If the gas were primarily ionized hydrogen, the bremsstrahlung emission—radiation generated by the electrical interaction of the charged particles—would be easily detectable. In any event, without an external energy source the ionized gas would quickly recombine to form atoms. If the gas were predominantly molecular, it would be detectable in the ultraviolet portion of the spectrum by means of the Lyman absorption bands against the light of distant quasars. In the one case where such lines are believed to have been detected (*56*), the column density of the hydrogen gas is too small to account for a massive galactic halo. The halo cannot also be the result of micrometer-sized dust particles similar to those found in the disk of the Milky Way. In that case, the extinction of starlight caused by the dust would be so strong that other galaxies could not be detected with optical telescopes.

What remains? One possibility is that the halo consists of starlike objects not sufficiently massive to have initiated the nuclear reactions typical of ordinary stars. These objects would have masses less than 0.08 solar mass and might resemble the planet Jupiter. They might be detectable in the infrared, but with great difficulty. Another possibility is that the halo consists of an unknown type of astronomical body larger than that of the microscopic dust but smaller than that of the smallest stars. Still another possibility has been raised by the recent controversial measurement of the neutrino mass (*57, 58*). The halo could consist of a cold gas of massive neutrinos that condensed at the time of formation of the galaxies.

Whatever the form of the massive halos in the Milky Way and in other galaxies, at least two things seem clear: the halos are exceedingly transparent and nonluminous at all wavelengths and the matter that comprises them is probably the dominant form of matter in the universe. An understanding of the universe as a whole, as well as the origin and formation of galaxies, demands an understanding of this new and strange component of the Milky Way, and it is sure to be one of the major challenges for astronomy in the coming years.

References and Notes

1. C. C. Lin and F. H. Shu, *Astrophys. J.* **140**, 646 (1964).
2. B. Bok has written two recent popular articles discussing some aspects of the new Milky Way not included in this article [*Sci. Am.* **244** (No. 3), 92 (1981); *Mercury* **10** (No. 5), 130 (1981)]. The general reader is also referred to B. J. Bok and P. F. Bok [*The Milky Way* (Harvard Univ. Press, Cambridge, Mass., ed. 5, 1981] for a fuller development of the history of Milky Way studies which have led up to this chapter.
3. A parsec is 3.26 light-years and is slightly smaller than the distance from the sun to the nearest star. It is also roughly equal to the mean distance between stars in our portion of the Milky Way. A parsec is equal to about 200,000 times the distance from the earth to the sun.
4. A number of recent observations indicate that the circular velocity of the sun is closer to 220 km/sec, and the distance to the center closer to 8.5 to 9 kpc. These values do not substantially alter any of the following discussion.
5. K. B. Jansky, *Proc. IRE* **20**, 1920 (1932).
6. D. Crampton and Y. M. Gerogelin, *Astron. Astrophys.* **40**, 317 (1975); R. M. Humphreys, in *Large Scale Characteristics of the Galaxy*, W. B. Burton, Ed. (Reidel, Dordrecht, 1979), p. 93; G. Lynga, in *Star Clusters*, J. E. Hesser, Ed. (Reidel, Dordrecht, 1980), p. 13.
7. H. C. van de Hulst, C. A. Muller, J. H. Oort, *Bull. Astron. Inst. Neth.* **12** 117 (1954); M. Schmidt, *ibid.* **13**, 247 (1957); F. J. Kerr and G. Westerhout, in *Galactic Structure*, A. Blaauw and M. Schmidt, Eds. (Univ. of Chicago Press, Chicago, 1965), p. 167.
8. W. B. Burton, *Astron. Astrophys.* **10**, 76 (1971); *ibid.* **19**, 51 (1972).
9. A. Toomre [*Annu. Rev. Astron. Astrophys.* **15**, 437 (1977)] presents an excellent review of the Lin-Shu theory as well as other theories of spiral structure.
10. G. Westerhout, *Bull. Astron. Inst. Neth.* **14**, 215 (1958).
11. D. M. Rank, C. H. Townes, and W. J. Welch [*Science* **174**, 1083 (1971)] provide an extensive review of the early molecular observations and their implications for understanding the interstellar medium.
12. J. Dickey, E. E. Salpeter, Y. Terzian, *Astrophys. J.* **228**, 465 (1980); *Astrophys. J. Suppl.* **36**, 77 (1978).
13. L. E. Snyder, D. Buhl, B. Zuckerman, P. Palmer, *Phys. Rev. Lett.* **22**, 679 (1969).
14. G. R. Carruthers, *Astrophys. J.* **161**, L81 (1970).
15. T. N. Gautier, V. Fink, R. R. Treffers, H. P. Larson, *Astrophys. J. Lett.* **207**, L129 (1976).
16. B. E. Turner [in *Galactic and Extragalactic Radio Astronomy* (Springer, New York, 1974), p. 199] reviews many of the arguments.
17. S. R. Federman, A. E. Glassgold, E. B. Jenkins, E. J. Shaya, *Astrophys. J.* **242**, 545 (1980).
18. J. G. A. Wouterloot, thesis, University of Leiden (1981).
19. L. Blitz, thesis, Columbia University (1978).
20. C. F. McKee and J. P. Ostriker, *Astrophys. J.* **218**, 148 (1977).
21. P. M. Solomon and D. B. Sanders, in *Giant Molecular Clouds in the Galaxy*, P. M. Solomon and M. G. Edmunds, Eds. (Pergamon, Oxford, 1980), p. 41.
22. G. E. Miller and J. M. Scalo, *Astrophys. J. Suppl.* **41**, 513 (1979).
23. M. A. Gordon and W. B. Burton, *Astrophys. J.* **208**, 346 (1976).

24. J. Lequeux, *Comments Astrophys.* **9**, 117 (1981).
25. W. B. Burton, in *Galactic and Extragalactic Radio Astronomy*, G. L. Verschuur and K. I. Kellerman, Eds. (Springer, Berlin, 1974), p. 82.
26. K. C. Freeman, *Astrophys. J.* **160**, 811 (1970).
27. M. Schmidt, in *Galactic Structure*, A. Blaauw and M. Schmidt, Eds. (Univ. of Chicago Press, Chicago, 1964), p. 513.
28. F. D. Kahn and L. Woltjer, *Astrophys. J.* **130**, 705 (1959).
29. J. P. Ostriker and P. J. E. Peebles, *ibid.* **186**, 467 (1973).
30. P. O. Vandervoort, *Astrophys. J. Lett.* **256**, L41 (1982).
31. J. Einasto, A. Kraasik, E. Saar, *Nature (London)* **250**, 309 (1973).
32. J. P. Ostriker, P. J. E. Peebles, A. Yahil, *Astrophys. J. Lett.* **193**, L1 (1974).
33. D. N. C. Lin and D. Lynden-Bell, *Mon. Not. R. Astron. Soc.* **181**, 59 (1977).
34. V. C. Rubin, W. K. Ford, N. Thonnard, *Astrophys. J. Lett.* **225**, L107 (1978).
35. A. Bosma, thesis, University of Leiden (1978).
36. F. D. A. Hartwick and W. L. W. Sargent, *Astrophys. J.* **221**, 512 (1978).
37. G. R. Knapp, S. D. Tremaine, J. E. Gunn, *Astron. J.* **83**, 1585 (1978).
38. P. D. Jackson, M. P. FitzGerald, A. F. J. Moffat, in *The Large Scale Characteristics of the Galaxy*, W. B. Burton, Ed. (Reidel, Dordrecht, 1979), p. 221.
39. L. Blitz, *Astrophys. J. Lett.* **231**, L115 (1979).
40. A. F. J. Moffat, M. P. FitzGerald, P. D. Jackson. *Astron. Astrophys. Suppl.* **38**, 197 (1979).
41. L. Blitz, M. Fich, A. A. Stark, *Astrophys. J. Suppl.* **49**, 183 (1982).
42. ______, in *Interstellar Molecules*, B. Andrews, Ed. (Reidel, Dordrecht, 1980), p. 213.
43. D. Lynden-Bell, in *Kinematics, Dynamics and Structure of the Milky Way*, W. L. H. Shuter, Ed. (Reidel, Dordrecht, 1983), p. 349.
44. H. Weaver and D. R. W. Williams, *Astron. Astrophys. Suppl.* **8**, 1 (1973).
45. A. P. Henderson, P. D. Jackson, F. J. Kerr, *Astrophys. J.*, in press.
46. S. Kulkarni, L. Blitz, C. E. Heiles, *ibid.* **259**, L63 (1982).
47. H. Weaver, in *Galactic Radio Astronomy*, F. J. Kerr and S. C. Simonson, Eds. (Reidel, Dordrecht, 1974), p. 573.
48. R. Sills, thesis, University of California, Berkeley (1981).
49. G. L. Verschuur, *Astron. Astrophys.* **27**, 73 (1973).
50. A. Bosma, *Astron. J.* **86**, 1825 (1981).
51. F. J. Kerr [*Annu. Rev. Astron. Astrophys.* **7**, 39 (1969)] reviews the early determinations.
52. J. E. Gunn, *Philos. Trans. R. Soc. London Ser. A* **296**, 313 (1980).
53. P. C. van der Kruit, *Astron. Astrophys.* **99**, 298 (1981).
54. J. N. Bahcall and R. N. Soneira, *Astrophys. J. Suppl.* **47**, 337 (1981).
55. H. Spinrad, J. P. Ostriker, R. P. S. Stone, L. T. G. Chiu, G. A. Bruzual, *Astrophys. J.* **225**, 56 (1978).
56. D. A. Varshalovich and S. A. Levshakov, *Comments Astrophys.* **9**, 199 (1982).
57. F. Reines, H. W. Sobel, E. Pasierb, *Phys. Rev. Lett.* **45**, 1307 (1980).
58. V. A. Lyubimov, E. G. Novikov, V. Z. Nozik, E. F. Tretyakov, V. S. Kosik, *Phys. Lett. B* **94**, 266 (1980).
59. L. Blitz and P. Thaddeus, *Astrophys. J.* **241**, 676 (1980).

This material originally appeared in *Science* 220, 17 June 1983.

6. The Most Luminous Stars

Roberta M. Humphreys and Kris Davidson

The intrinsically brightest stars are also the most massive. Is there a limit to stellar masses and luminosities?

Not long ago, most astronomers supposed that stars with much more than 60 times the mass of the sun were exceedingly rare or even nonexistent. Such objects, which we call "very massive" stars, would be more luminous than the sun by factors of a million or more, and were thought to be unstable. But today many very massive stars are known in our Galaxy and in several nearby galaxies. A few especially impressive objects may have masses of the order of 1000 $M_\odot$ (where $M_\odot$ is the mass of the sun, 2×10^{33} g) and luminosities of $10^7\ L_\odot$ or more ($L_\odot$, the sun's luminosity or radiated power, is nearly 4×10^{26} W). Very roughly, one star in a billion may be above 60 $M_\odot$.

Theoretical developments have kept apace with the new observational perspectives. We now know that a massive star's evolution is modified by internal mixing as well as by mass loss. Quite unlike the standard doctrine of only a decade ago, the star may remain "quasi-homogeneous" in chemical composition during most of its lifetime; and dramatic instabilities occur at particular stages of evolution. These processes are consequences of turbulent fluid physics in the presence of radiation.

In this chapter we review some of the recent observational and theoretical discoveries about very massive stars. We discuss the broader perspective now available from observations of stars in other galaxies and also a few individual stars of special interest (*1*). Such important topics as stellar winds and mass loss, which are now crucial in stellar astronomy (*2*), cannot be explained in detail here. We only briefly mention the Wolf-Rayet stars, which have very likely evolved from massive stars and which figure prominently in current research (*3*). We must also neglect the chemical enrichment of the interstellar gas and theories of massive supernovae. Even so, there is much to relate about massive stars and their evolution.

Historical Development

It has been known since the 1920's that radiation accounts for most of the pressure in a sufficiently massive star and that this reduces the star's stability (*4*). A self-gravitating ideal gas sphere supported entirely by radiation pressure would have zero net binding energy; so a very massive star is stable only because some of its pressure is due to ordinary gas rather than to radiation. A relatively small amount of energy can cause large-amplitude pulsations in a weakly bound object. Moreover, the CNO cycle of nuclear reactions, which liberates energy in massive stars, is very temperature-sensitive, and this helps motivate a pulsating heat engine. Stars above a critical mass—the Ledoux-Schwarzschild-Härm limit, estimated in 1959 to be about 60 $M_\odot$—are therefore vibrationally unstable (*5*). Before 1970 it was supposed that such a star's pulsation amplitude must grow until its outer layers were ejected, quickly reducing the mass or even destroying the star. Theory seemed to preclude the existence of observable very massive stars.

But a few known objects nevertheless appeared to be very massive. The total mass of one famous double star system had long been known to exceed 110 $M_\odot$ and very likely 150 $M_\odot$ (*6*). By 1960 Feast *et al.* (*7*) had observed the brightest stars in the Large and Small Magellanic Clouds, which are small satellite galaxies to our own Milky Way Galaxy. Luminosities of massive stars within our Galaxy tend to be uncertain, because their distances are usually uncertain and also because interstellar dust obscures most of our galactic disk at visual and ultraviolet wavelengths. The Magellanic Clouds, on the other hand, have well-determined distances [55 and 70 kpc, where 1 kpc (kiloparsec) is about 3260 light-years] and are not heavily obscured. Feast *et al.* found that many stars in the Clouds are luminous enough to require masses close to 100 $M_\odot$. By about 1970, Eta Carinae (η Car) was thought to be a very massive star; at the same time Walborn (*8*) introduced a new spectral classification system for the hottest stars and assigned a new type, O3, to several stars in our Galaxy, implying very high temperatures and luminosities.

Also around 1970, theorists found that the Ledoux-Schwarzschild-Härm limit may be closer to 100 $M_\odot$ than to 60 $M_\odot$ and that the vibrational instability can be self-limiting (*9*). The pulsation amplitude is limited by nonlinear effects, that is, by dissipation in shock waves. Some continuous, noncatastrophic mass loss may result from the pulsation (this is still quantitatively uncertain) but, at least, very massive stars are theoretically permitted to exist. During the past decade, further observations of the brightest stars in our Galaxy and in other nearby galaxies have revealed many very massive stars where few were previously believed to exist.

A very massive star is near catastrophe not only because of interior effects but also because of a direct consequence of radiation pressure in the star's outer layers. There the ratio between opposing forces of radiation pressure and of gravity depends simply on the star's luminosity/mass ratio, L/M, multiplied by an opacity coefficient. If L/M exceeds a critical value, generally called the Eddington limit, then emergent radiation will progressively blow the outer layers away. For very hot stars, where opacity is mainly due to scattering by free electrons, the Eddington limit is

$$(L/M)_{\max} = 4 \times 10^4 \, L_\odot/M_\odot$$

and additional causes of opacity tend to reduce this value. For moderately large stellar masses, 10 to 40 $M_\odot$, the mass-luminosity relation is such that L/M is roughly proportional to M^2. If extrapolated, this relation would violate the Eddington limit for $M \gtrsim 60 \, M_\odot$. Very massive stars must evidently be affected in their structures by radiation pressure, in such a way as to keep their L/M ratios safely below the Eddington limit—so that the mass-luminosity relation is modified above 40 $M_\odot$. For extremely large masses, the mass-luminosity relation may even be determined by the Eddington limit; L/M then becomes almost constant. Along with internal vibrational instability, surface radiation pressure may

Table 1. Summary of the physical properties (*47*) of the luminous stars of different temperatures.

Property	Hot	Intermediate	Cool
Spectral types (*48*)	O, B, A	F, G, K	M
Surface temperature (K)	50,000 to 10,000	10,000 to 4,000	$< 4{,}000$
Luminosity range ($L/L_\odot$)	10^4 to $> 5 \times 10^6$?	10^4 to 8×10^5	10^4 to 5×10^5
Mass range ($M/M_\odot$)	20 to 300?	20 to 50	20 to 50
Size range ($R/R_\odot$)	10 to 100	30 to 1,000	300 to 2,000

help to provoke continuous mass ejection (*2*).

The Hertzsprung-Russell Diagram

The physical characteristics of the very luminous stars (also called supergiants) with different surface temperatures are summarized in Table 1. The most luminous, most massive stars have surface temperatures above 30,000 K and therefore radiate mostly at ultraviolet wavelengths. At the visual wavelengths traditionally used in ground-based astronomy, the intrinsically brightest stars are somewhat cooler, less massive supergiants with temperatures around 10,000 K, like Cygnus OB2 #12 (see Fig. 1). These moderate-temperature supergiants will be the visually brightest ones in other galaxies; they are the most conspicuous stars in photographs of other nearby spiral galaxies.

Figure 1 is the Hertzsprung-Russell (H-R) diagram—that is, a plot of luminosity versus surface temperature with temperature increasing toward the left—for the most luminous known stars ($L \geq 5 \times 10^5\ L_\odot$) in our Galaxy (*10*), the Magellanic Clouds (*11*), the nearby small spiral galaxy M33 (*12*), and the nearby small irregular galaxies NGC 6822 and IC 1613 (*13*). (Because of its large size and high tilt angle there are no extensive surveys for the brightest stars in the giant spiral galaxy M31, the Andromeda nebula.)

The most significant feature of this H-R diagram is an observed upper envelope to the luminosities of normal stars. This luminosity boundary declines with decreasing temperature for the hotter stars, but becomes essentially constant for the cooler supergiant stars. The most luminous, most massive stars occupy the left part of Fig. 1, while the upper right is empty.

Our empirical knowledge of the brightest stars in our own Galaxy is largely restricted to the "solar neighborhood," a region 6 kpc in diameter centered on the sun which is only a few percent of the galactic disk. Stars at greater distances in the galactic disk are badly obscured at visual and ultraviolet wavelengths by interstellar dust. The galactic stars in Fig. 1 are members of star clusters and stellar associations, whose distances are reasonably known, and they are representative of the stellar population in the spiral arms of the Milky Way.

A more complete sample of luminous stars is available for the Large and Small Magellanic Clouds. The H-R diagram for the Large Cloud looks very much like that for the most luminous stars in the solar neighborhood, revealing very similar distributions of stellar temperatures and luminosities in the two galaxies (*14*).

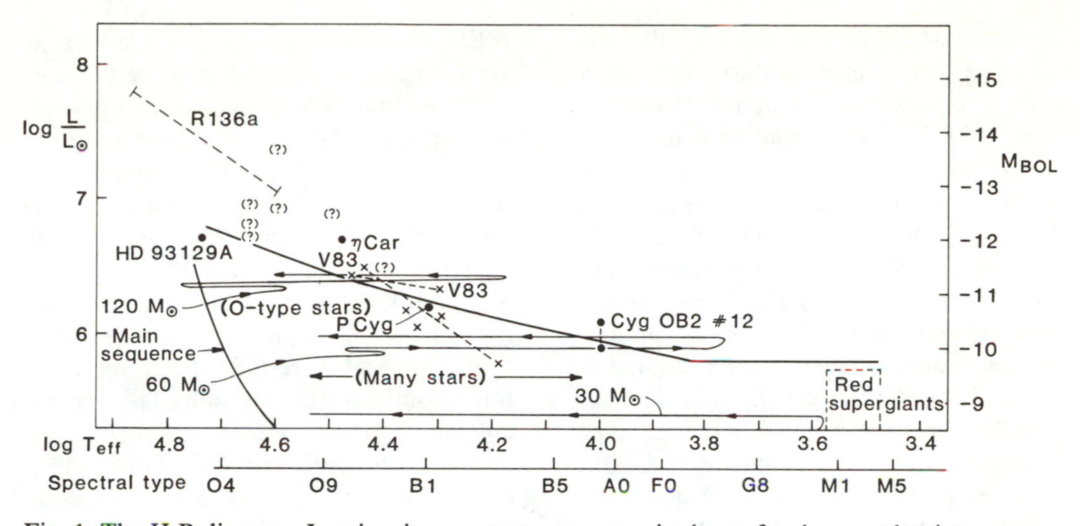

Fig. 1. The H-R diagram. Luminosity versus temperature is shown for the most luminous stars ($L > 5 \times 10^5 L_\odot$); M_{BOL} is bolometric magnitude. The position of the hydrogen-burning main sequence and the empirical upper luminosity boundary for normal stars are shown by solid lines. The locations of the O-type stars and the red supergiants are marked, as well as those of individual stars discussed in the text; the Hubble-Sandage variables are shown by x and the suspected supermassive Wolf-Rayet stars in M33 by (?). The evolutionary tracks for stars of 30, 60, and 120 $M_\odot$ are from Maeder (*39, 41*).

In the Small Magellanic Cloud the luminosities of the brightest hot stars are less than those of their counterparts in our Galaxy and the Large Cloud (*11*); this may be a size-of-sample effect, although it is also possible that star formation produces a different statistical distribution of initial stellar masses in the Small Cloud (*11*). In two other small nearby galaxies, NGC 6822 and IC 1613, there appear to be few hot stars with masses above 80 $M_\odot$ (*13*). Among the several galaxies that we have mentioned, the primary difference appears to be that smaller galaxies have fewer of the most luminous, hot stars. Otherwise, the H-R diagram for each galaxy resembles Fig. 1. It is significant that the upper luminosity envelope for the cool supergiants is the same in all of these galaxies.

"Main sequence" stars, which still have hydrogen at their centers, should be near the left (hot) side of Fig. 1; and as core hydrogen is exhausted, evolution should carry massive stars to the right across the diagram along nearly horizontal tracks as they expand and become cooler at roughly constant luminosity. Stars with masses less than 60 $M_\odot$, in the lower part of Fig. 1, apparently can evolve all the way across the diagram. Why, then, are the most luminous stars restricted to the left side of the H-R diagram? What prevents a very massive star from evolving into a yellow or red supergiant whose surface is cooler than 15,000 K but whose luminosity exceeds $10^6 L_\odot$? We will return to the theoretical implications of this question later; the answer is not simple.

Individual Very Massive Stars

We now describe a few remarkable objects whose characteristics and behavior provide important clues to the natures of very massive stars. We begin with a remarkable association of stars in the Milky Way constellation Carina, located in a prominent spiral arm of our Galaxy and close enough for us to study in detail. The Carina nebula, NGC 3372, is a region of star formation about 2.8 kpc (9000 light-years) away. Associated with this nebula are the famous variable star η Car and many other luminous stars, including several O3 stars. The O3 type denotes the hottest classified stars that have not yet evolved far from the main sequence (*8*). Their surface temperatures may be close to 50,000 K (*15*). Of the ten O3 stars known in 1982, six are associated with the Carina nebula (*16*). These, plus η Car, may all be within a region less than 15 pc across—an amazing concentration of very luminous stars. The brightest of the Carina O3 stars, HD 93129A, probably has a luminosity close to $5 \times 10^6 L_\odot$ (*15*, *17*), about the same as that of η Car and indicative of an initial mass around 200 $M_\odot$. These are the two most luminous known stars in the solar neighborhood.

Eta Carinae has been famous since the middle of the 19th century. Except for statements that it was visible to the naked eye and probably variable, little is known about its history prior to 1830. (It can be observed only from southern latitudes, where there were few astronomers at that time.) Then, between 1836 and 1858, η Car was seen to undergo a spectacular and prolonged outburst (*18*). In 1843 it was briefly the second brightest star in the sky. Later, between 1858 and 1870, its apparent visual brightness faded by a factor of several hundred. For a long time afterward it remained constant in visual brightness, except for one or two brief episodes; but since 1940 it has been gradually brightening.

Eta Carinae is now surrounded by a small nebula of gas and dust, called the "homunculus" because of its appearance (Plate III). The homunculus is expanding at several hundred kilometers per second and was obviously ejected during the explosion (*19*). The mass of the ejected material is most likely 0.1 to 1 $M_\odot$, additional outlying nebulosities suggest that other explosions occurred before 1800, and there is evidence for continued outflow from the star. The apparent fading between 1858 and 1870 was due to the formation of dust grains in the expanding ejected gas. These dust grains now absorb most of the star's visual and ultraviolet luminosity and are thereby heated enough to reradiate the luminosity at infrared wavelengths. Infrared radiation from the homunculus now provides us with a measurement of the total luminosity, almost $5 \times 10^6 L_\odot$ (*20*).

This same luminosity can nearly account for the maximum visual brightness seen in the 1840's and also seems reasonable for the ultraviolet luminosity before the explosion. From these and other considerations involving its spectrum, η Car is thought to have an effective surface temperature around 30,000 K (*21*).

Even though η Car and HD 93129A were formed in the same region and have similar luminosities, η Car appears to be cooler and less stable. Why? The most likely explanation is that their evolutionary states are different. Either η Car is so young that it has not yet reached the main sequence in the H-R diagram, or it is an evolved very massive star near the end of its lifetime. The latter alternative was favored by Burbidge (*22*) more than 20 years ago, but only lately have relevant and incisive observations been made. Ground-based spectra taken by A. D. Thackeray and by N. R. Walborn, and ultraviolet data taken by K. Davidson and T. R. Gull with the International Ultraviolet Explorer (IUE) satellite, have shown that the ejected gas is nitrogen-rich but carbon- and oxygen-poor (*23*). This is a natural outcome of the CNO cycle of nuclear reactions (*24*). The implication is that η Car has had enough time to mix the nuclear products from its core to its surface, from which the observed ejecta came. The star is therefore evolved. Moreover, the fact that mixing has occurred lends dramatic support to some theoretical ideas by Maeder, discussed later.

The precise reason why η Car is unstable—the cause of the observed outburst—is not yet understood. We later mention some possible single-star explanations and why they are relevant to Fig. 1. Alternatively, η Car may be a double star, with one component dynamically interfering with the surface of the other; this idea probably requires that at least one member of the hypothetical pair has evolved and is attempting to expand.

In any case, during recent centuries η Car has probably lost mass unsteadily but at an average rate between 0.001 and 0.1 $M_{\odot}$ per year (*14, 25*). The star may therefore lose most of its mass in less than 10^5 years. A very massive star's total lifetime, however, is much longer than this—of the order of 3×10^6 years. Eta Carinae has evidently reached a critical stage in its evolution.

Another well-known, unstable luminous star in our galaxy is P Cygni (P Cyg). This star temporarily brightened during the 17th century but has not fluctuated much since. Its total luminosity is now around 10^6 $L_{\odot}$ and its surface temperature is 20,000 K or somewhat cooler. It is now losing mass at a rate variously estimated to be 3×10^{-5} $M_{\odot}$ per year (*26*) or 3×10^{-4} $M_{\odot}$ per year (*27*). P Cygni may be qualitatively like η Car but cooler, less luminous, and less massive.

Several very luminous blue variable stars in other galaxies are similar to η Car and P Cyg. These include S Doradus in the Large Magellanic Cloud and the Hubble-Sandage variables in the nearby spiral galaxies M31 and M33 (*28*). One of these, Var A in M33, has a record of variability reminiscent of η Car as well as infrared radiation suggestive of dusty circumstellar gas. Ultraviolet spectra of five Hubble-Sandage variables were recently obtained with the IUE (*29*). Their ultraviolet fluxes, together with ground-based data, imply luminosities and temperatures intermediate between those of P Cyg and η Car. All of these are near or even above the empirical upper luminosity envelope for normal stars in Fig. 1.

Conspicuous in the upper left of the H-R diagram is Radcliffe 136a (R136a), a possible supermassive star. R136a, also called HD 38268, is the central object of the giant nebula 30 Doradus in the Large Magellanic Cloud (Fig. 2). Several years ago Schmidt-Kaler and Feitzinger (*30*) proposed that R136a accounts for most of the ultraviolet radiation that ionizes the nebula; this would entail a huge luminosity. After obtaining ultraviolet data with the IUE satellite, Cassinelli *et al.* (*31*) in 1981 concluded (along with Schmidt-Kaler and Feitzinger) that R136a is a single star, with a surface temperature above 60,000 K, whose luminosity exceeds that of η Car or HD 93129A by a factor of 10 or 20. The suggested mass was close to 3000 $M_{\odot}$. According to Savage *et al.* (*32*), newer data on the ultraviolet continuum are consistent with a somewhat lower mass of about 2000 $M_{\odot}$.

R136a is clearly a significant and extreme object, but there are two crucial uncertainties regarding its parameters. First, is it really a single object? This is

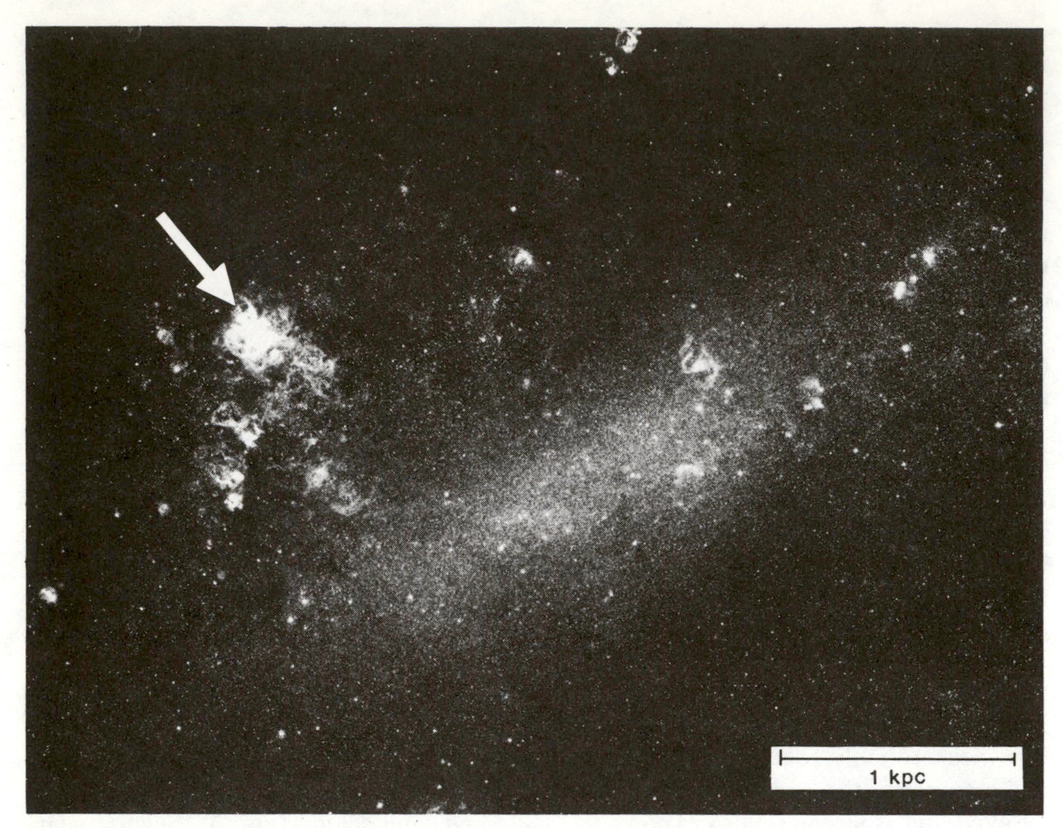

Fig. 2. The Large Magellanic Cloud showing the 30 Doradus or Tarantula nebula. [National Optical Astronomy Observatories]

uncertain because it is so far away. Second, what is its surface temperature? Most of the luminosity is at unobservably short wavelengths ($\lambda < 1000$ Å) and therefore must be extrapolated from the IUE and ground-based data (1000 Å $< \lambda < 3$ μm). The derived luminosity is roughly proportional to the cube of the assumed temperature.

Regarding the first question, direct photographs and visual observations show that R136a is smaller than 0.1 pc. Speckle interferometry (a new specialized technique for obtaining good angular resolution despite blurring by the earth's atmosphere) and the Space Telescope are capable of placing a limit of 0.005 pc on the size of R136a (*33*). But even 0.005 pc is large by some standards—it is 1000 times the earth-sun distance—and does not prove that R136a is a single star. A plausible cluster of many (more than ten) stars cannot be fit into 0.005 pc, but a double, triple, or quadruple system of stars is easily possible within that size. (The period for a 0.005 pc orbit would be of the order of a thousand years.) Long before the current brouhaha, Walborn (*34*) suggested that R136a is analogous to the very compact set of luminous stars in NGC 3603, the most massive nebula that has been optically observed in our Galaxy.

The appropriate temperature for

R136a is also uncertain. Schmidt-Kaler and Feitzinger (*30*) derived high temperatures by assuming that R136a photoionizes the entire 30 Doradus nebula. Cassinelli *et al.* (*31, 32*) argued from the general appearance of the ultraviolet spectrum that a temperature of 60,000 K or more seemed likely, and noted that this also sufficed for the ionization. However, various astronomers have noted that 30 Doradus is an extremely large nebula and contains many hot stars, which are not well studied and which may account for most of the ionization (*35*). The temperature of R136a should therefore be estimated more directly, from its spectral features. The shape of the observed continuum resembles a 40,000 K blackbody (*32*) and we take this to be a lower limit to the actual relevant temperature. Spectral lines at visual and near-infrared wavelengths resemble those in an O3-type spectrum, but not perfectly, and are not easy to analyze (*36*). The best temperature indicator may be the near absence of certain ultraviolet Si IV lines (*32*), implying that silicon is more than triply ionized. However, sophisticated atmosphere models—at least as good as those by Kudritzki (*15*) but specifically adapted to R136a—will be needed to quantify this. At present, it is quite conceivable that the relevant temperature is as low as 45,000 K. This would imply a luminosity around $1.5 \times 10^7\ L_\odot$, only three times as great as that of η Car. One star with a mass less than 1000 $M_\odot$ can achieve this.

Intuitively, and with little proof, we suspect that R136a will turn out to be either a single star with a mass between 500 and 1200 $M_\odot$, or else two or three stars in the range 200 to 1000 $M_\odot$. Such a "conservative" view is appealing because it does not leave a large gap between R136a and other known objects. Even so, R136a is a wonderfully extreme object and the recent observations constitute a significant and fascinating advance in our knowledge of the most massive stars.

Several objects with Wolf-Rayet–like spectra (*3*) have been identified in the large nebulae in M33 (*37*). Some of them may resemble R136a (*38*). Their proposed locations in the H-R diagram, denoted by question marks in Fig. 1, are less extreme than that of R136a. Unfortunately, since M33 is much farther away (700 kpc) than the Large Cloud, we cannot tell whether any of them are single stars.

In summary, Table 2 is a list of the physical parameters for several of the objects we have mentioned.

Table 2. Physical characteristics of some very luminous stars.

Characteristic	η Car	P Cyg	R136a	Var 83 in M33 (Hubble-Sandage variable)
Mass ($M/M_\odot$)	~ 200	~ 80 to 100	≤ 2,000	~ 100 to 150
Luminosity ($L/L_\odot$)	5×10^6	$\leq 1.5 \times 10^6$	$\leq 6 \times 10^7$	2×10^6
Temperature (K)	30,000	≤ 21,000	45,000 to 75,000	20,000 to 30,000
Mass loss rate ($M_\odot$ per year)	10^{-1} to 10^{-3}	10^{-4} to 10^{-5}	$\sim 5 \times 10^{-4}$	$\sim 3 \times 10^{-5}$

Theoretical Considerations

Several aspects of theory are influenced by the observational discoveries that we have described. In this chapter we can discuss only a few important theoretical ideas, but first we should outline some background. Nuclear reactions in a massive star, during most of its lifetime, occur within a central core which is continually stirred by convection. The hydrogen in this core is gradually changed into helium. Outside the core, according to the simplest view held a few years ago, material should remain mostly hydrogen because convective mixing does not generally occur there. During the past 15 years it has been recognized that massive stars tend to lose mass continuously (*2*). The outflowing "stellar winds" can be studied through infrared-, visual-, and ultraviolet-wavelength spectroscopy and even through radio and x-ray observations. Calculations of stellar evolution must therefore include mass loss, while theorists also attempt to explain why the stellar winds occur. (Radiation pressure plays a leading role but is not the only important process.) Some objects known as Wolf-Rayet stars appear to be relevant (*3*). A Wolf-Rayet star typically has a strong stellar wind and a hot surface which is hydrogen-poor but helium-rich and (relative to most stars) either carbon- or nitrogen-rich (spectral types WC and WN). If a moderately massive or very massive star can lose enough exterior mass while evolving a hydrogen-poor core, the star is eventually peeled down to its inner, processed material; it is then a Wolf-Rayet star. Hence, the stellar evolution and mass loss theorists have been greatly motivated by studies of Wolf-Rayet stars (*2*, *3*).

While keeping this background situation in mind, we think that very massive stars provide remarkable insights that are not obvious from studies of ordinary stellar winds or of Wolf-Rayet stars. Consider the upper envelope to luminosities in the H-R diagram (Fig. 1). What prevents a very massive star from crossing this boundary while evolving rightward (expanding and cooling) across the upper left-hand part of Fig. 1? In 1979 we drew attention to this question and proposed an intuitive answer, which naturally involved mass loss (*14*). Evolutionary calculations by de Loore, Chiosi, Stothers, Maeder, and others (*39*) show that continuous high mass loss rates do indeed prevent very massive stars from evolving far to the right or to cooler temperatures in the H-R diagram, but the average observed mass loss rates are inadequate to make normal loss the full explanation. Our 1979 hypothesis was somewhat different because we emphasized unsteady mass loss. Note that η Car, P Cyg, and at least some of the Hubble-Sandage variables, all lying near the critical boundary in Fig. 1, are thought to suffer spectacular episodes of mass ejection. These examples inspired our scenario, wherein some particular instability causes a drastic increase in mass loss just as the evolving star reaches the critical line in the H-R diagram. (Incidentally, the line drawn in Fig. 1 is an empirical envelope to the sample of "normal" stars. We expect the true boundary line for instability to lie slightly above this; perhaps η Car marks its location.) A star may even "bounce" recurrently on the critical line. When the star has evolved to this limit, perhaps the sudden instability causes an η Car–like outburst which ejects a fraction of a percent of the star's mass. This moves the star slightly away from the critical line and temporarily relieves the instability, but then, in a few centuries or decades, the star evolves back to the limit and suffers another explosion; and so on, perhaps until the star is reduced to a Wolf-Rayet star (unless it becomes a supernova first).

What causes the instability? There are several alternatives. An internal effect—possibly reminiscent of the classical vibrational instability—may be responsible [see a recent discussion by Stothers and Chin (*40*)]. But more likely, surface radiation pressure is involved. The temperature suspected for the surface of η Car, 30,000 K, is just low enough to allow the mostly ionized gas to contain a perceptible concentration of neutral hydrogen and low-ionization heavy ions (*21*). These raise the opacity noticeably above the value that applies at higher temperatures, which is due to scattering by free electrons. Lower temperature therefore means higher opacity, which means that the Eddington limit $(L/M)_{max}$ is decreased. For a given stellar L/M ratio, the star's surface then becomes less stable than it would be at higher temperatures. Thus, unless η Car has a mass well above 200 $M_{\odot}$, its surface temperature cannot evolve much below 30,000 K without some drastic change in its atmosphere. This is not really an explanation of the instability—a model atmosphere may be able to meet the opacity-related difficulty by expanding to lower densities—but the temperature dependence of the opacity seems to hint that the left-hand, sloping part of the luminosity envelope in the H-R diagram is at a critical location for the opacity-dependent Eddington limit.

Another promising cause for the hypothetical instability has been discussed by de Jager (*1*, pp. 11–14) and by Maeder (*41*). This intriguing but complicated mechanism involves turbulence. Supergiant stars with surface temperatures below 10,000 K or so are convective in some outer layers. This entails a pressure due to turbulent motions. Near the star's surface, dissipation of convective kinetic energy gives rise to a gradient of turbulent pressure. A pressure gradient, however, is in effect a volume force, which can oppose gravity. According to de Jager, the turbulent pressure gradient fully counteracts gravity if the star's luminosity is sufficiently high. (This may be analogous to the Eddington limit, with turbulent pressure instead of radiation pressure.) The expected luminosity limit is said to be at about the same location as the cooler, flat part of the empirical line in Fig. 1. Maeder remarks that this result is insensitive to chemical composition, which means that the luminosity limit for cool stars should be about the same in all star-formation regions and in all galaxies—which is what the observations suggest (*10–14*). It is not clear, though, whether the turbulent pressure mechanism is applicable to stars much hotter than 10,000 K. Stothers and Chin (*40*) discussed a few other processes that may contribute to the hypothetical instability.

Some of the most ambitious evolutionary calculations for massive stars have been done by Maeder (*41, 42*). A novel part of his work concerns mixing within very massive stars. Until recently, it was supposed that the chemical composition at the surface of a massive star (excluding Wolf-Rayet stars) is largely unaffected by the nuclear reactions in the core. Proposed mixing mechanisms (convective overshooting, meridional circulation due to rotation) did not seem to be very effective. Recently, though, Maeder found that turbulent diffusion incited by differential rotation can be important. Differential rotation is expected to occur inside a typical massive star. But the rotation-speed gradient (the shear) cannot be extremely large, for if it were, strong turbulence would develop and would redistribute the angular momentum. Hence, there is a tendency for differential rotation to adjust itself so that turbulence is only marginally induced (*43*). The Reynolds number is thus automatically of the order of 100. Turbulent mixing occurs and behaves like diffusion of material; the effective diffusion coefficient is roughly equal to the viscosity multiplied by the Reynolds number. In most stars this product is too small to have much effect. Maeder pointed out, however, that radiative viscosity, like radiation pressure, becomes large in massive stars, so that the turbulent diffusion coefficient is large—large enough for the mixing time scale in a very massive star to be less than the star's lifetime (*42*). This enables the star to remain quasi-homogeneous in its chemical composition even though nuclear processing occurs in the convective core. Maeder's turbulent mixing theory has apparently been confirmed by the discovery, mentioned earlier in this chapter, that nitrogen-rich material exists at the surface of η Car (*23, 41*).

In Maeder's models, which include mixing as well as mass loss, very massive stars become hot Wolf-Rayet stars after losing mass during their η Car–like encounters with the critical luminosity envelope in the H-R diagram. Mixing is important in this scenario. An interesting consequence concerns supernova events. A star that starts out moderately massive is probably a cool red supergiant when it explodes as a supernova, but a star which is initially very massive becomes a supernova while it is a hot Wolf-Rayet object.

It is not easy for very massive stars to form. One difficulty is that a massive protostar may be self-limiting in its "cocoon" stage. The densest part of a protostellar cloud should contract fastest, to form a sort of condensation nucleus. This is essentially a star, which grows as the outer parts of the surrounding cloud—the cocoon—continue to fall inward. As Larson and Starrfield and later Kahn noted (*44*), this growth should stop when the central object develops sufficient luminosity for radiation pressure to reverse the infall of dusty gas, because dust in the gas makes the effective opacity very large. At the same time, ultraviolet radiation may ionize and heat the same gas, making it less susceptible to gravitational infall. These limiting effects may occur when the star has acquired mass of the order of 60 $M_{\odot}$, and they become progressively more likely with increasing mass. However, according to Wolfire and Cassinelli (*45*), larger stellar masses are possible if the dust is more easily destroyed than Kahn assumed. The situation is not clear. We also do not know what special conditions occurred in the Carina nebula to create such a large and unusual concentration of very massive stars. This question is obviously relevant to theories of the "initial mass function," that is, of the statistical distribution of stellar masses throughout our Galaxy and in other galaxies.

Finally, we mention an unconventional idea by Bath (*46*) in which many of the very luminous objects we observe are really accretion disks around moderately massive stars in binary systems. One star in a very close binary system may lose mass, which is then accreted onto the other star; this accretion flow forms a luminous disk because of angular momentum and turbulent viscosity. The Eddington limit is modified in this context, so that a given luminosity requires smaller individual stellar masses. One objection to this type of model is that the part of Fig. 1 containing O3 stars, η Car, and the Hubble-Sandage variables can be explained without invoking accretion disks. It is difficult to assess the probability that there are enough massive, compact binary star systems in the necessary stage of their evolution. Bath's suggestion is perhaps most appealing for the most extreme objects. In the case of R136a, one naturally wonders whether the central object might be a massive black hole, with an accretion disk, rather like a miniature version of an active galactic nucleus. Savage *et al.* (*32*), on the other hand, remark that theoretically one does not expect an accretion disk to produce the observed O3-like spectrum.

The past decade has produced surprising revelations about the evolution of the most massive stars, and we expect that the study of very massive stars will figure prominently in the research programs of future very large telescopes and space telescopes. The angular resolution expected with the Space Telescope will be ten times better than with current ground-based telescopes. A good high-resolution Space Telescope image of the homunculus of η Car should be spectacular. The high spatial and spectral resolution planned for the telescopes of the future will be crucial for observations of individual stars in other galaxies.

References and Notes

1. For an extensive review of pre-1979 work see C. de Jager, *The Brightest Stars* (Reidel, Dordrecht, 1980).
2. V. G. Fessenkov [*Astron. Zh.* **26**, 67 (1949)] and A. G. Massevitch (*ibid.*, p. 207) suggested long ago that radiation-driven mass loss affects the evolution of massive stars; but it was not until the late 1960's that the topic really began to develop vigorously. See P. S. Conti, *Annu. Rev. Astron. Astrophys.* **16**, 371 (1978); J. P. Cassinelli, *ibid.* **17**, 275 (1979); P. S. Conti and R. McCray, *Science* **208**, 9 (1980); C. de Jager, in *The Most Massive Stars, ESO Workshop Proceedings*, S. D'Odorico, D. Baade, K. Kjär, Eds. (European Southern Observatory, Garching, West Germany, 1981), p. 67.
3. P. S. Conti, in *Effects of Mass Loss on Stellar Evolution*, C. Chiosi and R. Stalio, Eds. (Reidel, Dordrecht, 1981), p. 1; C. W. H. de Loore and A. J. Willis, Eds., *Wolf-Rayet Stars, IAU Symposium 99* (Reidel, Dordrecht, 1982); K. A. van der Hucht, P. S. Conti, I. Lundström, B. Stenholm, *Space Sci. Rev.* **28**, 227 (1981).
4. A. S. Eddington, *The Internal Constitution of the Stars* (Cambridge Univ. Press, Cambridge, 1926); S. Chandrasekhar, *An Introduction to the Study of Stellar Structure* (Univ. of Chicago Press, Chicago, 1939).
5. P. Ledoux, *Astrophys. J.* **94**, 537 (1941); M. Schwarzschild and R. Härm, *ibid.* **129**, 637 (1959). Strictly speaking, the quoted result is valid only for main sequence stars which burn hydrogen by the CNO cycle at their centers. The critical mass can be raised by readjusting the chemical composition. Schwarzschild and Härm noted that some stars were observed to be more massive than their theoretical limit of about 65 $M_\odot$.
6. This is "Plaskett's star," HD 47129. See J. S. Plaskett, *Publ. Dom. Astrophys. Obs.* **2**, 147 (1922); J. B. Hutchings and A. P. Cowley, *Astrophys. J.* **206**, 490 (1976).
7. M. W. Feast, A. D. Thackeray, A. J. Wesselink, *Mon. Not. R. Astron. Soc.* **121**, 337 (1960).
8. N. R. Walborn, *Astrophys. J. Lett.* **167**, L31 (1971); *Astrophys. J.* **179**, 517 (1973). Spectral type O3 required higher surface temperatures than the previous classification limit O5 in the decreasing-temperature O, B, A, F, G, K, M scheme. For main-sequence stars, burning hydrogen at their centers, higher temperatures correspond to higher luminosities.
9. I. Appenzeller, *Astron. Astrophys.* **5**, 355 (1970); *ibid.* **9**, 216 (1970); N. R. Simon and R. Stothers, *ibid.* **6**, 183 (1970); K. Ziebarth, *Astrophys. J.* **162**, 947 (1970); R. J. Talbot, *ibid.* **165**, 121 (1971). The relative pulsation amplitude may be small at the stellar surface even if it is large inside the star. One of us (K.D., unpublished results) has attempted to detect photometric variations, with expected periods less than a day, in several massive O3 stars in Carina; it appears that photometric amplitudes are typically less than 1 or 2 percent.
10. R. M. Humphreys, *Astrophys. J. Suppl.* **38**, 309 (1978).
11. ——, *ibid.* **39**, 389 (1979) for the Large Cloud; *Astrophys. J.* **265**, 176 (1983) for the Small Cloud.
12. ——, *Astrophys. J.* **241**, 587 (1980).
13. ——, *ibid.* **238**, 65 (1980).
14. —— and K. Davidson, *ibid.* **232**, 409 (1979).
15. R. P. Kudritzki, *Astron. Astrophys.* **85**, 174 (1980); in *The Most Massive Stars, ESO Workshop Proceedings*, S. D'Odorico, D. Baade, K. Kjär, Eds. (European Southern Observatory, Garching, West Germany, 1981), p. 49; P. S. Conti and S. A. Frost, *Astrophys. J.* **212**, 728 (1977).
16. N. R. Walborn, *Astrophys. J. Lett.* **254**, L15 (1982).
17. P. S. Conti and M. L. Burnichon, *Astron. Astrophys.* **38**, 467 (1975). The luminosities of O3 stars are uncertain partly because semitheoretical extrapolations into the far ultraviolet are required, and partly because the distances are

uncertain. The former uncertainty does not apply to η Car.
18. J. F. W. Herschel, *Results of Astronomical Observations . . . at the Cape* (Smith, Elder, London, 1847), p. 32; R. T. A. Innes, *Cape Ann.* **9**, 75B (1903); B. J. Bok, *Harv. Rep. No. 77* (1932); L. Gratton, *Star Evolution* (Academic Press, New York, 1963), p. 297; N. R. Walborn and M. H. Liller, *Astrophys. J.* **211**, 181 (1977).
19. A. D. Thackeray, *Observatory* **69**, 31 (1949); E. Gaviola, *Astrophys. J.* **111**, 408 (1950); A. E. Ringuelet, *Z. Astrophys.* **46**, 276 (1958); N. R. Walborn, B. M. Blanco, A. D. Thackeray, *Astrophys. J.* **219**, 498 (1978).
20. As seen from the earth, η Car is the brightest extra-solar-system object in the sky in the wavelength range 10 to 20 μm. G. Neugebauer and J. A. Westphal, *Astrophys. J. Lett.* **152**, L89 (1968); *ibid.* **156**, L45 (1969); G. Robinson, A. R. Hyland, J. A. Thomas, *Mon. Not. R. Astron. Soc.* **161**, 281 (1973); R. D. Gehrz, E. P. Ney, E. E. Becklin, G. Neugebauer, *Astrophys. J. Lett.* **13**, 89 (1973); A. R. Hyland, G. Robinson, R. M. Mitchell, J. A. Thomas, E. E. Becklin, *Astrophys. J.* **233**, 145 (1979).
21. K. Davidson, *Mon. Not. R. Astron. Soc.* **154**, 415 (1971); B. E. J. Pagel, *Astrophys. Lett.* **4**, 221 (1969).
22. G. R. Burbidge, *Astrophys. J.* **136**, 304 (1962). Burbidge's conjecture involved far too low a surface temperature but was otherwise much better than competing pre-1968 theories about η Car. Burbidge remarked that "η Car is a likely candidate for becoming a supernova, though whether this will occur in the next 100 or in the next 100,000 years we have no means of estimating."
23. K. Davidson, N. R. Walborn, T. R. Gull, *Astrophys. J. Lett.* **254**, L47 (1982); see also K. Davidson, *Mercury* **11**, 138 (1982).
24. In the CNO cycle, carbon, nitrogen, and oxygen nuclei serve as catalysts to convert hydrogen into helium. The relative abundance ratios of C, N, and O approach an equilibrium state in which ^{14}N is prevalent. See G. R. Caughlan and W. A. Fowler, *Astrophys. J.* **136**, 453 (1962); G. R. Cauglan, *ibid.* **141**, 688 (1965); C. de Jager (*1*, pp. 162–164).
25. C. D. Andriesse, B. D. Donn, R. Viotti, *Mon. Not. R. Astron. Soc.* **185**, 771 (1977).
26. D. van Blerkom, *Astrophys. J.* **221**, 186 (1978).
27. J. B. Hutchings, *ibid.* **203**, 438 (1976).
28. E. P. Hubble and A. Sandage, *ibid.* **118**, 353 (1953); R. M. Humphreys, *ibid.* **200**, 426 (1975) and **219**, 445 (1978); L. Rosino and A. Bianchini, *Astron. Astrophys.* **22**, 453 (1973); A. Bianchini and L. Rosino, *ibid.* **42**, 289 (1975).
29. R. M. Humphreys, C. Blaha, S. D'Odorico, T. R. Gull, P. Benvenuti, *Astrophys. J.*, in press.
30. T. Schmidt-Kaler and J. V. Feitzinger, in *The Most Massive Stars, ESO Workshop Proceedings*, S. D'Odorico, D. Baade, K. Kjär, Eds. (European Southern Observatory, Garching, West Germany, 1981), p. 105; and *Astrophys. Space Sci.* **41**, 357 (1976); J. V. Feitzinger, W. Schlosser, T. Schmidt-Kaler, C. Winkler, *Astron. Astrophys.* **84**, 50 (1980).
31. J. P. Cassinelli, J. S. Mathis, B. D. Savage, *Science* **212**, 1497 (1981).
32. B. D. Savage, E. L. Fitzpatrick, J. P. Cassinelli, D. C. Ebbets, *Astrophys. J.*, in press.
33. Visual observations [R. T. A. Innis, *Catalogue of Double Stars* (1927); C. E. Worley, unpublished results; Y. H. Chu and M. Wolfire, *Bull. Am. Astron. Soc.* **15**, 644 (1983)] reveal that R136a is double with a separation of 0.46 arc sec (≃ 0.1 pc). Recent speckle interferometric measurements by G. Weigelt [quoted in (*30, 32*)] are consistent with the visual data, but those by J. Meaburn, J. C. Hebden, B. L. Morgan, and H. Vine [*Mon. Not. R. Astron. Soc.* **200**, 1 (1982)] are not.
34. N. R. Walborn, *Astrophys. J.* **182**, 121 (1973).
35. H. Shapley and J. S. Paraskevolpoulos, *ibid.* **86**, 340 (1937); J. Melnick, in *Wolf-Rayet Stars, IAU Symposium 99*, C. W. H. de Loore and A. J. Willis, Eds. (Reidel, Dordrecht, 1982), p. 545.
36. D. C. Ebbets and P. S. Conti, *Astrophys. J.* **263**, 108 (1982); J. M. Vreux, M. Dennefeld, Y. Andrillat, *Astron. Astrophys.* **113**, L10 (1982).
37. P. S. Conti and P. Massey, *Astrophys. J.* **249**, 471 (1981); S. D'Odorico and M. Rosa, *ibid.* **248**, 1015 (1981); S. D'Odorico and P. Benvenuti, *Mon. Not. R. Astron. Soc.*, in press; P. Massey and P. S. Conti, *Astrophys. J.*, in press.
38. P. Massey and J. B. Hutchings, *Astrophys. J.*, in press.
39. C. de Loore, J. P. De Greve, H. Lamers, *Astron. Astrophys.* **61**, 251 (1977); C. Chiosi, E. Nasi, S. R. Sreenivasan, *ibid.* **63**, 103 (1978); R. Stothers and C.-W. Chin, *Astrophys. J.* **233**, 267 (1979); A. Maeder, *Astron. Astrophys.* **92**, 101 (1980); *ibid.* **99**, 97 (1981); *ibid.* **102**, 401 (1981).
40. R. Stothers and C.-W. Chin, *Astrophys. J.* **264**, 583 (1983).
41. A. Maeder, *Astron. Astrophys.* **120**, 113 (1983).
42. ———, *ibid.* **105**, 149 (1982).
43. E. Schatzman, *ibid.* **56**, 211 (1977).
44. R. B. Larson and S. Starrfield, *ibid.* **13**, 190 (1971); F. D. Kahn, *ibid.* **37**, 149 (1974). See also K. Davidson and M. Harwit [*Astrophys. J.* **148**, 443 (1967)], where the term "cocoon star" first appeared.
45. M. G. Wolfire and J. P. Cassinelli, *Bull. Am. Astron. Soc.* **15**, 644 (1983).
46. G. T. Bath, *Nature (London)* **282**, 274 (1979).
47. For comparison, the sun has the following parameters: $M_\odot = 2 \times 10^{33}$ g, $L_\odot = 4 \times 10^{26}$ W, $R_\odot = 7 \times 10^5$ km, and $T_\odot = 5800$ K.
48. Spectral type refers to the appearance of the absorption lines in a star's spectrum and is an indicator of the surface temperature. To some extent luminosity as well as temperature can be inferred from spectroscopy.
49. We are grateful to N. R. Walborn, P. Conti, and R. Kennicutt for many discussions about massive stars. In our work on this topic, we have been guest observers in the NASA-supported IUE program, as well as visiting astronomers at Kitt Peak National Observatory and Cerro Tololo Inter-American Observatory, which are operated by the Association of Universities for Research in Astronomy, Inc., under contract with the National Science Foundation. R.M.H.'s work is partially supported by an NSF grant.

This material originally appeared in *Science* 223, 20 January 1984.

7. Chromospheres, Transition Regions, and Coronas

Erika Böhm-Vitense

As long as total solar eclipses have been observed, people have seen the extended region around the sun, called the corona, which gives off a dim ghostly light during the total eclipse. The corona is not homogeneous but has a radial structure, or streamers (see Fig. 1). The resemblance to a crown gave it its name. The total amount of light emitted by the corona is only about one millionth of the light emitted by the solar disk itself, that is, photospheric spectrum. Analysis of the coronal light reveals that it consists of three parts:

1) The Fraunhofer corona, which consists of photospheric light scattered by "dust" particles at large distances from the sun.

2) The continuum corona, which is also due to photospheric light, but scattered by rapidly moving electrons in the solar corona. Because of the very rapid motions of the scattering particles, the Fraunhofer lines are washed out by the Doppler effect.

3) Some fairly strong emission lines, which differ so much from the photospheric spectrum that they cannot be scattered photospheric light but must originate in the corona itself. In discussing the coronal spectrum in this chapter, I mean this emission line spectrum of the corona proper. The coronal line at wavelength λ = 5303 Å can be seen in Fig. 2.

During the first seconds of a total solar eclipse, when the solar disk is just covered by the moon, another kind of spectrum, the "flash" spectrum, is visible (see Fig. 2). This emission occurs in a layer just above the photosphere. Its brightness is about 10 to 100 times that of the total corona, but it remains visible only for a few seconds, indicating that it is emitted by a layer whose vertical extent is of the order of 10,000 km. As shown in Fig. 1, the extent of the corona is several solar radii (the solar radius is 700,000 km). The layer that emits the colorful flash spectrum is called the chromosphere.

Observed Temperature Stratification in the Outer Layers of the Sun

The corona. The source of the coronal emission line spectrum remained a mystery for many years until Grotrian (*1*) and Edlén (*2, 3*) identified most of the observed corona lines with transitions in very highly ionized heavy elements. The identification of the lines implied that some elements had to be ionized 15 times. Removal of the last electron would require energies of about 500 eV. Such energies cannot be supplied by the

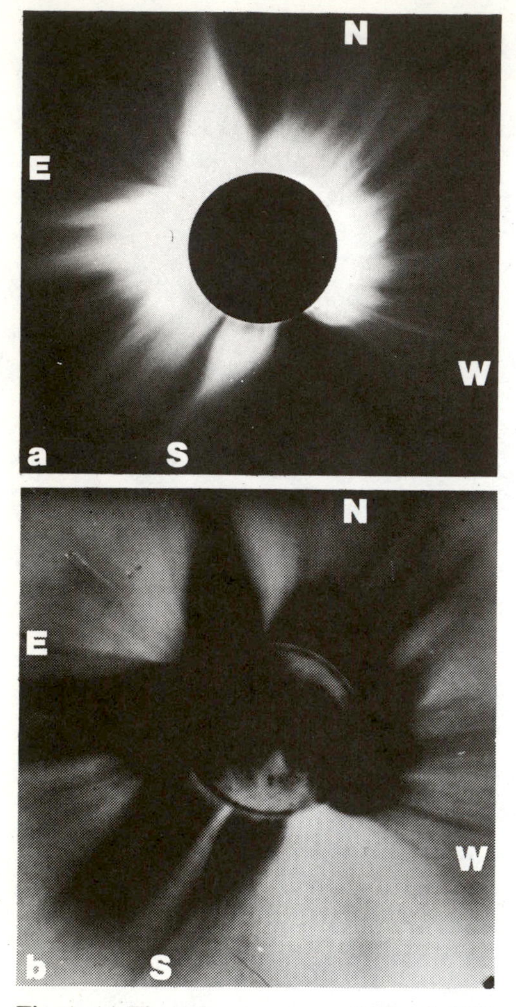

Fig. 1. (a) The solar corona on 7 March 1970 during total solar eclipse [Courtesy of G. Newkirk, National Center for Atmospheric Research]. (b) X-ray picture of the solar disk, taken shortly after the eclipse, photomontaged on the disk of the moon [Courtesy of the Solar Physics Group, American Science and Engineering, Inc.].

photospheric radiation field; they can only be supplied from the kinetic energy of particles. Grotrian inferred that the corona must have temperatures of the order of 1 million degrees, which came as a big surprise. It led to questions of how such high temperatures can be sustained above the solar photosphere, which has an average temperature of 5800 K and a surface temperature of about 4500 K; how the photosphere can remain cool in between the hot solar interior and the hot corona outside; and whether this contradicts the second law of thermodynamics. In fact, if the very high coronal temperatures had not been so difficult to believe, they might have been discovered much earlier from the large Doppler effects implied by the washed out Fraunhofer lines of the continuous corona.

Modern x-ray observations confirm the high coronal temperatures. Figure 1b shows a photomontage of an x-ray picture of the sun on the disk of the moon occulting the sun. The structure in the x-ray picture of the corona reflects the magnetic field strength and also the topology of the field lines. X-ray dark regions, the so-called coronal holes, coincide with regions where the magnetic field lines are "open" (this means that they close at very large distances from the sun). Bright coronal regions coincide with regions where the magnetic field lines are closed in the corona. Loop structures can often be seen (Fig. 3).

The chromosphere. Studies of emission line strengths and degree of ionization can also be used to determine the temperature in the chromosphere. The chromospheric line emission is mainly due to elements that are ionized only once, indicating a much lower temperature. For a quantitative determination of the temperature it is necessary to study how much of the energy needed for the ionization and light emission comes from absorption of photospheric radiation and how much must ultimately come from collisions (including the reabsorbed energy from chromospheric emission). From such a study we can, in principle,

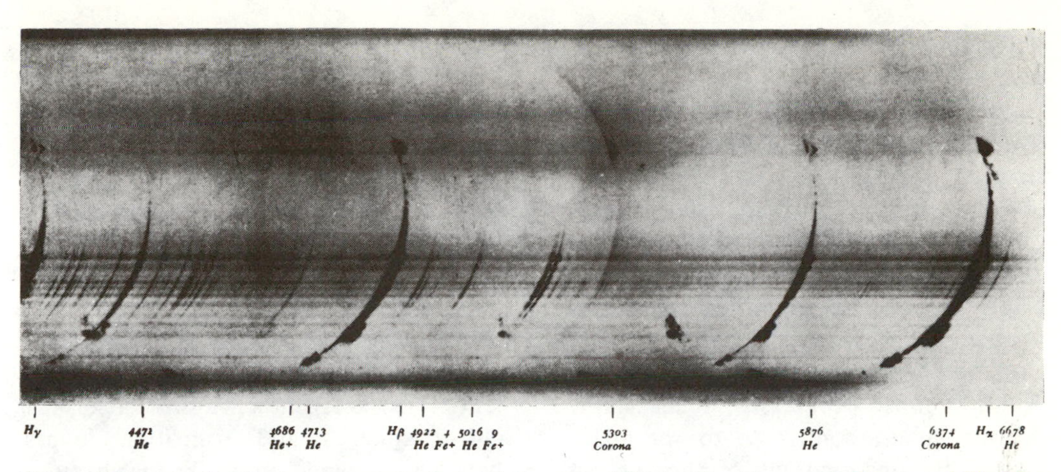

Fig. 2. Slitless solar flash spectrum taken 14 January 1926 by Davidson and Stratton (*41*). The emission spectrum shows lines of hydrogen, helium, and ions of the heavy elements. The curved solar limb is seen in the light of the different lines. A coronal emission line is seen at 5303 Å. [A. Unsöld, *Physik der Sternatmosphären* (Springer-Verlag, Berlin, 1938), reprinted with permission]

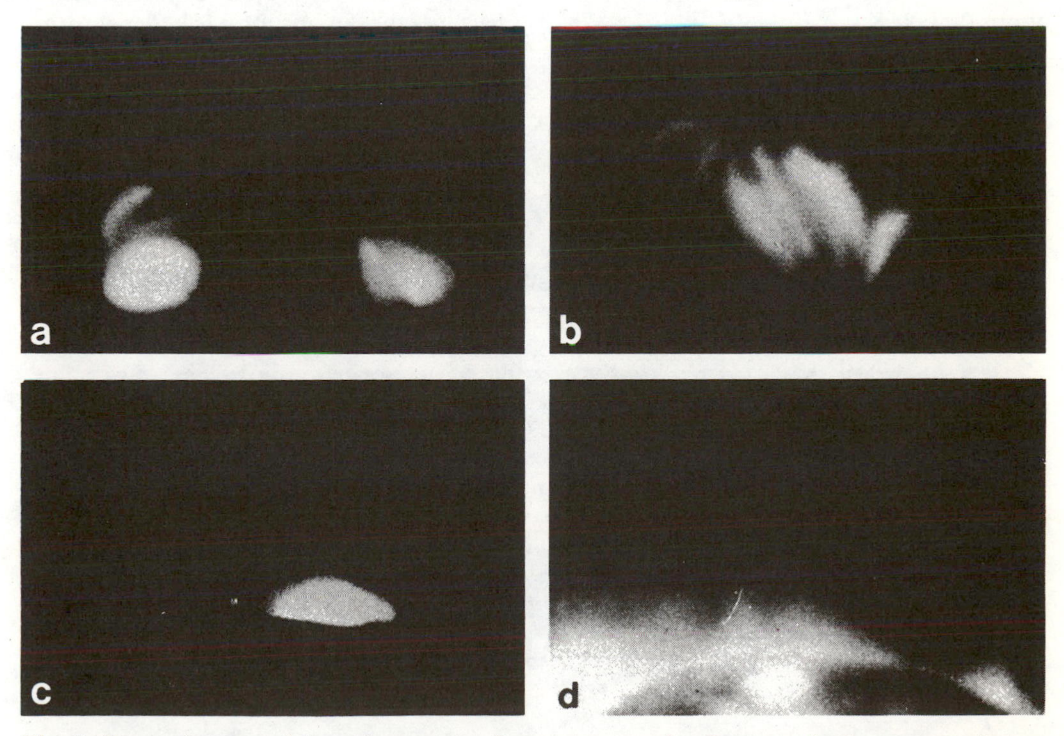

Fig. 3. X-ray pictures of the solar corona. Loop structure is often seen in the x-ray emission, outlining the geometry of the magnetic lines of force. [Reprinted courtesy of R. Rosner, *Astrophysical Journal* 220, 643 (1978), published by the University of Chicago, © American Astronomical Society]

determine the temperature necessary to give the observed chromospheric spectrum, but many of the details of the processes involved are still under investigation.

Solar eclipse observations can show directly the stratification of temperature and density. During a total solar eclipse, a few seconds before the disk becomes visible again, the light comes mainly from the highest chromospheric layers. As the eclipse progresses, successively deeper layers contribute to the light. Using the observations of the highest layers first, we can see how much light is added for successively deeper layers. It is generally found that in the chromosphere the temperature increases slowly outward to about 8500 K; thereafter it increases somewhat more rapidly to about 20,000 K.

The transition region. Since the corona has temperatures of about 2 million degrees and the chromosphere temperatures up to 20,000 K, there must be a layer with temperatures between 20,000 and 1 million degrees. This layer is called the transition region. Woolley and Allen (*4*) pointed out on theoretical grounds that the transition layer might be very thin, and the temperature might rise from 20,000 to 500,000 K over a distance of only 5000 km.

With the availability of rocket and satellite observations in the far ultraviolet ($\lambda < 1700$ Å), new ways of observing chromospheres and transition layers have been opened up. In the far ultraviolet the photospheric radiation of the sun is so reduced that the intensities of the emission lines of the chromosphere and transition layer are greater than the intensity of the photospheric light even when there is no eclipse. Figure 4 shows part of the solar ultraviolet spectrum. From such observations it is found that the temperature increase is even steeper in the lower transition region than sug-

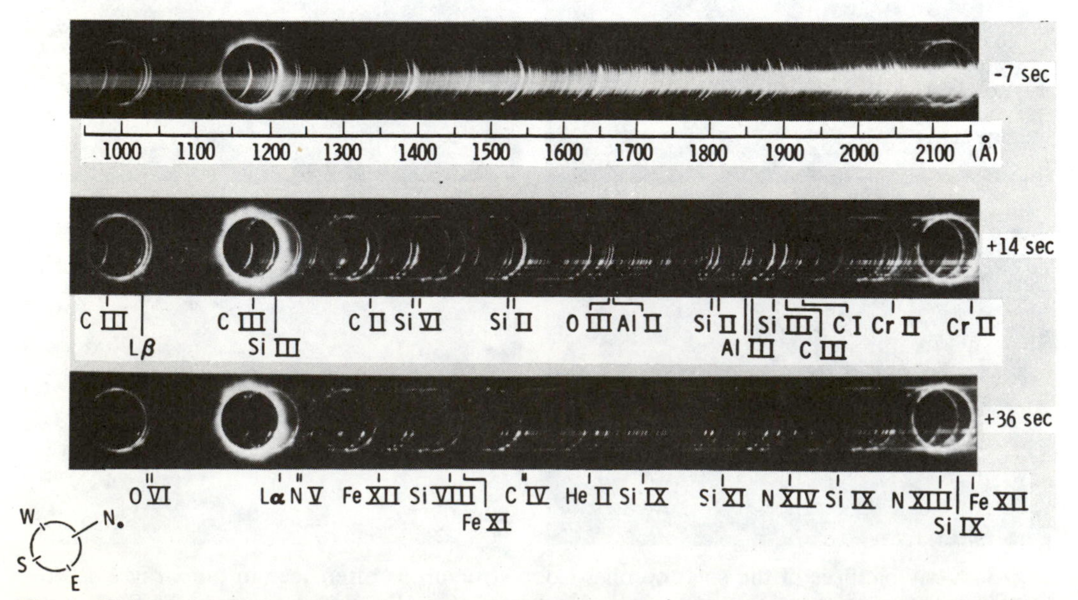

Fig. 4. Slitless ultraviolet spectrum of the transition layer taken from a rocket during total solar eclipse. Images of the solar limb are seen for the different spectral lines. The ions responsible for the line emission are indicated. [Courtesy of Harvard College Observatory]

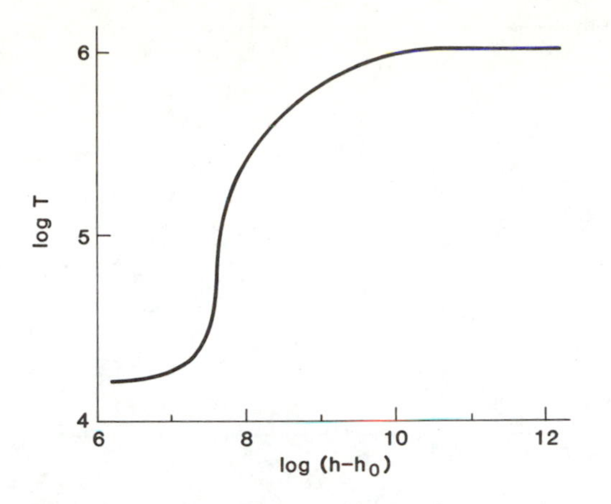

Fig. 5. Temperature stratification in the transition layer derived from the ultraviolet emission line spectrum by Böhm-Vitense (*5*). The height h is measured in centimeters and the temperature in kelvins; h_0 refers to a point in the upper photosphere and is chosen arbitrarily.

gested by Woolley and Allen (*4*). Figure 5 shows the temperature stratification as derived by Böhm-Vitense (*5*).

Understanding the Temperature Stratification

How can a temperature stratification such as that shown in Fig. 5 be maintained in nature? How can the temperature minimum at the surface of the photosphere ($h = 0$ in Fig. 5) be maintained between the hot solar interior and the hot corona? Keeping in mind that outside the solar corona there is the cool interstellar medium, we should ask instead how the corona can be maintained at a temperature between 1 million and 2 million degrees in between the relatively cool photosphere and the cold interstellar medium. This is the question addressed below.

The observational fact that the solar photosphere maintains its temperature in spite of its radiative energy loss at the surface is very important for understanding the photospheric temperature stratification. If the photosphere does not cool off, it must be heated from below at the same rate at which it loses energy at the surface. Such a heat flow requires a decreasing temperature outward. How can the photospheric heat flow go from the cool photosphere to the hot corona?

Actually, it does not go to the corona, it only goes through the corona. There is so little material in the corona (about 10^{-5} g in a column of 1 cm^2, compared to 1 g/cm^2 in the photosphere) that the corona is transparent to the photospheric radiation, and the photosphere can cool off in spite of the hot corona.

Furthermore, the hot corona transports very little heat into the photosphere to increase its temperature because the radiation from the corona is about one millionth of the photospheric radiation. Absorption of that much radiation by the photosphere results in heating that is negligible compared with the radiative loss from the photosphere. In addition, heat conduction is not important. The conductive heat flow downward is about one ten-thousandth of the photospheric radiation loss. In essence, the photosphere does not notice the corona; it barely notices the chromosphere.

Energy Balance in the Outer Layers of the Sun

Similar considerations apply in trying to understand the high temperatures of the outer layers of the sun. The chromosphere and corona are also constantly losing energy by radiation and by heat conduction. In order to stay hot, they also must constantly be supplied with

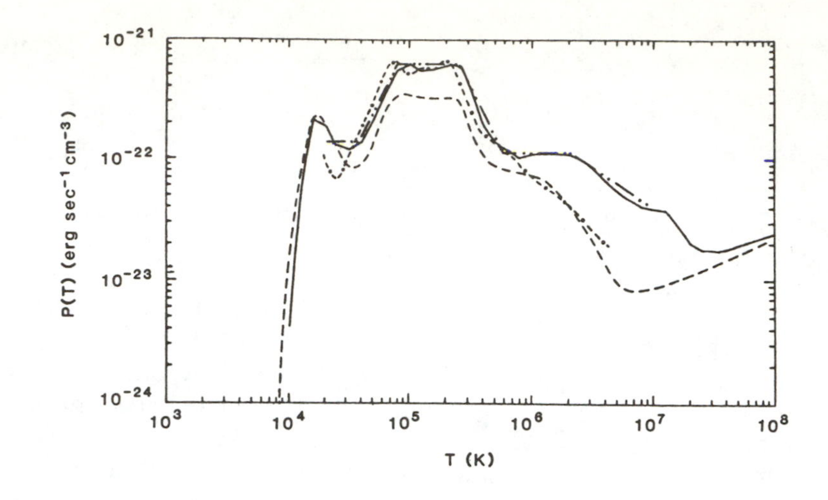

Fig. 6. Radiative energy loss as a function of temperature for a given density according to Pottasch (*8*) (·‑‑‑·), McWhirter *et al.* (*7*) (‑‑‑‑), Raymond (*9*) (——), and the approximation of Rosner *et al.* (*10*) (-···-). [Reprinted courtesy of R. Rosner, *Astrophysical Journal* 220, 643 (1978), published by the University of Chicago, © American Astronomical Society.

energy. The amount of energy a hot gas loses by radiation can be calculated, and at least that much energy must be supplied to keep the gas hot.

For light to be emitted electrons must be excited into a higher energy level. In the outer solar layers excitation occurs mainly through collisions between electrons and atoms or ions. Photon emission therefore drains kinetic energy from the gas. Higher densities mean more frequent collisions and excitations and larger energy losses by radiation. Locally, energy losses increase with the square of the density. At very low densities the gas finds it hard to get rid of its energy.

With increasing temperature, electrons can be excited to higher energy states, from which they can emit more energetic photons. Radiative energy losses increase with increasing temperature up to about 10,000 K. At higher temperatures other effects become important. Hydrogen ionizes around 10,000 K. Helium ionizes above 20,000 K the first time and above 40,000 K the second time. Heavier atoms become fully ionized at higher temperatures. Complete ionization of abundant elements reduces radiative energy losses, since no electrons are left to be excited.

Several authors have calculated radiative energy losses [for instance, Cox and Tucker (*6*)]. Figure 6 shows the results of McWhirter *et al.* (*7*), Pottasch (*8*), and Raymond *et al.* (*9*) as shown by Rosner *et al.* (*10, 11*). The differences indicate the uncertainties in such calculations. In a qualitative sense they all agree on some basic features. At constant density and a given chemical composition radiative losses increase with increasing temperature up to about 15,000 K. At higher temperatures hydrogen ionizes and the losses drop until helium becomes important and the losses increase again. When helium ionizes, heavier elements become responsible for the energy loss. At temperatures above 200,000 K the material is highly ionized and radiative losses drop finally, at least up to 6 million or 7 million degrees.

Energy Balance in the Chromosphere

In the following discussion we consider the density stratification as equal to the observed one. Clearly, a temperature increase in deeper layers will increase the density scale height and thereby increase the density in higher layers, and vice versa, but this is only a secondary effect resulting from the temperature changes. Our main concern here is to understand the temperature increase outward, which is the primary effect for the creation of the chromosphere and corona. Starting with the resulting density stratification permits us to consider the energy balance locally.

In the chromosphere the temperature increase is gradual. Heat conduction is therefore not very important. The energy balance is determined by the energy input and the radiative losses.

Suppose we start with a star whose temperature is as expected from our discussion of the photosphere: cool at the outside with the temperature increasing inward. In radiative equilibrium the amount of radiative energy absorbed in each layer is equal to the amount of energy reemitted. If we now put additional energy into the outer layers (assuming for simplicity that the energy input into each cubic centimeter is independent of height), the additional energy leads to an increase in temperature, T. For $T < 15{,}000$ K this leads to increased emission. When the increase in emission equals the increase in energy input, the heating stops. A new equilibrium temperature is reached. Figure 7 shows the radiative energy loss per unit volume for

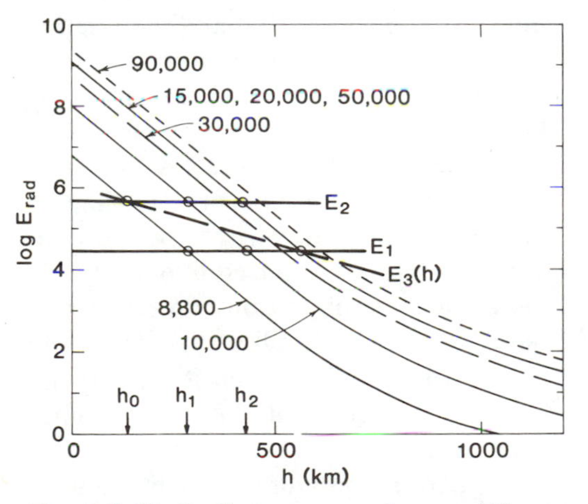

Fig. 7 (left). Radiative energy loss at different temperatures as a function of height (h) in the solar chromosphere. [The density and ionization stratification is from Vernazza *et al.* (*44*) and is assumed, for this discussion, to be the same for all temperatures.] For an energy input E_1 the temperature at h is 8,800 K. At h_2 it is 10,000 K and at h_3 15,000 K. A larger energy input E_2 leads to higher temperatures at each height, or a particular temperature occurs at a lower height with a higher density. The energy input as a function of height, for instance $E_3(h)$, determines the temperature as a function of height.

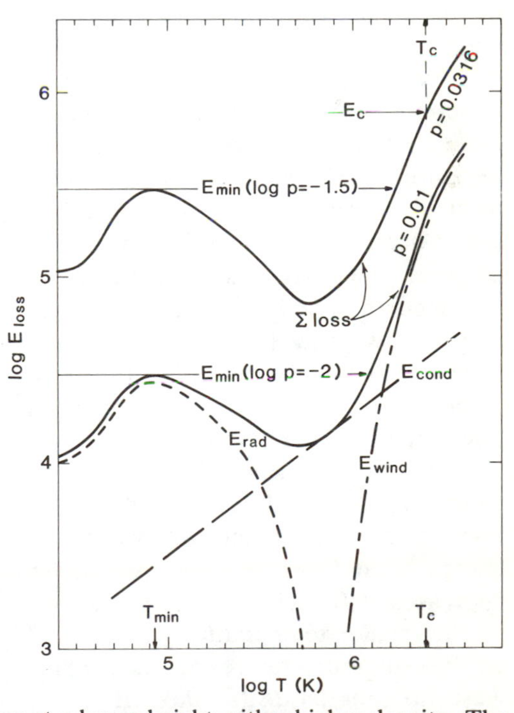

Fig. 8 (right). Total energy loss of a corona as a function of temperature for different pressures in the corona. The pressure is determined by the pressure at the base of the transition zone. For $p = 0.0316$ dyne/cm^2 and an energy input of E_c into the corona, the corona temperature must be T_c. For an energy input less than E_{min} the temperature stabilizes at $T < T_{min}$ and no corona is observed.

different temperatures as a function of height in the solar chromosphere (*12*). With increasing height the radiative losses per unit volume decrease approximately in proportion to the square of the density. We start out with a cool chromosphere, say 4300 K. At this temperature photon absorption and emission balance. We now put an additional amount of energy E into the chromosphere such that the total energy input per cubic centimeter is E_1 at each layer. At height h_1 the temperature now stabilizes at 8800 K because the energy loss for this temperature equals the energy input E_1. At height h_2 the temperature keeps rising until it is 10,000 K, and at height h_3 it stabilizes only at 15,000 K. For a given energy input we obtain a temperature increasing outward because the radiative energy losses decrease with decreasing density.

Suppose we now increase the energy input to the higher value E_2. The temperature now rises everywhere, and a particular temperature is now found at a lower depth with a higher density. For instance, 8000 K is now found at the lower depth h_0, where we have a larger density; we find 10,000 K at h_1 and 15,000 K at h_2. The density of the chromosphere at a given value of T tells us how large the energy input is.

So far, it has been assumed that the energy input is independent of height. Figure 7 also shows the case where the energy input $E_3(h)$ per cubic centimeter decreases with height. In this case we find a smaller temperature gradient dT/dh, but still temperatures increasing outward, as long as $dE_3(h)/dh < dE_{loss}/dh$, that is, as long as the energy input decreases more slowly with height than the energy loss for a constant temperature. The temperature gradient in the chromosphere can tell us how the energy input depends on height.

Energy Balance in the Transition Layer

Figures 6 and 7 show that at temperatures just above 20,000 K the energy loss decreases with increasing temperature. If energy balance cannot be achieved at a temperature less than 20,000 K, the temperature keeps rising until about 50,000 K is reached. This leads to a steep temperature increase. At a slightly greater height an energy input E_1 is even larger than the energy loss at 90,000 K. Beyond this temperature the energy loss finally decreases with increasing temperature, as seen in Fig. 6. The assumed energy input E_1 is now larger than the radiative energy loss at any temperature. The temperature keeps rising and the energy loss keeps falling, which makes the temperature increase even more. A runaway temperature increase occurs, building up the transition layer. The temperature increase stops only when additional energy losses become important. An energy input $E_2 > E_1$ would already surpass the energy loss at higher densities, and the transition layer would occur at higher densities.

A steep temperature gradient leads to heat conduction in the direction of decreasing temperature. In the case of the transition layer this constitutes energy transport back to the photosphere, which means that it drains energy from this layer. In the chromosphere the temperature gradient is modest, but in the transition layer it becomes so steep that the temperature stratification is determined essentially by heat conduction. The conductive flux through the transition layer serves to drain the surplus energy from the corona. Such transition layer stratifications were first discussed by Woolley and Allen (*4*) and by Unsöld (*13*).

Energy Balance in the Corona

When studying the corona we must look at yet another energy loss mechanism. Observations of comet tails led Biermann (*14, 15*) to conclude that a wind must be blowing from the direction of the sun with particle velocities of several hundred kilometers per second. Parker (*16*) clarified the physical processes that lead to the solar wind: at a distance of several solar radii gravitation becomes too weak to keep hot gas bound. If hot coronal gas extends to such large distances, the pressure forces make the gas escape at high speeds. (For low temperatures the density at such distances would be too low to cause any noticeable wind.) The pressure forces increase with increasing temperature, causing a rapidly increasing energy loss. For high coronal temperatures the solar wind is a very efficient cooling agent. The same considerations apply to other stars.

In the following discussion I will assume a spherically symmetrical homogeneous corona of constant temperature. This is a grossly simplified picture, as seen from Fig. 1, but it helps in understanding the major effects. For such a schematic corona, Hearn (*17*) gave approximate equations for calculation of the total energy loss.

Figure 8 shows the energy loss of a corona by radiation, E_{rad}, and by heat conduction, E_{cond}, plotted as a function of coronal temperature, assuming that the stratification in the transition layer is determined by heat conduction. The energy loss due to the stellar wind, E_{wind}, is also plotted as a function of temperature. Curves are drawn for different pressures in the corona. As discussed in the preceding section, the density in the transition zone, and therefore also in the corona, is determined by the energy input at the base of the transition zone. For a given energy input, E_c, into the corona, the temperature is determined by the equilibrium condition $E_c = E_{wind} + E_{cond} + E_{rad}$. The larger the energy input the larger the temperature. The increase in temperature with increasing energy input is, however, very small, since the energy loss due to the stellar wind increases so steeply with temperature.

In Fig. 8 we see that for a given pressure the total coronal energy loss has a minimum at a temperature, T_{min}, which depends on the density. There have been discussions in the literature, starting with a paper by Hearn (*17*), whether coronal density and temperature would adjust in such a way as to obtain this temperature of minimum energy loss for a coronal density which would adjust according to the energy input. I do not see how it could. If the energy input equaled the energy loss at this minimum, the corona would actually stabilize at the lower temperature T_0. However, T_{min} is a lower limit to the actual temperature. It appears that the actual coronal temperature cannot be very much higher, since the energy loss increases so steeply with temperature.

Energy Input into the Chromosphere

The discussions above show how we can understand the temperature stratification in the outer layers by considering the equilibrium between energy loss and energy input. We must still consider what causes the energy input. It cannot be radiation alone, because if it were the chromosphere would be in radiative equilibrium, which would generally give a temperature decreasing outward. Also, there can be no other kind of heat transport from the cool photospheric gas to the hotter chromosphere and corona. The only way to create a hot layer of gas within a cool surrounding is by heating the gas with mechanical energy (for instance, by acoustic waves or shock waves) or with some form of magnetohydrodynamic energy. A goal of modern studies of chromospheric and coronal temperature stratifications is to try to decide which of these mechanisms is the more important one. The next question is how such energies can be created in a hot ball of gas like the sun.

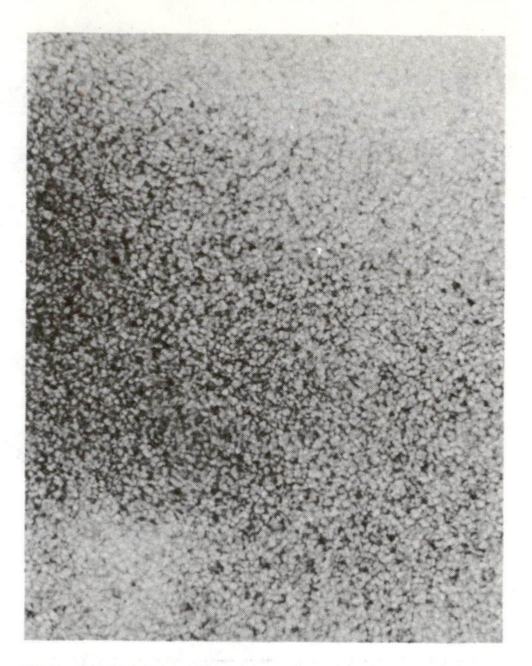

Fig. 9. High-resolution image of the solar surface showing solar granulation. Bright regions are moving upward, dark areas downward.

Mechanical or kinetic energy can be generated by means of temperature differences. There is a natural temperature gradient in a star due to the heat transport from the inside out. In layers where the radiative conductivity is very small, a very steep temperature gradient is required to transport the energy by radiation. This leads to convection. In the sun, the radial temperature gradient leads to rising and falling gas streams, which we can see in the photosphere as the solar granulation (see Fig. 9). The bright granules appear to rise. The dark areas generally fall, as can be seen from the Doppler shifts of the spectral lines in the different regions. Velocities up to 7000 km/hour are observed; for comparison, the velocity of sound in the photosphere is about 18,000 km/hour. At such high velocities large amounts of noise (acoustical waves) are generated (*18*). Some of these are absorbed in the photosphere, but a not very well-determined fraction of the acoustic waves are able to travel outward into the low-density chromosphere, where they steepen to form shock waves. We know from experience that shock waves dissipate their energy rapidly and generate large amounts of heat, which may possibly heat the chromosphere (*19*). If this is indeed the main source of energy for heating the chromosphere, then stars with larger convective velocities than the sun may be expected to have stronger chromospheric emission than the sun. This can be tested by observation.

If convection alone causes chromospheric heating, we may expect stars with outer convection zones to have chromospheres, while those without outer convection zones do not. Overall, stellar properties are best visualized in

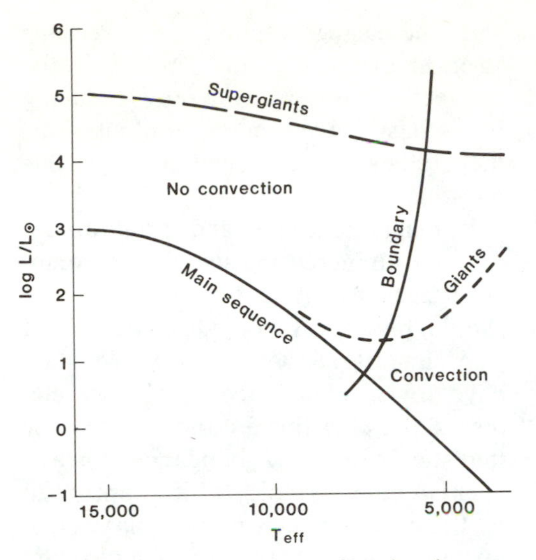

Fig. 10. Schematic diagram of luminosity versus T_{eff}, showing the positions of the main sequence, the giants, and the supergiant branches. Cool stars have a surface convection zone; hot ones do not. The dividing line between convective and nonconvective stars is shown as a solid line labeled *Boundary*. Stars to the right of this line show chromospheric emission lines; stars to the left do not.

the Hertzsprung-Russell diagram (Fig. 10), where stars are entered according to their temperature and luminosity, or intrinsic brightness. In the H-R diagram most stars fall along the main sequence. Larger and hence brighter stars are called giants or supergiants. Stars to the right of the solid line in Fig. 10 are expected to have outer convection zones; stars to the left, which have higher temperatures, are not. The change is expected to be quite abrupt (*20*).

Since the launching of the International Ultraviolet Explorer (IUE) satellite (*21*), chromospheres of many cool stars have been observed. It was confirmed that stars with outer convection zones have chromospheres, while for stars without outer convection zones no chromospheric emission lines were visible. In fact, chromospheric emission decreases rather abruptly at the boundary line for outer convection zones (*22*). As expected theoretically, it is generally observed to increase for stars with increasing convective velocities, although there are still quantitative disagreements between theory and observations (*23*). There are, however, many stars with enhanced chromospheric emission, referred to as stars with active chromospheres, which shows that other effects are also important.

Stars may inherit a magnetic field at birth from the interstellar medium. At the photospheres and outer convection zones of stars these magnetic fields are expected to decay within 100 million to 1 billion years. Older stars with surface magnetic fields, like our sun, must have regenerated their fields. It is generally thought that convective mass flows, subject to Coriolis forces on a rotating star, in combination with differential rotation (the sun rotates faster at the equator than at the pole) can result in a dynamo action which builds up a magnetic field in the surface layers of a star. In the case of the sun, this turns out to be a periodically varying magnetic field leading to the 22-year sunspot cycle. More rapidly rotating stars, possibly with stronger differential rotation, might be expected to have stronger magnetic fields than the sun, but fields still too weak to be measured on a distant star.

Weak chromospheric emission lines can be found in the solar spectrum in the center of the strong Fraunhofer line due to the calcium ion (the Ca II *K* line) and also in the ultraviolet lines of the magnesium ion (the Mg II *h* and *k* lines). Wilson (*24*) made extensive observations of stellar Ca II *K* lines. Based on these observations, Skumanich (*25*) concluded that for ages up to the age of the sun,

4.5×10^9 years, chromospheric emission for stars of a particular type decreases with increasing age of the stars. Stellar rotation also decreases with increasing age of the stars. Satellite observations of the Mg II *h* and *k* lines confirmed that young stars show greater emission than older stars (*26*).

Solar observations also showed that the Ca II *K* emission is stronger in regions of stronger magnetic fields (*27*), indicating that either more energy is fed into the chromosphere in regions of stronger magnetic fields, or the energy is preferentially dissipated in regions of stronger magnetic fields. Since the overall Ca II *K* emission increases at sunspot maximum, when large magnetic fields occur at the solar surface, it appears that more energy is delivered to the chromosphere if the magnetic field is stronger. Calculations indicate that more acoustic noise will be generated in regions with increased magnetic fields.

With this in mind, Skumanich concluded that the observed decreasing chromospheric emission with increasing stellar age is probably caused by a decreasing magnetic field due to decreasing rotation. In the case of young stars, a decreasing magnetic field may also be due to decay of the inherited fossil field.

Observations of the Ca II *K* line (*28*) and the Mg II *k* line (*29*) also indicate enhanced emission for close binary stars, which probably means that they too have enhanced magnetic fields due to either rapid rotation or tidal interaction.

The Ca II *K* line emission has been observed only for stars cooler than about 6500 K. These stars generally are slow rotators. For hotter stars, which are frequently rapid rotators, the Mg II $h + k$ lines have been observed. Such observations show that for very rapid rotation the emission does not increase further with increasing rotation (*26*). Perhaps stronger magnetic fields can reduce the convective motions, thereby reducing the acoustic wave energy output. Further studies will be needed to answer this question.

If magnetic fields and rotation decrease with increasing age, then chromospheric emission from very old stars should be very weak. Observations of star clusters show that very old stars have lower abundances of heavy elements. Turning this around, we can say that stars with a low abundance of heavy elements are very old, which appears to be the case for stars in our galaxy. Observations indicate that for such old stars the total chromospheric emission probably decreases little, if at all, with increasing age, where age is measured by decreasing heavy element abundance (*30*). It seems that once the fossil magnetic field has decayed, after about 10^9 years, the magnetic fields decrease very slowly, or perhaps for very weak fields the field strength becomes unimportant for chromospheric heating. Further studies are needed to confirm this result.

In conclusion, acoustic noise generation in stellar outer convection zones appears to be necessary for the heating of solar-type chromospheres. Some additional mechanism, probably related to magnetic fields, may enhance chromospheric energy input, except possibly for very weak and very strong fields.

Heating of the Corona

Since the corona loses less energy than the chromosphere, it needs much less energy to stay hot. Nevertheless, the heating of the corona poses more difficult problems to the theoreticians. Shock waves appear to dissipate their energy in the chromosphere and have no energy left by the time they reach the corona. In addition, the steep temperature and density gradient in the transition zone would reflect acoustic waves. Nevertheless, the convection zone seems to be very important for heating of the corona.

I pointed out earlier that it has been possible to take x-ray pictures of the solar corona. It is not possible to take similar x-ray pictures of other stars, but the Einstein satellite, equipped with an x-ray telescope, has been able to measure the total x-ray emission of many stars (*31*). Surprisingly, x-ray emission was observed for almost all kinds of stars close enough to us that the flux was not too much diluted by distance. Only for cool, large stars—the giants and supergiants, with temperatures less than about 4000 K—has x-ray emission not yet been detected (*32*).

For the problem of the heat source of the corona, it is a significant observation that the x-ray emission of normal single stars increases abruptly by a factor of about 10 as the stellar temperature decreases from about 8000 to about 7500 K. The much higher x-ray emission of the cooler stars shows that they have a well-developed corona, while the hotter stars have either a low-temperature or a low-density corona. Stars with surface temperatures $T \leq 7800$ K are expected to have strong outer convection, while stars with $T \geq 8000$ K are not (Fig. 10). The abrupt increase in x-ray flux in this temperature range shows the importance of convection for coronal heating.

On the other hand, the fact that stars with higher temperatures also emit some x-rays shows that hot regions may also exist in the outer layers of stars that do not have an outer convection zone. In fact, for stellar surface temperatures above 8000 K the x-ray emission increases with increasing surface temperatures. The strong x-ray flux from hot stars is one of the many unsolved puzzles presented by the data from the Einstein observatory.

Why the cool giants and supergiants do not seem to have coronas in spite of their strong convection is another puzzle which astronomers are trying to solve. Observations by Simon *et al.* (*33*) showed that for these stars lines originating in transition regions cannot be seen either. It was suggested that this might be due to cooling by a very strong stellar wind. But in the absence of a hot corona, what would drive such a strong wind? No satisfactory solution has been found yet for the winds blowing from the cool luminous stars, though radiative acceleration due to photon absorption in the hydrogen Lyman alpha line (*34*) and Alfvén wave pressure (*35*) have been discussed.

A look at the x-ray picture of the sun shows rays (or streamers) and loops with enhanced x-ray emission. The only explanation of these streamers and loops is that they outline the direction of the magnetic lines of force in the corona. The loops appear to be confined by magnetic flux tubes. X-ray emission is enhanced along these structures which apparently outline stronger magnetic fields. Regions above sunspot groups with strong magnetic fields also show enhanced x-ray emission. Studies of line intensities indicate that these regions of enhanced x-ray emission are regions of higher temperature or density than their surroundings (*36*). Stronger magnetic fields appear to lead to higher radiative energy losses, which probably indicates higher energy input.

If it is correct that stellar magnetic fields are constantly regenerated by the interaction of convection and rotation, then we might expect the x-ray emission to increase with increasing stellar rotation for otherwise similar stars. Pallavicini *et al.* (*37*) observed increasing x-ray emission with increasing rotation, but their sample did not contain any stars rotating faster than 35 km/sec (actually, only the product v_{rot} sin i can be measured, where v_{rot} is the equatorial rotational velocity and i is the angle of inclination between the rotation axis and the line of sight). For stars with $T \gtrsim 7500$ K and rotational velocities larger than 50 km/sec, no further increase was found with increasing rotational velocity (*30*). Again, there seems to be a saturation effect.

Strongly enhanced x-ray emission is also observed for close binaries. Walter and Bowyer (*38*) and Walter (*39*) observed an increasing x-ray flux with decreasing separation of the two stars. Coronal temperatures appear to increase presumably due to increasing magnetic fields for close binaries either due to increasing rotation, expected for close binaries, or due to increasing tidal effects. More observations will be needed to determine which effect causes the increased emission and presumably an increased magnetic field.

In any case, these observations show a strong dependence of coronal x-ray emission on effects that are unrelated to convection and acoustic waves. The coronal structures seen in x-ray emission indicate the importance of magnetic fields. On the other hand, the observations also show that the presence of convection also increases coronal emission.

The variation in chromospheric emission from different stars is rather small, about a factor of 4, compared to the variation in x-ray flux, which can amount to a factor of 10^4. The range in chromospheric energy output can easily be explained in terms of the expected range of acoustic noise generation; in fact, we would expect larger variations. But the large increase in x-ray emission from some stars must be due to a much more efficient means of transporting energy into the corona. For strong x-ray emitters, a large fraction of the generated mechanical energy must be transformed into a form of energy that can reach the corona and heat it.

At present we do not know the mechanism of heating of the corona. We do not even know whether the heating takes place mainly in the loops. It is not possible here to discuss this in detail, but I would like to point out one possible mechanism for heating of the corona.

As discussed above, any acoustic energy reaching the chromosphere is dissipated rapidly in the chromosphere and cannot reach the corona. The rapid convective motions in the photosphere and chromosphere may, however, displace and distort the magnetic fields. This would generate magnetohydrodynamic waves (Alfvén waves), which are damped very little and can therefore travel along the lines of force into the corona. But this presents another problem: since the waves are not damped they will not dissipate their energy in the corona either. In fact, satellite observations have shown that Alfvén waves exist in the solar wind, and thus they may persist to very large distances from the sun. Looking at the corona, we see that only near the poles do the streamers indicate magnetic field lines that go almost radially outward. In most areas the field lines are bent with rather small radii of curvature. In such curved and tangled magnetic fields, Alfvén waves could regenerate waves related to acoustic waves (fast and slow modes, which may also form shocks), which can dissipate their energy in the corona. It is possible that Alfvén waves are the vehicle for transporting energy into the corona, where dissipation may occur due to shocks generated through fast and slow modes [for example, see (*15*) and (*20*)]. However, at present this suggestion is only speculative.

References and Notes

1. W. Grotrian, *Naturwissenschaften* **27**, 214 (1939).
2. B. Edlén, *Ark. Mat. Astron. Fys.* **28B** (No. 1) (1941).
3. ______, *Z. Astrophys.* **22**, 30 (1942).
4. R. v. d. R. Woolley and C. W. Allen, *Mon. Not. R. Astron. Soc.* **110**, 358 (1950).
5. E. Böhm-Vitense, unpublished results.
6. D. P. Cox and W. H. Tucker, *Astrophys. J.* **157**, 1157 (1969).
7. R. W. P. McWhirter, P. C. Thonemann, R. Wilson, *Astron. Astrophys.* **40**, 63 (1975).
8. S. R. Pottasch, *Bull. Astron. Inst. Neth.* **18**, 7 (1965).
9. J. C. Raymond, D. P. Cox, B. W. Smith, *Astrophys. J.* **204**, 290 (1976).
10. R. Rosner *et al.*, *ibid.* **220**, 643 (1978).
11. These calculations give the amount of energy emitted per cubic centimeter. The amount of radiative energy that might be absorbed from the surrounding must be considered as energy input.
12. The radiative energy absorbed by this volume is considered as energy input.
13. A. Unsöld, *Physik der Sternatmosphären* (Springer-Verlag, Berlin, ed. 2, 1955), p. 670.
14. L. Biermann, *Z. Astrophys.* **29**, 274 (1951).
15. ______ and R. Lüst, in *The Moon, Meteorites and Comets*, B. M. Middelhurst and G. P. Kuiper, Eds. (Univ. of Chicago Press, Chicago, 1963), p. 618.
16. E. N. Parker, *Astrophys. J.* **128**, 664 (1958).
17. A. Hearn, *Astron. Astrophys.* **40**, 355 (1975).
18. M. Lighthill, *Proc. R. Soc. London Ser. A* **211**, 564 (1952).
19. P. Ulmschneider, *Astron. Astrophys.* **12**, 297 (1971).
20. E. Böhm-Vitense, *Z. Astroph.* **46**, 108 (1958).
21. A. Boggess *et al.*, *Nature (London)* **275**, 372 (1978).
22. E. Böhm-Vitense and T. Dettmann, *Astrophys. J.* **236**, 560 (1980).
23. R. Stein and J. Leibacher, in *Stellar Turbulence*, D. F. Gray and J. L. Linsky, Eds. (Springer-Verlag, Berlin, 1981), p. 225.
24. O. C. Wilson *Astrophys. J.* **144**, 695 (1966).
25. A. Skumanich, *ibid.* **171**, 565 (1972).
26. E. Böhm-Vitense, *ibid.*, in press.
27. A. Skumanich, C. Smythe, E. N. Frazier, *ibid.* **200**, 747 (1975).
28. R. Glebocki and A. Stawikowski, *Acta Astron.* **27**, 225 (1977).
29. F. Middelkoop and C. Zwaan, *Astron. Astrophys.* **101**, 26 (1981).
30. E. Böhm-Vitense, *Advances in Ultraviolet Astronomy* [NASA conference publication 2238 (1982)], p. 231.
31. G. S. Vaiana *et al.*, *Astrophys. J.* **245**, 163 (1981).
32. T. R. Ayres *et al.*, *ibid.* **250**, 293 (1981).
33. T. Simon, J. Linsky, R. Stencel, *ibid.* **257**, 225 (1982).
34. B. M. Haisch, J. L. Linsky, G. S. Basri, *ibid.* **235**, 519 (1980).
35. L. Hartmann and K. B. MacGregor, *ibid.* **242**, 260 (1980).
36. J. C. Raymond and J. G. Doyle, *ibid.* **247**, 686 (1981).
37. R. Pallavicini, L. Golub, R. Rosner, G. C. Vaiana, T. Ayres, J. L. Linsky, *ibid.* **248**, 279 (1981).
38. F. M. Walter and S. Bowyer, *ibid.* **245**, 671 (1981).
39. F. M. Walter, *ibid.*, p. 677.
40. E. G. Gibson, *NASA Spec. Publ. SP-303* (1972), p. 256.
41. C. R. Davidson and F. J. M. Stratton, *Mem. R. Astron. Soc.* **64**, IV (1927).
42. A. Unsöld, *Physik der Sternatmosphären* (Springer-Verlag, Berlin, ed. 1, 1938), p. 391.
43. E. G. Gibson, *NASA Spec. Publ. SP-303* (1972), p. 280.
44. J. E. Vernazza, E. H. Avrett, R. Loezer, *Astrophys. J. Suppl.* **45**, 635 (1981).
45. E. G. Gibson, *NASA Spec. Publ. SP-303* (1972), p. 156.

This material originally appeared in *Science* **223**, 24 February 1984.

8. Interstellar Matter and Chemical Evolution

M. Peimbert, A. Serrano, S. Torres-Peimbert

The chemical composition of the interstellar medium can be determined by studying the emission line spectra of gaseous nebulae. Hydrogen, helium, carbon, nitrogen, oxygen, and neon are the six most abundant elements in our galaxy and in those galaxies for which accurate abundances have been determined. The relative abundance of these elements in the interstellar medium can be derived from H II regions, planetary nebulae, and supernova remnants.

Accurate abundance determinations of H II regions have made it possible to find small differences in the abundance of elements among various galaxies and among different regions of the same galaxy. These small differences occur because material produced in stellar interiors and rich in elements heavier than hydrogen has been injected into the interstellar medium. The enrichment of the interstellar medium is due to loss of mass from stars of intermediate mass that become planetary nebulae before turning into white dwarfs and from massive stars that explode as supernovae. By studying H II regions where almost no star formation has occurred, it is possible to determine pregalactic chemical abundances; and by studying H II regions in galaxies where a substantial fraction of the gas has gone into stars, it is possible to study the chemical evolution of galaxies.

Abundance determinations of supernova remnants are not discussed in this chapter because their accuracy is considerably smaller than that of abundance determinations of H II regions and planetary nebulae. Similarly, stellar surface abundances are not discussed because, with the exception of a few results for the Magellanic Clouds, there are no individual determinations for stars in other galaxies. Even in our galaxy, the accuracy with which the relative abundances of some key elements can be derived is smaller for stars than for H II regions and planetary nebulae. This is particularly the case for the He/H ratio.

Observational Data on Abundances

H II regions are conglomerations of gas and dust where stars are being formed at present; the most massive H II regions are hot enough to be able to ionize hydrogen atoms (see Fig. 1). The chemical composition of their gaseous component has been affected by stellar enrichment throughout the life of the galaxy. They are called H II regions because most of the gas is in the form of ionized hydrogen. In H II regions of the galactic neighborhood about 90 percent of the atoms are of hydrogen, about 10 percent are of helium, and about 0.1

Fig. 1. Galactic H II region, called Messier 8 or the Lagoon Nebula. From spectra of its bright regions it is possible to obtain accurate abundances of hydrogen, helium, carbon, nitrogen, oxygen, neon, and heavier elements. This is a reproduction of a photograph taken with the Lick Observatory Crossley 0.9-meter telescope.

percent are of all the other elements combined. The fractions of these three constituents, normalized by mass, are called *X, Y*, and *Z*, and for the interstellar medium in the solar neighborhood their values amount to 0.70, 0.28, and 0.02, respectively.

Planetary nebulae are shells of gas ejected from and expanding about extremely hot central stars. Stars of intermediate mass go through the planetary nebulae stage after the red giant phase and before they become white dwarfs. The ultraviolet flux of the hot central star ionizes the ejected shells. The physical processes responsible for the emission line spectrum in the outer envelopes of planetary nebulae and in H II regions are very similar.

The emission line intensity of a given ion in a gaseous nebula (planetary nebulae and H II regions are called gaseous nebulae) depends on the distribution along the line of sight of the electron density (N_e), the electron temperature (T_e), and the ionic concentration, $N(X^{+i})$, as well as on atomic parameters that represent the likelihood of photon emission. Values of N_e can be determined from the ratio of two emission lines of the same ion with almost the same excitation energy, and values of T_e can be determined from the ratio of two emission lines of the same ion with very different excitation energies (*1*).

The relative ionic abundances are determined from the ratio of two line intensities of the ions involved. The lines chosen are not affected by radiative transfer effects, and the accuracy of the abundance determinations is given by the accuracy of the line intensity determinations and the precision of atomic constants. There are two types of general processes that produce emission line photons in a gaseous nebula: recombination, which in the case of hydrogen is responsible for the Balmer lines, and collisional excitation of atomic energy levels followed by radiative de-excitation. The recombination line intensities depend weakly on T_e, whereas the intensities of the collisionally excited lines in the optical and ultraviolet electromag-

Table 1. Chemical abundances expressed as log N(X) and scaled to log N(H) = 12.00.

Type	Element						References
	H	He	C	N	O	Ne	
			H II regions				
Orion	12.00	11.01	8.57	7.68	8.65	7.80	*(9)*
LMC	12.00	10.92	7.86	7.03	8.34	7.44	*(2, 7, 10)*
SMC	12.00	10.89	7.00	6.41	7.89	7.03	*(2, 8, 10)*
			Planetary nebulae				
$(He\text{-}N)_{rich}$	12.00	11.18	8.3–9.1	8.6	8.6	7.9	*(19)*
Intermediate	12.00	11.04	8.4–9.2	8.1	8.7	8.1	*(17)*
Halo	12.00	10.99	8.9	6.7–8.3	7.9	6.2–8.0	*(18, 40)*

netic domains depend strongly on T_e. Therefore, high-quality determinations of the He^+/H^+ abundance ratio, which are based on recombination lines, have a typical accuracy of $10^{0.04}$ (≡ 0.04 dex), and high-quality ionic determinations of carbon, nitrogen, oxygen, and neon, relative to hydrogen, based on collisionally excited lines have typical accuracies of about 0.15 dex as a result of their higher dependence on T_e and the lower precision of the atomic constants.

For some of the atoms we do not observe lines corresponding to all the stages of ionization present in the nebula. Obtaining the total abundances requires evaluation of the fraction of atoms in the unobserved stages of ionization. This estimation is usually based on the degree of ionization and on ionization structure models. Observations at different positions of a given nebula often permit a better determination of this correction.

Some of the more accurate abundance determinations for galactic and extragalactic H II regions are given in Table 1, where the Orion Nebula is representative of the solar neighborhood, and average values for several H II regions in the Large Magellanic Cloud (LMC) and the Small Magellanic Cloud (SMC). Table 1 also shows the abundances for planetary nebulae of different types; the $(He\text{-}N)_{rich}$ and intermediate planetary nebulae are solar neighborhood objects whose parent stars are relatively young (1×10^8 to 3×10^9 years), whereas halo planetary nebulae have very old parent stars, with ages similar to the age of the galaxy (about 10^{10} years). Most of the abundances in Table 1 were obtained from observations made with ground-based optical telescopes; however, abundances of carbon and some ions of nitrogen and neon were obtained from observations made with the *International Ultraviolet Explorer* satellite.

Pregalactic Chemical Abundances

Observational studies of planetary nebulae and supernovae, as well as stellar evolution models have shown that during their lifetimes stars enrich the interstellar medium with elements heavier than hydrogen. Therefore, to find the pregalactic chemical abundances, it is necessary to observe regions where star formation has not been important—that is, to study H II regions located in galaxies where the fraction of mass in the form of interstellar gas (M_{gas}/M_{total}) is large.

From abundances determined for extragalactic H II regions, it has been found that generally the higher the values of M_{gas}/M_{total} in the galaxies where the H II regions are located, the smaller the abundances of carbon and heavier elements (*2, 3*). This can also be seen in Table 1, since the values of M_{gas}/M_{total} ratios for the LMC and the SMC are 0.12 and 0.42, respectively (*2*). This result implies that the Z value with which galaxies were formed is less than one-tenth of the value for the Orion Nebula or for the sun. Stars of masses smaller than about one solar mass (1 $M_{\odot}$) would not be expected to show variations in elements heavier than neon because nuclear reactions in their central regions do not produce or destroy them; therefore their abundances are indicative of the prestellar values. Measuring the abundances of elements heavier than neon relative to the abundance of hydrogen in very old stars has shown that such elements are underabundant compared to their abundances in the Orion Nebula and the sun by about two orders of magnitude. This is the case for sulfur and argon in halo planetary nebulae (*4*), and for iron in globular cluster stars (*5*). From the previous discussion, it follows that the pregalactic Z value is smaller than 0.002 and possibly smaller than 0.0002.

All of the well-observed H II regions in our galaxy and other galaxies show similar He/H abundance ratios, about 0.07 to 0.11, which correspond to Y values between 0.22 and 0.30. One of the main uncertainties in the determination of the He/H abundance ratio is due to the correction for the amount of neutral helium. Fortunately for objects with low O/H values, this correction is small and amounts to 0.01 (or less) of the Y value. From observations of H II regions with O/H values similar to that of the Orion Nebula, Y is found to be between 0.27 and 0.30, whereas from those H II regions with O/H values smaller by a factor of 2 or more than the O/H value of the Orion Nebula, Y is found to be between 0.22 and 0.25 (*2, 6–12*). These values indicate that a very general process occurred before the formation of galaxies that produced $X \approx 0.77$, $Y \approx 0.23$, and almost no heavy elements. This result together with the 3 K background radiation field and the recession of the galaxies (expansion of the universe) are the three strongest pieces of evidence in favor of Big Bang cosmology.

All of the observed H II regions have $Z \neq 0$ and belong to galaxies in which a substantial fraction of the mass has already been converted into stars. The Z value is probably due to enrichment of the interstellar medium by loss of mass from previously formed stars. For a very accurate pregalactic helium abundance (Y_p) to be determined, the observed Y values have to be extrapolated to the case with $Z = 0$. As a first approximation it can be assumed that, as a result of star formation, for any increase in X_{16} there is a corresponding increase in freshly made Y; that is, $\Delta Y/\Delta X_{16} = \beta$, where X_{16} is the oxygen abundance by mass and constitutes about 45 percent of Z (*9*). While one group of observers finds that for the Orion Nebula and for metal-poor H II regions $\Delta Y/\Delta X_{16} \sim 6 \pm 2$ (*2, 7–10*), another group finds that for metal-poor H II regions $\Delta Y/\Delta X_{16} \sim 3 \pm 8$ (*12*). Adopting different determinations of the initial mass function for the solar neighborhood (which gives the relative proportion at birth of stars of different masses) and different stellar evolution models has resulted in the prediction that $\Delta Y/\Delta X_{16}$ is between 2 and 7 (*13*).

Values of Y versus O/H presented in Fig. 2 are based on observations of H II

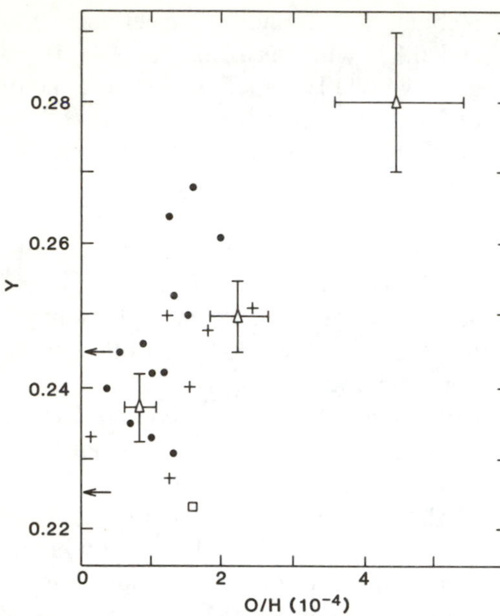

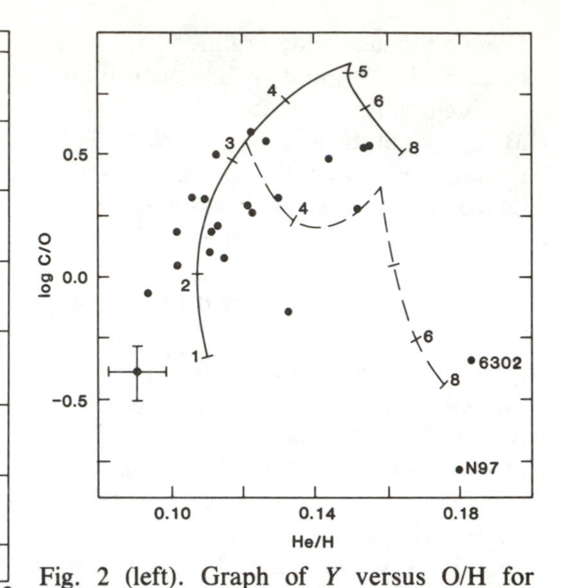

Fig. 2 (left). Graph of Y versus O/H for galactic and extragalactic H II regions. Crosses are the data from (*2*); the square corresponds to NGC 5471 in M101 (*11*), and the filled circles are the data from (*12*). For these points, the errors are not labeled to avoid crowding. Two possible values of the pregalactic helium abundance, Y_p, are marked by arrows. Fig. 3 (right). C/O versus He/H abundance ratios. From stellar evolution models, the predicted values for the planetary nebula shell are presented for two values of α, the ratio of the convective mixing length to the pressure scale height. The solid line is for $\alpha = 0$, and the broken line is for $\alpha = 2$. The numbers in the figure denote the mass of the progenitor stars. The filled circles are the observed values for planetary nebulae. Standard deviation errors (1σ) shown for one object are typical for the sample.

regions in spiral, irregular and blue compact galaxies. In this figure, we show the three best Y determinations that have been reported (*7–9*), with estimated standard deviation errors (1σ), together with additional very good data reported by others (*2, 11, 12*). We consider our Y values (*7–9*) to be better than those of the other three groups of observers (*2, 11, 12*) because the procedure used to estimate the amount of neutral helium inside the H II regions was based on an empirical ionization correction formula that included O^+ and O^{2+} and, in addition, was based on the condition that the dispersion of the He/H ratio would be minimum for a set of H II regions of a given galaxy. The Mexican group (*2, 7–9, 11*), which finds $\Delta Y/\Delta X_{16} \sim 6$, used a subset of the data in Fig. 2 to obtain $Y_p = 0.225$ by extrapolating to $Z = 0$. Kunth and Sargent (*12*) used another subset of the data in Fig. 2, and letting $\Delta Y/\Delta X_{16} = 0$, obtained $Y_p = 0.245$. An exhaustive discussion of the internal and external errors of these determinations has not been reported, but from unpublished work the error in Y_p is estimated to be about 0.01 at the 1σ level (*14*). Furthermore, it is not known if ΔY should be linear with ΔZ for values smaller than the minimum observed O/H ratio. The point is very important because $Y_p = 0.225$ is consistent with a standard (that is, the simplest) hot Big Bang model with only two families of two component light neutrinos ($m_\nu \ll 1$ million electron volts, where m_ν is the rest mass of the neutrinos), whereas $Y_p = 0.245$ is consistent with the standard model with three families of two component light neutrinos (*15*). If $Y_p < 0.23$, either the τ neutrino might be heavy ($m_\nu > 1$ MeV) or a nonstandard Big Bang model would have to be adopted.

To increase the accuracy of the Y_p determinations from H II region observations, we need (i) more accurate measurements of several He/H emission line ratios in a given H II region, (ii) measurements in several areas of a given H II region, (iii) better H II region ionization structure models, and (iv) observations of several H II regions in a galaxy with a well mixed interstellar medium.

Interstellar Medium Enrichment

The previous discussion suggests that galaxies are formed with $X \approx 0.77$, $Y \approx 0.23$, and almost no heavy elements.

The theory of stellar evolution predicts that in the lifetime of the galaxy, only stars more massive than $\sim 1\ M_\odot$ have evolved and ejected material that has modified the chemical composition of the interstellar medium. The stars more massive than $\sim 8\ M_\odot$ produce supernovae that enrich the interstellar medium with oxygen and heavier elements such as neon, sulfur, argon, iron, and nickel. The intermediate mass stars produce planetary nebulae that enrich the interstellar medium with helium, nitrogen, and carbon; this can be seen by comparing the abundances of these elements in $(\text{He-N})_{\text{rich}}$ and intermediate planetary nebulae of the solar neighborhood with the Orion Nebula values (Table 1).

The stellar evolution predictions for the chemical composition of planetary nebulae shells by Iben and his collaborators (*16*) have been compared with observations (*17–20*). For example, a comparison of theory with observations is shown in the plot of log C/O versus He/H in Fig. 3. In general, the agreement is good and indicates that intermediate mass stars produce large amounts of carbon and helium. Moreover, these stars are responsible for most of the carbon and nitrogen and a substantial amount of the helium enrichment of the interstellar medium (*13, 21*).

The location of NGC 6302 and N97 in Fig. 3 indicates that, in the stellar models, the ratio of the convective mixing length to the pressure scale height, α, has to be about 2. There are two other arguments in favor of this value of α; a smaller value of α would produce a C/H ratio larger than that observed in the solar neighborhood (*21*); a smaller value of α would be inconsistent with the decrease of the O/H ratio with increasing N/H found in $(\text{He-N})_{\text{rich}}$ planetary nebulae (*18, 21, 22*).

Another result that can be derived from Fig. 3 is that the predicted mass for the observed chemical composition of NGC 6302 and N97 is around $8\ M_\odot$. This result is important because it indicates that at least some stars with initial masses around $8\ M_\odot$ do not become supernovae. However, before this result is accepted, other tests of the stellar evolution theory of intermediate mass stars should be made. An independent observation that supports this result is the detection, in galactic clusters, of white dwarfs that had progenitors with masses around $7\ M_\odot$ (*23*); white dwarfs and nuclei of planetary nebulae have masses smaller than $1.4\ M_\odot$, and therefore the excess mass was lost by stellar winds and planetary nebulae shell ejection.

In spiral galaxies, like ours, the $M_{\text{gas}}/$

M_{total} ratio increases from the center outward (*24*), and therefore the amount of heavy elements relative to hydrogen should decrease from the center outward. This is indeed what is seen in our galaxy and other galaxies (*11, 25*). The quantitative study of these gradients is related to the problem of the chemical evolution of galaxies.

Chemical Evolution of Galaxies

To explain the observed abundances in the interstellar medium of our galaxy and other galaxies, it is necessary to construct models for the chemical evolution of galaxies (*2, 26, 27*); in these models the abundance at a given time in a given region of a galaxy depends on (i) the initial mass function, (ii) the birth rate of stars and the chemical composition of mass lost by the stars during their evolution, and (iii) the existence of large-scale mass flows, like infall from the halo or radial flows across the disk of a galaxy. Since these physical parameters are not generally known in galaxies, the observed abundances can be used as constraints of the galactic chemical evolution models.

In closed models, in which there are no mass flows, the total mass (gas plus stars) is conserved and abundances depend only on the fraction of the initial mass of the galaxy that has been converted into stars independently of the details of the birth rate function (*28*). This result holds as long as the time scale for gas consumption in the galaxy is much longer than the lifetime of the stars that produce predominantly a given element; this is the instantaneous recycling approximation (IRA). Such a simple model (closed plus IRA) results in abundances that increase logarithmically with the gas fraction M_{gas}/M_{total}. All information about the contribution of individual stars is condensed in the "yield," which is the mass of the newly formed elements that stars eject to the interstellar medium in units of the mass that is never returned to it (mass permanently trapped in "remnants"). The yield is defined for a stellar generation and weighted by the initial mass function, and it depends only on the assumed stellar mass loss; that is, the amount and chemical composition. In a diagram of abundance versus the logarithm of the gas fraction, such as Fig. 4,

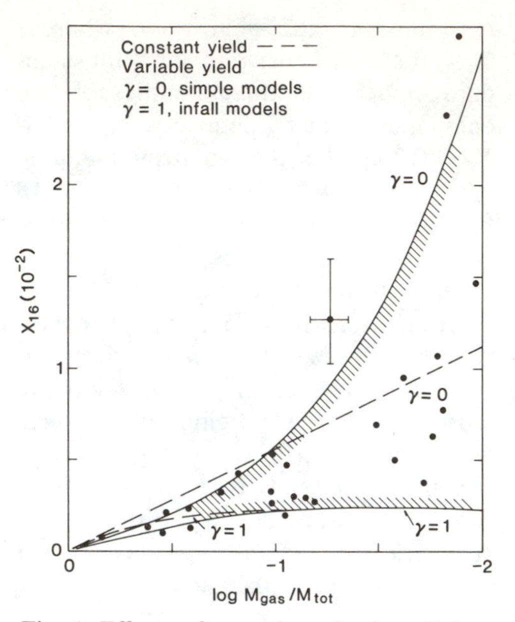

Fig. 4. Effects of accretion of primordial gas and of yields increasing with metallicity on the relation between the metallicity, represented by X_{16}, and the gas fraction M_{gas}/M_{total}. Filled circles represent observations of galactic and extragalactic H II regions (*3*). Also presented are models with a constant yield ($p_{16} = 0.0023$) and models with a yield increasing with metallicity ($p_{16} = 0.0009 + 0.6\ X_{16}$). The infall models coincide for log M_{gas}/M_{total} smaller than -1. Standard deviation errors (1σ) presented for one object are typical for the sample; note that the vertical error is proportional to the oxygen abundance.

the yield of a simple model corresponds to the slope of the galactic evolution line. Furthermore, the simple model predicts that if the yield is the same for all galaxies, they should lie on a straight line in Fig. 4.

Theoretical estimates of the yield of "heavy elements" (carbon and heavier) give values between 0.003 and 0.01 (*2, 26, 27*). The same range is obtained when simple models with a yield independent of metallicity are used to explain observed abundances and gas fractions in galactic and extragalactic H II regions; this is the "great success of the simple model." But although the order of magnitude of the heavy elements predicted by this model is correct, there are galaxies with the same mass fraction and with different values for abundances of heavy elements (represented in Fig. 4 by the abundance of oxygen with respect to hydrogen). Therefore, unless we accept an arbitrary yield that varies from galaxy to galaxy, the simple model has to be dropped. Models with a yield independent of metallicity and infall of gas with primordial composition were proposed in order to solve this problem (*29*). Models with infall can be characterized by the parameter γ, the ratio of accretion to the rate of star formation. For example, if the model is closed, $\gamma = 0$, and if infall (or accretion) exactly balances star formation so that the mass in the form of gas is constant, then $\gamma = 1$. These IRA models with a yield independent of metallicity are shown in Fig. 4 as dashed lines. In the case of the extreme infall model, $\gamma = 1$, the oxygen abundance quickly saturates to the value given by the yield (a horizontal line in Fig. 4) because an equilibrium is reached in which dilution due to accretion of primordial material does not allow abundances to increase. If galaxies diminish their gas fraction as the result of star formation, they should lie between the simple model ($\gamma = 0$) and the extreme infall model ($\gamma = 1$) in Fig. 4. Thus, by assuming different values of the ratio of infall to star formation rates, it would be possible to explain a range of O/H values at a given value of M_{gas}/M_{total}.

However, as can be seen in Fig. 4, this permitted region does not agree with observations, since there is a much larger observed range of X_{16} than constant-yield models predict for small gas fractions. To explain this fact it is necessary for the yield to increase with the abundance of heavy elements. In Fig. 4, the variable yield presented (p) is an expression of the form $p = p_0 + aZ$ (*3*). This in turn implies that galaxies with low O/H are more efficient in the formation of low mass stars (less than 1 $M_\odot$). These models are shown as continuous lines in Fig. 4. In this case the simple model ($\gamma = 0$) curves upward in contrast with closed models with a yield independent of metallicity, which are straight lines. On the other hand, extreme infall models ($\gamma = 1$) are very similar in both cases. The high metallicity objects are better explained with a yield increasing with metallicity.

In the simple model, there is a clear distinction between primary and secondary elements, where primary elements are those that are directly synthesized from hydrogen and helium, and secondary elements are those synthesized from heavy elements that were already present in the star when it was formed. For example, in an N/O versus O/H diagram (Fig. 5), primary nitrogen production corresponds to a horizontal line (N/O is constant), whereas secondary nitrogen production corresponds to a 45° line (N/O $\propto$ O/H) (*30*). It is clear from Fig. 5 that there is no unique correlation of either

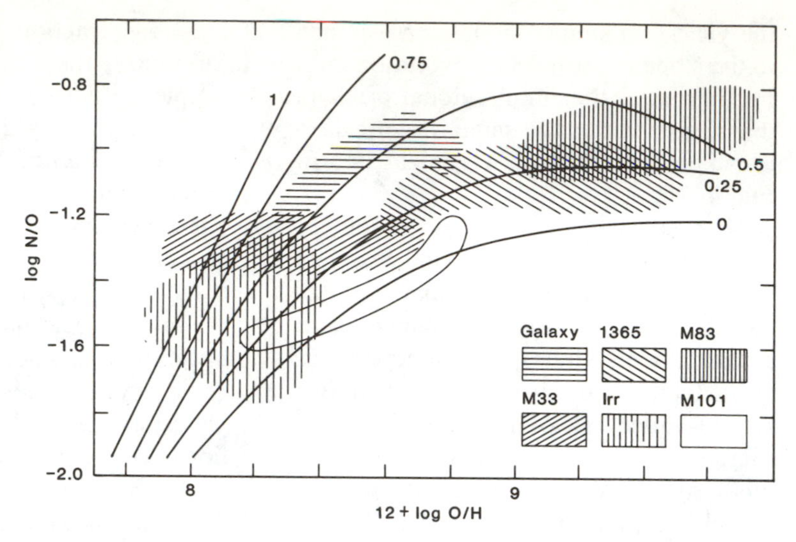

Fig. 5. Graph of N/O versus O/H, assuming secondary production of nitrogen and a yield increasing with metallicity. Observations of galactic and extragalactic H II regions are schematically represented by shaded areas. As in Fig. 4, curves are labeled by γ. Accretion increases with γ, defining the region between the curves $\gamma = 0$ and $\gamma = 1$ as the permitted region. Lines of constant age are nearly horizontal in this diagram, with age increasing with N/O. Lines of constant gas fraction are nearly vertical, with M_{gas}/M_{total} decreasing with O/H. Galaxies identified in the boxes are our own galaxy; spiral galaxies NGC 1365, M33, M83, and M101; and a group of irregular and blue compact galaxies designated Irr.

type between these two abundance ratios. The spread in Fig. 5 makes it more difficult to find out whether nitrogen is a primary or secondary element. The assumption of IRA breaks down for nitrogen because the average star that produces nitrogen has a smaller initial mass (around 3 $M_\odot$) than the average star that produces oxygen (around 25 $M_\odot$). Since the lifetime of a star is a strongly decreasing function of its mass, this means that there is a delay in the injection of nitrogen into the interstellar medium with respect to oxygen. Moreover, there are usually no strong gradients of N/O across disks of spiral galaxies, whereas O/H usually shows well-defined gradients. This led to the suggestion that nitrogen is primary, so that each galaxy is represented by a horizontal line in Fig. 5, and that there is a vertical spread in this figure caused by the effect of delays in the ejection of nitrogen (*31*).

Since the oxygen content in a simple model (closed plus IRA) is determined by the gas fraction, the spread in Fig. 5 requires a second physical parameter whose variation from galaxy to galaxy explains the dispersion in N/O at a given O/H. Such a second parameter is also needed to explain the variation of O/H, at N/O nearly constant, from place to place across the disks of spiral galaxies. It has been proposed that the spread in Fig. 5 is due to variations in (i) the initial mass function, (ii) the star formation rate, (iii) the history of the star formation rate, and (iv) the accretion rate. It has been found that Fig. 5 is best explained by models with accretion of uncontaminated gas, secondary nitrogen, and a yield increasing with O/H (*32*). Each of these three assumptions is necessary: primary nitrogen would fill up the forbidden region at high O/H and low N/O; closed models result in a too narrow diagram; and, finally, a constant yield would not predict galaxies with high O/H at all. In these models a given galaxy has the same age across its disk; O/H gradients can exist under small differences in the local gas fractions; and much smaller N/O gradients are explained by gradients in the ratio of infall to the star formation rate. Under these assumptions, the stars that produce nitrogen have masses of about 1.3 $M_{\odot}$ and larger.

If the explanation of Fig. 5 (*32*) is correct, most of the nitrogen of the interstellar medium is of secondary origin and has been produced by intermediate mass stars, progenitors of planetary nebulae. This is contrary to the suggestion that novae inject most of the nitrogen present in the interstellar medium (*33*), because the nitrogen produced by novae is of primary origin.

The models of chemical evolution of galaxies require infall of gas to be important in order to explain the observations presented in Figs. 4 and 5. Moreover, the chemical history of the solar neighborhood requires also that the rate of infall of gaseous material from outside the disk has been approximately equal to half the star formation rate as indicated by the age-metallicity relationship in stars (*34*) and by isotopic abundance ratios (*35*). The existence of a substantial infall rate in galaxies is a subject of great controversy, although recent observations have shown that it exists in both irregular and spiral galaxies (*36*).

A study of a sample of irregular galaxies shows that there is no relation between the mass gas fraction and the O/H abundance ratio. This also indicates that even if the Z value of a given galaxy is consistent with the closed model, the galaxies as a group are not; in addition, the observations are best explained by an infall model (*37*). Moreover, the constancy of the star formation rate determined for isolated irregular galaxies cannot be explained by the simple model; an alternative explanation could be steady accretion from a bound gaseous halo or other types of accretion. Other explanations are also possible (*38*).

The suggestion that formation of low mass stars is more efficient in low-metallicity environments is also controversial, since a simple gravitational instability argument would suggest exactly the opposite. Other theories, however, predict that the relative number of low mass stars decreases with metallicity (*39*). A yield increasing with metallicity, as proposed, would produce a positive gradient in the observed mass-to-luminosity ratio in spiral galaxies and would be consistent with a substantial amount of hidden mass in the form of low mass stars in galactic halos.

References and Notes

1. D. E. Osterbrock, *Astrophysics of Gaseous Nebulae* (Freeman, San Francisco, 1974).
2. J. Lequeux, M. Peimbert, J. F. Rayo, A. Serrano, S. Torres-Peimbert, *Astron. Astrophys.* **80**, 155 (1979).
3. M. Peimbert and A. Serrano, *Mon. Not. R. Astron. Soc.* **198**, 563 (1982).
4. T. Barker, *Astrophys. J.* **237**, 482 (1980); *ibid.* **270**, 641 (1983).
5. H. L. Helfer, G. Wallerstein, J. L. Greenstein, *ibid.* **129**, 700 (1959); G. Wallerstein and H. L. Helfer, *Astron. J.* **71**, 350 (1966).
6. M. Peimbert, *Annu. Rev. Astron. Astrophys.* **13**, 113 (1975); B. E. J. Pagel, *Philos. Trans. R. Soc. London Ser. A* **307**, 19 (1982); T. D. Kinman and K. Davidson, *ibid.*, p. 37. A thorough discussion on the pregalactic helium abundance is given in *Primordial Helium*, P. A. Shaver, D. Kunth, K. Kjar, Eds. (European Southern Observatory, Garching, 1983).
7. M. Peimbert and S. Torres-Peimbert, *Astrophys. J.* **193**, 327 (1974).
8. _____, *ibid.* **203**, 581 (1976).
9. _____, *Mon. Not. R. Astron. Soc.* **179**, 217 (1977); S. Torres-Peimbert, M. Peimbert, E. Daltabuit, *Astrophys. J.* **238**, 133 (1980).
10. R. J. Dufour, G. A. Shields, R. J. Talbot, Jr., *Astrophys. J.* **252**, 461 (1982); R. J. Dufour, *Int. Astron. Union Symp.* **108**, 353 (1984).
11. J. F. Rayo, M. Peimbert, S. Torres-Peimbert, *Astrophys. J.* **255**, 1 (1982).
12. D. Kunth and W. L. W. Sargent, *ibid.* **273**, 81 (1983).
13. A. Serrano and M. Peimbert, *Rev. Mex. Astron. Astrofis.* **5**, 109 (1981); A. Maeder, *Astron. Astrophys.* **101**, 385 (1981); C. Chiosi and F. M. Matteucci, *ibid.* **105**, 140 (1982).

14. K. Davidson and T. D. Kinman, *Astrophys. J.*, in press; M. Peimbert and S. Torres-Peimbert, in preparation.
15. J. Yang, M. S. Turner, G. Steigman, D. N. Schramm, K. A. Olive, *Astrophys. J.* **281**, 493 (1984).
16. A. Renzini and M. Voli, *Astron. Astrophys.* **94**, 175 (1981); I. Iben, Jr., and A. Renzini, *Annu. Rev. Astron. Astrophys.* **21**, 271 (1983).
17. L. H. Aller, *Int. Astron. Union Symp.* **103**, 1 (1983); J. B. Kaler, *ibid.*, p. 245; M. Peimbert, in *Physical Processes in Red Giants* (Reidel, Dordrecht, Netherlands, 1981), p. 409.
18. S. Torres-Peimbert, in *Stellar Nucleosynthesis* (Reidel, Dordrecht, Netherlands, 1984), p. 3.
19. M. Peimbert, *Int. Astron. Union Symp.* **76**, 215 (1978); ______ and S. Torres-Peimbert, *ibid.* **103**, 233 (1983).
20. M. J. Barlow, S. Adams, M. J. Seaton, A. J. Willis, A. R. Walker, *ibid.*, p. 538.
21. M. Peimbert, *Int. Astron. Union Symp.* **108**, 363 (1984).
22. A. Renzini, in *Stellar Nucleosynthesis* (Reidel, Dordrecht, Netherlands, 1984), p. 99.
23. D. Reimers and D. Koester, *Astron. Astrophys.* **116**, 341 (1982).
24. M. Peimbert, *Int. Astron. Union Coll.* **45**, 149 (1977).
25. L. Searle, *Astrophys. J.* **168**, 327 (1971); M. Peimbert, *Int. Astron. Union Symp.* **84**, 307 (1979); P. A. Shaver, R. X. McGee, L. M. Newton, A. C. Danks, S. R. Pottasch, *Mon. Not. R. Astron. Soc.* **204**, 53 (1983); H. E. Bond, in preparation.
26. R. J. Talbot, Jr., and W. D. Arnett, *Astrophys. J.* **186**, 51 (1973).
27. B. M. Tinsley, *Fundam. Cosmic Phys.* **5**, 287 (1980); B. E. J. Pagel and M. G. Edmunds, *Annu. Rev. Astron. Astrophys.* **19**, 77 (1981); B. A. Twarog, *Int. Astron. Union Symp.*, in press.
28. L. Searle and W. L. W. Sargent, *Astrophys. J.* **173**, 25 (1972).
29. R. B. Larson, *Nature (London) Phys. Sci.* **236**, 21 (1972).
30. R. J. Talbot, Jr., and W. D. Arnett, *Astrophys. J.* **190**, 605 (1974).
31. M. G. Edmunds and B. E. J. Pagel, *Mon. Not. R. Astron. Soc.* **185**, 77 (1978).
32. A. Serrano and M. Peimbert, *Rev. Mex. Astron. Astrofis.* **8**, 117 (1983).
33. R. E. Williams, *Astrophys. J.* **261**, L77 (1982).
34. B. A. Twarog, *ibid.* **242**, 242 (1980).
35. M. Tosi, *ibid.* **254**, 699 (1982).
36. I. F. Mirabel, *ibid.* **256**, 112 (1982); J. M. Van der Hulst *et al.*, *ibid.* **264**, L37 (1982); I. F. Mirabel and R. Morras, *ibid.* **279**, 86 (1984).
37. D. A. Hunter, J. S. Gallagher, D. Rautenkranz, *Astrophys. J. Suppl. Ser.* **49**, 53 (1982).
38. J. S. Gallagher, D. A. Hunter, A. V. Tutukov, *Astrophys. J.*, in press.
39. J. I. Silk, *ibid.* **214**, 718 (1977); V. C. Reddish, *Stellar Formation* (Pergamon, Oxford, 1978), p. 208.
40. S. Adams, M. J. Seaton, I. D. Howarth, M. Aurriere, J. R. Walsch, *Mon. Not. R. Astron. Soc.* **207**, 471 (1984).

This material originally appeared in *Science* **224**, 27 April 1984.

9. The Formation of Stellar Systems from Interstellar Molecular Clouds

Robert D. Gehrz, David C. Black, Philip M. Solomon

Perhaps the greatest fascination in man's age-old quest to understand the forces that order the cosmos has centered around the origin of the stars, sun, planets, and life itself. What complex physical processes have conspired to produce life and the solar system? How many of the countless stellar systems in our own galaxy and the universe have experienced a similar evolutionary process? Such questions, which challenge the limits of understanding, have intrigued generations of scientists, philosophers, and laymen. Never before has astronomical science had such great potential to provide answers. Observational and theoretical results of modern astronomical research have given insight into the chain of events that led from the early expansion of the universe to life on the planet Earth. Star formation, a crucial link in the chain, is a continuous process in the disk of our galaxy. Studies of regions where star formation is occurring provide us with an opportunity to observe firsthand the physical conditions and events that must have attended the formation of the solar system.

From Primordial Material to Stellar Systems

The process of star formation can be placed in the context of our current understanding of cosmic evolution (*1*, *2*). As the Big Bang fireball expanded at nearly the speed of light, matter cooled from $\geq 10^{12}$ to 10^8 K within ~ 1000 seconds. Nearly all primordial nucleosynthesis must have occurred during this brief interval. Theory suggests that nuclei heavier than ^{4}He did not form in significant quantities before the cessation of Big Bang nucleosynthesis (*3*), so that the expanding primordial medium consisted almost entirely of isotopes of hydrogen and helium ($\approx$ 90 and 10 percent, respectively, by number). Elements basic to life, such as carbon, nitrogen, and oxygen, were highly deficient in the primordial material as compared to their abundance in the solar system. Very heavy elements were virtually nonexistent.

As the expanding universe cooled, localized fragments of the primordial medium collapsed by self-gravitation to form protogalaxies. Spiral galaxies such as our own are thought to have condensed from relatively rapidly rotating protogalaxies which fragmented into stars, clusters, and interstellar clouds. A halo of stars and globular clusters was left behind the contracting protogalaxy, but most of the material collapsed into a thin rotating disk containing newly formed stars and interstellar matter (*4*). Although galaxy formation appears to have been confined to an era about (10 to 15) $\times 10^9$ years ago, star formation in active regions of the interstellar medium (ISM) in the disks of spiral galaxies is a continuing process during which the by-products of nucleosynthesis in massive short-lived stars are incorporated into successive stellar generations. This stellar processing of primordial material in evolving galaxies has produced all the other elements in the periodic table.

During the 1950's, progress in our understanding of stellar evolution and nuclear reaction rates led to a theory of nucleosynthesis that successfully accounts for the production of heavy elements from primordial hydrogen and helium inside evolving massive stars (*5, 6*). These massive stars return some of their processed matter back into the interstellar medium. Several different ejection mechanisms seem required to account for all the elements found in the solar system. Injection of processed material from aging massive stars into the ISM was documented during the past decade by infrared astronomers (*7*). Red giant and supergiant stars have extended circumstellar shells containing dust grains composed of heavy elements. The dust is probably silicates (for example, $MgSiO_3$) in the shells of oxygen-rich stars (*8*), while carbon and silicon carbide (SiC) grains form in the shells of carbon-rich stars (*9*). Circumstellar dust grains are driven into the ISM by pulsations and stellar radiation pressure, carrying the shell gas with them by momentum coupling (*7*). This steady-state process alone does not satisfactorily account for the presence of elements heavier than iron in the interstellar medium since such elements cannot form in the interiors of even the most massive stars. These elements appear to be formed mainly by rapid neutron capture in the shells ejected in supernova eruptions of very massive [$\approx$ 30 to 100 times the mass of the sun ($M_\odot$)] stars (*10*). Thus, the presence of elements heavier than iron in the solar system implies that supernova ejecta were at some time mixed into the presolar nebula (*10a*).

The ISM is, then, the crucible in which primordial material and heavy elements ejected from generations of massive short-lived stars are mixed. Star formation is the critical process that incorporates this new mixture, vastly different in chemical composition from protogalactic material, into later generations of stars like the sun. Existing observational knowledge about star formation is confined to the interstellar and early post-protostellar collapse configurations. New astronomical facilities proposed for the 1980's (*11*) may enable astronomers to examine the elusive protostellar collapse phase.

Molecular Clouds in the Interstellar Medium

Early 21-cm radio observations of atomic hydrogen (HI) traced the diffuse component of the ISM in our galaxy (*12*). High-density regions of the nearby ISM were originally recognized because they obscured visible radiation from stars in the galactic plane, appearing as dark structures against the bright background of the Milky Way. Although numerous nearby dark clouds were identified by star counting, quantitative measurements of their composition, density, mass, and galactic distribution were not possible with optical astronomical techniques because these regions emit no visible light. The advent of millimeter-wave detection techniques enabled astronomers to observe emission from dozens of interstellar molecules and to identify the chemical composition and structure of the dense interstellar clouds (*13*).

Millimeter-wavelength spectral line surveys of the emission from the carbon monoxide (CO) molecule have substantially revised our view of the density and distribution of interstellar matter in the galaxy (*14, 15*). The CO fundamental rotational emission at 2.6 mm traces the more abundant hydrogen molecule (H_2), which has no permanent electric dipole moment and no radio or millimeter transitions. Although H_2 emission at 28.2 μm is believed to be an efficient cooling mechanism for clouds hotter than 100 K (*16*), the line is blocked by atmospheric water vapor. Moreover, most of the matter in interstellar molecular clouds is at much lower temperatures (*15*). Thus, CO emission reveals the spatial morphology of the higher density regions of the ISM where almost all the hydrogen is molecular.

High-spatial-resolution CO maps of the inner portion of our galaxy show that the ISM is dominated by large massive clouds (*15*) (giant molecular clouds or GMC's), which are the primary sites of current star formation in the galaxy. Relatively nearby (to the sun) GMC's such as the one associated with the Orion Nebula (*17*) are invariably found near associations of young stars and ionized gas. Plate IV is a CO map of the galaxy between 290° and 90° longitude, and Plate V shows CO emission in a cross section of part of the galactic plane. Clumping of the emission in both spatial and velocity coordinates is characteristic of discrete clouds rather than of a diffuse medium. The strongest emission comes from clouds or cloud complexes associated with HII regions (ionized gas surrounding recently formed massive stars). Most of the clouds producing the emission in Plates IV and V are at distances between 3 and 13 kiloparsecs (kpc) (9,000 to 39,000 light-years) from the sun in the highly flattened disk of the galaxy.

Contours of CO emission (Plate IV) can be used to measure the size and number of molecular clouds (*15*). Cloud size along the galactic plane is determined from knowledge of the distance (inferred from rotation models of the galaxy) and angular extent. Internal gas velocities implied by CO line widths show that large-scale motions within these clouds are highly supersonic. GMC's are relatively dense, having 10^2 to 10^3 H_2 per cubic centimeter, as compared to 1 to 10 hydrogen atoms per cubic centimeter in atomic hydrogen clouds and much less than 1 atom per cubic centimeter between clouds. Although there are numerous small molecular clouds, the mass in the ISM of the inner galaxy is dominated by a few thousand GMC's. In addition, these clouds tend to cluster into complexes. Plate V shows a complex about 150 pc (450 light-years) at galactic longitude $\sim 35°$. A typical GMC is 20 to 60 pc in diameter with an average density of 300 H_2 per cubic centimeter (*15, 18*) and a mass of 10^5 to 10^6 $M_\odot$ ($1\ M_\odot = 2 \times 10^{33}$ g). GMC's are the most massive objects in the galaxy. The molecular mass throughout the galaxy is about $3 \times 10^9\ M_\odot$. The ringlike nature of the distribution (*14*), which peaks at a distance of

5.5 kpc from the galactic center (the sun is 10 kpc from the center), is the most obvious large-scale morphological feature. The tendency for emission to occur approximately along lines in Plate IV suggests the presence of enhanced emission from spiral arms or segments of spiral arms, but there is currently no general agreement about the degree of confinement of molecular clouds to spiral arms. The research group at the Goddard Institute for Space Studies maintains (*19*) that virtually all CO-emitting molecular clouds are in two or three spiral arms in the inner galaxy. The Stony Brook–Massachusetts group finds (*15, 20*) that, although the hottest and the largest GMC's are in spiral-like patterns, similar to the patterns of ionized hydrogen, a substantial fraction of molecular clouds including giant clouds exists between spiral arms (*15, 20*) indicating that star formation is widespread in the galactic disk.

Physical conditions within the clouds are apparently favorable for promoting chemical reactions required to form complex molecules. More than 50 different molecules have been identified in the gas phase of GMC's, primarily by millimeter-wave observations (*21*). These include such exotic organic molecules as ethyl cyanide (CH_3CH_2CN), methyl cyanide (CH_3CN), methyl mercaptan (CH_3SH), acetaldehyde (CH_3CHO), and ethyl alcohol (CH_3CH_2OH). Clearly, the chemistry of the galaxy is much richer than originally expected. Some investigators have argued that the infrared spectrum of interstellar dust shows evidence for the presence of organic or even biological particles (*22*). It is apparent that a very complex chemical broth is incorporated into protostellar condensations within molecular clouds.

Several fascinating problems have emerged from our study of GMC's. Densities and temperatures derived from CO spectra show that thermal gas pressure in clouds is insufficient to oppose their gravitational collapse (*23*). However, the current star formation rate in our galaxy is much lower than would be predicted if molecular clouds were freely collapsing, even if we were to assume a star formation efficiency of less than 1 percent (*24*). Comparison of the gravitational potential energy with the internal pressure derived from the velocities of large-scale nonthermal internal motions (deduced from Doppler-velocity broadening of molecular lines in the clouds) shows that the clouds are in or near equilibrium. It is difficult, however, to understand how such motions are maintained. Turbulence would be highly supersonic and should rapidly dissipate by shock heating.

Highly energetic outflows recently observed in the star-forming cores of many molecular clouds may bear on this problem. For example, millimeter and infrared CO and H_2 observations (*25*) reveal a shock front propagating away from a central source in the core of the Orion Molecular Cloud (OMC). The speed of the shock (30 to 50 km/sec) and its distance (0.3 pc) from the central source suggest that the events that precipitated the shock occurred only 1000 to 3000 years ago. The energy in the outflow exceeds 10^{47} ergs. Presumably, the source of the outflow is a highly luminous newborn star that is driving material away from its outer envelope or protocloud. Although there is no general agreement on the specific dynamical mechanism responsible for such outflows, they appear to be a normal occurrence accompanying star formation. Sufficient amounts of kinetic energy may be deposited into the molecular cloud by many young stars to oppose gravitational collapse. An intriguing possibility is that star formation in a given molecular cloud is a self-regulating process. The fact that interstellar matter is still plentiful (not all transformed to stars) in our galaxy is due to this inefficiency or self-regulation of the star formation process. By contrast, star formation may have been much more efficient in the early stages of the evolution of elliptical galaxies or in globular clusters that are left with no interstellar matter (*26*).

Star Formation in Molecular Cloud Cores

A typical GMC is more massive by four orders of magnitude than individual massive stars and more massive by two orders of magnitude than an entire galactic cluster. Thus, star formation must proceed in fragments or core regions of the GMC's (*27*). Processes within a molecular cloud that cause the high-density cores to form and that subsequently lead to the condensation of protostars are not well understood. Compression of matter by shocks may induce gravitational collapse. Mechanisms proposed to date include supernova shocks (*28*), ionization fronts (*29*), cloud-cloud collisions (*30*), and, on a larger scale, galactic density waves (*31*). It is not clear from the observations whether kinetic energy input by the shocks observed around young stars on the average disrupts or encourages star formation.

High-density molecular cloud cores form when portions of a GMC fragment and condense by self-gravitation. A typical cloud core may contain 10^2 to 10^3 $M_\odot$ of material at densities of 10^4 to 10^5 H_2 per cubic centimeter and temperatures of 50 to 100 K in a region less than 1 pc in diameter. A small molecular cloud might develop only a single core near its geometric center, whereas GMC's containing roughly 100 to 1000 cloud core masses might be expected to form a number of core regions throughout their volume, some of them relatively near the GMC boundaries (*32*). Astronomers have observationally identified core regions of molecular clouds and recently formed stars embedded within them. The actual protostellar collapse phase, discussed below, is predicted to be exceedingly short-lived and has thus far escaped observation.

The morphology of and physical conditions in cloud cores vary enormously, depending upon the degree to which star formation has progressed. Very young cloud cores in which stars have not yet formed appear merely as hot spots in the CO emission from the parent molecular cloud and emit no significant optical, near-infrared, or radio-continuum radiation (*32*). Several cool young cloud cores have also recently been identified in the OMC by their far-infrared ammonia (NH_3) line emission (*33*). On the other hand, cloud cores containing recently evolved young stars become observable throughout the electromagnetic spectrum. The youngest stars are still embedded in dense cocoons of gas and dust that are remnants of the protostellar collapse phase (*34*). Such objects, like the Becklin-Neugebauer (BN) star in Orion, appear as compact infrared hot spots within the cloud core. The inner regions of these cocoons may be as hot as 300 to 1000 K, and therefore they emit strongly in the 2- to 30-μm spectral region. Infrared spectroscopy has shown that dust, similar in composition to the material being ejected from aging massive stars, is a major constituent of these cocoons, and the associated gas phase contains such molecules as H_2O, CO, and H_2.

Another signpost of dense fragments, maser emission from H_2O, SiO, and OH, has been observed in the immediate vicinity of many molecular cloud cores.

As hot young stars (OB stars) in cloud cores evolve toward the main sequence (hydrogen burning) phase, they can dissipate their cocoons by heating, radiation pressure, and stellar winds. The initial stages of such a process may account for the high-velocity CO outflows. During the dissipation of the remnant protostellar cocoon, the circumstellar material can become optically thin to ionizing radiation from the central star; this causes a compact ionized hydrogen (HII) region to propagate outward through the molecular cloud core material surrounding the young star (*35*). At this point, the cloud core emits both a rich optical recombination spectrum and thermal bremsstrahlung radio-continuum radiation from the ionized gas and thus becomes identifiable by a wide variety of detection techniques. An evolving young OB star may eventually vaporize and blow away nearly all the remnant material. It can then be observed as an optical object surrounded by a large HII region. OB stars that form near the edge of a molecular cloud appear able to break out of the cloud, sweeping away the surrounding molecular medium by radiation pressure and stellar winds (*36*), and are observed to be in relatively low-density regions of the ISM just outside the molecular cloud boundary. Because highly evolved cloud cores emit most of their radiation at optical and near-infrared wavelengths, they are most readily identified when they lie near the GMC boundaries facing the sun. More deeply embedded cores suffer large amounts of extinction at short wavelengths. Thus, far-infrared, millimeter, and radio observations are necessary to delineate the distribution of cores within GMC's.

The OMC is an excellent example of a GMC that has fragmented into several cloud cores containing recently formed stars in various stages of evolution (*35, 37*). Nearly all the cloud core and post-protostellar configurations described above are represented. In **Plate VI, a, b,** and **c**, we show three views of the OMC that describe active cloud cores and young stellar stages of the star formation process. The central 0.3° (3 pc) of the OMC, shown in ^{12}CO **(Plate VIa),** is fragmented into several smaller cloud core regions each ~ 0.5 pc in diameter. The active star formation region centered on the Trapezium, a small cluster of optically visible OB stars **(Plate VIb),** contains $\approx 10^3\ M_\odot$ of material at densities of 10^4 to 10^6 H_2 per cubic centimeter. The Trapezium stars and the OB stars to the southeast are examples of recently formed stars that have substantially dissipated most of their remnant protostellar clouds. They provide the ionizing photons that cause the fluorescence and radio-continuum emission from the well-developed HII region and are driving HII shock fronts into the surrounding medium. The Becklin-Neugebauer-Kleinmann-Low (BNKL) object in the cloud core **(Plate VIc)** is a very young star or cluster of stars still enveloped by post-protostellar material (*25*). Thus, the OMC complex shows fragmentation on several scales, the largest representing cloud cores and the smallest representing individual stellar masses.

Current observational knowledge about extremely young stellar systems comes from infrared observations of the compact condensations in molecular cloud cores, but it is highly unlikely that any of the objects observed thus far are in the state of protostellar collapse. They probably represent the immediate post-protostellar contraction phases during which the initial stages of nuclear fusion have begun. The high-energy outflows might be expected for young stars beginning to dissipate their protoclouds. There is as yet no conclusive evidence for the presence of infalling material that would indicate a collapse phase. However, observations of infrared sources embedded in molecular clouds are crucial since they may provide indirect evidence about the physical processes that occurred in the immediately preceding but less easily observable phases of rapid protostellar collapse.

Recent high-resolution spatial maps acquired using new infrared imaging techniques on objects such as Sharpless 106 (S106) and W3 (see Plate VII) are providing insight into the geometric morphology of the immediate postprotostellar collapse phase (*38*). Images of S106 show that the central source is surrounded by a flattened ring of dense gas and dust similar to that expected to form in the collapse of a rotating protostellar cloud (*39*). The resulting biconical nebula contains a number of dense subcondensations that may be formation sites for companion stars. The outflow observed optically in the biconical lobes of S106 (*40*) suggests that the newly formed star is beginning to dissipate the remnant cocoon.

Protostellar Collapse

Since the observations described above refer only to the initial and final configurations in star formation, they provide only circumstantial evidence about crucial physical details of the star formation process. The missing observational link is the true "protostar," a gravitationally bound collapsing cloud fragment whose dynamical evolution and luminosity are driven by gravitational contraction with negligible contribution from nucleosynthesis (*34*). Detailed knowledge about protostellar collapse would be gratifying because the physical events occurring during this stage evidently determine the important morphological characteristics of the final stellar configuration. Formation of binary stars and planets, spatial fractionation of elements within the collapsed disk, and the initial phases of molecular chemistry in the circumstellar material are probably determined during the protostellar phase. Recent theoretical studies of protostellar evolution provide insight into why this crucial evolutionary stage of star formation has escaped direct detection (*41*). The protostellar collapse is believed to progress in three phases: phase I, an initial quiescent contraction; phase II, a rapid near–free-fall collapse; and phase III, an accretion phase at the onset of contraction of the condensed core to the main sequence.

Phase I begins when the molecular cloud core fragments into gravitationally bound stellar mass–sized systems less than a few tenths of a parsec in diameter with densities of 10^3 to 10^4 H_2 per cubic centimeter. Presumably, this phase could be precipitated either by some of the shock mechanisms described above or by gravitational instabilities in the molecular material. Phase I is relatively quiescent, and internal mass motions of a few kilometers per second are typical. During phase I, the cloud density increases about a hundredfold and the cloud temperature decreases as increasing density promotes efficient cooling from far-infrared spectral lines. Cloud temperature may fall as low as ≈10 K. Magnetic fields remain frozen into the protostellar gas as it collapses (*42*), so that the magnetic field strength increases roughly as the square root of the gas density in the cloud (*43*). It is possible that external material, such as ejecta from nearby stars or material from the parent cloud, may be mixed into the protostar at this time. Thus, this phase of evolution may represent the point at which considerable amounts of heavy elements were incorporated within the presolar nebula. There is speculation (*10a, 44*), based upon isotopic anomalies in solar system meteorites, that material from supernovae may have been mixed into the presolar nebula. Evidently, detection of phase I clouds requires high spatial resolution and high sensitivity in the far-infrared (wavelength $\gtrsim$ 100 μm).

After the initial quiescent contraction, the protocloud enters a phase of rapid gravitational collapse (phase II) at a density of ≈ 10^5 to 10^6 molecules per cubic centimeter (*41*). This rapid infall slows at a density of ≈ 10^{10} molecules per cubic centimeter when the central regions of the cloud become optically thick to the radiation released by the material as it gains internal energy during the collapse. Efficient far-infrared cooling mechanisms within the cloud maintain its temperature at around 10 to 20 K. Although the protostar's density increases by a factor of 10^5 during phase II, the evolution time is only 10^4 to 10^5 years. The cloud is cold and small so that it must be observed at high spatial resolution with the use of far-infrared, millimeter, and submillimeter detection techniques. These conditions render the rapid protostellar collapse the most difficult of all the stages of star formation to study observationally. If angular momentum is conserved during the free-fall collapse, strong rotational forces may cause the protostar to fragment into a central ring-like structure that can in turn fragment into several subcondensations orbiting the original center of mass (*45*). Thus, this stage of collapse is crucial to our understanding of the formation of both binary stellar systems and planetary systems. The theoretical prediction of breakup of the cloud during rapid collapse by rotation is encouraging. Perhaps only 30 percent of observed evolved stellar systems are thought to involve single stars; the vast majority are binary or multiple star systems.

The final phase of the protostellar collapse (phase III) begins when the central region of the cloud becomes optically thick to its own radiation, causing the central temperature to increase. Gas pressure is soon sufficient to halt the gravitational free fall, and an equilibrium core (or cores if the system has become a binary star) forms in the center of the protostar (*41*). This core, which may be small if the system's rotation rate is high, appears to accrete material from the outer layers of the protostellar cloud, a process that may lead to the formation of a preplanetary nebula in which additional planet-mass condensations will later fragment (*39*). The central core contracts adiabatically during phase III until the dissociation of H_2 occurs at a temperature of 1800 to 2000 K. The central condensation at this time has a density of ≃ 10^{20} atoms per cubic centimeter and a radius of less than half an astronomical unit (1 A.U. = 1.5×10^{13} cm, the Earth-sun distance). Additional dissociation and atomic ionization barriers are encountered as the density and temperature increase, each acting as an energy reservoir for the nearly isothermal absorption of gravitational energy generated by additional compression. After all ionization barriers have been passed, the core contracts slowly and adiabatically as it accretes mass until its central temperature and density are sufficient for nucleosynthesis to begin (*41*).

Viewed from the distance of the OMC, the 2000 K core would subtend an angle of only 0.8 milliarc second. Dynamical motions are of the order of 10 km/sec. The entire protostellar nebula, with outer regions having temperatures on the order of 100 K, would be smaller than several hundred astronomical units (2 to 5 arc seconds at a distance of 500 pc). Physical conditions during this phase must be studied with near-infrared (10 to 30 μm) and mid-infrared (2 to 5 μm) spectroscopic and photometric techniques that have high spatial and spectral resolution. There are a number of important issues to be addressed regarding the physics of phase III. The angular momentum of the protostar determines to a great extent how much of the material ends up in the central condensation. In the case of our solar system, for example, the angular momentum per unit mass of the sun's rotation is very low compared to that of the planets' orbital motion. Mechanisms such as the transfer of central angular momentum to the outer regions of the protostellar cloud by magnetic braking appear plausible but, to our knowledge, have received no definitive observational or theoretical test. The development of the central condensation may have important consequences for the early nucleosynthetic history of stars.

During the final phases of protostellar collapse, the outer cool regions of the protostar initially contain vast amounts of material condensed in dust grains. The intense radiation field generated as the infall energy of material accreting onto the central condensation may be intense enough to vaporize the dust grains, a process that could alter the chemical composition, structure, and size distribution of the grains and therefore affect the processing of biogenic elements. Unfortunately, there is little observational evidence concerning these processes at present. A final question of great importance concerns the spatial fractionation of the elements within the protostellar nebula during late contraction stages. Here also the theoretical models are severely constrained by the paucity of observational tests.

Planetary Systems

There is currently firm observational evidence for the existence of only one planetary system: our own. The recent discovery of cold matter associated with the nearby star Vega (*46*), made with the new infrared astronomical satellite (IRAS), will no doubt accelerate the search for other planetary systems. If the accepted theoretical picture of star formation is correct, planetary systems should form in nearly every stellar system. A significant exception to this expectation involves binary stars. It is suspected that the process by which binaries are formed somehow precludes the formation of planetary companions. If so, the fact that at least 70 percent of stars are in binary or multiple star systems would drastically reduce the expected number of planetary systems. Recent studies (*47*) have shown that a wide variety of stable orbits for planets do exist in binary systems if the planets can be formed there. Results from a comprehensive search for other planetary systems would provide an important check on our current theories of star formation (*48*).

Future Research

We have outlined the observationally known properties of regions where star formation is occurring and have presented a theoretical point of view concerning the way in which stars and planetary systems are thought to form. This picture of star formation is only a beginning. Much remains to be learned, both observationally and theoretically.

Observational research during the past decade has been successful in determining the physical and chemical environment of star formation and in identifying very young, recently formed stars. The interstellar molecular clouds and the newly formed stars have been studied primarily at infrared, submillimeter, and millimeter wavelengths. Thus far, these observations have been carried out with instruments that are primitive when compared with those used in optical astronomy and centimeter-meter radio astronomy. In particular, the limits due to lack of high spatial resolution in infrared and millimeter observations and the blanketing of the earth by an absorbing atmosphere in much of the infrared will have to be overcome in order to provide images of star formation comparable in clarity to the optical pictures of normal stars or galaxies now obtained with ordinary medium-sized, ground-based telescopes. For the most part, the spatial resolution of existing far-infrared and millimeter observation is only slightly better than the naked-eye resolution for yellow light. Thus, the universe of molecules, giant molecular clouds, young stars, and interactions between stars and clouds has been revealed, but the picture is still out of focus. A partial remedy has been achieved by the IRAS, but only a beginning. Some basic questions remain. On what scale do GMC's fragment when forming individual stars or star clusters? What are the properties of a true protostellar fragment? What is the energy output from a star or protostar back to the parent cloud? What does star formation look like in other galaxies where we can get a "global" picture? Does the formation of an ordinary star like the sun proceed in a fundamentally different way from that of more massive stars? What does star formation look like in active "star burst" galaxies where the process is faster by one or two orders of magnitude than in our own galaxy?

Several new ground-based and orbiting astronomical facilities that could provide high spatial and spectral resolution appear technologically ready for development during the next several decades. They include the cryogenically cooled shuttle infrared telescope facility, the 15-m ground-based optical-infrared new technology telescope, an orbiting 10-m large deployable reflector for far-infrared–submillimeter measurements, and several ground-based infrared-radio interferometers. Part of a comprehensive plan for the development of new astronomical facilities during the 1980's, these facilities are discussed in detail in the "Report of the Astronomy Survey Committee" (*11*). Another very powerful instrument recently suggested by millimeter-wave astronomers is a high-resolution millimeter-wave synthesis telescope that would resolve giant molecular clouds in other galaxies and that could be used to study nearby cloud cores with a resolution of 0.01 pc. The timely development of these facilities is a basic prerequisite for continued progress in studies of the process of planet and star formation and consequently for understanding the evolution of spiral galaxies, whose beautiful appearance is almost totally determined by recent star formation.

Much of the research that must be done if we are to understand the formation of stars and planetary systems is theoretical. It includes both the modeling of observed phenomena and the calculation of quantities that have not yet been observed. The complexity and diversity of processes involved in star formation, as well as the extreme range of physical parameters, will make it necessary for theoretical studies to rely increasingly upon detailed numerical simulations based on the use of the most advanced computers available (*11*). Although a number of important theoretical questions must be answered, the following are of particular significance: How do relatively small objects (stars) form from massive parent GMC's? During what phases of collapse do fragments form? What are the various ways in which binary systems evolve? What physical conditions and parameters (such as angular momentum and magnetic field strength) have a major influence on the effectiveness of star formation mechanisms? How are planetary systems like our solar system formed? What is their structure, and the range of physical conditions within them? These questions and others will play a central role in our efforts to determine how cosmically insignificant, yet breathtakingly beautiful objects like Earth were formed.

References and Notes

1. A. W. Wolfendale, Ed., *Progress in Cosmology* (Reidel, Dordrecht, Netherlands, 1981).
2. J. Billingham, Ed., *Life in the Universe* (MIT Press, Cambridge, Mass., 1981), and references therein.
3. R. V. Wagoner, *Astrophys. J.* **179**, 343 (1973; ——, W. A. Fowler, F. Hoyle, *ibid.* **148**, 3 (1964).
4. K. O. Thielhelm, in (*1*), p. 305; W. H. McCrea, in (*1*), p. 239; W. W. Roberts and W. B. Burton, in *Topics in Interstellar Matter*, H. Van Woerden, Ed. (Reidel, Dordrecht, Netherlands, 1977), p. 195.
5. F. Hoyle, *Astrophys. J. Suppl. Ser.* **1**, 121 (1954); E. M. Burbidge, G. R. Burbidge, W. A. Fowler, F. Hoyle, *Rev. Mod. Phys.* **29**, 547 (1957).
6. C. A. Barnes, D. D. Clayton, D. N. Schramm, Eds., *Essays in Nuclear Astrophysics* (Cambridge Univ. Press, Cambridge, 1982).
7. R. D. Gehrz and N. J. Woolf, *Astrophys. J.* **165**, 285 (1971).
8. N. J. Woolf and E. P. Ney, *ibid.* **155**, L181 (1969).
9. R. Treffers and M. Cohen, *ibid.* **188**, 545 (1974).
10. W. D. Arnett, *Annu. Rev. Astron. Astrophys.* **11**, 73 (1973); D. N. Schramm, in (*6*), p. 325.

10a. See D. D. Clayton [*Q. J. R. Astron. Soc.* **23**, 174 (1982)] for a stimulating analysis of the scenarios whereby stellar ejecta are mixed into the ISM and presolar nebula.

11. *Astronomy and Astrophysics for the 1980's*: (report of the Astronomy Survey Committee, National Academy Press, Washington, D.C., 1982), vol. 1.
12. F. J. Kerr and G. Westerhaut, in *Galactic Astronomy*, A. Blaauw and M. Schmidt, Eds. (Univ. of Chicago Press, Chicago, 1965), p. 167.
13. The first few radio-wavelength discoveries of interstellar molecules were actually at centimeter wavelengths prior to the use of millimeter-wave technology [D. M. Rank, C. H. Townes, W. J. Welch, *Science* **174**, 1083 (1971); P. M. Solomon, *Phys. Today* **26**, 32 (1973)].
14. N. Z. Scoville and P. M. Solomon, *Astrophys. J.* **199**, L105 (1975); M. A. Gordon and W. B. Burton, *ibid.* **208**, 346 (1976). For a presentation of the two-dimensional axisymmetric distribution of molecular clouds in the galaxy, see: D. B. Sanders, thesis, State University of New York, Stony Brook (1982); ——, P. M. Solomon, N. Z. Scoville, *Astrophys. J.*, in press.
15. P. M. Solomon, D. B. Sanders, N. Z. Scoville, in *The Large Scale Characteristics of the Galaxy*, W. B. Burton, Ed. (Reidel, Dordrecht, Netherlands, 1979), p. 35; A. A. Stark, thesis, Princeton University (1980).
16. S. Drapatz and K. W. Michel, *Astron. Astrophys.* **36**, 211 (1974).
17. M. Kutner, K. D. Tucker, G. C. Chin, P. Thaddeus, *Astrophy. J.* **215**, 521 (1977); L. Blitz, in *Giant Molecular Clouds in the Galaxy*, P. M. Solomon and M. G. Edmunds, Eds. (Pergamon, New York, 1980), p. 1.
18. H. S. Liszt, D. L. Xiang, W. B. Burton, *Astrophys. J.* **249**, 532 (1981).
19. R. S. Cohen, H. I. Cong, T. M. Dame, P. Thaddeus, *ibid.* **239**, L53 (1980); P. Thaddeus, in *Molecules in Interstellar Space*, A. Carrington, Ed. (Royal Society, London, 1981), p. 3.
20. P. M. Solomon, in *Molecules in Interstellar Space*, A. Carrington, Ed. (Royal Society, London, 1981), p. 16; D. B. Sanders, N. Z. Scoville, P. M. Solomon, *Astrophys. J.*, in press; D. B. Sanders, P. M. Solomon, N. Z. Scoville, D. B. Clemens, IAU Symposium 106, *The Milky Way*

Galaxy, H. Van Wouden, Ed. (Reidel, Dordrecht, Netherlands, 1984), for Fig. 2.

21. W. Watson, *Rev. Mod. Phys.* **48**, 513 (1976); E. Herbst, in (*41*), p. 88. Theoretical interstellar ion-molecule chemistry was first introduced by E. Herbst and W. Klemperer, *Astrophys. J.* **185**, 505 (1973).
22. F. Hoyle, M. C. Wickramasinghe, A. H. Olanesen, S. Al-Mafti, D. T. Wickramasinghe, *Astrophys. Space Sci.* **83**, 405 (1982).
23. P. Goldreich and J. Kwan, *Astrophys. J.* **189**, 441 (1974); N. Z. Scoville and P. M. Solomon, *ibid.* **187**, L67 (1974).
24. B. Zuckerman and P. Palmer, *Annu. Rev. Astron. Astrophys.* **12**, 279 (1974); N. Z. Scoville and P. M. Solomon, *Astrophys. J.* **199**, 105 (1975).
25. R. Genzel and D. Downes, in *Regions of Recent Star Formation*, R. S. Roger and P. E. Dewdney, Eds. (Reidel, Dordrecht, Netherlands, 1982), p. 251.
26. K. M. Strom and S. E. Strom, *Science* **216**, 571 (1982), and references therein.
27. A comprehensive reference regarding regions of star formation within molecular clouds is *Proceedings, International Astronomical Union Symposium 92, Infrared Astronomy*, C. G. Wynn-Williams and D. P. Cruileshauk, Eds. (Reidel, Dordrecht, Netherlands, 1981).
28. W. Herbst and G. E. Assousa, in (*41*), p. 368.
29. B. G. Elmegreen and C. J. Lada, *Astrophys. J.* **214**, 725 (1977).
30. J. Smith, *ibid.* **238**, 842 (1980).
31. F. H. Shu *et al.*, *ibid.* **173**, 557 (1972); P. R. Woodward, *ibid.* **207**, 484 (1976).
32. N. J. Evans II, in (*27*), p. 107, and references therein.
33. A. Harris, C. H. Townes, D. N. Matsakis, P. Palmer, *Astrophys. J.* **265**, L63 (1983).
34. C. G. Wynn-Williams [*Annu. Rev. Astron. Astrophys.* **20**, 587 (1982)] has reviewed the properties of objects embedded in molecular cloud cores.
35. G. G. Fazio, in *Infrared Astronomy*, G. Setti and G. G. Fazio, Eds. (Reidel, Dordrecht, Netherlands, 1978), p. 25.
36. A. Whitworth, *Mon. Not. R. Astron. Soc.* **186**, 59 (1979); T. J. Mazurek and G. E. Brown, *Astron. Astrophys.* **81**, 382 (1980).
37. G. L. Grasdalen, R. D. Gehrz, J. A. Hackwell, in (*27*), p. 179; N. Z. Scoville, in (*27*), p. 187.
38. R. D. Gehrz *et al.*, *Astrophys. J.* **254**, 550 (1982).
39. P. Cassen and A. Summers, *Icarus*, in press.
40. H. Hippelein and G. Munch, *Astron. Astrophys.* **99**, 248 (1981).
41. A general review of protostellar processes is contained in *Protostars and Planets*, T. Gehrels, Ed. (Univ. of Arizona Press, Tucson, 1978).
42. J. Nittmann, in (*25*), p. 123.
43. E. H. Scott and D. C. Black, *Astrophys. J.* **239**, 166 (1980); D. C. Black and E. H. Scott, *ibid.* **263**, 686 (1982).
44. A. G. W. Cameron and J. W. Truran, *Icarus* **30**, 447 (1977).
45. D. C. Black and P. Bodenheimer, *Astrophys. J.* **206**, 138 (1976).
46. F. C. Gillett, personal communication; H. H. Aumann, F. C. Gillett, C. A. Beichman, T. de Jong, J. R. Houck, F. Low, G. Neugebauer, R. G. Walker, P. Wesselius, IRAS preprint (1984).
47. D. C. Black, *Astron. J.* **87**, 1333 (1982).
48. ———, *Space Sci. Rev.* **25**, 35 (1980).
49. N. Scoville, R. P. Schloerb, P. Goldsmith, personal communication.
50. Optical photograph by G. Herbig; 10-μm contours from R. D. Gehrz, J. A. Hackwell, J. R. Smith, *Astrophys. J.* **202**, L33 (1976).
51. Infrared data from R. D. Gehrz, G. L. Grasdalen, J. A. Hackwell, unpublished data; H_2 contours from S. Beckwith, S. E. Persson, G. Neugebauer, E. E. Becklin, *Astrophys. J.* **223**, 464 (1978).
52. R. D. Gehrz, G. L. Grasdalen, J. A. Hackwell, unpublished data.
53. D. B. Sanders, N. Z. Scoville, P. M. Solomon, D. Clemens, *Astrophys. J.*, in press.
54. Publication costs for colored images in the original article were supported in part by NSF grants to the University of Wyoming and contracts at Stony Brook. This chapter summarizes only a limited aspect of the great promise and excitement of galactic research in the years to come. We owe a great debt to our fellow members of the Galactic Astronomy Working Group, who contributed in many ways to the material discussed here. They were W. B. Burton, D. F. Carbon, J. G. Cohen, P. DeMarque, F. K. Lamb, B. Margon, S. Van Den Bergh, and P. O. Vandervoort. Consultants to the working group were R. McCray, C. F. McKee, and L. Searle.

This chapter is a review of the status of research in the field of star formation and interstellar matter based in part upon the deliberations of the Galactic Astronomy Working Group of the National Academy of Sciences Astronomy Survey Committee. The paper originally appeared in *Science* **224**, 25 May 1984.

10. Binary Stars

Bohdan Paczyński

The star closest to us, our sun, is single. This singularity is an exception rather than the rule. Most stars form either double or multiple systems held together by their gravitational interaction. For example, among solar type stars the observed ratios of single:double:triple:quadruple systems is 45:46:8:1 (*1*). This seems to be a typical situation in the galactic disk, close to the plane of symmetry of our Milky Way, whereas in the galactic halo a majority of stars are single (*1*). The reason for this difference is not known.

Multiple systems usually have hierarchical structure: at a much larger distance from a close pair of stars, there might be another star or another close pair. It is easy to understand why all the components of a multiple system cannot have comparable mutual separations—such systems are known to be unstable unless the number of components is very large. Rich and very rich systems do exist, and they are known as clusters of stars and galaxies.

Various binaries may differ enormously in their extent. There are many pairs of stars touching each other. These are the contact systems of the W Ursae Majoris or SV Centauri type (*2*). A whole contact binary is only slightly larger than the sun, some 3×10^{11} cm in diameter. At the other extreme there is the nearby triple system, α Centauri ABC. The component α Centauri C, also known as Proxima Centauri, is a distant companion to a binary, α Centauri AB. The separation between components A and B is 3×10^{14} cm, whereas component C is about 10^{17} cm away from A and B.

Various binaries populate fairly uniformly a whole range of separations from about 3×10^{11} to 3×10^{17} cm, which corresponds to orbital periods of 1 to 10^9 days (that is, up to about 3 million years). Roughly 10 percent of all stars are binaries with orbital periods between 1 and 10 days, another 10 percent have orbital periods between 10 and 100 days, and so on (*1*). There is a rapid cutoff in the number of systems with separations larger than about 3×10^{17} cm, most likely due to gravitational perturbations caused by other stars belonging to our galaxy and randomly passing by (*3*). The average distance between separate stars or stellar systems in the solar neighborhood is about 4×10^{18} cm.

Ordinary Stars in Binary Systems

Binary and multiple stars are so common that an understanding of their origin and evolution is important for an understanding of the origin and evolution of stars in general. Binaries, like single stars, spend most of their "lives" burning hydrogen in their deep interiors. That phase of evolution a star spends on the main sequence. This is a relatively narrow band on the Hertzsprung-Russell or color-magnitude diagram, where the majority of known stars are located. The band runs diagonally across this diagram from faint and cool (that is, red) stars all the way to luminous and blue (that is, hot) stars. This confinement of a vast majority of known stars to an almost one-dimensional domain on a two-dimensional diagram implies that stellar properties are largely determined by one parameter. It is firmly established now that this parameter is the stellar mass. Low-mass stars are faint and cool, whereas massive ones are luminous and hot. The empirical relations between stellar masses, luminosities, and radii were obtained through studies of binaries. For the main sequence these relations may be well approximated as

$$R/R_{\odot} \approx (M/M_{\odot})^{0.8},\ L/L_{\odot} \approx (M/M)_{\odot})^{3.5}$$

where $M_{\odot} = 1.99 \times 10^{33}$ g, $R_{\odot} = 6.96 \times 10^{10}$ cm, and $L_{\odot} = 3.86 \times 10^{33}$ erg sec^{-1} are the solar mass, radius, and luminosity, respectively. These empirical relations are well understood and reproduced theoretically with numerical models of stars calculated with modern computers.

Many other parameters can be measured more conveniently or more accurately in binaries than in single stars. The following example may illustrate how this is done.

ζ Phoenicis, also known as HD 6882 or HR 338, is a bright binary with an orbital period of 1.67 days. Its light intensity (*4*) as well as the radial velocities of the two components (*5*) vary with this period. The variations are shown in Fig. 1. The light curve (Fig. 1A) displays two eclipses: the primary at phase 0 (or 1), when the cool component partly eclipses the hotter one and the secondary at phase 0.5 (or 1.5), when the hot component is in front and the cool component is hidden behind it. A detailed analysis of the light changes during the eclipses allows accurate determination of relative stellar radii, expressed in units of their mutual separation. A schematic presentation of the binary is given in Fig. 2. The extent of stellar motion and the size of the binary orbit follow from an analysis of the radial velocity curves (Fig. 1B). Using Kepler's law, one may determine the stellar masses as well. The results for ζ Phoenicis are as follows (*4*):

$$M_1/M_{\odot} = 3.85,\ R_1/R_{\odot} = 2.84,\ L_1/L_{\odot} = 320$$

$$M_2/M_{\odot} = 2.50,\ R_2/R_{\odot} = 1.85,\ L_2/L_{\odot} = 32$$

There are several dozen systems for which a similar analysis has been done with high accuracy, that is with errors less than 10 percent and sometimes significantly less (*6*).

A binary like ζ Phoenicis is very useful because all the parameters are known very accurately and the two stars have significantly different masses. This allows one to estimate the age and chemical composition of this system by comparing stellar models with the observations.

A main sequence star generates energy by burning hydrogen in the core. The energy diffuses from the core to the surface, and finally it is radiated away. Gradually, hydrogen is processed through a sequence of thermonuclear reactions into helium. As a result of this process, stellar luminosity and radius increase with time (*7*). The best quantitative comparison between models and real stars is possible for binaries like ζ Phoenicis. Calculations of models for a star of a given mass and a given chemical

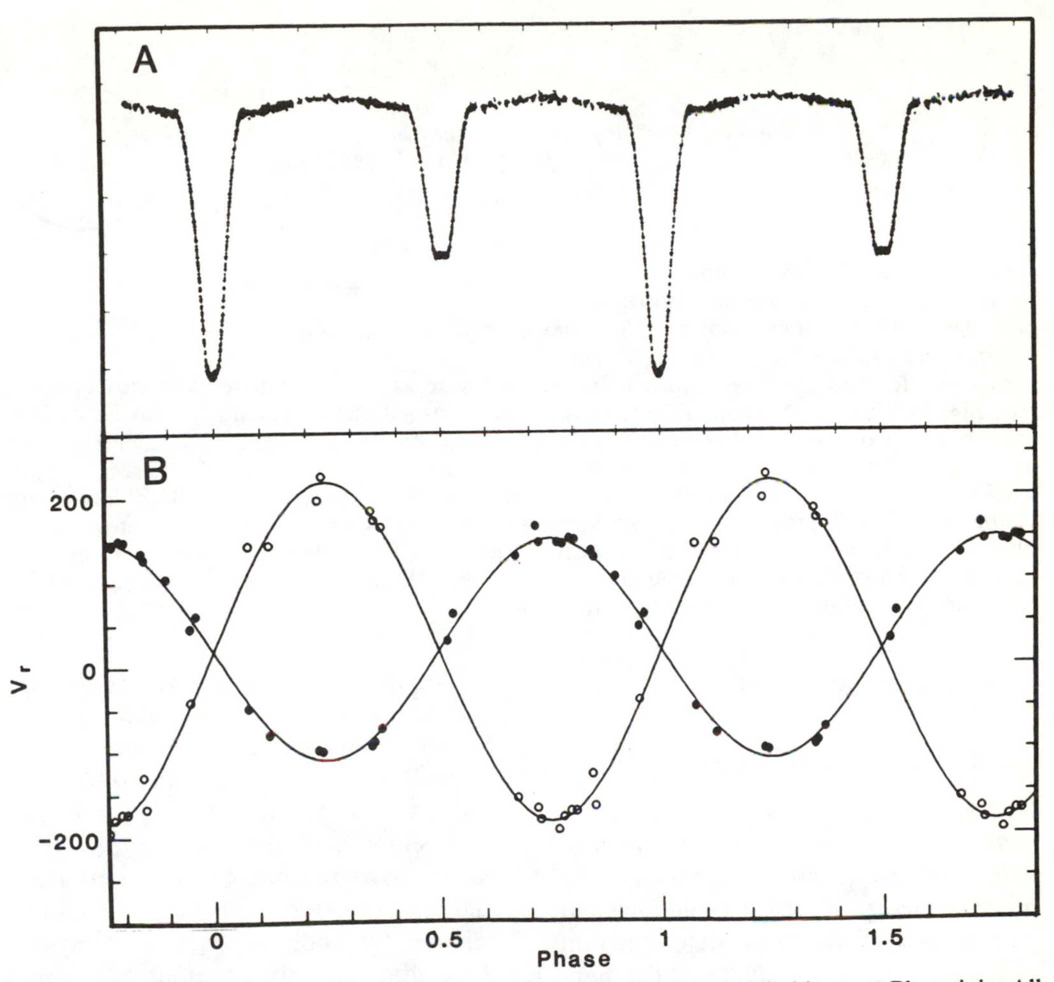

Fig. 1. Light variations (A) and radial velocity variations (B) for a bright binary, ζ Phoenicis. All the variations have the 1.67-day binary period. At phase 0 (or 1) the hotter of the two stars is eclipsed by the cooler companion (primary eclipse). At phase 0.5 (or 1.5) the cooler star is eclipsed by the hotter one (secondary eclipse). The radial velocities, expressed in kilometers per second, are displayed for the hot primary with filled circles and for the cool secondary with open circles. Radial velocity is positive when the source is receding from the observer.

composition provide the variations of stellar radius and luminosity as a function of time. Chemical composition may be conveniently described with just two parameters: helium content Y and heavy element content Z. All the remaining matter is hydrogen. It is natural to assume that both components of a binary system were formed simultaneously from condensations of interstellar gas, and that the initial chemical composition was the same for the two stars. As we know the masses of the two stars from observations, there are only three parameters that we may vary in the models: age (t), Z, and Y. We want to reproduce four observed quantities: the present radii and luminosities of the two stars (R_1, R_2, L_1, L_2). Therefore, we may even check the models for self-consist-

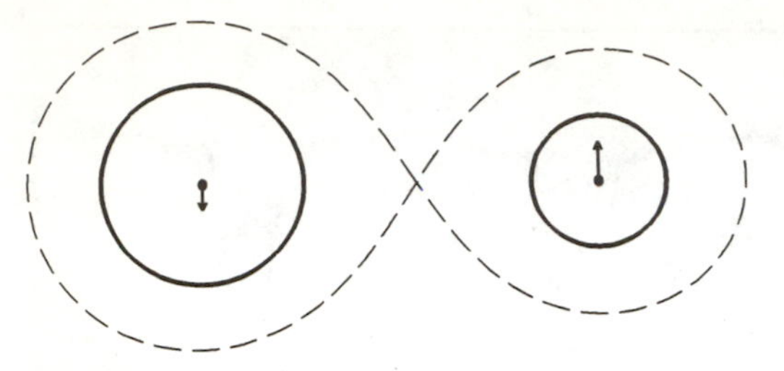

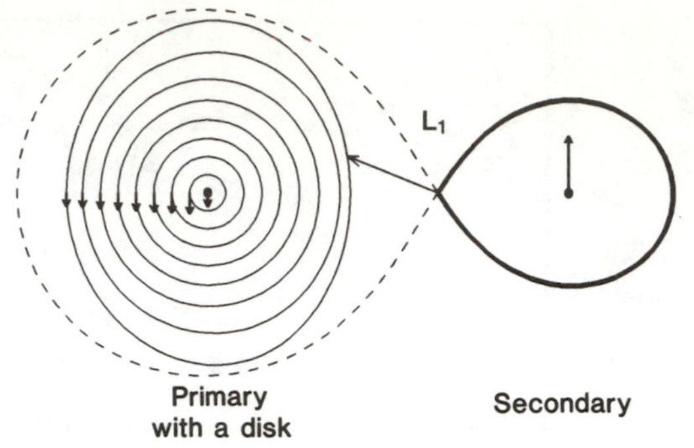

Fig. 2 (left). The two components of the binary system ζ Phoenicis are shown as circles. Their relative dimensions and the separation were deduced from the light curve shown in Fig. 1A. Their velocities, indicated with arrows, were measured spectroscopically (see Fig. 1). The infinity-shaped figure drawn with a dashed line represents the Roche surface around the two stars. Fig. 3 (right). A schematic picture of a cataclysmic binary. The primary is a very compact star: a white dwarf, a neutron star, or a black hole. The secondary is a solar type (that is, main-sequence) star. It slightly overflows its Roche lobe. Tidal forces due to the primary pull matter through the vicinity of the inner Lagrangian point, L_1, toward the primary. Coriolis forces deflect the stream of gas and let it collide with the outer rim of a gaseous accretion disk rotating around the primary. Arrows indicate the orbital velocities of the two stars and the rotational velocity of the disk.

ency (*4*). This type of analysis gives for ζ Phoenicis

$$t = 8 \times 10^7 \text{ years}, Y = 0.28, Z = 0.015$$

Similar modeling has been done for about half a dozen systems and has provided the best quantitative checks of the modern theory of stellar evolution, and perhaps the most accurate determination of the helium content in stars. The helium abundance obtained in this way is in good agreement with determinations from models of hot stellar atmospheres and from observations of the interstellar medium. It is now believed that helium was produced mainly during the first 3 minutes after the Big Bang; hence, a determination of helium abundance is of considerable cosmological importance.

White Dwarfs, Neutron Stars, and Black Holes

Not only main-sequence stars but also more exotic objects may have their masses measured if they are members of binary systems. The three white dwarfs for which masses are well known are Sirius B (0.94 $M_\odot$), Procyon B (0.65 $M_\odot$), and σ^2 Eridani B (0.43 $M_\odot$). All are members of multiple systems (*6*). Their radii are also reasonably well determined and are consistent with the mass-radius relation for cool, degenerate stars (*8*). According to this relation, no white dwarf can exist with a mass exceeding the Chandrasekhar limit of 1.4 $M_\odot$. Indeed, none is known. For many decades white dwarfs were the most compact stars known to astronomers.

Even more compact objects, neutron stars, were discovered as single radio pulsars (*9*). But it was only after the discovery of x-ray pulsars, which are neutron stars in close binaries, that their masses could be measured (*10, 11*). Most available estimates have rather large errors, but all are in the range of 1 to 2 $M_\odot$, as expected theoretically. The best available estimates for the neutron star masses are provided by a binary radio pulsar,

PSR 1913 + 16. Both components have almost identical masses of 1.4 $M_{\odot}$ with an accuracy of a few percent (*12*). Unfortunately, existing observations do not allow reliable determinations of neutron star radii. Some attempts were made for x-ray bursters, a subclass of binary neutron stars undergoing thermonuclear flashes in freshly accreted matter (*11*). The masses and radii of x-ray bursters agree reasonably well with the prediction of models, but the available estimates are not very accurate. It is not yet possible to use them to select the correct equation of state at supernuclear densities (*13*).

No matter what equation of state is adopted at supernuclear densities, there can be no cool stars above 3 $M_{\odot}$, provided that the theory of general relativity is correct in the strong field limit (*13*). Anything more massive must ultimately collapse and form the most compact object so far imagined, a black hole. Discovery of a black hole would be one of the most spectacular achievements of modern astronomy. There are two general areas where the search is going on: active galactic nuclei and quasars (*14*) and massive x-ray binaries (*10*). So far there is no proof that a black hole has been found, but there are two very good candidates, the compact components of two binary systems: Cygnus X-1 in our galaxy (*10*) and LMC X-3 in the nearby galaxy, Large Magellanic Cloud (*15*). Observations indicate that in both binaries one component is a fairly normal, massive, hot and luminous star, probably close to the main sequence, whereas the second component is a very strong x-ray source. Each component has a mass of about 10 $M_{\odot}$.

It is generally believed that there are only two types of stars that can be strong emitters of x-rays, and both are very compact. These are neutron stars and black holes powered by the accretion of matter flowing from the nearby companion. A majority of strong x-ray sources have masses between 1 and 2 $M_{\odot}$ and display strong short-period variations in their x-ray power. These are the x-ray pulsars, believed to be rotating neutron stars with strong magnetic field channeling the accretion flow toward the magnetic poles, where the two bright spots are formed. Rotation periodically displays the spots and hides them behind the star, hence the so-called pulsations. It is believed that a black hole must be axially symmetric, and accretion onto a black hole cannot give rise to strictly periodic rapid oscillations. Indeed, neither of the two black hole candidates is a pulsar. Unfortunately, present theoretical models of x-ray emission cannot predict with any accuracy the x-ray spectrum to be expected from an accreting black hole, and the evidence for the existence of black holes in Cygnus X-1 and LMC X-3 is still rather circumstantial. In fact, it is difficult to formulate a realistic criterion that would convince everybody of the presence of black holes in these two systems. One of the problems is that black holes are the most extreme objects predicted by general relativity. Their convincing discovery would be a very important step in demonstrating that general relativity is the correct theory of gravitation in strong-field conditions. However, there is no independent observational evidence that this theory is really valid when the gravitational field becomes very strong. So far, almost all tests of this theory were restricted to the weak field and the so-called post-Newtonian approximation. With one important exception.

The post-Newtonian approximation for gravitation and motion allows for terms of the order $(v/c)^2$, where v is the characteristic velocity of objects in the system under consideration and c is the speed of light. Higher order terms are too small to produce detectable effects under most astronomical conditions. The only exception is gravitational radiation, which is supposed to carry away energy at a rate proportional to $(v/c)^5$. This is a post-post-Newtonian effect, and it is usually too small to notice, as there are so many other competing astrophysical processes. However, there is one ideal case of a binary radio pulsar, PSR 1913 + 16, which has an eccentric orbit with a period of 7 hours and 45 minutes. The pulsar itself is an excellent clock, rotating once every 59 msec. Timing the radio pulses arriving at the radio telescope in Arecibo, Puerto Rico, can be done with an accuracy of a few tens of microseconds. As a result, this became the binary system with the best known orbital parameters. It seems that the companion is also a neutron star, and the two objects may be treated as two point masses while one is analyzing the dynamics of their orbit. Gravitational radiation should remove energy from the binary and should reduce the binary period P at the dimensionless rate $dP/dt = -2.403 \pm 0.005 \times 10^{-12}$. The best recent observations give the empirical value of $dP/dt = -2.30 \pm 0.22 \times 10^{-12}$ (*12*). It is likely that in a few years the observational errors will be reduced by an order of magnitude. If the observed rate agrees with that predicted by general relativity to an accuracy of 1 percent, then it will be very difficult to imagine that the period change is an unrelated phenomenon and that the agreement is just a coincidence. The present agreement at the level of 10 percent is highly suggestive evidence for gravitational radiation. For agreement at the 1 percent level that is likely to be accepted as a proof, there would be increasing confidence in general relativity and indirect support for the inferred existence of black holes in Cygnus X-1 and LMC X-3.

Accretion Disks

Some relatively nearby binaries exhibit phenomena similar to those displayed on a much larger scale by very distant active galactic nuclei and by quasars. The characteristic linear scale of galactic nuclei is larger by many orders of magnitude than the scale of close binaries. Therefore, the characteristic time scale for any variations is orders of magnitude longer in galactic nuclei than in binaries. This makes binaries much easier to study and to understand within the limited lifetime of the astronomers. Among many possible phenomena and structures that may be similar in these two different types of objects, there are two that are most interesting: accretion disks and jets.

There is a large class of very compact binaries, called cataclysmic because of their violent activity (*16, 17*), ranging from minor flares to fair-sized dwarf nova eruptions to powerful explosions of novae. It is also possible that type I supernovae are the end products of the evolution of some cataclysmic binaries. Supernovae are the most powerful stellar explosions known, energetic enough to destroy a whole star. A standard geometrical model of a cataclysmic binary is shown in Fig. 3. Their typical orbital periods are a few hours. The shortest of all known binary periods is 17.5 minutes. This record belongs to a cataclysmic star, AM Canum Venaticorum (*18*).

The more massive component of a cataclysmic system is a compact star of about 1 $M_{\odot}$. The less massive secondary is usually a main-sequence star. It overflows its Roche lobe slightly and loses matter through the vicinity of inner Lagrangian point (L_1) because of tidal forces induced by the primary. The outflowing gas is deflected by the Coriolis forces and collides with an outer rim of gaseous disk rotating around the primary. A disk is like a thin star. It is rotationally supported against the gravitational pull of the centrally located primary. Ordinary stars are supported against their own gravity with a pressure gradient.

Most of the energy radiated by a cataclysmic binary is generated by viscous dissipation in a differentially rotating disk. The inner parts of the disk rotate faster than the outer parts. This gives rise to a shear flow. Viscosity generates a lot of heat throughout the disk, and it also transfers angular momentum outward. From the outer rim of the disk angular momentum is transferred back to the binary orbital motion through the tidal forces induced by the secondary. Outward transport of angular momentum forces disk matter to spiral toward the primary. The ultimate source of energy is gravitational. Disk luminosity may be calculated as a product of the rate of mass accretion and the gravitational binding energy at the surface of the primary. The smaller the stellar radius, the larger the binding energy and the more efficient the accretion disk "engine." This is why it is most efficient to have a compact star at the center.

The most popular model for the central parts of an active galactic nucleus or a quasar is an accretion disk around a supermassive black hole (*14*). This makes it fairly similar to the popular model of a cataclysmic binary. There is one major problem with both models. It is not known what physical processes give rise to disk accretion (*19*). Ordinary "molecular" viscosity is too small by many orders of magnitude. Magnetic fields and turbulent motions are likely to be important, but it is not known how to calculate their effect on disk evolution. In practice, one is forced to compensate for one's lack of knowledge by introducing some free parameters in the model.

Studies of cataclysmic binaries are easier than studies of active galactic nuclei. Binaries are smaller, and therefore all variations have shorter time scales. For example, a few decades of observations contributed to the development of a phenomenological model of disk instabilities in dwarf novae. Recently, many groups more or less independently developed a disk model that reproduces the observed instability and relates it to the partial ionization of hydrogen and helium in the outer parts of an accretion disk (*20*). Rapid progress in the development of those models and the availability of a large number of easily observable dwarf novae leave little doubt that the understanding of processes involved in disk accretion will increase rapidly in the near future. Application of models to the more difficult stellar and extragalactic objects will become possible. Certainly, the most important development would be a realistic model of a quasar. But there are less dramatic opportunities, too. For example, there are reasons to believe that many stars that are forming now, and have not yet reached their location on the main sequence, have extended disks around them. Some disks may be predecessors of planetary systems (*21*). Some, like FU Orionis stars, may have unstable accretion disks, just like dwarf novae (*22*). However, those disks are a few hundred times larger than those found in dwarf novae. Therefore, eruptions of FU Orionis stars may last more than 10 years, while the eruption of a dwarf nova lasts only a week.

Jets

Radio jets, some expanding with highly relativistic velocity, are known to be ejected from some active galactic nuclei and quasars (*14*). These are among the most spectacular phenomena in the universe. They are studied on scales as small as 1 parsec (3×10^{18} cm), and as large as 1 megaparsec (3×10^{24} cm). Jets are found very close to the "central engine," most likely streaming along the rotation axis of an accretion disk that spins around a supermassive black hole. Enormous energies are transmitted with jets to the distant "radio lobes" where jets collide with tenuous intergalactic matter. In spite of considerable effort the formation of jets and their dynamics are poorly understood. Similar phenomena are observed on a much smaller scale in two binaries in our own galaxy. These are Scorpio X-1 (*23*) and SS 433 (*24*).

Scorpio X-1 is the brightest x-ray source in the night sky. It is a binary with an orbital period of 0.787 days and a prominent disk accreting onto a neutron star or a black hole. Scorpio X-1 is also a triple radio source, pretty much like a typical radio galaxy, but on a scale millions of times smaller. The central radio source coincides with optical and radio images of Scorpio X-1. On the two sides there are two additional radio sources, presumably formed in a collision between the jets and interstellar matter. It is most puzzling that recent observations failed to detect any motion of the outer radio sources (*23*). In fact, a surprisingly small upper limit of 32 km sec^{-1} has been put on their velocity. There is no hope to achieve a comparable accuracy of measurement for any extragalactic radio source in the foreseeable future.

The binary system SS 433 has an orbital period of 13 days and an accretion disk that precesses with a period of 164 days. The most spectacular phenomenon is a pair of gaseous jets outflowing along the rotation axis of the disk in opposite directions, each with a velocity of 80,000 km sec^{-1}, that is, just over a quarter of the speed of light. The twin jets precess with a period of 164 days, just like the disk, and over thousands of years they have produced a huge bubble in the interstellar medium.

The two binaries with jets are very different from each other, and their jets are very different also. There is no good model for them, and this gives little credibility to models of much more powerful jets generated by active galactic nuclei and quasars. It is clear that a lot of work must be done before these phenomena will be understood. Scorpio X-1 and SS 433 offer a very convenient testing ground for theories of jets. We may hope that by understanding them we may come to understand their powerful extragalactic relatives.

Evolution of Close Binaries

Binaries with orbital periods in excess of about 10 years have their components so far apart that they evolve as two single stars. Neither star can affect the internal structure or the evolution of the companion. The evolution of single stars is reasonably well understood (*7*). While on the main sequence, they burn hydrogen into helium. When hydrogen is exhausted in the core, the core contracts and heats up. After some time helium ignites and burns into carbon. The sequence of nuclear reactions in the cores of massive stars can lead all the way to iron. In low mass stars this chain is terminated with carbon or oxygen. While this nuclear evolution proceeds in the deep interior, the outer part of the star, its envelope, gradually expands. The star becomes first a red giant and later a supergiant, almost as big as our planetary system. The evolution may be terminated in two ways. A star initially less massive than about 8 $M_{\odot}$ loses its envelope upon becoming a red supergiant. The loss is gradual and supplies fresh matter into the interstellar medium. A hot stellar core is left. It slowly cools and finally becomes a white dwarf. More massive stars die more violently. Their cores are too massive to support themselves against gravity when they run out of nuclear fuel. They collapse and become neutron stars. The collapse of a very massive star is likely to produce a black hole. Tremendous energy released in the process is believed to give rise to a supernova explosion and the violent ejection of the outer layers of the star.

How is this scheme changed if a star has a nearby companion? To be effective the companion must be at a distance not larger than the largest extent of the primary. The primary is the star that is initially the more massive of the two (the companion is called the secondary). As a result of its nuclear evolution, the primary expands, fills its Roche lobe, and starts transferring matter to the secondary (*25, 26*). Observations and model calculations show that well over 50 percent of the primary's mass is transferred, and the secondary becomes the more massive of the two stars. There is an enormous variety of possible evolutionary patterns, depending on the initial masses of the two stars and their initial separation (*2*). It seems that the qualitative features of binary evolution and mass transfer between the components are reasonably well understood but there are only a few binaries for which a detailed successful model has been calculated (*27*). There is a major uncertainty about the mass loss rates of binaries undergoing vigorous mass transfer. Also, it is not known how much angular momentum may be carried away from a binary. This problem is as difficult as the problem of mass outflow from single stars. It appears from observations that the high rate of loss of matter by luminous stars is important for their evolution. Unfortunately, there is no generally accepted theory for the phenomenon (*21*). The rate at which angular momentum may be lost is even more difficult to estimate, not only theoretically but also observationally.

It is very likely that the concept of a Roche lobe, which turned out to be so fruitful for understanding the evolution of many binary systems, may not be useful when the mass ratio is very different from unity. As an example, a schematic picture of a binary with a mass ratio 20:1 is shown in Fig. 4. The moment of inertia due to orbital motion is smaller than the primary's moment of inertia due to internal structure. The concept of Roche lobes is useful as long as we may assume that the two components rotate synchronously with their binary motion. The synchronism is easy to maintain if the internal stellar moments of inertia are small as compared to the orbital moment of inertia. This is certainly so when the two stars have comparable masses. It is no longer true when their mass ratio is 20:1. While the massive primary gradually expands, its moment of inertia increases. The tides due to the secondary component try to

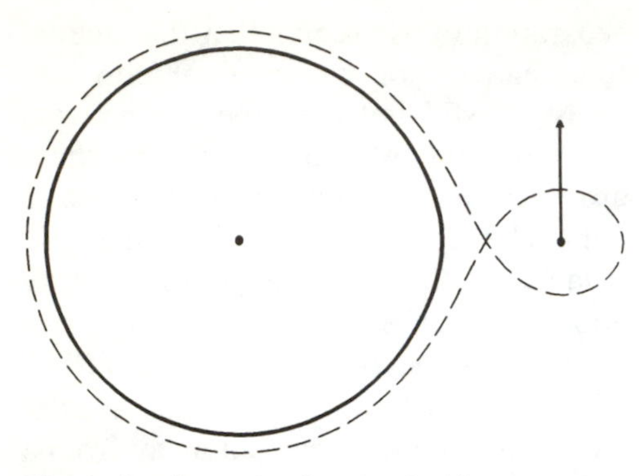

Fig. 4. A schematic picture of a binary system with a mass ratio 20:1. The big star, the more massive primary, almost fills its Roche lobe, the dashed line. A low-mass secondary is much smaller than its Roche lobe.

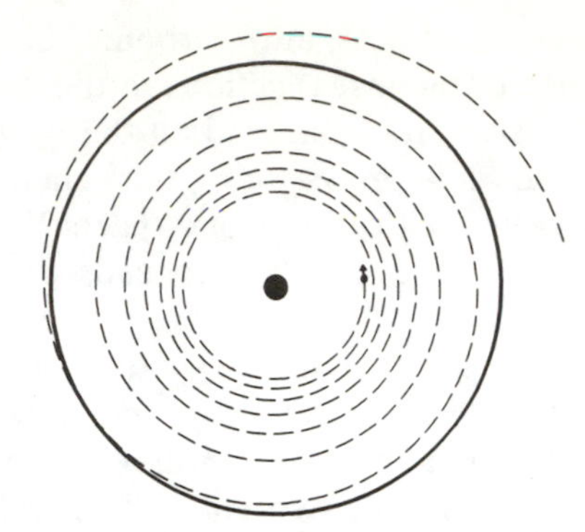

Fig. 5. A schematic picture of the evolution of a binary system with a very large initial mass ratio. The low-mass secondary spirals through the expanding common envelope toward the dense core of the primary.

enforce the synchronous rotation of the primary. A large part of the orbital angular momentum is used to spin up the expanding primary. This transfer of angular momentum forces the binary orbit to shrink. The secondary star spirals toward the primary, sinks in the extended envelope, and later spirals down toward the primary's core as shown in Fig. 5. Because of its angular momentum, the secondary cannot just fall straight down onto the core. While moving through the extended envelope, it experiences a strong drag. This drag transfers angular momentum from the secondary to the envelope, allows the secondary to spiral in, and dissipates a lot of energy.

We may expect two different outcomes from this type of evolution. In the first case the energy released during the spiral-in phase is so large that it cannot be radiated fast enough from the stellar surface. The common envelope first expands, and later it is driven away. The binary is left with much reduced mass and separation, and with a much smaller orbital angular momentum and period. The bulk of initial mass and angular momentum has been carried away with the expelled envelope. In the second case the energy released during the spiral-in phase is not sufficient to disrupt the envelope. The process continues until the secondary merges with the primary's core. The binary fuses, and a single star is left. It is very difficult to model this type of evolution, as the problem is really three-dimensional and time-dependent. However, it is to be expected that the first scenario may be valid when the secondary mass is not too small, whereas the second scenario is more likely when the secondary mass is really tiny. Ultimately, theory and observations should make more quantitative statements possible.

Binary evolution through a common envelope stage should be about as frequent as binary evolution for which Roche lobes and the mass transfer concept are useful. However, the common envelope scenario is much more difficult to deal with. It has been invented rather late, and it has not yet received sufficient study (*28*). Nevertheless, it is generally accepted that common envelope evolution is needed to explain the formation of cataclysmic binaries. Even though not a single case of a common envelope binary has been observed so far, there are binaries that have recently emerged from their common envelopes. These are the short-period systems discovered in the centers of some planetary nebulae (*29*). The nebulae are the original common envelopes ejected some 10,000 years ago. It is very likely that future progress in this area will be semiobservational and semitheoretical, with theory organizing a rich "zoo" of observed objects into a plausible evolutionary pattern.

References and Notes

1. H. A. Abt, *Annu. Rev. Astron. Astrophys.* **21**, 343 (1983).
2. V. Trimble, *Nature (London)* **303**, 137 (1983).
3. J. M. Retterer and I. R. King, *Astrophys. J.* **254**, 214 (1982).
4. J. V. Clausen, K. Gyldenkerne, B. Gronbech, *Astron. Astrophys.* **46**, 205 (1976).
5. D. M. Popper, *Astrophys. J.* **162**, 925 (1970).
6. ——, *Annu. Rev. Astron. Astrophys.* **18**, 115 (1980).
7. I. Iben, Jr., *ibid.* **12**, 215 (1974).
8. S. Chandrasekhar, *An Introduction to the Study of Stellar Structure* (Univ. of Chicago Press, Chicago, 1939); J. Liebert, *Annu. Rev. Astron. Astrophys.* **18**, 363 (1980).
9. R. N. Manchester and J. H. Taylor, *Pulsars*, (Freeman, San Francisco, 1977); W. Sieber and R. Wielebinski, Eds., International Astronomical Union Symposium No. 95, *Pulsars* (Reidel, Dordrecht, Netherlands, 1980).
10. J. N. Bahcall, *Annu. Rev. Astron. Astrophys.* **16**, 241 (1978).
11. P. C. Joss and S. A. Rappaport, *ibid.*, in press.
12. J. H. Taylor and J. M. Weisberg, *Astrophys. J.* **253**, 908 (1982).
13. S. L. Shapiro and S. A. Teukolsky, *Black Holes, White Dwarfs, and Neutron Stars* (Wiley, New York, 1983).
14. M. J. Rees, *Q. J. R. Astron. Soc.* **18**, 429 (1977); J. B. Hutchings, *Publ. Astron. Soc. Pac.* **95**, 799 (1983).
15. A. P. Cowley, D. Crampton, J. B. Hutchings, R. Remillard, J. Penfold, *Astrophys. J.* **272**, 118 (1983).
16. E. L. Robinson, *Annu. Rev. Astron. Astrophys.* **14**, 119 (1976).
17. J. S. Gallagher and S. Starrfield, *ibid.* **16**, 171 (1978).
18. J. Smak, *Acta Astron.* **17**, 255 (1967); J. Patterson, R. E. Nather, E. L. Robinson, F. Handler, *Astrophys. J.* **232**, 819 (1979).
19. J. E. Pringle, *Annu. Rev. Astron. Astrophys.* **19**, 137 (1981).
20. R. Hoshi, *Prog. Theor. Phys.* **61**, 1307 (1979); F. Meyer and E. Meyer-Hofmeister, *Astron. Astrophys.* **104**, L10 (1981); J. Smak, *Acta Astron.* **32**, 199 (1982); J. K. Cannizzo, P. Ghosh, J. C. Wheeler, *Astrophys. J.* **260**, L83 (1982); for a review, see J. Smak, *Publ. Astron. Soc. Pac.*, in press.
21. D. Sugimoto, D. Q. Lamb, N. D. Schramm, Eds., International Astronomical Union Symposium No. 93, *Fundamental Problems in the Theory of Stellar Evolution* (Reidel, Dordrecht, Netherlands, 1980).
22. B. Paczyński, paper presented at the First European Conference on Astronomy, Leicester, England, 1975 [reported by V. Trimble, *Q. J. R. Astron. Soc.* **17**, 31 (1976)].
23. E. W. Gottlieb, E. L. Wright, W. Liller, *Astrophys. J.* **195**, L33 (1975); E. B. Fomalont, B. J. Geldzahler, R. M. Hjellming, C. M. Wade, *ibid.* **275**, 802 (1983).
24. B. Margon, H. C. Ford, S. A. Grandi, R. P. Stone, *ibid.* **233**, L63 (1979); J. I. Katz, S. F. Anderson, B. Margon, S. A. Grandi, *ibid.* **260**, 780 (1982); K. Davidson and R. McCray, *ibid.* **241**, 1082 (1980).
25. H.-C. Thomas, *Annu. Rev. Astron. Astrophys.* **15**, 127 (1977).
26. F. H. Shu and S. H. Lubow, *ibid.* **19**, 277 (1981).
27. S. Refsdal, M. L. Roth, A. Weigert, *Astron. Astrophys.* **36**, 113 (1974).
28. J. P. Ostriker, personal communication; B. Paczyński, in International Astronomical Union Symposium No. 73 (Reidel, Dordrecht, Netherlands, 1976), p. 75; H. Ritter, *Mon. Not. R. Astron. Soc.* **175**, 279 (1976); R. E. Taam, P. Bodenheimer, J. P. Ostriker, *Astrophys. J.* **222**, 269 (1978); F. Meyer and E. Meyer-Hofmeister, *Astron. Astrophys.* **78**, 167 (1979); M. Livio, J. Saltzman, G. Shaviv, *Mon. Not. R. Astron. Soc.* **188**, 1 (1980); M. Livio, *Astron. Astrophys.* **105**, 37 (1982).
29. H. E. Bond, in *Cataclysmic Variables and Low-Mass X-Ray Binaries*, J. Patterson and D. Q. Lamb, Ed. (Reidel, Dordrecht, Netherlands, in press).

30. This work was supported by NSF grant AST-8317116.

This material originally appeared in *Science* **225**, 20 July 1984.

11. Dynamics of Globular Clusters

Lyman Spitzer, Jr.

Globular clusters are nearly spherical stellar systems associated with many galaxies and generally containing from 10^5 to 3×10^6 stars. Figure 1 is a photograph of Messier 19 (No. 19 in a catalog of some hundred diffuse objects compiled by Charles Messier late in the 18th century), a conspicuous such cluster in our own Galaxy, at a distance of some 3000 parsecs (1 parsec = 3.26 light years). Typically the relatively dense central core of a globular cluster, with a radius of about 1 parsec, contains some 10^4 stars, while the outer regions of the cluster extend with much diminished density out to distances of roughly 25 to 100 parsecs. Studies of the stellar spectra indicate that these systems within our own Galaxy were formed early in the life of the Universe, about 10^{10} years ago, not very long after the initial Big Bang.

The dynamical evolution of these beautifully symmetrical, very ancient systems has provided astrophysicists with an intriguing and challenging problem, which so far is only partly solved. Even in the cores, the average distance between neighboring stars is generally more than 10^4 times the radii of even the giant stars and direct collisions between stars are extremely rare. Thus the cluster stars move as mass points under their mutual gravitational attraction, with random velocities of some tens of kilometers per second. The large value of N, the total number of stars, can be expected to average out any large statistical fluctuations, and the way a spherical cluster evolves with time, as a result solely of Newton's laws of motion, appears deceptively simple.

Although the general principles underlying this evolution have been known for some time, it is only within the last decade that theoretical analyses, supported by high-speed computers, have provided a detailed understanding of the later evolutionary phases. As we shall see below, these involve actual collapse of the central core and lead to the occurrence of new physical processes not important at the earlier stages. During all this activity the outer regions of the cluster gradually expand. The evolution of the cluster in the post-collapse phase is an active research field. The x-ray sources observed in the cores of some of the more centrally condensed clusters may well result from processes occurring during and after the core collapse.

In this chapter, as in several general surveys (*1, 2*), first the physical principles affecting the early evolution of the cluster and the detailed evolutionary models based on these principles are outlined. Then additional physical processes that become important during the collapse phase, such as formation of binary systems, both by tidal capture in a close two-body encounter and by direct three-body encounters, are discussed.

Physical Principles

In discussing stellar motions in a globular cluster, we first separate the gravitational potential energy, $\phi(\mathbf{r}, t)$ into the sum of two terms. The first is a smoothed, spherically symmetric potential obtained by averaging $\phi(\mathbf{r}, t)$ over a time interval including several orbital periods of the stars. A star moving at 10 kilometers per second goes 1 parsec in 10^5 years, and the time required to travel back and forth across a cluster is generally less than 10^6 years, which in turn, is a small fraction of the evolution time. The average of $\phi(\mathbf{r}, t)$, over roughly 10^6 years we denote by $\phi_A(r)$, assumed to be spherically symmetric. This smoothed potential will change slowly as the cluster evolves. In a zero-order approximation each star moves in this spherical potential, with constant energy E and angular momentum $\mathbf{J}$, both measured per unit mass.

On this approximation no evolution occurs. A basic constraint on a cluster in this approximation is that the average smoothed stellar density, $\rho_A(r)$, must be consistent with Poisson's equation

$$\nabla^2 \phi_A(r) = 4\pi\rho_A(r) \quad (1)$$

where $\rho_A(r)$ is averaged over the same time interval used in determining $\phi_A(r)$. Many equilibrium solutions are possible.

The difference in potential $\phi(\mathbf{r}, t) - \phi_A(r)$ results from the granularity of the gravitational field. It is generally assumed that this granularity can be represented in a first approximation by two-body encounters between stars, and that the effects of such encounters in altering

Fig. 1. Photograph of globular cluster M19 (NGC 6273) with the 3.9-m Anglo-Australian telescope. [Courtesy of D. Malin]

E and $\mathbf{J}$ of each star can be computed as though the two stars involved were moving in a hyperbolic path relative to each other, unaffected by other stars. The effects of such encounters have been computed in detail (*3, 4*). The results can be used to follow the way in which the distribution of stars among different orbits is changed and thus how the cluster evolves. It should be emphasized that, for the cluster as a whole, the effects produced by stellar encounters occur very slowly in comparison with the time for a star to move across the cluster. Essentially the mean free path is many orders of magnitude greater than the dimensions of the cluster, and thousands of cluster crossings are required for appreciable evolution.

One important feature of two-body gravitational encounters is that the cumulative effect of many distant encounters, each of which produces only a small change in stellar velocities, tends to outweigh the less frequent close encounters, in which the stars are deflected by some 90° or more. Thus the velocity of a star is subject to a diffusion process, and similar diffusion occurs in E and $\mathbf{J}$. Changes in the velocity distribution of stars are governed by the integrodifferential Fokker-Planck equation (*5*).

While the details of these dynamical interactions are somewhat complex, the general physical tendency is clear. Encounters between stars will tend to increase entropy, evolving the stellar system toward a state of higher probability. The distribution of stars can be described by the density $f(\mathbf{r}, \mathbf{v})$ in phase space; $f(\mathbf{r}, \mathbf{v})$, multiplied by the phase space volume element $dxdydzdv_x dv_y dv_z$ is the number of stars within this volume element centered at $\mathbf{r}$, $\mathbf{v}$. In the local state of highest probability, toward which the cluster evolves, the phase space density $f(\mathbf{r}, \mathbf{v})$, which we designate simply by f, is given by

$$f = Ke^{-\beta E} \quad (2)$$

where E is the energy of each star per unit mass, β is inversely proportional to the average energy, and K is a normalization constant. For stars within a small region of space the potential energy is constant, and only the kinetic energy, $mv^2/2$, need be considered; Eq. 2 then gives f_M, the usual Maxwellian distribution function

$$f_M = \frac{n}{v_m{}^3}\left(\frac{3}{2\pi}\right)^{3/2} e^{-3v^2/2v_m{}^2} \quad (3)$$

where v_m is the root-mean-square (rms) velocity, and n is the number of stars per unit volume of physical space. Multiplication of f_M by $4\pi v^2 dv$ gives the number of stars per unit volume whose total velocity lies between v and $v + dv$.

It is this tendency toward a more probable state, as in thermodynamic equilibrium, which leads the cluster straight to catastrophe. The volume of accessible phase space per energy increment is greatest for stars which are at the greatest distances from the cluster center, especially those which escape the cluster entirely and have an entire galaxy to roam around in. On the other hand, f in Eq. 2 is maximized if some of the cluster stars are very close together, giving a large negative potential energy E. Thus velocity perturbations lead to an expansion of some regions of the cluster and contraction of others. Analysis of the various ways in which these processes occur in a spherical star cluster provides a challenging task, whose status is summarized in this chapter.

Catastrophes with Simple Models

The simultaneous processes of expansion and contraction to which star clusters are subject can be understood physically from very simple models. While precise numerical results can be obtained only from the realistic, detailed calculations discussed later in this chapter, these simple models are helpful in understanding and interpreting the more complex calculations. Three of these simple models, each of which leads the cluster to catastrophe in a different way, are presented below.

In the first model, discussed some 40 years ago (*6*, *7*), the cluster is regarded as a uniform sphere, whose density ρ and rms velocity v_m are constant. The total mass is M, and all stars are taken to have the same mass, m. We make use of the virial theorem, which states that for an isolated system of self-gravitating mass points in equilibrium

$$2T = Mv_m{}^2 = -W \quad (4)$$

where T is the total kinetic energy and W is the total gravitational energy. Thus the average kinetic energy per star is half the corresponding average gravitational binding energy. However, the average change of potential energy involved in removing one star initially from the cluster is twice the average potential binding energy of all the stars; this may be seen if one computes the energy required to disassemble the entire cluster—the energy required per star declines steadily as the remaining mass decreases, with the initial value twice the average value. It follows that the average energy for escape of the first few stars is not twice but four times the average kinetic energy; if we denote the escape velocity by v_{esc}, we obtain the general result for any isolated stellar system

$$<v_{esc}^2> = 4v_m^2 \equiv 4<v^2> \quad (5)$$

where the brackets denote average values over all the stars.

The Maxwellian distribution in Eq. 3 can be used to compute the fraction, ξ_e, of stars for which $v^2 > 4v_m^2$, giving $\xi_e = 7.4 \times 10^{-3}$. Encounters between stars will tend to establish a Maxwellian velocity distribution during some time interval, which is called the time of relaxation and is denoted by t_r. If we assume that for velocities exceeding $4v_m^2$, f approaches its Maxwellian value in the time t_r and that all particles escape if their kinetic energy exceeds four times the average, we obtain

$$\frac{1}{M}\frac{dM}{dt} = -\frac{\xi_e}{t_r} \quad (6)$$

For the relaxation time we adopt the value (*8*)

$$t_r = \frac{v_m^3}{15.4G^2m^2n \ln(0.4N)} \quad (7)$$

where v_m is again the rms stellar velocity, m the stellar mass, n the density of stars per unit volume, and $N = M/m$, the total number of stars in the system. The general form of Eq. 7 follows from the fact that the cross section for a 90° deflection in the relative orbit is of order $\pi(Gm/v^2)^2$; the numerical constant and the logarithmic term are obtained from the detailed theory of stellar encounters. For the uniform sphere considered here t_r is independent of position in the cluster.

The assumption that the fraction of stars escaping during the time t_r is given so directly by f_M is, of course, a simplification. A solution of the Fokker-Planck equation for a system of stars in a hypothetical square-well spherical potential (constant inside the cluster and zero outside) gives (*9*) Eq. 6 with the constant ξ_e now equal to 8.5×10^{-3}. The stars which diffuse to values exceeding v_{esc} leave the cluster with very little excess energy. As a result, the total energy of the cluster, proportional to M^2/R, where R is the cluster radius, remains constant as M decreases. Hence v_m^2 varies as $1/M$, n varies as M^{-5}, and Eqs. 6 and 7 may be integrated approximately to yield (*10*)

$$M(t) = M(0)\left(1 - \frac{7}{2}\frac{\xi_e t}{t_r(0)}\right)^{2/7} \quad (8)$$

where $M(0)$ and $t_r(0)$ are the initial values of M and t_r. Evidently, evaporation of stars produces a collapse of the cluster, with cluster mass M and radius R approaching zero together after a time interval equal to $2t_r(0)/7\xi_e$.

Equation 8 is applicable not only to this idealized homogeneous cluster but also to any cluster of constant total energy, E_T, which undergoes homologous contraction; that is, a cluster in which the smoothed density is a function of $r/$

$r_c(t)$, where $r_c(t)$ is some characteristic time-dependent cluster dimension, either the outer radius of a uniform cluster or the radius of a compact central core.

For homologous contraction the structure of the system, including the spatial variation of ρ and v_m, remains constant except for time-dependent scale factors. If the evaporating stars carry away appreciable energy, diminishing E_T, and if ζ is the ratio of the fractional loss of energy to the fractional loss of mass, then

$$\zeta = \frac{dE_T}{E_T dt} \bigg/ \frac{dM}{Mdt} \tag{9}$$

and Eq. 8 is replaced (*10*) by

$$1 - (3.5 - 1.5\zeta)\frac{\xi_e t}{t_r(0)} = \left[\frac{M(t)}{M(0)}\right]^{3.5 - 1.5\zeta} \tag{10}$$

In addition, r_c becomes proportional to $M^{2 - \zeta}$, and v_m to $M^{(\zeta - 1)/2}$.

We turn now to a second model, in which the cluster is replaced by an isothermal sphere, whose equilibrium structure has been extensively studied. Since in such a sphere $\rho(r)$ varies asymptotically as $1/r^2$, the mass is infinite if the radius is infinite. To give a model with finite mass, the sphere is truncated at some radius R with a hypothetical rigid, confining shell. For large R, the phenomena of interest occur well inside this confining surface, which does not much affect the results. Since the central regions of a cluster are in fact nearly isothermal, this model is much more realistic than the first.

This model is subject to the remarkable "gravothermal" instability (*11*), associated with the negative specific heat of self-gravitating stellar systems. According to the virial theorem in Eq. 4, the total energy $T + W$ is, of course, negative and equal to $-T$. Thus if the total energy is increased (becomes less strongly negative), T will decrease. For example, if a small satellite loses energy as it orbits around the Earth (from frictional retardation by the Earth's atmosphere), it spirals inward, accelerating its motion, so that the centrifugal force remains nearly in balance with the gravitational force.

Since Eq. 4 applies to isolated systems confined by their self-gravitational attraction, it is not strictly valid for a system confined by a rigid wall. Nevertheless, results based on this equation provide a good first approximation for the compact core of a bounded isothermal sphere, in view of the dominant self-attraction of this core.

Consequently, the core of an isothermal sphere can contract, heat up, and release energy, which flows to the outer regions. The outer regions, being less bound gravitationally, will tend to have a positive specific heat; but if the sphere is sufficiently condensed at the center, the core temperature will increase faster than the temperature of the outer regions, and the temperature gradient will increase, accelerating the core collapse. The rate of the collapse will be limited only by the rate at which heat can flow outward. Analysis shows that if the velocity distribution is nearly isotropic, the gravothermal instability can occur (*11, 12*) if the density at the center exceeds the density at the assumed bounding shell by a factor of 709.

A detailed time-dependent solution for such a collapsing sphere has been found (*13*) on the assumption that the contraction is homologous, as defined above. The result does not apply exactly to actual clusters, since the mean free path is assumed to be short, and the velocity distribution is consequently nearly isotropic. In fact, stars on radial orbits, which pass frequently through the heated collapsing core, will have a higher kinetic energy than stars in outer circular orbits, which are less immediately affected by the process of collapse. In the short-mean-free-path approximation, Eq. 10 is valid, with M replaced by M_c, the mass in the core, which remains essentially isothermal. The value of ζ is found to be 0.74, giving v_m varying very slowly with the core density ρ_c (as $\rho_c^{0.047}$). The density distribution outside the core differs slightly from that of an isothermal sphere in equilibrium, with ρ varying asymptotically as $r^{-2.21}$ instead of r^{-2}. While the inner regions have an inward velocity, the outer regions move outward, with the velocity vanishing at the radius where $\rho(r)/\rho(0) = 0.0071$. As we shall see below, the properties of this theoretical model are in general agreement with those obtained from more detailed, more realistic models.

A third simplified model considers effects associated with stars of two different masses, which tend toward equipartition of energy as a result of mutual encounters. In this model, the system of heavier stars must inevitably collapse if their relative number exceeds a small limiting value. We omit the detailed analysis but derive this result from simplified physical arguments. First we assume that $\rho_2(0)$, the smoothed density of heavy stars at the cluster center, is small compared to $\rho_1(0)$, the corresponding density for the lighter stars. We can then assume that the gravitational potential is entirely produced by the lighter stars. We take this potential, $\phi(r)$, to be zero at $r = 0$. Then in equilibrium, the radial distance attained by stars of each type is determined by the condition that the mean potential energy is proportional to the mean kinetic energy. If $\rho_1(r)$ is constant with r, the gravitational potential varies as r^2 according to Eq. 1, and we may write

$$\frac{v_{2m}^2}{v_{1m}^2} = k\frac{r_{2m}^2}{r_{1m}^2} \tag{11}$$

where v_{2m}^2 and v_{1m}^2 are the mean square velocities for stars of the two types, and r_{2m}^2 and r_{1m}^2 are the mean square distances from the center. The numerical constant k is needed because in determining r_{1m}^2 one cannot neglect the decrease of ρ_1 with increasing r, and the effect of this change on $\phi(r)$. This effect is negligible for the heavier stars, provided the mass m_2 of such a star appreciably exceeds m_1.

In equipartition, v_{2m}^2/v_{1m}^2 equals m_1/m_2. Evidently as m_2/m_1 becomes larger, the equilibrium condition (Eq. 11) requires that r_{2m}^2/r_{1m}^2 decrease; as the velocities of the heavier stars become smaller, because of equipartition, the radial distance out to which they can rise, against the gravitational attraction of the lighter stars, decreases in proportion.

However, equilibrium becomes impossible if the ratio $\rho_2(0)/\rho_1(0)$ becomes too great, since in this circumstance the self-attraction of the heavier stars becomes appreciable, and the value of v_{2m}^2 required for equilibrium consequently increases as r_{2m} decreases. Thus if the total mass M_2 of the heavier stars is sufficiently large compared to M_1, the total mass of the lighter stars, there is no equilibrium distribution of heavy stars in which v_{2m} is much less than v_{1m}. From Eq. 11, plus the assumed equipartition of kinetic energies, we find that

$$\frac{\rho_2(0)}{\rho_1(0)} = \kappa k^{3/2}\frac{M_2}{M_1}\left(\frac{m_2}{m_1}\right)^{3/2} \tag{12}$$

where κ is another numerical constant relating $\rho(0)$ to M/r_m^3. Determination of the critical value of $\rho_2(0)/\rho_1(0)$, below which equilibrium is possible, and of the constants κ and k shows (*14*) that $(M_2/M_1)(m_2/m_1)^{3/2}$ must be less than 0.16 for equilibrium. For higher values, the loss of kinetic energy to the lighter stars will lead to continuing contraction of a dense system of the heavier stars, which, as we have seen above, will heat up as they lose energy, another example of the negative specific heat of a self-gravitating system.

In the realistic models described below, the three effects shown here separately all occur together, each contributing to the cluster collapse.

Detailed Models of Globular Clusters

To follow the dynamical evolution of a spherical cluster a number of detailed numerical calculations have been made. While the procedures have varied, all have considered the motion of point-mass stars in the smoothed potential, ϕ_A, given by Eq. 1, with perturbations of these motions by two-body encounters. We discuss first the analyses of systems that are (i) isolated from other gravitating masses and (ii) composed of stars all of the same mass. While these two assumptions are unrealistic, they simplify the problem and provide a clear indication of the physical processes involved.

Two different approaches have been followed. In the first, the orbits of stars in the smoothed potential field are considered, and the changes in energy, E, and angular momentum, J, resulting from stellar encounters are considered. This approach has been adopted in Monte Carlo computations (*15, 16*), with a number of representative stars followed through time, with frequent small changes in E and J computed in accordance with the appropriate probability distributions. The Fokker-Planck equation, transformed to give the diffusion of stars in E, J space, has also been solved numerically (*17*). In the second approach (*8, 18*), the motions of 1000 representative stars in the potential field $\phi_A(r)$ are followed by numerical integration; frequent small changes in velocity, produced by two-body stellar encounters, are obtained with the usual Monte Carlo techniques. In both approaches, changes of the smoothed potential with changing density are, of course, taken into account.

The results obtained by these different methods are in close agreement. The various models show that whatever its initial origin, the spherically symmetrical system develops a core-halo structure, with a nearly isothermal central region surrounded by a halo in which the orbits are mostly radial. The resultant structure is shown in Fig. 2, where the computed values of the smoothed density, ρ, are plotted against radius, r (both in dimensionless units). This particular system began as a homogeneous sphere, shown by the dotted line, with all stars in circular orbits about the cluster center but with random orientation. Evidently for r less than about 50, the density profile in the evolved system is close to that of an isothermal sphere. At larger r, the orbits are more nearly radial, and the density approaches the theoretically anticipated (*19*) relation $\rho \propto r^{-3.5}$ for an isolated cluster, shown by the dashed line.

Before discussing further the results obtained with these numerical models, we introduce the reference relaxation time t_{rh}, which is a convenient measure

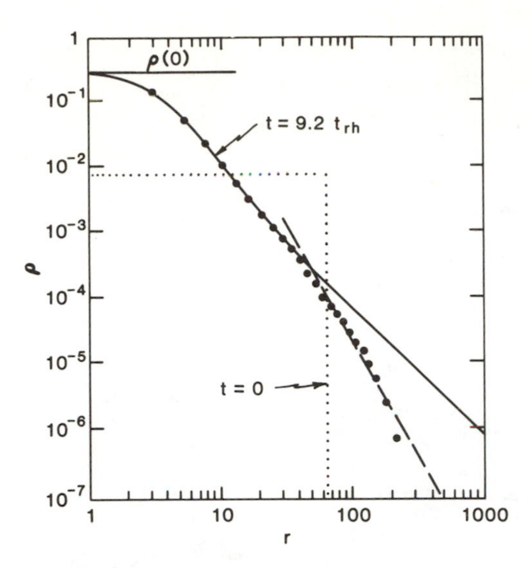

Fig. 2. Structure of an evolving globular cluster. The dotted line shows the initial density, ρ, as a function of radius, r, at $t = 0$, when the sphere is in equilibrium, with all orbits circular. The plotted points (*18*) show $\rho(r)$ at $t = 9.2\ t_{rh}$, where t_{rh} is the reference relaxation time (see text). The solid curve represents the theoretical relationship for an isothermal sphere, and the dashed straight line represents ρ varying as $r^{-3.5}$, the behavior predicted theoretically in the halo, where the orbits are predominantly radial.

of a cluster's evolutionary age; the quantity t_{rh} is defined as the value of Eq. 7 when $\rho \equiv mn$ is set equal to the mean density inside the radius r_h, containing half the cluster mass, and v_m is set equal to the rms velocity for the entire cluster. During the evolution of the cluster, t_{rh} usually remains relatively constant. If the virial theorem is used to equate $v_m{}^2$ and $0.4\ GM/r_h$ (a reasonably accurate approximation for $-W/M$), we obtain

$$t_{rh} = \frac{0.060\ M^{1/2} r_h{}^{3/2}}{mG^{1/2} \log(0.4N)} \tag{13}$$

If we express r_h in parsecs and m in terms of the solar mass, $M_{\odot}$, we obtain for t_{rh} in years

$$t_{rh} \frac{0.90(r_h)^{3/2} N^{1/2}}{(m/M_{\odot})^{1/2} \log(0.4N)} \times 10^6 \tag{14}$$

For most clusters t_{rh} is from 10^8 to 10^{10} years. For comparison, the ratio of the period of a circular orbit at the half radius, r_h, to t_{rh} equals $148 \log(0.4N)/N$ and is less than 1/147 for $N > 10^5$. This small ratio is an example of the general result, referred to earlier, that the mean free path for a star, before encounters strongly modify its velocity, much exceeds the dimensions of a typical cluster.

The system shown in Fig. 2 is rather advanced in its evolution, which has proceeded for a time interval $9.2\ t_{rh}$ since the origin of the uniform system at $t = 0$. As we see below, the collapse of the core to a central singularity occurs at $12.1\ t_{rh}$ for this particular model, only $2.9\ t_{rh}$ after the state shown in Fig. 2.

Another important characteristic of a cluster, in addition to its density profile, is the variation of the phase space density function $f(r, v)$ and, in particular, how this differs from Eq. 2, valid for an isothermal sphere. In the central regions, the velocity distribution is isotropic and f is a function of E only. Values of $f(E)$ near the center of a model cluster (*17*), relatively late in its evolution, are plotted in Fig. 3; both f and E are in dimensionless units. Since the relaxation time much exceeds the orbital period of a cluster star, the phase space density is relatively constant along each orbit in the cluster. Hence these values of $f(E)$ refer to all radial orbits, at any distance from the cluster center. The figure shows that, as expected, for appreciable $-E$, encounters between stars establish the exponential form of $f(E)$ in Eq. 2. However, $f(E)$ must clearly vanish for positive energy, since unbound stars escape

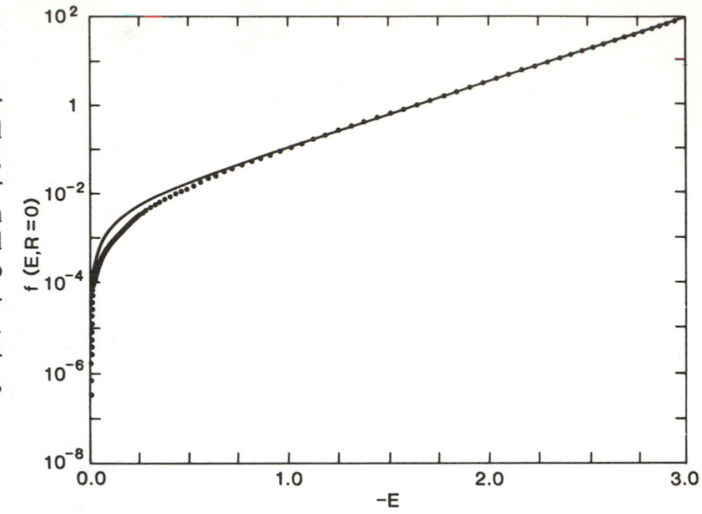

Fig. 3. Phase space density for radial orbits in an evolved cluster. The points represent f, the density in phase space for stars at zero distance R from the center of a highly evolved cluster (*17*). These values also apply to all radial orbits passing through the central core. The light upper curve represents a lowered Maxwellian, given by Eq. 15 with $E_0 = 0$.

rapidly. As shown by the light upper curve in Fig. 3, the computed values of f can be fitted reasonably well with the "lowered Maxwellian" distribution, given by

$$f(E) = K(e^{-\beta E} - e^{-\beta E_0}) \text{ for } E < E_0 \quad (15)$$

and vanishes otherwise. For an isolated cluster the energy E_0 at which $f(E)$ is taken to vanish is zero.

Last, we discuss the evolutionary changes shown by these numerically derived globular clusters. Not surprisingly, the numerical models demonstrate the three effects found earlier with simpler models—escape of stars, gravothermal instability, and mass segregation; the relative importance of each in the final collapse is demonstrated by these more realistic models. The observed rate of escape from the cluster may be used in Eq. 6, with the reference relaxation time t_{rh} replacing t_r, to determine an effective value of ξ_e. The escape of stars from an isolated cluster results from the energy change of a halo star, with an energy only slightly negative, in its passage through the high-density central region where encounters are important. Thus the period of these halo stars plays an important part in the escape rate. As a result, models for computing the diffusion of orbits in E, J space which ignore the values of the orbital periods give no escape at all from an isolated cluster. The models integrating the detailed equations of motion yield "observed" values of ξ_e of about 3×10^{-3}. Although this evaporation process, together with the gradual accumulation of stars in the far halo, is responsible for the initial contraction of the cluster core, as in the simple model, it can apparently not explain the later evolutionary stages of isolated clusters.

The nature of this final evolution is shown in Fig. 4, where the radii containing indicated percentages of the total initial mass are shown plotted against the time (*1*). The initial state of the system was the same uniform sphere in equilibrium whose structure is plotted in Fig. 2. The radii are expressed in units of r_h, the radius initially containing half the mass, while one unit of the dimensionless time

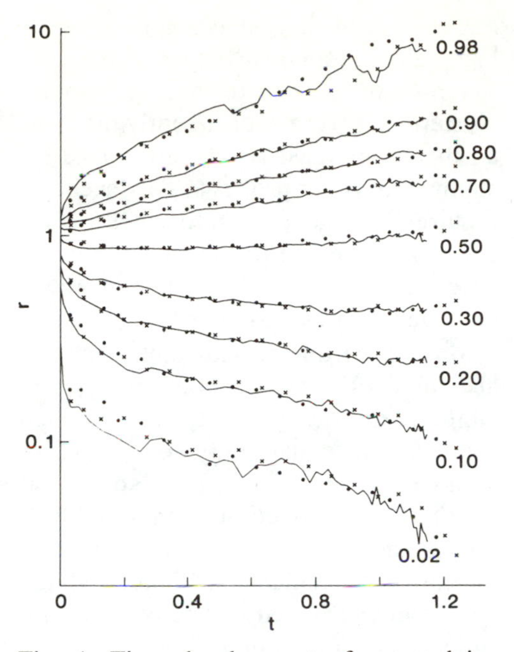

Fig. 4. Time development of an evolving globular cluster. The ordinate shows the values of the radius r, containing the fractions of the total mass on the right. The model is the same as the one portrayed in Fig. 2. The solid line represents the results of Monte Carlo computations based on the diffusion of orbits in E, J space, while the dots and crosses represent results obtained from direct integration of stellar orbits, with diffusion in velocity (*1*).

indicated equals about 9.1 t_{rh}. For comparison, extrapolation of the results indicates that the core approaches a singularity at a "collapse time," t_{coll} equal to 12.1 t_{rh}. The agreement between the solid line (obtained from the diffusion of orbits in E and J) and the plotted points (obtained from integration of dynamical trajectories) is excellent. The figure shows that the outer half of the cluster mass expands almost from the very beginning, while the radii containing less than half the mass first contract, then expand; the late expansion for the radii containing 2 and 10 percent of the mass is shown in calculations (*16*) that extend closer to the final collapse.

The general evolutionary behavior of these models is remarkably close to the behavior predicted for the gravothermal instability in a gaseous sphere. In the contraction of the central core, r_c^2 is observed to decrease nearly linearly with $t_{coll} - t$, permitting an accurate determination of the time t_{coll} at which the density becomes infinite; in the theoretical model $r_c^{1.9}$ varies linearly with $t_{coll} - \mathrm{t}$. Similarly, v_m increases as $\rho_c^{0.05}$, in agreement with a theoretical variation as $\rho_c^{0.047}$. Finally the density at the radius where contraction stops and expansion begins is about two orders of magnitude less than the central density, as compared with the theoretical ratio 0.0071 noted above. Exact agreement is not to be expected, in view of the approximations in the instability theory, especially the assumption of a short mean free path and a consequently isotropic velocity distribution. However, there seems little question that the collapse found in the model clusters must be due to the gravothermal instability of an isothermal sphere.

The numerical models also show the expected mass segregation when stars of differing masses are assumed to be present. This segregation occurs relatively rapidly, within a time of 1 to 2 t_{rh}. For example, a two-component model was computed with a stellar mass ratio of 5 to 1, and with 10 percent of the cluster mass in the heavier stars, uniformly distributed initially. After a time interval of only 0.81 t_{rh}, at the cluster center the more massive stars provide 62 percent of the smoothed stellar density, a dramatic increase in the relative densities in the two stellar components. While the initial collapse rate for these models results from the tendency towards equipartition, at a later time this process slows down, since the relative number of lighter stars near the center becomes progressively smaller. Hence it seems likely, though not yet proven, that the final collapse of multicomponent models is due to the gravothermal instability.

Several other effects must be taken into account before a detailed comparison can be made between any of these models and the observations. The most important of these is the gravitational force of the Galaxy. The tidal force produced by the mass in the inner regions of our Galaxy can draw some stars out of globular clusters if their distance from the cluster center exceeds the "tidal cutoff," r_t. While the dynamics of the cluster stars in the presence of such a tidal force form a complex problem, a simple first approximation is that the cluster remains spherical, with vanishing density for $r > r_t$. A theoretical model for such a system can be computed (*20*) if a lowered Maxwellian distribution function, given in Eq. 15, is assumed, with E_0 (per unit stellar mass) set equal to $-GM/r_t$, where M is again the cluster mass. Such "King models" have been widely used for comparison with observed cluster data; with two parameters, r_c and r_t, they usually provide a good fit to the observed surface density profiles. The evaporation probability ξ_e per time interval t_{rh} for these tidally truncated models can exceed by more than an order or magnitude the corresponding value for an isolated cluster, with a major effect on the evolution.

Other important physical effects are the gravitational perturbations produced when a cluster crosses the galactic plane; such perturbations heat the cluster, leading to the escape of halo stars and usually (if somewhat paradoxically) a more rapid collapse of the central core (*1*). It has even been suggested that most of the high-velocity subdwarf stars in the Galaxy may have been formed in globular clusters that were subsequently dissipated by the increased exporation rate resulting from this process. This scenario requires (*21*) that most of these early clusters had mean densities one or two orders of magnitude smaller than those observed in present clusters.

Mass loss from individual cluster stars also affects the dynamics of the system, although at the present epoch, when giant stars lose mass only shortly before their death, this effect may be somewhat minor. The presence of a massive black hole at the center would certainly affect the cluster dynamics (*2, 16*), although the x-ray evidence discussed below does not support this possibility. Finally, binary star effects may be important.

Formation and Evolution of Binaries

Binaries are potentially very important in cluster evolution because they can give up energy to passing stars and become more and more tightly bound. The energy available is more than enough to slow down or even reverse the core collapse discussed above. If we regard the two stars in a binary system as mass points, with masses m_A and m_B, the total energy, denoted by $-x$, can be written as

$$x = Gm_Am_B/2a \qquad (16)$$

where a is the semimajor axis of the binary orbit. The factor 1/2 in Eq. 16 results from Eq. 4, according to which the average kinetic energy is half the average negative gravitational energy.

When a single star encounters a binary, the net binding energy, x, may be changed. The result depends critically on whether the mean stellar kinetic energy of translation is large or small compared to x. If the former, the binary is called soft; the velocities of the stars in the binary orbit are less than those of passing stars, and one would expect from equipartition arguments that encounters will impart energy to the binary, decreasing x on the average (*22*). If x exceeds the mean stellar kinetic energy, the binary is called hard; the equipartition argument is now less directly applicable, since a passing star is itself accelerated as it approaches the binary, but x will, in fact, increase on the average (*23, 24*). Thus there is a "watershed" value of x, comparable with the mean kinetic energy of single stars; binaries with greater x will become more tightly bound, on the average, giving up energy to the system.

While the energy absorbed by soft binaries is negligibly small, that given up by hard binaries may strongly influence cluster evolution. In any one encounter, a hard binary may change its binding energy by any amount, from zero to infinity, but on the average the rate of change of x is given by

$$<\frac{dx}{dt}> = \frac{4 \times 3^{1/2}A}{35} \times \frac{G^2 m_s{}^3 n_s}{v_{sm}} \quad (17)$$

where m_s is the stellar mass, here assumed the same for all stars, while n_s and v_{sm} are the particle density and the random rms velocity of the single stars. The constant A, found by averaging some 10^6 numerical orbits, is between 30 and 35 for x someone to two orders of magnitude greater than $mv_{sm}{}^2/2$ (*25*). While $<dx/dt>$ is nearly independent of x, the increase of x per close encounter averages $0.4x$. As a result, when a binary gradually hardens, it loses energy less and less frequently but in progressively larger increments. When x becomes substantially greater than the energy required to escape from a globular cluster, interaction with a passing star can result in the ejection of both the star and the binary.

Hard binaries, with their important dynamical effects, may appear in globular clusters through three different routes. (i) They may be primordial—that is, present in the initial stellar population from which the cluster formed. (ii) They may be formed by three-body encounters between cluster stars. (iii) They may be produced by dissipative two-body collisions between these stars. Detailed computations of formation rates have been carried out for the second (*24*) and third (*26*) of these processes. For globular clusters these formation rates are negligible under normal conditions but can become important during core collapse.

The effect of primordial binaries on the evolution of a cluster has been analyzed in several models. Computations for a cluster in a square-well potential, the simple model used originally in computation of the evaporation rate, show (*27*) that the energy released from hard binaries will about cancel the contraction produced by evaporation if N, the total number of stars, is in the range 10^4 to 10^5, and if about 35 percent of the mass of the system is in binaries, with all stars of the same mass. However, this result is not supported in a group of Monte Carlo models that consider the detailed density profile of the cluster (*28*). In these models, the binaries, each with twice the mass of a single star, settle toward the center and outnumber single stars in the inner regions, driving the mass segregation instability discussed above. Binary-binary reactions become dominant and have been taken into account approximately. The energy released by binaries is mostly carried away by energetic reaction products, which travel away from the center and give up relatively little of their energy to the central core. As a result, the cores finally collapse, even if 50 percent of the system mass is in primordial binaries. The final stage of this collapse may result from the gravothermal instability shown in systems of single stars.

The details of the evolution shown by these models would certainly have been altered if a distribution of stellar masses had been taken into account. In particular, exchange reactions, in which a heavy single star displaces a lighter star in a binary system, will occur frequently (*29, 30*). Even if only a few binaries are present in the core, these will probably end up containing the heaviest stars, or possibly some even more massive black holes, formed long ago by supernovae.

Unfortunately the relative number of hard primordial binaries in globular clusters is uncertain. The orbital velocity in a hard binary should substantially exceed the random velocities of single stars, and the variable Doppler shift should be easily measurable spectroscopically with modern techniques. In a recent study (*31*) of the globular cluster M3, 33 stars (all of them red giants, the brightest cluster stars) were measured two or more times and showed no variation of radial velocity significantly greater than the measuring error of about 1 km/sec. There is a similar scarcity (*32*) of spectroscopic binaries for other old stars distributed through the galactic disk. Like the globular clusters themselves, these stars were formed early in the life of our Galaxy, and have large velocities, over 100 km/sec relative to the Galaxy.

The chief evidence for the presence of some binaries in globular clusters is the presence of x-ray sources in several of these systems. In the galactic disk, most strong x-ray sources are believed to be

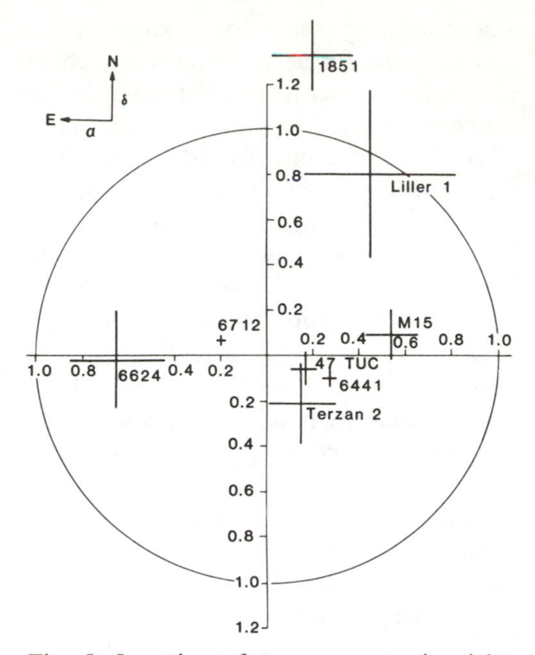

Fig. 5. Location of x-ray sources in eight globular clusters. The circle defines one core radius (see text). The plotted points (*33*) show preliminary Einstein Observatory results for the position of the observed strong sources; the error bars show approximate $1 = \sigma$ values. Final positions (*34*) to be published.

binaries, with gas expelled from a relatively normal star falling on its dense companion, generally a neutron star. The position of eight high-luminosity x-ray sources with respect to the visible cores of globular clusters is shown (*33*) in Fig. 5. The circle represents the radius at which the apparent surface brightness in each core is half its central value. If the visible stars are assumed all to have the same mass (0.8 $M_{\odot}$ for the giant stars in question), the distribution of source positions is consistent (*34*) with a mass of about 1.5 ± 0.5 $M_{\odot}$, supporting the assumption that these are normal x-ray binaries.

In addition to these strong sources, which are believed to be neutron stars, with normal less massive stars as components, weaker x-ray sources, with luminosities less by some three orders of magnitude, are also seen in several clusters (*35*). Some of these are found well outside the cluster cores, suggesting a binary mass less than 1 $M_{\odot}$. The evidence suggests that these sources are binaries in which gas from a normal dwarf star falls onto a lower-mass white dwarf. Thus, some binaries containing neutron stars and some containing white dwarfs seem to be present in a number of globular clusters.

As we shall see later these relatively compact binaries may have been formed during core collapse, especially since the neutron-star binaries are found mostly in the most compact clusters. Thus they provide no evidence for the presence of primordial binaries. While it seems unlikely that primordial binaries will avert the initial collapse of cores, binaries formed early or late in the cluster's history almost certainly play a major role in the final stages of this collapse, a complex process to which we now turn.

To Collapse and Beyond

The final stages of core collapse and the subsequent dynamical fate of the surrounding cluster have been explored only partially. The discussion of this fascinating problem must rely partly on approximate models with homogeneous spherical cores and partly on some provisional, more realistic, calculations.

Interesting guidance is provided by precise dynamical computations for systems with a small value of N (*36*). For a cluster with 250 stars and a distribution of stellar masses, binaries form by three-body encounters in the contracting core, and two of the most massive stars generally end up as a central binary, whose binding energy may exceed half the total binding energy of the cluster. The energy released by the hardening of this massive binary goes into the expansion of the cluster, with escape of many stars. The system gradually dissipates at a slower and slower rate as its density falls. In the final, most probable, accessible state—beyond the range of the computations—the central binary is presumably surrounded by an unbound, expanding aggregation of single stars, binaries, and perhaps some triple systems.

The effect of three-body binary formation on a system with a large N has been computed (*37*) in some detail with a Monte Carlo program, following the diffusion of particle orbits in E and J. Among the modifications made to earlier programs of this type were the following: assumption of a suitable distribution of stellar masses; inclusion of binary formation by three-body encounters (*24*); dynamical integration of encounters between single stars and binaries; inclusion of binary-binary encounters (*28*); and modification of the E, J diffusion calculations to give an approximate value for the escape rate. The results show that binaries are formed in the very central region, harden and then escape because of their recoil energy. The central density reaches a maximum, about 10^5 to 10^6 times its initial value, and then decreases systematically. The total number of binaries also decreases after peak density is reached. The general behavior is qualitatively similar to that of the small N systems. Further studies of such models will be needed to clarify the details of what is happening.

Binaries formed by tidal capture are also important during the collapse phase. The process involved is a close encounter between two stars, with a distance of closest approach between stellar centers equal to roughly three stellar radii. The tides each star produces in the other absorb sufficient energy that the two stars are bound in a highly elliptical orbit. At each successive close approach, further energy is lost, decreasing the semimajor axis of the orbit and the eccentricity, until the orbit becomes circular with a radius roughly twice the original distance of closest approach.

The theory of tidal capture receives some confirmation from the presence of x-ray sources in globular clusters. In particular, the number of weaker sources can be accounted for by this process (*35*) if white dwarfs, thought to be the compact objects in these systems, constitute about 10 percent of the stellar population. The number of binaries containing neutron stars, and constituting the stronger sources, seems also consistent with preliminary calculations based on the tidal capture theory. However, a more conclusive comparison of theory with observation requires more detailed models, taking into account the strong concentration of the relatively massive neutron stars toward the cluster center and the concentration of neutron stars in binaries as a result of exchange reactions with lower-mass binaries.

A binary produced by tidal capture will have an orbital velocity exceeding 100 km/sec. Such a binary is so hard that if it interacts with a single star, both the binary and the star will likely be ejected from the cluster with appreciable energy, and only a small part of the reaction energy will be available to heat the cluster by reducing the mass M of the bound system. On the other hand, tidal dissipation involves a loss of translational kinetic energy by the interacting stars (including the interactions that dissipate only a fraction of this kinetic energy) and therefore cools the system, accelerating the collapse. Detailed calculations (*38, 39*) indicate that this cooling by two-body tidal interactions tends to exceed the heating by three-body binaries, at least during the early stages of collapse.

There are no detailed models that take the spatial structure of the cluster into account and that include the effects of tidal energy loss and tidal capture. Hence the discussion of these effects must be tentative. The newly formed binaries, each with a mass of two stars, will tend to settle to the center of the collapsing cluster. At sufficiently high densities they will interact with each other and with single stars, and will be ejected. This loss of mass from the center of the system tends to produce an expansion of the system, offsetting, perhaps, the tidal cooling. Formation of quadruple systems, with two very hard binaries bound together, will also occur through encounters involving two of these binaries and a third mass. Direct collisions of stars will also occur (*2*), leading either to single stars or to contact binaries, with two stellar nuclei in a common envelope. Some black holes may be formed by coalescence of several stars.

While we do not yet know which processes are most important late in the collapse of a globular cluster, the accelerating collapse must clearly end. Certainly the gravothermal instability cannot produce continued collapse when there are only a handful of stars in the dense central core. The equations used for deriving this instability, either in the short- or long-mean-free-path limit, are not applicable when the number of stars in the core is as small as 10 to 100. After the collapse terminates, whatever processes occur near the center will presumably maintain a quasi-steady state, with a continuing release of energy from binaries leading to a gradual expansion of the cluster as a whole.

The nature of this post-collapse phase has been explored in some half dozen investigations, most of them based on the assumption that some unspecified source of energy is available at the center. Solutions involving homologous expansion have been obtained, again assuming stars all of the same mass. These solutions are very similar to those for collapse, except that conditions are more nearly isothermal (*40*). In one investigation (*41*) large-amplitude oscillations were found under some conditions, with the cluster expanding after collapse to a normal pre-collapse configuration, and then repeating the earlier gravothermal collapse.

The nature of the collapse and post-collapse phases is likely to be relevant for at least some globular clusters. According to the model computations, the time remaining at any instant until the singularity is reached equals about 200 t_{r0} at that instant, where t_{r0} is the central relaxation time. A detailed tabulation of globular cluster properties (*42*) shows that of 41 values of t_{r0}, 7 are less than 10^8 years; 4 of these are less than 3×10^7 years. The distribution of these t_{r0} values among clusters is consistent with the view that a few of these clusters have already collapsed and are now expanding (*43*). However, these are not readily distinguished from clusters which are still collapsing.

In any case clusters not far from collapse should show a density increase far into the cluster. In fact, precise profile measurements (*44*) have recently shown that several compact clusters do show radial density gradients extending into radii of about 1 arcsecond, the limit set by variable refraction through our inhomogeneous atmosphere. In a few years, the Hubble Space Telescope, yielding 0.1-arcsecond images, may give more conclusive data on the inner structure of these systems and should help to unveil the final evolutionary history of globular clusters.

References and Notes

1. L. Spitzer, in *Dynamics of Stellar Systems* (IAU Symposium Number 69), A. Hayli, Ed. (Reidel, Dordrecht, Netherlands, 1975), p. 3.
2. A. P. Lightman and S. L. Shapiro, *Rev. Mod. Phys.* **50**, 437 (1978).
3. S. Chandrasekhar, *Principles of Stellar Dynamics* (Univ. of Chicago Press, Chicago, 1942).
4. ———, *Astrophys. J.* **97**, 255 (1943).
5. M. N. Rosenbluth, W. M. MacDonald, D. L. Judd, *Phys. Rev.* **107**, 1 (1957).
6. V. A. Ambartsumian, *Ann. Leningrad State Univ. (Astron. Ser.), No. 22* (1938); *Tr. Astron. Obs.*, Issue 4, p. 19.
7. L. Spitzer, *Mon. Not. R. Astron. Soc.* **100**, 396 (1940).
8. L. Spitzer and M. H. Hart, *Astrophys. J.* **164**, 399 (1971).
9. L. Spitzer and R. Harm, *ibid.* **127**, 544 (1958).
10. I. R. King, *Astron. J.* **63**, 114 (1958).
11. V. A. Antonov, *Vestn. Leningr. Univ. Mat. Mekh. Astron.* **7**, 135 (1962).
12. D. Lynden-Bell and R. Wood, *Mon. Not. R. Astron. Soc.* **138**, 495 (1968).
13. D. Lynden-Bell and P. P. Eggleton, *ibid.* **191**, 483 (1980).
14. L. Spitzer, *Astrophys. J. Lett.* **158**, L139 (1969).
15. M. Hénon, *Astrophys. Space Sci.* **13**, 284 (1971); *ibid.* **14**, 151 (1971).
16. A. B. Marchant and S. L. Shapiro, *Astrophys. J.* **239**, 685 (1980).
17. H. Cohn, *ibid.* **234**, 1036 (1979).
18. L. Spitzer and T. X. Thuan, *ibid.* **175**, 31 (1972).
19. L. Spitzer and S. L. Shapiro, *ibid.* **173**, 529 (1972).
20. I. R. King, *Astron. J.* **71**, 64 (1966).
21. J. P. Ostriker, L. Spitzer, R. A. Chevalier, *Astrophys. J. Lett.* **176**, L51 (1972).
22. L. E. Gurevich and B. Yu. Levin, *Astron. Zh.* **27**, 273 (1950); *NASA Technical Translation TT F-11, 541* (1968).
23. D. C. Heggie, in *Dynamics of Stellar Systems* (IAU Symposium Number 69), A. Hayli, Ed. (Reidel, Dordrecht, Netherlands, 1975), p. 73; see also the discussions on pp. 93 and 94.
24. ———, *Mon. Not. R. Astron. Soc.* **173**, 729 (1975).
25. P. Hut, *Astrophys. J. Lett.* **272**, L29 (1983).
26. W. H. Press and S. A. Teukolsky, *Astrophys. J.* **213**, 183 (1977).
27. J. G. Hills, *Astron. J.* **80**, 1075 (1975).
28. L. Spitzer and R. Mathieu, *Astrophys. J.* **241**, 618 (1980).
29. J. G. Hills, *Astron. J.* **82**, 626 (1977).
30. ——— and L. W. Fullerton, *ibid.* **85**, 1281 (1980).

31. J. E. Gunn and R. F. Griffin, *ibid.* **84**, 752 (1979).
32. H. Abt, *ibid.*, p. 1591.
33. J. E. Grindlay, in *X-Ray Data with the Einstein Satellite*, R. Giacconi, Ed. (Astrophysics and Space Science Library, Reidel, Dordrecht, Netherlands, 1981), vol. 87, pp. 79–109.
34. ——, P. Hertz, J. E. Steiner, S. S. Murray, A. P. Lightman, *Astrophys. J. Lett.* **282**, L13 (1984).
35. P. Hertz and J. E. Grindlay, *ibid.* **267**, L83 (1983).
36. S. J. Aarseth and M. Lecar, *Annu. Rev. Astron. Astrophys.* **13**, 1 (1975).
37. J. S. Stodolkiewicz, *Publ. Astron. Inst. Czechoslovak Acad. Sci.* **56**, 103 (1983).
38. L. M. Ozernoy and V. I. Dokuchaev, *Astron. Astrophys.* **111**, 1 and 16 (1982).
39. S. Inagaki, *Mon. Not. R. Astron. Soc.* **206**, 149 (1984).
40. —— and D. Lynden-Bell, *ibid.* **205**, 913 (1983).
41. E. Bettwieser and D. Sugimoto, *ibid.* **208**, 493 (1984).
42. C. J. Peterson and I. R. King, *Astron. J.* **80**, 427 (1975).
43. H. Cohn and P. Hut, *Astrophys. J. Lett.* **277**, L45 (1984).
44. S. Djorgovski and I. R. King, *ibid.*, p. L49.
45. Grateful acknowledgements are due to various colleagues, especially J. Goodman, D. C. Heggie, P. Hut, J. P. Ostriker, and S. Tremaine, for helpful comments on this paper and to D. Malin for the photograph of M19, obtained with his unsharp masking technique to show both the bright core and the faint outer region.

This material originally appeared in *Science* **225**, 3 August 1984.

12. The Magnetic Activity of Sunlike Stars

Arthur H. Vaughan

The roughly 11-year periodicity in the appearance of sunspots was first recognized about 140 years ago by the German amateur astronomer Heinrich Schwabe. This discovery met with little notice for nearly a decade until Alexander Humboldt, "next to Napoleon Bonaparte the most famous man in Europe" (*1*), drew attention to Schwabe's work and pointed out a corresponding cyclic trend in the amplitudes of small daily variations of the magnetic field at the surface of Earth (*2*). In 1908 the astronomer George Ellery Hale showed, from observations of the Zeeman splitting of lines in the Sun's spectrum, that sunspots are regions of intense magnetic field, several thousand times stronger than Earth's field. He found that sunspots usually appear in pairs of opposite magnetic polarity and that in the hemisphere above the solar equator the leading spots in the direction of solar rotation nearly always have one polarity, opposite to that of the leading spots in the hemisphere below the equator. In the course of his studies from 1912 to 1925 (*3*), Hale showed that in successive sunspot cycles the sense of the polarity of the leading spots in the two hemispheres is interchanged, so that the solar cycle is actually a magnetic cycle with a period of about 22 years.

A great deal of knowledge about magnetic fields on the solar surface has been gained from decades of intensive observation of the Sun, the only star whose surface we can examine in detail. But, as we approach the end of the 20th century, the ultimate origin of the cycle remains an intriguing secret. There is reason to expect that future understanding will be guided by a study in which stellar analogs of solar activity have been kept under consistent observation for the past 18 years. As the study continues, important details are being added to what is known about the stars under observation (*4*).

A wide variety of phenomena on the Sun wax and wane in synchronism with the pattern of the cycle, including the excitation of the chromosphere (Fig. 1a)—a hot outer layer of the Sun's atmosphere first described and named for its "pinkish" color by the English astronomer Norman Lockyer in 1869. From spectroscopic evidence it has long been known that chromospheres are a feature of sunlike, or dwarf, stars less massive than about 1.2 solar masses, as well as of virtually all giants and supergiants of spectral type G0 and cooler. The chromospheres of these classes of stars cover a wide range of excitation. It was appar-

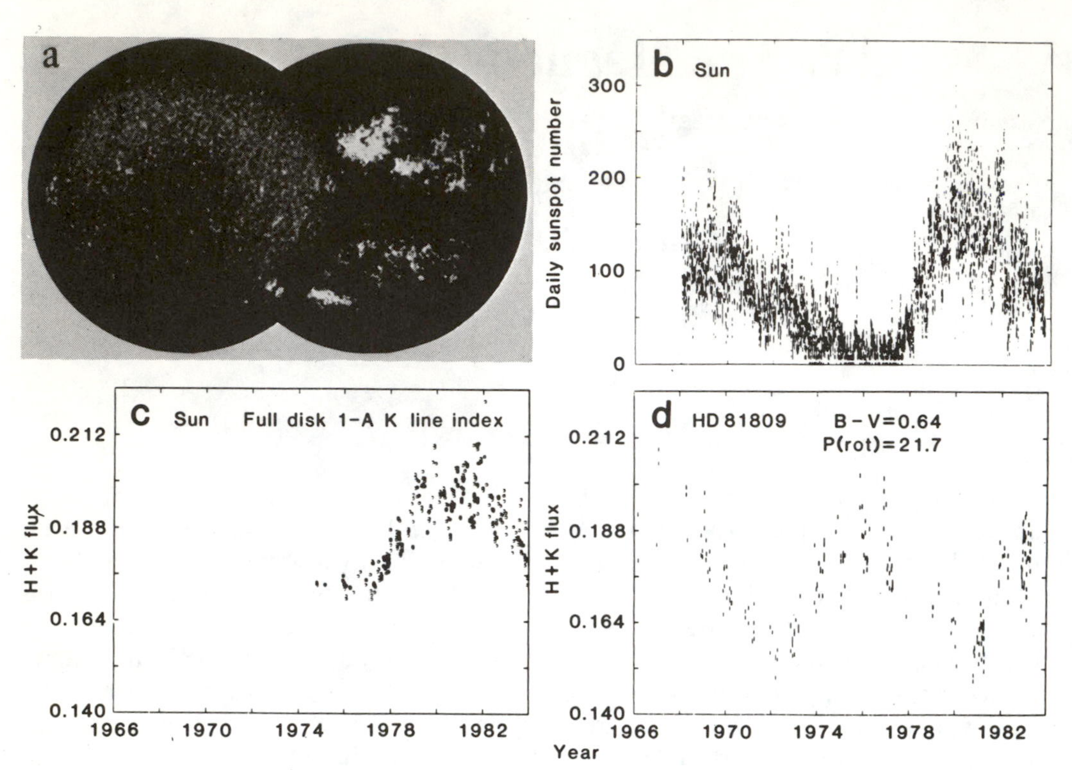

Fig. 1. (a) Chromospheric active regions on the solar surface photographed at the wavelength of the calcium K line near minimum in January 1976 (left) and in April 1979 (right) near the last maximum of solar activity. The bright patches are known to be associated with regions of enhanced magnetic flux on the solar surface. (b) Daily values of the international sunspot index from 1968 to 1983. (c) An index of full-disk calcium K line emission measured at Kitt Peak National Observatory with a solar K line photometer (*22*) in the same period, converted to the Mount Wilson scale. (d) Measurements of an analogous index of H + K line emission for the star HD 81809 (see text). This record exhibits variations similar to those of the Sun over a corresponding time interval. The Sun's activity cycle observed in this way would be clearly discernible at stellar distances.

ent even in the 1930's that detection of stellar analogs of the solar cycle would be of great interest and that a search might best be undertaken by looking for a spectroscopic signature of the chromospheric variations that conceivably would occur in the course of time in such stars (*5*). But an investigation of this kind became practical only with the development of a suitably precise photoelectric measuring technique whose calibration could be maintained over a period of many years.

Such a long-term study was initiated in mid-1966 at the Mount Wilson Observatory in California by O. C. Wilson. From 11 years of observation of 91 stars at monthly intervals he showed that variations in the chromospheric excitation of sunlike stars can be observed and systematically studied and that many such stars exhibit cyclic changes strongly resembling the activity cycle of the Sun (*6*). This discovery was of revolutionary importance in the field of astronomy concerned with explaining the solar cycle, a field in which progress had long been hindered by the fact that knowledge of the phenomenon was limited to the single example provided by the Sun. Moreover, the discovery occurred in an era of newly acquired ability to make very large-scale numerical computations, which had led to renewed interest on the part of theoreticians in modeling the complex processes underlying the solar cycle. The past decade has also seen remarkable advances in our knowledge and understanding of the atmospheric physics of sunlike stars and the Sun itself, brought about by observations in the ultraviolet and x-ray spectral regions made with spaceborne observatories.

Differential Rotation

No one would have sought to discover evidence for the existence of magnetic activity in stars if it were not for the example of the Sun, which must remain the source of all detailed knowledge of the phenomenon. From solar studies extending over centuries [for a recent general review, see (*7*)] it is evident that solar activity manifests a process that is organized on a grand scale.

Many of the most important features of the pattern had become evident before the end of the 19th century. Among the most striking features is the tendency of the sunspot zone to drift toward the equator in the course of each cycle and for the first spots belonging to a new cycle to appear at high solar latitudes even before the last spots of the previous cycle have dissipated. The trend is especially evident in an updated version of the "butterfly" diagram, first published by E. W. Maunder in 1922 (see Fig. 2b). From such a diagram it is apparent that successive cycles sometimes differ markedly from each other in intensity and duration. This is also evident from an examination of the mean annual number of sunspots over an interval of not quite four centuries (Figs. 1b and 2a).

By using sunspots as tracers, Galileo was able to show in 1610 that the Sun rotates with a period of about 25 days. Scheiner noted in 1630 that low-latitude spots indicate a more rapid rotation than high-latitude spots: in other words, the Sun does not rotate as a rigid body. This effect, now referred to as differential rotation, was thoroughly and systematically studied in the 19th century by Carrington (*8*).

Modern investigators use a variety of tracers as well as highly sensitive spectroscopic measurements of the Doppler effect to study the differential rotation of the Sun (*9*). The precise magnitude and latitude dependence of the differential rotation is found in general to depend upon the method of measurement, and this is partly because the differential rotation is a function not only of latitude but also of depth in the Sun's outer layers. Systematic deviations in the photospheric pattern of differential rotation occur in synchronism with the 22-year magnetic cycle (*10*). At a fixed latitude, the velocity in the direction of rotation of the Sun varies by an average of about 5 meters per second from the mean velocity in a period of about 11 years. At any given time, the velocity deviation is found to be a function of latitude, symmetrical in both hemispheres. With the passage of time a given zone of excess velocity, and the corresponding zone in the opposite hemisphere, drift toward the equator as traveling "waves." There are two cycles of the wave present in each hemisphere at all times, so that the time required for a wave to travel from

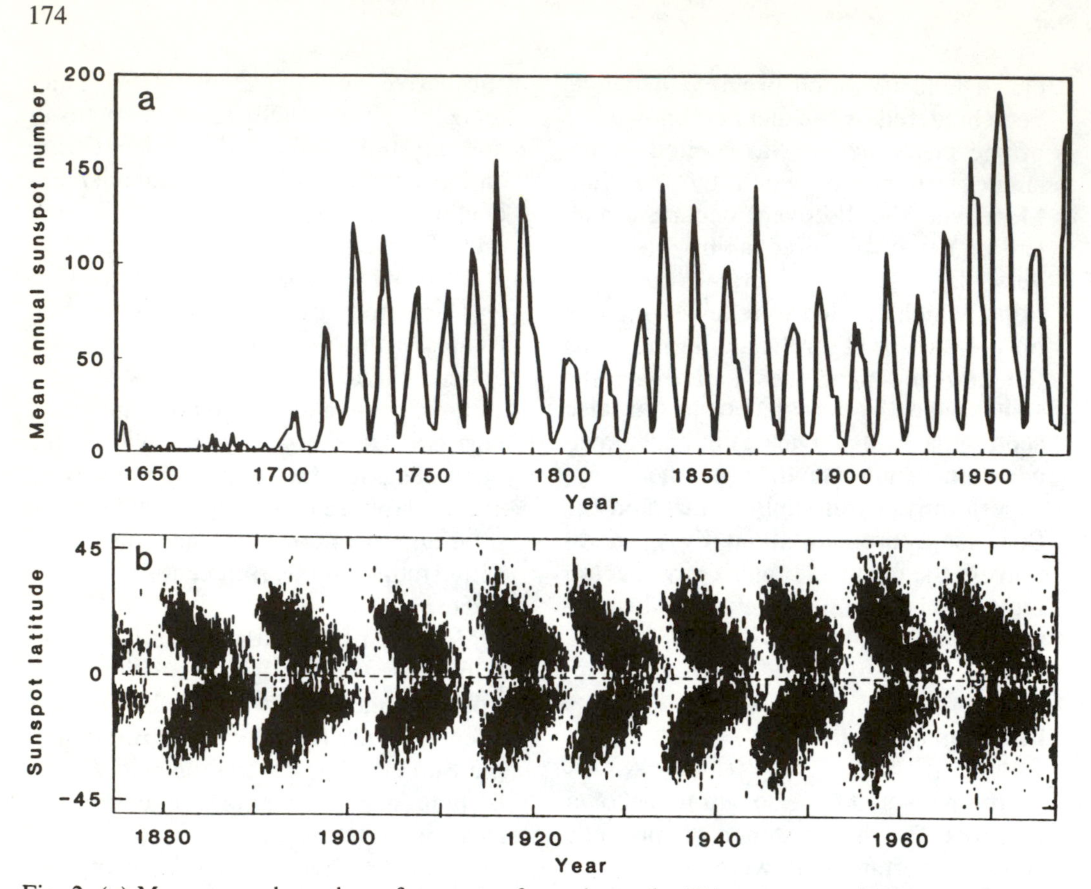

Fig. 2. (a) Mean annual number of sunspots from the early 17th century to 1980. Pronounced differences from one cycle to the next are evident both in intensity and in the timing of maxima. The 60-year period of low activity extending to about 1710 is the Maunder minimum. (b) "Butterfly" diagram in which the latitudes of emerging sunspot groups are plotted monthly; such a plot illustrates the tendency of the sunspot zone to migrate equatorward during each cycle. The first spots from a new cycle usually appear at higher latitudes before the last spots of the previous cycle have dissipated. [Reprinted with permission of G. Newkirk and K. Frazier, *Physics Today* 35, 25 (April 1982), and courtesy of the Science and Engineering Research Council of Great Britain]

pole to equator on the Sun is 22 years. The pattern of these waves is that of a torsional oscillation of wave number 2 per hemisphere. Other modes of torsional oscillation have also been reported. The waves are correlated with the butterfly pattern of emerging spots and magnetic flux on the solar surface: most of the activity emerges in the shear zone that lies just equatorward of a zone of accelerated rotation. Although the intensity of magnetic activity can differ strongly from one cycle to the next, no corresponding variations appear in the torsional oscillation; this result has led investigators to suspect that the torsional oscillation might have a fundamental causal role in the solar cycle. Cycle-related variations in the rotation measured from sunspots are also seen (*11*).

A feature of solar activity described in 1894 by Maunder but not generally appreciated until relatively recently (*12*) is the fact that the activity fell to a very low level in the latter part of the 17th century. This 60-year interval of low activity, now called the "Maunder minimum," has been found to coincide with the most recent of several known past epochs for which an above-average content of the radioactive isotope ^{14}C relative to the stable isotope ^{12}C is found in ancient tree rings of known age. Carbon-14 is one of several isotopes produced by galactic cosmic-ray particles entering Earth's atmosphere. It is known that there is a negative correlation between the flux of cosmic rays impinging upon the atmosphere and solar activity. This is attributed to the bending of the trajectories of these electrically charged particles by the magnetic fields carried into interplanetary space by the solar wind, although other factors affecting the cosmic-ray fluxes are also known. Carbon dioxide containing ^{14}C resides in the atmosphere and oceans for at least a decade before being incorporated in sediment, polar ice, and living plants, so that variations on a time scale as short as the solar cycle are erased. But variations on time scales of a century or longer are apparent in the radiocarbon record in ancient wood of the last 8000 years (*13*). This record suggests that, at the current stage of its evolution, the Sun may have spent as much as 30 to 40 percent of the last 1000 years in a "Maunder minimum" state of low activity (*14*).

About 20 years ago it was discovered from precise Doppler measurements that small areas of the solar surface rise and fall at a rate of a few hundred meters per second in a period of about 5 minutes. About 10 years later it was recognized that these 5-minute oscillations are associated with global motions in the Sun, analogous to the seismic waves that propagate along the surface of Earth and through its interior (*15*). Just as geologists make use of seismic waves to derive information about Earth's crust and core, solar scientists are beginning to use highly sensitive Doppler measurements to probe the inner structure, composition, and dynamics of the Sun. From recent studies has come evidence that the core of the Sun may be rotating several times faster than the surface (*16*). This would have obvious importance, not only for theories of solar magnetic activity but for theories of differential rotation and angular momentum loss in stars and perhaps for the understanding of the low flux of neutrinos produced by nuclear reactions in the solar core.

The most generally accepted explanation for the existence and large-scale behavior of the magnetic field of the Sun depends upon the fact that the sun rotates, and the fact that vertical circulation of convective eddies takes place in an outer shell that extends from the photosphere—the apparent surface—down to a depth that is an appreciable fraction (about a quarter to a third) of the Sun's radius. The thermonuclear fusion of hydrogen into helium that supplies the Sun's radiant energy is confined to a core whose radius is also about a quarter of the solar radius. This energy diffuses radiatively from the core outward to the base of the outer shell, where high opacity begins to force the onset of convection as the dominant mechanism of energy transport to the surface. The convection, in combination with the rotation of the Sun, gives rise to a pattern of circulation in which the fluid at different latitudes and depths in the convective shell is forced to rotate at appreciably different rates.

The fluid of which the Sun is composed is a good conductor of electric current. A magnetic field embedded in such a fluid behaves as if it is "frozen" into the fluid, moving with it and becoming deformed, twisted, and amplified, feeding on the energy of motion of the fluid. Investigators have made extensive studies of the generation of magnetic fields by such "dynamo" processes (*17, 18*). A number of models have been considered in attempts to understand the Sun as a magnetic oscillator. Although a great deal of new insight into the processes at work is emerging, thus far no model has succeeded in accounting for all the known features of the solar magnetic cycle. It is evident that a successful model will depend intimately upon physical conditions and processes that are hidden in the solar interior, and it is reasonable to expect that understanding the solar cycle will bring fundamental advances in knowledge about these processes in stars, and vice versa.

Strong Emission Lines

The convective outer shell is believed to be necessary for the existence of the hot chromosphere of the Sun and for the corona and the solar wind; the stars in which chromospheres are found are those in which outer convection zones are expected. The structure of these extremities of the solar atmosphere is modified by the magnetic field, producing a variety of transitory effects that can be observed and studied in detail at the wavelengths of the emission lines of abundant metallic ions and hydrogen when the white light of the underlying photosphere is filtered out.

Although most of the strong emission lines produced by the chromosphere and corona are located in the ultraviolet part of the spectrum that is absorbed by Earth's atmosphere and is observable only from space, the Fraunhofer H and K lines of ionized calcium are important exceptions. They carry about 15 percent of the energy radiated into space by the chromosphere of the Sun. Light at and near the wavelengths of these lines, at 3968.470 and 3933.664 Å in the near-ultraviolet, is strongly absorbed in the photosphere, so that bright chromospheric features above the photosphere stand out in sharp contrast (see Fig. 1, a and c). Their brightness is highly correlated point by point with the magnetic flux present in the underlying photosphere, and thus calcium H and K line core emission can be used to infer the existence and character of magnetic activity at stellar distances.

The method of observation introduced by Wilson consists of measuring, with a photoelectric spectrometer, the light flux in two spectral bands 1 Å wide centered at the H and K lines, relative to the flux in a second pair of bands about 20 Å wide located on either side. From these measurements one computes a relative HK flux index by dividing the sum of the fluxes at H and K by the sum of the fluxes in the reference bands. Since the fluxes in the numerator and denominator of this ratio are measured simultaneously by photon counters, the ratio is insensitive to fluctuations in the light caused by telescope guiding errors and air turbulence.

For the first 11 years, the measurements were made at the 100-inch Hooker reflector at Mount Wilson with a scanning spectrometer that had been developed in the late 1950's and early 1960's. After mid-1977, the measurements were made at the Mount Wilson 60-inch telescope with a specialized "HK photometer" devised by Vaughan *et al.* (*19*). The new instrument closely reproduces Wilson's original photometric system for the H and K lines but is simpler to calibrate and use than the apparatus it replaced. Its operation is indicated in Fig. 3.

The stars included in the first 11 years of the survey range in surface temperature from about 3000 to 7000 K, the latter being about 1000 K hotter than the Sun. All of them lie, with the Sun, in the lower

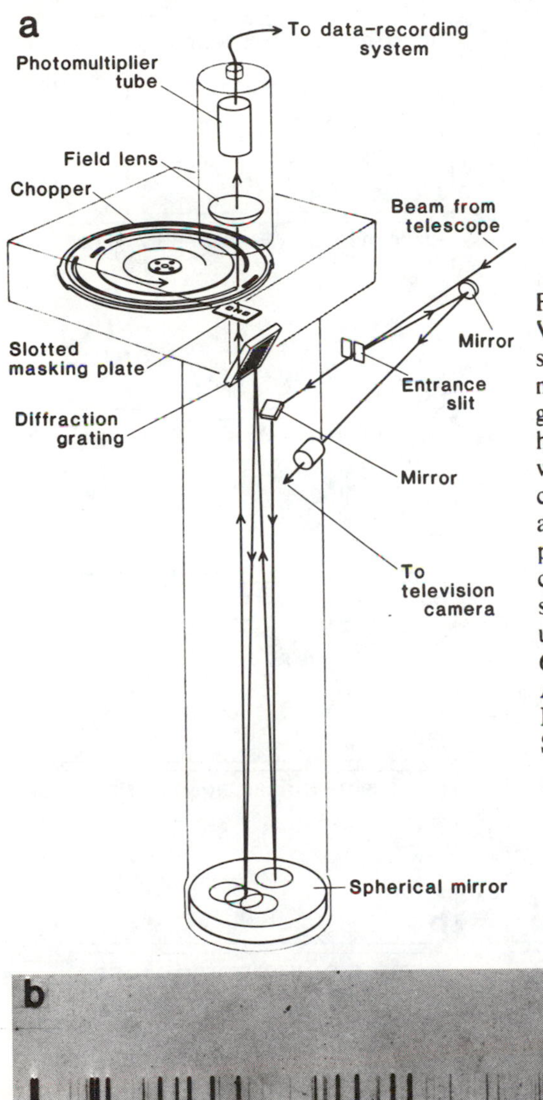

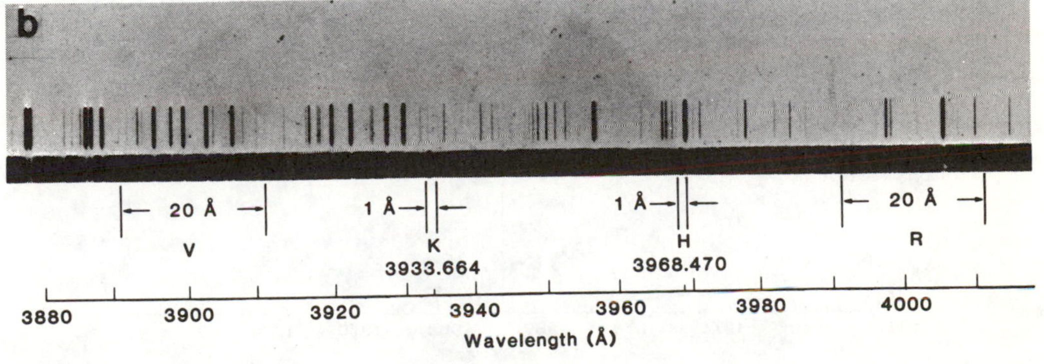

Fig. 3. (a) The HK photometer devised by Vaughan *et al.* (*19*) to measure the chromospheric H and K line emission associated with magnetic activity in stars. A grating spectrograph images the stellar spectrum onto a mask having narrow slots at the H and K lines and wider slots on either side, as shown in (b). A chopper wheel rotating at about 30 hertz admits light from only one slot at a time to a photon detector. A digital electronic system counts the photons from the four slots in separate registers. The instrument has been used in conjunction with the Mount Wilson 60-inch telescope since 1977. [From "The Activity Cycles of Stars," O. C. Wilson, A. H. Vaughan, D. Mihalas. Copyright © 1981 by Scientific American, Inc. All rights reserved]

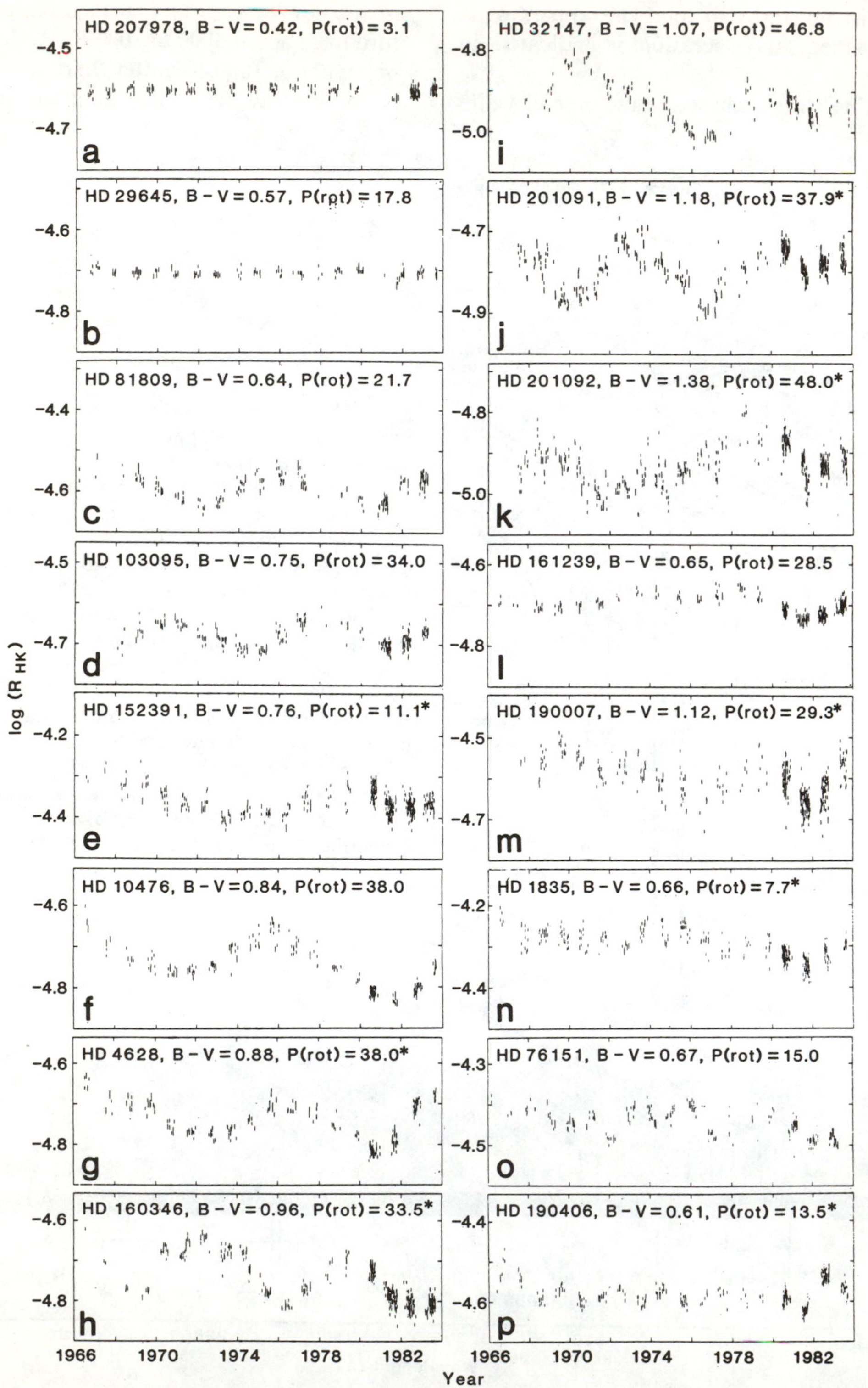
HD 207978, B - V = 0.42, P(rot) = 3.1
a
HD 29645, B - V = 0.57, P(rot) = 17.8
b
HD 81809, B - V = 0.64, P(rot) = 21.7
c
HD 103095, B - V = 0.75, P(rot) = 34.0
d
HD 152391, B - V = 0.76, P(rot) = 11.1*
e
HD 10476, B - V = 0.84, P(rot) = 38.0
f
HD 4628, B - V = 0.88, P(rot) = 38.0*
g
HD 160346, B - V = 0.96, P(rot) = 33.5*
h
HD 32147, B - V = 1.07, P(rot) = 46.8
i
HD 201091, B - V = 1.18, P(rot) = 37.9*
j
HD 201092, B - V = 1.38, P(rot) = 48.0*
k
HD 161239, B - V = 0.65, P(rot) = 28.5
l
HD 190007, B - V = 1.12, P(rot) = 29.3*
m
HD 1835, B - V = 0.66, P(rot) = 7.7*
n
HD 76151, B - V = 0.67, P(rot) = 15.0
o
HD 190406, B - V = 0.61, P(rot) = 13.5*
p
log (R HK)
1966
1970
1974
1978
1982

Year

main sequence of the Hertzsprung-Russell diagram in which the luminosity of stars is plotted against their surface temperature or color index (*20*), a function of stellar mass.

The relative HK flux scale of a star measured by the HK photometer depends not only upon the strength of the star's chromospheric emission but also upon the strength of its photospheric emission at the nearby wavelengths of the reference bands of the photometer. Thus, the relative HK flux is larger for a cool star than it would be for a hotter star with the same amount of chromospheric emission. To eliminate this effect, investigators often choose to express a star's HK flux as a fraction of its total energy output by applying a color correction (*21*). The resulting absolute HK flux indeed is denoted R_{HK}.

Examples of long-term stellar chromospheric HK flux records are shown in Fig. 4, in which R_{HK} is plotted on a logarithmic scale as a function of time, over an interval extending from 1966 to 20 September 1983.

To control the calibration of observations, some 14 stars having very weak and presumably nearly constant chromospheric emission were included from the beginning of the survey to serve as standards. The flux records of two standard stars (HD 207978 and HD 29645) are shown in Fig. 4, a and b. For these stars the seasonal scatter in R_{HK} is at a minimal level of about 2 percent and the seasonal average values of R_{HK} have remained constant to within about 1 percent, which must be regarded as the effective limit of stability of the measuring apparatus. Some of the standards exhibit seasonal scatter that is larger than in these examples, an indication that such scatter, even in the standards, is partly real and not the result of measuring errors.

The stars whose HK flux records are illustrated in Fig. 4 exhibit a wide variety of forms of behavior, from sunlike activity cycles to pronounced variations that are markedly different from the solar cycle as we know it. This variety in itself was a remarkable finding of Wilson's original study. The resemblance of some of these records to the solar cycle is made clear if one compares them to observations of solar K-line emission recorded daily since 1975 by Livingston (*22*) at the Kitt Peak National Observatory in Arizona by means of a solar K-line photometer (see Fig. 1c). The Kitt Peak index differs in scale by a factor of 2 from that used for stars at Mount Wilson; when this is taken into account, the solar record is seen to resemble that of the star HD 81809 (Figs. 1d and 4c), whose color index and rate of rotation

Fig. 4 (facing page). Long-term HK flux records for 16 of the 91 stars that have been kept under consistent observation at Mount Wilson since 1966. The index R_{HK} (see text) is plotted on a logarithmic scale. A change in 0.1 in log R_{HK} represents a 26 percent change in R_{HK}. Stars are identified by their number in the Henry Draper catalog. The average of a night's observation for each star is plotted as a short vertical bar whose length is roughly equal to the accuracy of measurement. The measurements appear in seasonal groups separated by gaps of a few months of each year when a given star is out of range of nighttime observation. Prior to 1980, observations were limited to an average of about four nights per month. After June 1980, the frequency of observation was increased to a near nightly basis, resulting in the increase in density of points for the last four seasons. The legend for each star gives its B–V color index and its period of rotation [P(rot)] in days, either measured from its rotational modulation (with asterisk) or predicted from the Rossby relation (without asterisk) as described in the text.

are also close to the solar values. It is clear that the solar cycle itself would be discernible at stellar distances, were the Sun observed in its H and K lines.

At the time of publication of Wilson's survey in 1978 (*6*), the rotation rates of dwarf stars exhibiting cycles were unknown with the single exception of the Sun. The classical method of measuring stellar rotation depends upon the Doppler broadening of photospheric absorption lines in the stellar spectrum. The broadening is proportional to the line-of-sight component of surface velocity resulting from the rotation. In the Sun, whose equatorial speed of rotation is about 2 kilometers per second, the rotational broadening is so small as to be almost entirely masked by the intrinsic line widths produced by thermal and convective motions in the photosphere.

A feature of all the stars under observation, noted quite early in Wilson's survey and evident in Fig. 4, is the seasonal scatter in R_{HK}, small in the standards but many times larger than the errors of measurement in chromospherically active stars. Some part of this scatter must arise from sporadic events in the surface activity of a star. However, in the case of the Sun, occasionally a large "complex" of activity emerges and persists for several rotations. If this were to occur in a star, one would expect to see, superimposed upon any long-term variation that might be present, a periodic modulation of the HK flux index as the activity complex is carried into and out of view by rotation, the period of the modulation being equal to the star's period of rotation. Unresolved in observations at monthly intervals, the modulation would give rise to just such seasonal scatter as is observed.

To resolve the rotational modulation effect it would be necessary to make observations at intervals short as compared to the period of rotation, over a span of several weeks or months. Just such an intensive observing program was undertaken by a team of researchers working at Mount Wilson, beginning in July 1980 (*23*). This work demonstrated that in fact rotational modulation is the principal cause of the seasonal scatter in the long-term survey. Moreover, from these synoptic observations periods of rotation have thus far been determined, with a precision of a few percent, for 41 of the stars in the study. The observed periods range from 2.5 to 48 days (see Fig. 5).

From an early stage in the observation of rotational modulation in stars, a close connection was clearly apparent between a star's rate of rotation and the strength of its chromospheric emission in the H and K lines, in the sense that the faster the star's rotation, the larger the average value of its HK flux index; the relation between rotation and emission is also dependent upon stellar color index or temperature. Far-ultraviolet and x-ray emission in many of the same stars is also known to be correlated with rotation rate (*24*).

Rossby Relation

The rotation-activity connection is regular enough to suggest that it might be represented by a simple empirical formula. Recently, Noyes and his colleagues (*25*) discovered a formulation that, in addition to adequately representing the observations, may have theoretical significance. Their formulation makes use of a fundamental parameter in hydromagnetic dynamo theory known as the Rossby number. The dimensionless number is the ratio between the rotation-

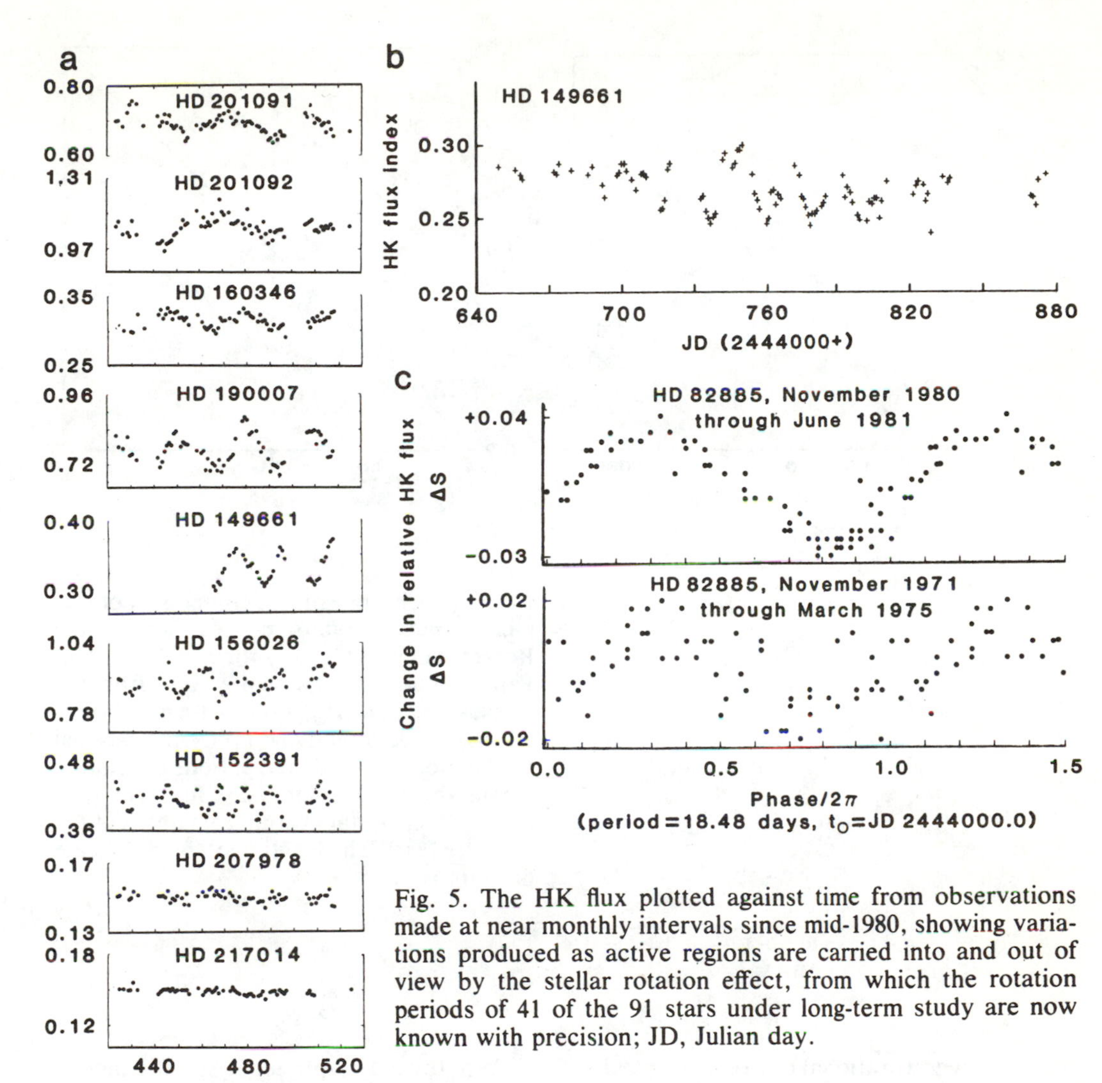

Fig. 5. The HK flux plotted against time from observations made at near monthly intervals since mid-1980, showing variations produced as active regions are carried into and out of view by the stellar rotation effect, from which the rotation periods of 41 of the 91 stars under long-term study are now known with precision; JD, Julian day.

al period of a star and the "convective turnover" time (t_c) required for a convective element to traverse the convection zone. The convective turnover time depends upon the (not precisely known) depth of the convection zone; this depth in turn depends upon a parameter α describing the efficiency of the convective heat transport process in the convection zone of a star. Values for the turnover time in the Sun computed (*26*) from an approximate theory range from 4 to 16 days for values of α between 1 and 3. Rossby number is a measure of the importance of rotation-dependent Coriolis forces in producing helicity in the motion of rising convective eddies within the convection zone, which is usually regarded in dynamo theory as the location of the field generation process.

Noyes and his colleagues found that, when the chromospheric component of the mean HK fluxes of stars is plotted against their Rossby numbers in which

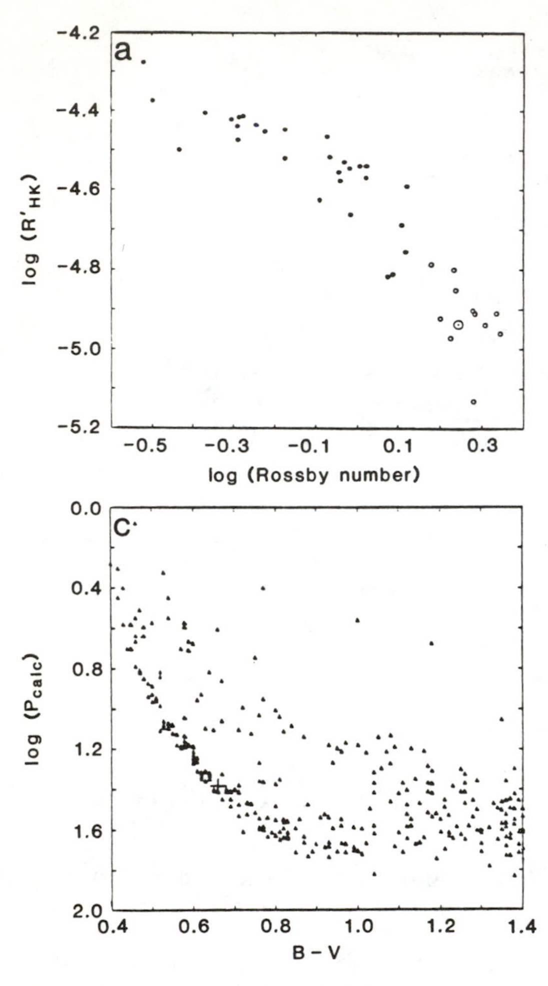

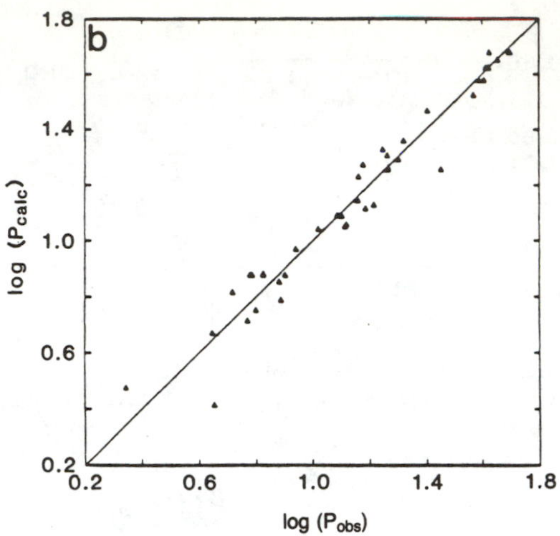

Fig. 6. (a) Chromospheric component of HK flux output for stars plotted against the Rossby number (P_{obs}/t_c) for $\alpha = 2$ (*24*). (b) Periods of stellar rotation predicted from the Rossby relation (P_{calc}) plotted against periods measured directly (P_{obs}) from rotational modulation. (c) Predicted periods of rotation versus the B−V color index for 311 lower main sequence stars within 25 parsecs of the Sun (see text). The location of the Sun in the diagram is indicated by +.

the observed rotational periods are used, the points fall with fairly small scatter along a curve (Fig. 6a). This occurs, however, only if α is taken to be larger than unity (1.9 is perhaps close to the best value). Such a value for α is not in conflict with recent estimates, based upon theoretical calculations and upon helioseismology, for the depth of the convection zone in the Sun. That α should have the same value for all lower main sequence stars is an assumption remaining open to question, as the discoverers of the Rossby relation have emphasized.

Whatever the physical significance of the Rossby relation may prove to be, it can be a valid empirical tool with which to predict the period of rotation of a lower main sequence star from knowledge only of its color index or temperature and the average value of its H and K emission. For the stars used to derive the relation, the predicted periods are accurate to within about 20 percent (see Fig. 6b). By this method it is possible to supply predicted periods of rotation for all the stars under long-term observation and to examine the connection between rotation and their long-term behavior.

Trends in the Stellar Records

It is widely noted in the literature that stellar activity variations most strongly reminiscent of the solar cycle are mainly found only in stars with rotation periods in excess of about 20 days. Indeed, with one exception (HD 152391) (Figs. 4e and 5a), this is true of the 16 stars in the study that exhibit exceptionally well-defined and obvious cycles in their HK flux records.

Several characteristics familiar in the solar cycle can be noted in the stellar records. The scatter associated with rotational modulation usually becomes greatest near the maximum of the cycle. Successive maxima often differ appreciably in amplitude; this is especially notable in HD 160346 (Fig. 4h). Successive minima may also differ. In many cases, as in the Sun, the cycle rises to a maximum more rapidly than it declines after the maximum; HD 32147 (Fig. 4i) provides an extreme example of this, but for HD 201092 (Fig. 4k) the reverse is true. The intervals between successive maxima in stars range from about 7 to 12.5 years (or longer), with an average close to 10 years. For the Sun the long-term average is near 11 years, but the average is closer to 10 years in this century. The time between individual maxima has ranged from 8 to 14 years. The cause of this dispersion in the Sun is not known. Since the Sun must be regarded as a typical star, it is reasonable to suppose that the dispersion in stars arises at least in part from the same cause.

From the limited evidence now available one can say that whether a clear cycle exists seems to depend upon the rate of rotation of a star. Once the cycle becomes established, however, there is no obvious correlation between its time scale and the rotation period of the star or its color index or mass (*27*). Investigators are hopeful that systematic trends of some sort may yet be uncovered in studies that are in progress. Such trends, or their absence, could be of crucial importance in testing the divergent predictions of competing theoretical models of the cycle.

There are some 37 stars in the study whose observed or expected periods of rotation exceed about 18 to 20 days, including the 15 slow rotators already noted as showing pronounced sunlike cycles. Of the remaining 22 slow rotators, nine or ten can be described as probably cyclic but having ambiguous time scales. For example, HD 161239 (Fig. 4l) could have a cycle of about 12 years, but a component of variation on a time scale of about 4 years is also present. More than one time scale appears in HD 190007 (Fig. 4m) and perhaps also in other examples shown in Fig. 4. The activity of HD 156026 (Fig. 5a) increased more or less steadily from 1968 until 1980 but began to decline slightly in 1981: it could be an example of a cyclic star with a very long period. If we give the ambiguous cases the benefit of the doubt, it is possible that altogether 25 of the slowly rotating dwarfs in the study are cyclic. Even the remaining 12 slow rotators show some hint of long-term variation that could be significant but whose amplitude is much smaller than that of the solar cycle in the last two centuries. It is entirely conceivable that these dozen or so stars with weak activity (one-third of the sample of slow rotators that "might" be cyclic) are in a Maunder minimum state, in which the Sun may have spent a comparable fraction of its time in past epochs. If this is the case, then sooner or later such a star will regenerate its cycle or a star that now exhibits cycles will cease to vary. Half a millennium might elapse before any one star—or the Sun—undergoes another such change. But it is statistically likely that the waiting time for such an event will be reduced in proportion to the number of slowly rotating stars under scrutiny.

Even among stars that rotate rapidly one finds variations that could be construed as possibly sunlike or periodic, apart from the example of HD 152391 (Fig. 4e) already mentioned. HD 1835 (Fig. 4n) has twice declined slowly in activity for several years and then recovered in the span of one or two seasons. Similar "sawtooth" behavior is evident in HD 25998 and to some extent in HD 18256. HD 20630 remained at a high level for the first 6 years of the Wilson survey and then began to vary on roughly a 4-year time scale. Thus researchers suspect that the dichotomy between stars with and without cycles is perhaps not completely sharp.

It is among the more rapidly rotating stars that one encounters behavior quite unlike that exhibited by the Sun. HD 76151 (Fig. 4o) and HD 190406 (Fig. 4p) show prominent variations on time scales as short as 2 or 3 years and occasional systematic changes within a single season. Although common characteristics can be found in the HK flux records of such stars, these characteristics seem to emerge, like biological traits, in differing combinations from one individual star to another or even from one interval of time to another in the same star. With few exceptions, one can say that the more rapidly a star rotates, the more chaotic its chromospheric variations are likely to appear. Are distinct time scales or "modes" shorter than the "fundamental" sunlike cycle present in these stars? Given the sparseness of the data collected thus far, it is not yet possible to say.

A surprising fact uncovered by the study of rotational modulation was that in many stars the modulation continues unchanged in phase for very long times, far longer than the typical lifetimes of even the largest activity complexes on the Sun. An extreme example is perhaps HD 82885 (Fig. 5c), found (*28*) to show rotational modulation with remarkable stability in amplitude and phase over an 8-month interval in 1980–1981, with a period of between 18.0 and 18.6 days. Analysis of Wilson's survey subsequently showed (*29*) that, if the period is assumed actually to be 18.49 ± 0.05 days, the variations observed in 1980–1981 are almost exactly in phase with the unresolved variations of about the same amplitude in the survey data for this star throughout the decade from 1971 to 1981. It is difficult to understand how the precise location of a surface feature on a differentially rotating star could be thus "remembered" unless it is in some way associated with long-lived magnetic field patterns deep inside the star. Studies have shown that, even in the Sun, there is some tendency for active regions to occur preferentially at certain longitudes over intervals of several years (*30*). From geomagnetic studies it was inferred in 1975 that large-scale irregularities in the Sun's general dipolar magnetic field had persisted for as long as five sunspot cycles or 47 years (*31*).

In the chromospherically active star HD 149661 (Fig. 5, a and b) two frequencies of HK flux variation are sometimes simultaneously present, producing a "beat" in the modulation and a corresponding double peak in the autocorrelation. It is tempting to suppose that this effect might arise from a pair of active regions at different latitudes on the star in the presence of differential rotation (*29*). The two periods, 18 and 21 days, differ by 15 percent, which is comparable to the solar differential rotation between the equator and regions near the poles but larger than the solar differential rotation within the sunspot zones. The differential rotation in HD 149661 could well be larger than the Sun's. It is also conceivable that a star could have more than one spot zone (or active shear zone) in each hemisphere, although the butterfly diagram clearly shows that the Sun has only one. Ten stars have thus far been identified in the Mount Wilson survey that occasionally show such double periodicities (*32*).

About 5 years ago the HK photometer was used to survey (*33*) the HK emission fluxes in a large fraction (several hundred) of the dwarf stars in the northern celestial hemisphere within a distance of 25 parsecs (82 light-years) of the Sun. In this survey each star was observed only once or a few times, but the range of variation of an individual star is usually small enough that even a "snapshot" suffices reasonably well as an approximation of a star's average level of activity. To this solar neighborhood survey it is now possible to apply the Rossby relation in order to estimate from the known HK fluxes and colors the previously unknown periods of rotation of these stars. The resulting periods are plotted, on a logarithmic scale (with periods increasing downward), against their B−V colors, in Fig. 6c. To minimize extrapolation beyond the range in which the Rossby relation is calibrated by observation, only stars with B−V less than 1.4 are plotted. Even so, the diagram should be regarded as tentative. Such a diagram is of interest for at least one reason, and possibly a second that can be mentioned only briefly.

It was established some years ago that both chromospheric activity and the rate of rotation of main-sequence stars decrease with advancing stellar age (*34*), presumably as the result of torque exerted by the stellar wind. Thus, the early Sun would have been represented by a point at the top of the diagram in Fig. 6c. In the course of time this point would have moved almost vertically downward, reaching its present position after about 4.6 billion years. But its present position is very close to the rather sharp lower boundary of the distribution of points in the diagram, along with many other stars of about the same color, some of which must be even older than the Sun. This circumstance suggests that a star with B−V between about 0.45 and 0.9 slows in the course of time to some definite minimal rotation rate that is a function of its mass. Thereafter, either no significant further spindown occurs or else thereafter, unlike the Sun, the much older stars presumably included in the sample do not "obey" the Rossby relation. There is independent evidence from satellite measurements of the solar wind in interplanetary space that the angular momentum now being carried away from the Sun by particle flow and magnetic fields in the solar wind is not sufficient to account for significant further deceleration of the Sun's present rate of rotation (*35*). Figure 6c may well provide a strong observational confirmation of this result and a demonstration that it is a universal phenomenon among dwarf stars, thus far unexplained by theory. Indeed, from Fig. 6c it can be inferred that this terminal rate varies as about the fourth power of stellar mass.

It is also of interest that in Fig. 6c the density of points, as far as can be determined from the limited number, decreases more or less smoothly upward, as would be expected if (as most investigations suggest) star formation has occurred in the Galaxy at an essentially uniform rate during the last few billion years, and if stellar spindown takes place smoothly as a function of time. The nonlinearity of the Rossby relation (in which the HK flux "saturates" at high rates of rotation) gives a plot of the HK flux against B−V an appearance rather different from that of Fig. 6c. Such a plot was discussed in *(33)*.

Sensitive modern techniques of observation are beginning to offer researchers the possibility of detecting other effects of magnetic activity besides the enhancement of chromospheric emission lines. These effects include the broadening of magnetically sensitive photospheric absorption lines (compared with insensitive lines) by the Zeeman effect in magnetically active stars *(36)*, from which the strength of the field can be directly inferred. Precise photometry can reveal the subtle variation in stellar brightness—or rotational modulation—caused by the presence of "starspots" in the case of very active stars *(37)*. From high-dispersion spectrograms of detailed features on the solar surface it is known that the profiles of the H and K lines differ systematically from one kind of feature to another; from study of changes in the line profiles as a star rotates, inferences might be made about the kinds of features present on its surface and their distribution in latitude and longitude. These and other detailed and systematic studies remain tasks for the future. It is reasonable to expect that the extension of solar physics into the domain of stars will continue to be an exciting venture.

References and Notes

1. Quoted from the article by A. M. Clerke in *Encyclopedia Brittanica*, ed. 14 (1929), p. 887.
2. A. Humboldt, *Kosmos* **1** (1851).
3. G. E. Hale and S. B. Nicholson, *Astrophys. J.* **62**, 270 (1925); list of errata, *ibid.* **63**, 72 (1926).
4. O. C. Wilson, A. H. Vaughan, D. Mihalas, *Sci. Am.* **244**, 104 (February 1981).
5. O. C. Wilson, *Science* **151**, 1487 (1966); an early attempt to uncover chromospheric variations in stellar spectra photographically was reported by D. Popper [*Astrophys. J.* **123**, 377 (1956)].
6. O. C. Wilson, *Astrophys. J.* **226**, 379 (1978).
7. G. Newkirk and K. Frazier, *Phys. Today* **35**, 25 (April 1982).
8. R. C. Carrington, *Observations of Spots on the Sun* (Williams and Norgate, London, 1863).
9. R. F. Howard, *Rev. Geophys. Space Phys.* **16**, 721 (1978).
10. ______ and B. J. LaBonte, *Astrophys. J.* **239**, L33 (1980).
11. P. A. Gilman and R. F. Howard, *ibid.*, in press.
12. J. A. Eddy, *Sci. Am.* **236**, 80 (May 1977); *Science* **192**, 1189 (1976).
13. M. Stuiver and P. D. Quay, *Science* **207**, 11 (1980); *Solar Phys.* **74**, 479 (1981); H. E. Suess, *Radiocarbon* **22**, 200 (1980); M. Stuiver and P. M. Grootes, in *The Ancient Sun*, R. O. Pepin, J. A. Eddy, R. B. Merrill, Eds. (Pergamon, New York, 1980), pp. 165–173.
14. J. A. Eddy, personal communication; *Solar Phys.* **89**, 195 (1983).
15. R. B. Leighton, R. W. Noyes, G. Simon, *Astrophys. J.* **135**, 474 (1962); R. Ulrich, *ibid.* **162**, 993 (1970).
16. A. Claverie, G. R. Isaak, C. P. McCleod, H. B. van der Raay, *Nature (London)* **293**, 443 (1981).
17. M. Schüssler, in International Astronomical Union Symposium 102, *Solar and Stellar Magnetic Fields: Origins and Coronal Effects*, J. O. Stenflo, Ed. (Reidel, Dordrecht, 1983), p. 213.
18. E. N. Parker, *Sci. Am.* **249**, 44 (August 1983).
19. A. H. Vaughan, G. W. Preston, O. C. Wilson, *Publ. Astron. Soc. Pac.* **90**, 276 (1978).
20. A measurement of surface temperature used by astronomers is the ratio of a star's radiant energy output at two selected wavelengths or, on a logarithmic scale, the difference in magnitudes at these wavelengths. In the most commonly used system the logarithmic color index B−V refers to the "blue" minus "visual" or yellow spectral regions. This index increases as temperature decreases and is close to 0.66 for the Sun.
21. F. Middelkoop, *Astron. Astrophys.* **107**, 31 (1982).
22. W. C. Livingston, personal communication.
23. A. H. Vaughan *et al.*, *Astrophys. J.* **250**, 276 (1981); S. L. Baliunas *et al.*, *ibid.* **275**, 752 (1983).
24. R. Pallavicini, L. Golub, R. Rosner, G. Vaiana, in International Astronomical Union Symposium 102, *Solar and Stellar Magnetic Fields: Origins and Coronal Effects*, J. O. Stenflo, Ed. (Reidel, Dordrecht, 1983), p. 77.
25. R. W. Noyes, L. W. Hartmann, S. L. Baliunas, D. K. Duncan, A. H. Vaughan, *Astrophys. J.*, in press.
26. P. A. Gilman, in International Astronomical Union Colloquium 51, *Stellar Turbulence*, D. Gray and J. Linsky, Eds. (Springer, Berlin, 1980), p. 19.
27. A. H. Vaughan, in International Astronomical Union Symposium 102, *Solar and Stellar Magnetic Fields: Origins and Coronal Effects*, J. O. Stenflo, Ed. (Reidel, Dordrecht, 1983), p. 113.

28. J. Frazer and H. Lanning, personal communication.
29. R. W. Noyes, *Second Cambridge Conference on Cool Stars, Stellar Systems, and the Sun*, M. S. Giampapa and L. Golub, Eds. (SAO Special Report 392, Smithsonian Astrophysical Observatory, Cambridge, Mass., 1982), vol. 2, p. 41.
30. J. W. Dodson and E. R. Hedeman, *Solar Phys.* **42**, 121 (1975).
31. L. Svalgaard and J. M. Wilcox, *ibid.* **41**, 461 (1975).
32. S. L. Baliunas *et al.*, in preparation.
33. A. H. Vaughan and G. W. Preston, *Publ. Astron. Soc. Pac.* **92**, 385 (1980); L. Hartmann, D. Soderblom, R. W. Noyes, N. Burnham, A. H. Vaughan, *Astrophys. J.* **276**, 254 (1984).
34. O. C. Wilson, *Astrophys. J.* **138**, 832 (1963); R. P. Kraft, *ibid.* **150**, 551 (1967); A. Skumanich, *ibid.* **171**, 565 (1972).
35. V. Pizzo *et al.*, *ibid.* **271**, 335 (1983).
36. G. W. Marcy, *ibid.*, in press.
37. R. R. Raddick *et al.*, *Publ. Astron. Soc. Pac.* **94**, 934 (1982).
38. Construction of the HK photometer was funded by National Aeronautics and Space Administration grant NSG 07148. The work at Mount Wilson was supported by the Carnegie Institution of Washington, by grants AST 79-21070 and AST 81-21726 from the National Science Foundation, and by grant 2548-82 from the National Geographic Society. The work reviewed owes its rapid recent development to the ideas and cooperative labors of many workers. I thank S. L. Baliunas, D. K. Duncan, J. A. Eddy, R. F. Howard, M. M. Neugebauer, G. Newkirk, R. W. Noyes, and O. C. Wilson for informative discussions and suggestions during the preparation of this paper and W. C. Livingston for providing solar K line data in advance of publication.

This material originally appeared in *Science* **225**, 24 August 1984.

13. On Stars, Their Evolution and Their Stability

S. Chandrasekhar

Introduction

When we think of atoms, we have a clear picture in our minds: a central nucleus and a swarm of electrons surrounding it. We conceive them as small objects of sizes measured in angstroms ($\sim 10^{-8}$ cm); and we know that some hundred different species of them exist. This picture is, of course, quantified and made precise in modern quantum theory. And the success of the entire theory may be traced to two basic facts: first, the Bohr radius of the ground state of the hydrogen atom, namely

$$\frac{h^2}{4\pi^2 me^2} \sim 0.5 \times 10^{-8} \text{ cm} \qquad (1)$$

where h is Planck's constant, m is the mass of the electron, and e is its charge, provides a correct measure of atomic dimensions; and second, the reciprocal of *Sommerfeld's fine-structure constant*,

$$\frac{hc}{2\pi e^2} \sim 137 \qquad (2)$$

gives the maximum positive charge of the central nucleus that will allow a stable electron orbit around it. This maximum charge for the central nucleus arises from the effects of special relativity on the motions of the orbiting electrons.

We now ask: Can we understand the basic facts concerning stars as simply as we understand atoms in terms of the two combinations of natural constants 1 and 2? In this lecture, I shall attempt to show that in a limited sense we can.

The most important fact concerning a star is its mass. It is measured in units of the mass of the sun, ⊙, which is 2×10^{33} g: stars with masses very much less than, or very much more than, the mass of the sun are relatively infrequent. The current theories of stellar structure and stellar evolution derive their successes largely from the fact that the following combination of the dimensions of a mass provides a correct measure of stellar masses:

$$\left(\frac{hc}{G}\right)^{3/2} \frac{1}{H^2} \simeq 29.2\odot \qquad (3)$$

where G is the constant of gravitation and H is the mass of the hydrogen atom. In the first half of the lecture, I shall essentially be concerned with the question: How does this come about?

This chapter is the lecture delivered by S. Chandrasekhar in Stockholm on 8 December 1983 when he received the Nobel Prize in Physics, which he shared with William A. Fowler.

The Role of Radiation Pressure

A central fact concerning normal stars is the role which radiation pressure plays as a factor in their hydrostatic equilibrium. Precisely the equation governing the hydrostatic equilibrium of a star is

$$\frac{dP}{dr} = -\frac{GM(r)}{r^2}\rho \tag{4}$$

where P denotes the total pressure, ρ the density, and $M(r)$ is the mass interior to a sphere of radius r. There are two contributions to the total pressure P: that due to the material and that due to the radiation. On the assumption that the matter is in the state of a perfect gas in the classical Maxwellian sense, the material or the gas pressure is given by

$$p_{\text{gas}} = \frac{k}{\mu H}\rho T \tag{5}$$

where T is the absolute temperature, k is the Boltzmann constant, and μ is the mean molecular weight (which under normal stellar conditions is ~ 1.0). The pressure due to radiation is given by

$$p_{\text{rad}} = \frac{1}{3}aT^4 \tag{6}$$

where a denotes Stefan's radiation-constant. Consequently, if radiation contributes a fraction $(1 - \beta)$ to the total pressure, we may write

$$P = \frac{1}{1-\beta}\,\frac{1}{3}aT^4 = \frac{1}{\beta}\,\frac{k}{\mu H}\rho T \tag{7}$$

To bring out explicitly the role of the radiation pressure in the equilibrium of a star, we may eliminate the temperature, T, from the foregoing equations and express P in terms of ρ and β instead of ρ and T. We find:

$$T = \left(\frac{k}{\mu H}\,\frac{3}{a}\,\frac{1-\beta}{\beta}\right)^{1/3}\rho^{1/3} \tag{8}$$

and

$$P = \left[\left(\frac{k}{\mu H}\right)^4 \frac{3}{a}\,\frac{1-\beta}{\beta^4}\right]^{1/3}\rho^{4/3} = C(\beta)\rho^{4/3} \text{ (say)} \tag{9}$$

The importance of this ratio, $(1 - \beta)$, for the theory of stellar structure was first emphasized by Eddington. Indeed, he related it, in a famous passage in his book on *The Internal Constitution of the Stars*, to the "happening of the stars" (*1*). A more rational version of Eddington's argument which, at the same time, isolates the combination 3 of the natural constants is the following.

There is a general theorem (*2*) which states that the pressure, P_c, at the center of a star of a mass M in hydrostatic equilibrium in which the density, $\rho(r)$, at a point at a radial distance, r, from the center does not exceed the mean density, $\bar{\rho}(r)$, interior to the same point r, must satisfy the inequality,

$$\frac{1}{2}G\left(\frac{4}{3}\pi\right)^{1/3}\bar{\rho}^{4/3}M^{2/3} \leqslant P_c \leqslant \frac{1}{2}G\left(\frac{4}{3}\pi\right)^{1/3}\rho_c^{4/3}M^{2/3} \tag{10}$$

where $\bar{\rho}$ denotes the mean density of the star and ρ_c its density at the center. The content of the theorem is no more than the assertion that the actual pressure at the center of a star must be intermediate between those at the centers of the two configurations of uniform density, one at a density equal to the mean density of the star, and the other at a density equal to the density ρ_c at the center (see Fig. 1). If the inequality 10 should be violated then there must, in general, be some regions in which adverse density gradients must prevail; and this implies instability. In other words, we may consider conformity with the

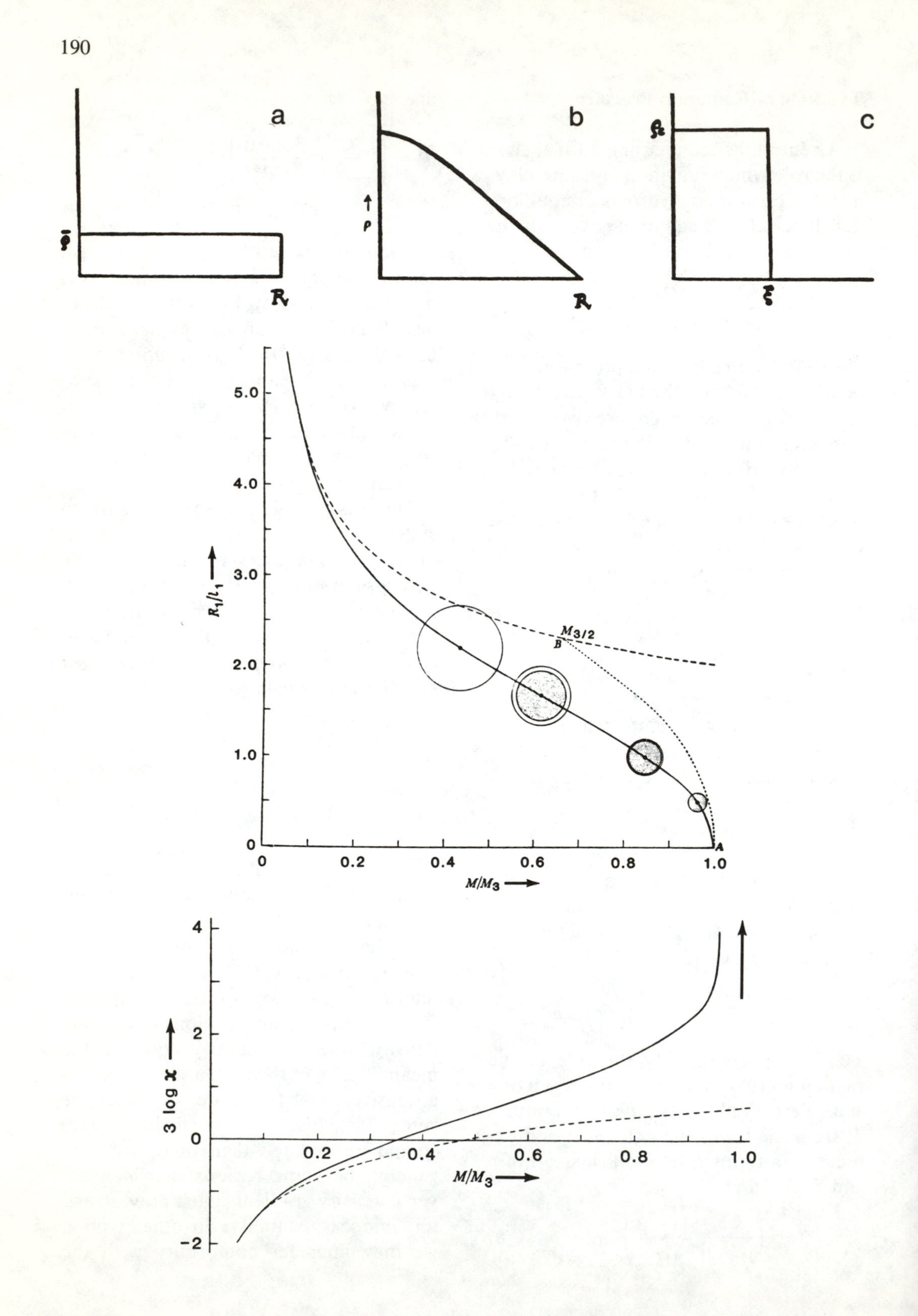
a
b
c
R
R
5.0
4.0
3.0
2.0
1.0
0
R_1/l_1
$M_{3/2}$
B
A
0
0.2
0.4
0.6
0.8
1.0
M/M_3
4
2
0
-2
3 log x
0.2
0.4
0.6
0.8
1.0
M/M_3

inequality 10 as equivalent to the condition for the stable existence of stars.

The right-hand side of the inequality 10 together with P given by Eq. 9, yields, for the stable existence of stars, the condition,

$$\left[\left(\frac{k}{\mu H}\right)^4 \frac{3}{a} \frac{1-\beta_c}{\beta_c^4}\right]^{1/3} \leq \left(\frac{\pi}{6}\right)^{1/3} GM^{2/3} \tag{11}$$

or, equivalently,

$$M \geq \left(\frac{6}{\pi}\right)^{1/2} \left[\frac{k}{\mu H}^4 \frac{3}{a} \frac{1-\beta_c}{\beta_c^4}\right]^{1/2} \frac{1}{G^{3/2}} \tag{12}$$

where, in the foregoing inequalities, β_c is a value of β at the center of the star. Now Stefan's constant, a, by virtue of Planck's law, has the value

$$a = \frac{8\pi^5 k^4}{15h^3c^3} \tag{13}$$

Inserting this value a in the equality 12 we obtain

$$\mu^2 M\left(\frac{\beta_c^4}{1-\beta_c}\right)^{1/2} \geq \frac{(135)^{1/2}}{2\pi^3} \left(\frac{hc}{G}\right)^{3/2} \frac{1}{H^2} = 0.1873\left(\frac{hc}{G}\right)^{3/2} \frac{1}{H^2} \tag{14}$$

We observe that the inequality 14 has isolated the combination 3 of natural constants of the dimensions of a mass; by inserting its numerical value given in Eq. 3, we obtain the inequality,

Table 1. The maximum radiation pressure, $(1 - \beta_*)$, at the center of a star of a given mass, M.

$1-\beta_*$	$M\mu^2/\odot$	$1-\beta_*$	$M\mu^2/\odot$
0.01	0.56	0.50	15.49
0.03	1.01	0.60	26.52
0.10	2.14	0.70	50.92
0.20	3.83	0.80	122.5
0.30	6.12	0.85	224.4
0.40	9.62	0.90	519.6

$$\mu^2 M\left(\frac{\beta_c^4}{1-\beta_c}\right)^{1/2} \geq 5.48\odot \tag{15}$$

This inequality provides an upper limit to $(1 - \beta_c)$ for a star of a given mass. Thus,

$$1 - \beta_c \leq 1 - \beta_* \tag{16}$$

where $(1 - \beta_*)$ is uniquely determined by the mass M of the star and the mean molecular weight, μ, by the quartic equation,

$$\mu^2 M = 5.48\left(\frac{1-\beta_*}{\beta_*^4}\right)^{1/2} \odot \tag{17}$$

In Table 1, we list the values of $1 - \beta_*$ for several values of $\mu^2 M$. From this table it follows in particular that, for a star of solar mass with a mean molecular weight equal to 1, the radiation pressure at the center cannot exceed 3 percent of the total pressure.

Fig. 1 (facing page, top). A comparison of an inhomogeneous distribution of density in a star (b) with the two homogeneous configurations with the constant density equal to the mean density (a) and equal to the density at the center (c). Fig. 2 (facing page, center). The full-line curve represents the exact (mass-radius) relation (l_1 is defined in Eq. 46 and M_3 denotes the limiting mass). This curve tends asymptotically to the dashed curve appropriate to the low-mass degenerate configurations, approximated by polytropes of index 3/2. The regions of the configurations which may be considered as relativistic [$\rho > (K_1/K_2)^3$] are shown shaded. Fig. 3 (facing page, bottom). The full-line curve represents the exact (mass-density) relation for the highly collapsed configurations. This curve tends asymptotically to the dashed curve as $M \rightarrow 0$.

What do we conclude from the foregoing calculation? We conclude that to the extent Eq. 17 is at the base of the equilibrium of actual stars, to that extent the combination of natural constants 3, providing a mass of proper magnitude for the measurement of stellar masses, is at the base of a physical theory of stellar structure.

Do Stars Have Enough Energy to Cool?

The same combination of natural constants 3 emerged soon afterward in a much more fundamental context of resolving a paradox Eddington had formulated in the form of an aphorism: "a star will need energy to cool." The paradox arose while considering the ultimate fate of a gaseous star in the light of the then new knowledge that white-dwarf stars, such as the companion of Sirius, exist, which have mean densities in the range 10^5 to 10^7 g cm^{-3}. As Eddington stated (*3*)

> I do not see how a star which has once got into this compressed state is ever going to get out of it. . . . It would seem that the star will be in an awkward predicament when its supply of subatomic energy fails.

The paradox posed by Eddington was reformulated in clearer physical terms by R. H. Fowler (*4*). His formulation was "The stellar material, in the white-dwarf state, will have radiated so much energy that it has less energy than the same matter in normal atoms expanded at the absolute zero of temperature. If part of it were removed from the star and the pressure taken off, what could it do?"

Quantitatively, Fowler's question arises in this way.

An estimate of the electrostatic energy, E_v, per unit volume of an assembly of atoms, of atomic number Z, ionized down to bare nuclei, is given by

$$E_v = 1.32 \times 10^{11} Z^2 \rho^{4/3} \qquad (18)$$

while the kinetic energy of thermal motions, E_{kin}, per unit volume of free particles in the form of a perfect gas of density, ρ, and temperature, T, is given by

$$E_{kin} = \frac{3}{2}\frac{k}{\mu H}\rho T = \frac{1.24 \times 10^8}{\mu}\rho T \qquad (19)$$

Now if such matter were released of the pressure to which it is subject, it can resume a state of ordinary normal atoms only if

$$E_{kin} > E_v \qquad (20)$$

or, according to Eqs. 18 and 19, only if

$$\rho < \left(0.94 \times 10^{-3}\frac{T}{\mu Z^2}\right)^3 \qquad (21)$$

This inequality will be clearly violated if the density is sufficiently high. This is the essence of Eddington's paradox as formulated by Fowler. And Fowler resolved this paradox in 1926 in a paper (*4*) entitled "Dense Matter"—one of the great landmark papers in the realm of stellar structure: in it the notions of Fermi statistics and of electron degeneracy are introduced for the first time.

The Degeneracy of the Electrons in White-Dwarf Stars

In a completely degenerate electron gas all the available parts of the phase space, with momenta less than a certain "threshold" value p_0—the Fermi threshold—are occupied consistently with the Pauli exclusion-principle, that is, with two electrons per "cell" of volume h^3 of the six-dimensional phase space. Therefore, if $n(p)dp$ denotes the number of electrons, per unit volume, between p and $p + dp$, then the assumption of complete degeneracy is equivalent to the assertion,

$$n(p) = \frac{8\pi}{h^3}p^2 \qquad (p \leq p_0)$$
$$= 0 \qquad (p > p_0) \qquad (22)$$

The value of the threshold momentum, p_0, is determined by the normalization condition

$$n = \int_0^{p_0} n(p)dp = \frac{8\pi}{3h^3}\, p_0^3 \qquad (23)$$

where n denotes the total number of electrons per unit volume.

For the distribution given by Eq. 22, the pressure p and the kinetic energy E_{kin} of the electrons (per unit volume), are given by

$$P = \frac{8\pi}{3h^3}\int_0^{p_0} p^3 v_{\mathrm{p}} dp \qquad (24)$$

and

$$E_{\mathrm{kin}} = \frac{8\pi}{h^3}\int_0^{p_0} p^2 T_{\mathrm{p}} dp \qquad (25)$$

where v_{p} and T_{p} are the velocity and the kinetic energy of an electron having a momentum p. If we set

$$v_{\mathrm{p}} = p/m \text{ and } T_{\mathrm{p}} = p^2/2m \qquad (26)$$

appropriate for nonrelativistic mechanics, in Eqs. 24 and 25, we find

$$P = \frac{8\pi}{15h^3m}\, p_0^5 = \frac{1}{20}\left(\frac{3}{\pi}\right)^{2/3}\frac{h^2}{m}\, n^{5/3} \qquad (27)$$

and

$$E_{\mathrm{kin}} = \frac{8\pi}{10h^3m}\, p_0^5 = \frac{3}{40}\left(\frac{3}{\pi}\right)^{2/3}\frac{h^2}{m}\, n^{5/3} \qquad (28)$$

Fowler's resolution of Eddington's paradox consists in this: at the temperatures and densities that may be expected to prevail in the interiors of the white-dwarf stars, the electrons will be highly degenerate and E_{kin} must be evaluated in accordance with Eq. 28 and *not* in accordance with Eq. 19; and Eq. 28 gives,

$$E_{\mathrm{kin}} = 1.39 \times 10^{13}\,(\rho/\mu)^{5/3} \qquad (29)$$

Comparing now the two estimates 18 and 29, we see that, for matter of the density occurring in the white dwarfs, namely $\rho \sim 10^5$ g cm^{-3}, the total kinetic energy is about two to four times the negative potential energy; and Eddington's paradox does not arise. Fowler concluded his paper with the following highly perceptive statement:

> The black-dwarf material is best likened to a single gigantic molecule in its lowest quantum state. On the Fermi-Dirac statistics, its high density can be achieved in one and only one way, in virtue of a correspondingly great energy content. But this energy can no more be expended in radiation than the energy of a normal atom or molecule. The only difference between black-dwarf matter and a normal molecule is that the molecule can exist in a free state while the black-dwarf matter can only so exist under very high external pressure.

The Theory of the White-Dwarf Stars: The Limiting Mass

The internal energy (= $3P/2$) of a degenerate electron gas that is associated with a pressure P is *zero-point energy*;

and the essential content of Fowler's paper is that this zero-point energy is so great that we may expect a star to eventually settle down to a state in which all of its energy is of this kind. Fowler's argument can be more explicitly formulated in the following manner (*5*).

According to the expression for the pressure given by Eq. 27, we have the relation,

$$P = K_1 \rho^{5/3}$$

where

$$K_1 = \frac{1}{20}\left(\frac{3}{\pi}\right)^{2/3} \frac{h^2}{m\,(\mu_e H)^{5/3}} \qquad (30)$$

where μ_e is the mean molecular weight per electron. An equilibrium configuration in which the pressure, P, and the density, ρ, are related in the manner,

$$P = K\rho^{1+1/n} \qquad (31)$$

is an *Emden polytrope* of index n. The degenerate configurations built on the equation of state 30 are therefore polytropes of index 3/2; and the theory of polytropes immediately provides the relation,

$$K_1 = 0.4242\,(GM^{1/3}R) \qquad (32)$$

or, numerically, for K_1 given by Eq. 30,

$$\log_{10}(R/R_{\odot}) = -\frac{1}{3}\log_{10}(M/\odot) - \frac{5}{3}\log_{10}\mu_e - 1.397 \qquad (33)$$

For a mass equal to the solar mass and $\mu_e = 2$, the relation 33 predicts $R = 1.26 \times 10^{-2}\ R_{\odot}$ and a mean density of 7.0×10^5 g cm^3. These values are precisely of the order of the radii and mean densities encountered in white-dwarf stars. Moreover, according to Eq. 32 and 33, the radius of the white-dwarf configuration is inversely proportional to the cube root of the mass. On this account, finite equilibrium configurations are predicted for all masses. And it came to be accepted that the white dwarfs represent the last stages in the evolution of all stars.

But it soon became clear that the foregoing simple theory based on Fowler's premises required modifications. For, the electrons at their threshold energies, at the centers of the degenerate stars, begin to have velocities comparable to that of light as the mass increases. Thus, already for a degenerate star of solar mass (with $\mu_e = 2$) the central density (which is about six times the mean density) is 4.19×10^6 g cm^{-3}; and this density corresponds to a threshold momentum $p_0 = 1.29\ mc$ and a velocity which is $0.63\ c$. Consequently, the equation of state must be modified to take into account the effects of special relativity. And this is easily done by inserting in Eqs. 24 and 25 the relations,

$$v_p = \frac{p}{m\,(1 + p^2/m^2c^2)^{1/2}}$$

and

$$T_p = mc^2\,[(1 + p^2/m^2c^2)^{1/2} - 1] \qquad (34)$$

in place of the nonrelativistic relations 26. We find that the resulting equation of state can be expressed, parametrically, in the form

$$P = Af(x) \text{ and } \rho = Bx^3 \qquad (35)$$

where

$$A = \frac{\pi m^4 c^5}{3h^3},\ B = \frac{8\pi m^3 c^3 \mu_e H}{3h^3} \qquad (36)$$

and

$$f(x) = x(x^2 + 1)^{1/2}(2x^2 - 3) + 3\sinh^{-1}x \qquad (37)$$

And similarly

$$E_{\text{kin}} = Ag(x) \quad (38)$$

where

$$g(x) = 8x^3 [(x^2 + 1)^{1/2} - 1] - f(x) \quad (39)$$

According to Eqs. 35 and 36, the pressure approximates the relation 30 for low enough electron concentrations ($x \ll 1$); but for increasing electron concentrations ($x \gg 1$), the pressure tends to (*6*)

$$P = \frac{1}{8}\left(\frac{3}{\pi}\right)^{1/3} hcn^{4/3} \quad (40)$$

This limiting form of relation can be obtained very simply by setting $v_p = c$ in Eq. 24; then

$$P = \frac{8\pi c}{3h^3}\int_0^{p_0} p^3 dp = \frac{2\pi c}{3h^3}\, p_0^4 \quad (41)$$

and the elimination of p_0 with the aid of Eq. 23 directly leads to Eq. 40.

While the modification of the equation of state required by the special theory of relativity appears harmless enough, it has, as we shall presently show, a dramatic effect on the predicted mass-radius relation for degenerate configurations.

The relation between P and ρ corresponding to the limiting form 41 is

$$P = K_2\, \rho^{4/3}$$

where

$$K_2 = \frac{1}{8}\left(\frac{3}{\pi}\right)^{1/3} \frac{hc}{(\mu_e H)^{4/3}} \quad (42)$$

In this limit, the configuration is an Emden polytrope of index 3. And it is well known that when the polytropic index is 3, the mass of the resulting equilibrium configuration is uniquely determined by the constant of proportionality, K_2, in the pressure-density relation. We have accordingly,

$$M_{\text{limit}} = 4\pi\left(\frac{K_2}{\pi G}\right)^{3/2}(2.018) = 0.197\left(\frac{hc}{G}\right)^{3/2}\frac{1}{(\mu_e H)^2} = 5.76\,\mu_e^{-2}\odot \quad (43)$$

(In Eq. 43, 2.018 is a numerical constant derived from the explicit solution of the Lane-Emden equation for $n = 3$.)

It is clear from general considerations (*7*) that *the exact mass-radius relation for the degenerate configurations must provide an upper limit to the mass of such configurations given by Eq. 43; and further, that the mean density of the configuration must tend to infinity*, while the radius tends to zero, and $M \rightarrow M_{\text{limit}}$. These conclusions, straightforward as they are, can be established directly by considering the equilibrium of configurations built on the exact equation of state given by Eqs. 35 to 37. It is found that the equation governing the equilibrium of such configurations can be reduced to the form (*8, 9*)

$$\frac{1}{\eta^2}\frac{d}{d\eta}\left(\eta^2\frac{d\phi}{d\eta}\right) = -\left(\phi^2 - \frac{1}{y_0^2}\right)^{3/2} \quad (44)$$

where

$$y_0^2 = x_0^2 + 1 \quad (45)$$

and mcx_0 denotes the threshold momentum of the electrons at the center of the configuration and η measures the radial distance in the unit

$$\left(\frac{2A}{\pi G}\right)^{1/2}\frac{1}{By_0} = l_1 y_0^{-1} \text{ (say)} \quad (46)$$

By integrating Eq. 44, with suitable boundary conditions and for various initially prescribed values of y_0, we can derive the exact mass-radius relation, as well as the other equilibrium properties, of the degenerate configurations. The

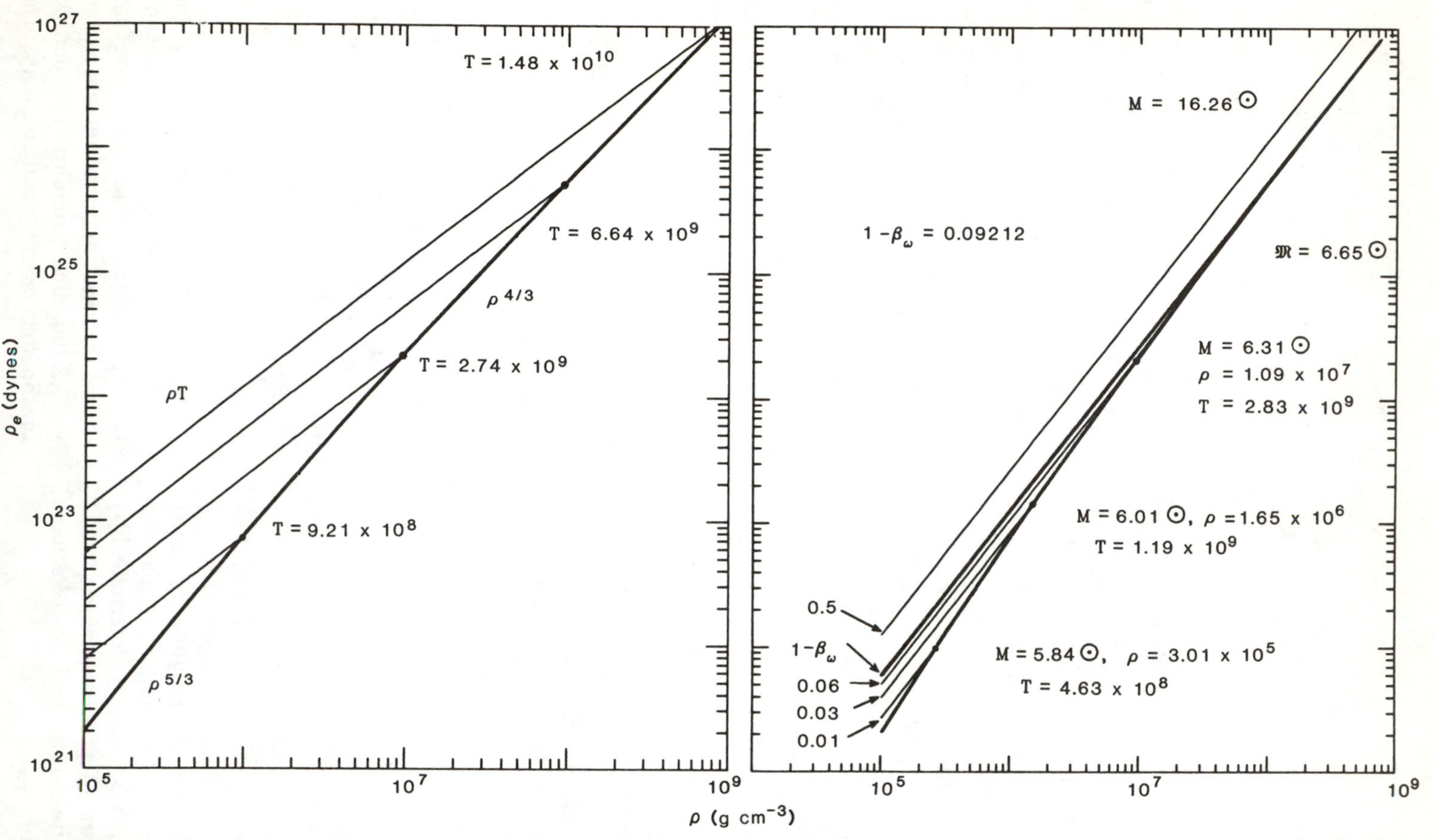

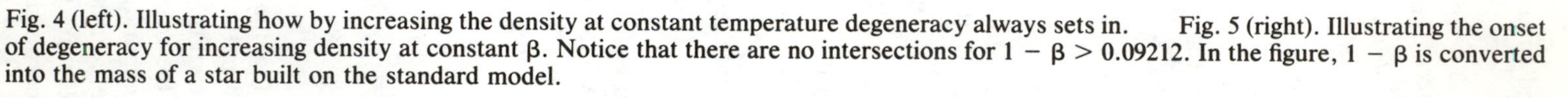
Fig. 4 (left). Illustrating how by increasing the density at constant temperature degeneracy always sets in. Fig. 5 (right). Illustrating the onset of degeneracy for increasing density at constant β. Notice that there are no intersections for $1 - \beta > 0.09212$. In the figure, $1 - \beta$ is converted into the mass of a star built on the standard model.

principal results of such calculations are illustrated in Figs. 2 and 3.

The important conclusions which follow from the foregoing considerations are: first, there is an upper limit, M_{limit}, to the mass of stars which can become degenerate configurations, as the last stage in their evolution; and second, that stars with $M > M_{limit}$ must have end states which cannot be predicted from the considerations we have presented so far. And finally, we observe that the combination of the natural constant 3 now emerges in the fundamental context of M_{limit} given by Eq. 43: its significance for the theory of stellar structure and stellar evolution can no longer be doubted.

Under What Conditions Can Normal Stars Develop Degenerate Cores?

Once the upper limit to the mass of completely degenerate configurations had been established, the question that required to be resolved was how to relate its existence to the evolution of stars from their gaseous state. If a star has a mass less than M_{limit}, the assumption that it will eventually evolve toward the completely degenerate state appears reasonable. But what if its mass is greater than M_{limit}? Clues as to what might ensue were sought in terms of the equations and inequalities of the second and third sections (*10, 11*).

The first question that had to be resolved concerns the circumstances under which a star, initially gaseous, will develop degenerate cores. From the physical side, the question, when departures from the perfect-gas equation of state 5 will set in and the effects of electron degeneracy will be manifested, can be readily answered.

Suppose, for example, that we continually and steadily increase the density, at constant temperature, of an assembly of free electrons and atomic nuclei, in a highly ionized state and initially in the form of a perfect gas governed by the equation of state 5. At first the electron pressure will increase linearly with ρ; but soon departures will set in and eventually the density will increase in accordance with the equation of state that describes the fully degenerate electron gas (see Fig. 4). The remarkable fact is that this limiting form of the equation of state is independent of temperature.

However, to examine the circumstances when, during the course of evolution, a star will develop degenerate cores, it is more convenient to express the electron pressure (as given by the classical perfect-gas equation of state) in terms of ρ and β_e defined in the manner (see Eq. 7),

$$p_e = \frac{k}{\mu_e H} \rho T = \frac{\beta_e}{1 - \beta_e} \frac{1}{3} aT^4 \qquad (47)$$

where p_e now denotes the electron pressure. Then, analogous to Eq. 9, we can write

$$p_e = \left[\left(\frac{k}{\mu_e H}\right)^4 \frac{3}{a} \frac{1 - \beta_e}{\beta_e}\right]^{1/3} \rho^{4/3} \qquad (48)$$

Comparing this with Eq. 42, we conclude that if

$$\left[\left(\frac{k}{\mu_e H}\right)^4 \frac{3}{a} \frac{1 - \beta_e}{\beta_e}\right]^{1/3} > K_2 = \frac{1}{8}\left(\frac{3}{\pi}\right)^{1/3} \frac{hc}{(\mu_e H)^{4/3}} \qquad (49)$$

the pressure p_e given by the classical perfect-gas equation of state will be greater than that given by the equation if degeneracy were to prevail, not only for the prescribed ρ and T, but for *all* ρ and T having the same β_e.

Inserting for a its value given in Eq. 13, we find that the inequality 49 reduces to

$$\frac{960}{\pi^4}\,\frac{1-\beta_e}{\beta_e} > 1 \qquad (50)$$

or, equivalently (see Fig. 5)

$$1 - \beta_e > 0.0921 = 1 - \beta_\omega \text{ (say)} \qquad (51)$$

For our present purposes, the principal content of the inequality 51 is the criterion that for a star to develop degeneracy, it is necessary that the radiation pressure be less than 9.2 percent of $(p_e + p_{rad})$. This last inference is so central to all current schemes of stellar evolution that the directness and the simplicity of the early arguments are worth repeating.

The two principal elements of the early arguments were these: first, that radiation pressure becomes increasingly dominant as the mass of the star increases; and second, that the degeneracy of electrons is possible only so long as the radiation pressure is not a significant fraction of the total pressure—indeed, as we have seen, it must not exceed 9.2 percent of $(p_e + p_{rad})$. The second of these elements in the arguments is a direct and an elementary consequence of the physics of degeneracy; but the first requires some amplification.

That radiation pressure must play an increasingly dominant role as the mass of the star increases is one of the earliest results in the study of stellar structure that was established by Eddington. A quantitative expression for this fact is given by Eddington's *standard model* which lay at the base of his early studies summarized in his *The Internal Constitution of the Stars*.

On the standard model, the fraction β (= gas pressure/total pressure) is a constant through a star. On this assumption, the star is a polytrope of index 3 as is apparent from Eq. 9; and, in consequence, we have the relation (see Eq. 43)

$$M = 4\pi \left[\frac{C(\beta)}{\pi G}\right]^{3/2} (2.018) \qquad (52)$$

where $C(\beta)$ is defined in Eq. 9. Equation 52 provides a quartic equation for β analogous to Eq. 17 for β_*. Equation 52 for $\beta = \beta_\omega$ gives

$$M = 0.197\, \beta_\omega^{-3/2} \left(\frac{hc}{G}\right)^{3/2} \frac{1}{(\mu H)^2} = 6.65\, \mu^{-2}\odot = \mathfrak{M} \text{ (say)} \qquad (53)$$

On the standard model, then, stars with masses exceeding $\mathfrak{M}$ will have radiation pressures exceeding 9.2 percent of the total pressure. Consequently, stars with $M > \mathfrak{M}$ cannot, at any stage during the course of their evolution, develop degeneracy in their interiors. Therefore, for such stars an eventual white-dwarf state is not possible unless they are able to eject a substantial fraction of their mass.

The standard model is, of course, only a model. Nevertheless, except under special circumstances, briefly noted below, experience has confirmed the essential qualitative correctness of the conclusions drawn from the standard model, namely that the evolution of stars of masses exceeding 7 to 8 ⊙ must proceed along lines very different from those of less massive stars. These conclusions, which were arrived at some 50 years ago, appeared then so convincing that assertions such as these were made with confidence:

> Given an enclosure containing electrons and atomic nuclei (total charge zero) what happens if we go on compressing the material indefinitely? (1932) (*10*)
>
> The life history of a star of small mass must be essentially different from the life history of a star of large mass. For a star of small mass the natural white-dwarf stage is an initial step towards complete extinction. A star of large mass cannot pass into the white-dwarf stage and one is left speculating on other possibilities. (1934) (*8*)

And these statements have retained their validity.

While the evolution of the massive stars was thus left uncertain, there was no such uncertainty regarding the final states of stars of sufficiently low mass (*11*). The reason is that by virtue, again, of the inequality 10, the maximum central pressure attainable in a star must be less than that provided by the degenerate equation of state, so long as

$$\frac{1}{2}G\left(\frac{4}{3}\pi\right)^{1/3} M^{2/3} < K_2 = \frac{1}{8}\left(\frac{3}{\pi}\right)^{1/3} \frac{hc}{(\mu_e H)^{4/3}} \tag{54}$$

or equivalently

$$M < \frac{3}{16\pi}\left(\frac{hc}{G}\right)^{3/2} \frac{1}{(\mu_e H)^2} = 1.74\ \mu_e^{-2}\odot \tag{55}$$

We conclude that there can be no surprises in the evolution of stars of mass less than 0.43 $\odot$ (if $\mu_e = 2$). The end stage in the evolution of such stars can only be that of the white dwarfs. (Parenthetically, we may note here that the inequality 55 implies that the so-called "mini" black holes of mass $\sim 10^{15}$ g cannot naturally be formed in the present astronomical universe.)

The Evolution of Massive Stars and the Onset of Gravitational Collapse

It became clear, already from the early considerations, that the inability of the massive stars to become white dwarfs must result in the development of much more extreme conditions in their interiors and, eventually, in the onset of gravitational collapse attended by the supernova phenomenon. But the precise manner in which all this will happen has been difficult to ascertain in spite of great effort by several competent groups of investigators. The facts which must be taken into account appear to be the following.*

In the first instance, the density and the temperature will steadily increase without the inhibiting effect of degeneracy since for the massive stars considered $1 - \beta_e > 1 - \beta_\omega$. On this account, "nuclear ignition" of carbon, say, will take place which will be attended by the emission of neutrinos. This emission of neutrinos will effect a cooling and a lowering of $(1 - \beta_e)$; but it will still be in excess of $1 - \beta_\omega$. The important point here is that the emission of neutrinos acts selectively in the central regions and is the cause of the lowering of $(1 - \beta_e)$ in these regions. The density and the temperature will continue to increase until the next ignition of neon takes place followed by further emission of neutrinos and a further lowering of $(1 - \beta_e)$. This succession of nuclear ignitions and lowering of $(1 - \beta_e)$ will continue until $1 - \beta_e < 1 - \beta_\omega$ and a relativistically degenerate core with a mass approximately that of the limiting mass ($= 1.4\ \odot$ for $\mu_e = 2$) forms at the center. By this stage, or soon afterward, instability of some sort is expected to set in (see the following section) followed by gravitational collapse and the phenomenon of the supernova (of type II). In some instances, what was originally the highly relativistic degenerate core of approximately 1.4 ⊙, will be left behind as a neutron star. That this happens sometimes is confirmed by the fact that in those cases for which reliable estimates of the masses of pulsars exist, they are consistently close to 1.4 ⊙. However, in other instances—perhaps, in the majority of the instances—what is left behind, after all "the dust has settled," will have masses in excess of that allowed for stable neutron stars; and in these instances black holes will form.

In the case of less massive stars ($M \sim$ 6 to 8 ⊙) the degenerate cores, which are initially formed, are not highly relativistic. But the mass of the core increases with the further burning of the nuclear fuel at the interface of the core and the mantle; and when the core reaches the limiting mass, an explosion occurs following instability; and it is believed that this is the cause underlying supernova phenomenon of type I.

From the foregoing brief description of what may happen during the late stages in the evolution of massive stars, it is clear that the problems one encounters are of exceptional complexity, in which a great variety of physical factors compete. This is clearly not the occasion for me to enter into a detailed discussion of these various questions.

Instabilities of Relativistic Origin:

Vibrational Instability of Spherical Stars

I now turn to the consideration of certain types of stellar instabilities which are derived from the effects of general relativity and which have no counterparts in the Newtonian framework. It will appear that these new types of instabilities of relativistic origin may have essential roles to play in discussions pertaining to gravitational collapse and late stages in the evolution of massive stars.

We shall consider first the stability of spherical stars for purely radial perturbations. The criterion for such stability follows directly from the linearized equations governing the spherically symmetric radial oscillations of stars. In the framework of the Newtonian theory of gravitation, the stability for radial perturbations depends only on an average value of the adiabatic exponent, Γ_1, which is the ratio of the fractional Lagrangian changes in the pressure and in the density experienced by a fluid element following the motion; thus,

$$\Delta P/P = \Gamma_1 \Delta\rho/\rho \qquad (56)$$

And the Newtonian criterion for stability is

$$\overline{\Gamma_1} = \int_0^m \Gamma_1(r)P(r)dM(r) \div \int_0^M P(r)dM(r) > \frac{4}{3} \qquad (57)$$

If $\overline{\Gamma_1} < 4/3$, *dynamical instability* of a global character will ensue with an *e*-folding time measured by the time taken by a sound wave to travel from the center to the surface.

When one examines the same problem in the framework of the general theory of relativity, one finds *(12)* that, again, the stability depends on an average value of Γ_1; but, contrary to the Newtonian result, the stability now depends on the radius of the star as well. Thus, one finds that no matter how high $\overline{\Gamma_1}$ may be, instability will set in provided the radius is less than a certain determinate multiple of the *Schwarzschild radius*,

$$R_S = 2\,GM/c^2 \tag{58}$$

Thus, if for the sake of simplicity, we assume that Γ_1 is a constant through the star and equal to 5/3, then the star will become dynamically unstable for radial perturbations, if $R_1 < 2.4\, R_S$. And further, if $\Gamma_1 \to \infty$, instability will set in for all $R < (9/8)\, R_S$. *The radius* $(9/8)\, R_S$ *defines, in fact, the minimum radius which any gravitating mass, in hydrostatic equilibrium, can have in the framework of general relativity*. This important result is implicit in a fundamental paper by Karl Schwarzschild published in 1916. [Schwarzschild actually proved that for a star in which the energy density is a uniform $R > (9/8)\, R_S$.]

In one sense, the most important consequence of this instability of relativistic origin is that if Γ_1 (again assumed to be a constant for the sake of simplicity) differs from and is greater than 4/3 only by a small positive constant, then the instability will set in for a radius R which is a large multiple of R_S; and, therefore, under circumstances when the effects of general relativity, on the structure of the equilibrium configuration itself, are hardly relevant. Indeed, it follows (*13*) from the equations governing radial oscillations of a star, in a first post-Newtonian approximation to the general theory of relativity, that instability for radial perturbations will set in for all

$$R < \frac{K}{\Gamma_1 - 4/3}\,\frac{2GM}{c^2} \tag{59}$$

where K is a constant which depends on the *entire*† march of density and pressure in the equilibrium configuration in the Newtonian framework. Thus, for a

Table 2. Values of the constant K in the inequality 59 for various polytropic indices, n.

n	K	n	K
0	0.452381	3.25	1.28503
1.0	0.565382	3.5	1.49953
1.5	0.645063	4.0	2.25338
2.0	0.751296	4.5	4.5303
2.5	0.900302	4.9	22.906
3.0	1.12447	4.95	45.94

polytrope of index n, the value of the constant is given by

$$K = \frac{5-n}{18}\left[\frac{2\,(11-n)}{(n+1)\,\xi_1^4\,|\theta_1'|^3}\int_0^{\xi_1}\theta\left(\frac{d\theta}{d\xi}\right)^2\xi^2 d\xi + 1\right] \tag{60}$$

where θ is the Lane-Emden function in its standard normalization ($\theta = 1$ at $\xi = 0$), ξ is the dimensionless radial coordinate, ξ_1 defines the boundary of the polytrope (where $\theta = 0$), and θ_1' is the derivative of θ at ξ_1.

In Table 2, we list the values of K for different polytropic indices. It should be particularly noted that K increases without limit for $n \to 5$ and the configuration becomes increasingly centrally condensed.‡ Thus, already for $n = 4.95$ (for which polytropic index $\rho_c = 8.09 \times 10^6\, \bar{\rho}$), $K \sim 46$. In other words, for the highly centrally condensed massive stars (for

†It is for this reason that we describe the instability as *global*.

‡Since this was written it has been possible to show [Chandrasekhar and Lebovitz, *Mon. Not. R. Astron. Soc.* **207**, 13P (1984)] that for $n \to 5$, the asymptotic behavior of K is given by

$$K \to 2.3056/(5-n)$$

and, further, that along the polytropic sequence, the criterion for instability 59 can be expressed alternatively in the form

$$R < 0.2264\,\frac{2GM}{c^2}\,\frac{1}{\Gamma_1 - 4/3}\left(\frac{\rho_c}{\bar{\rho}}\right)^{1/3} \quad (\rho_c > 10^6\,\bar{\rho})$$

which Γ_1 may differ from 4/3 by as little as 0.01),§ the instability of relativistic origin will set in, already, when its radius falls below $5 \times 10^3\ R_S$. Clearly this relativistic instability must be considered in the contexts of these problems.

A further application of the result described in the preceding paragraph is to degenerate configurations near the limiting mass (*14*). Since the electrons in these highly relativistic configurations have velocities close to the velocity of light, the effective value of Γ_1 will be very close to 4/3 and the post-Newtonian relativistic instability will set in for a mass slightly less than that of the limiting mass. On account of the instability for radial oscillations setting in for a mass less than M_{limit}, the period of oscillation, along the sequence of the degenerate configurations, must have a minimum. This minimum can be estimated to be about two seconds (see Fig. 6). Since pulsars, when they were discovered, were known to have periods much less than this minimum value, the possibility of their being degenerate configurations

§By reason of the dominance of the radiation pressure in these massive stars and of β being very close to zero.

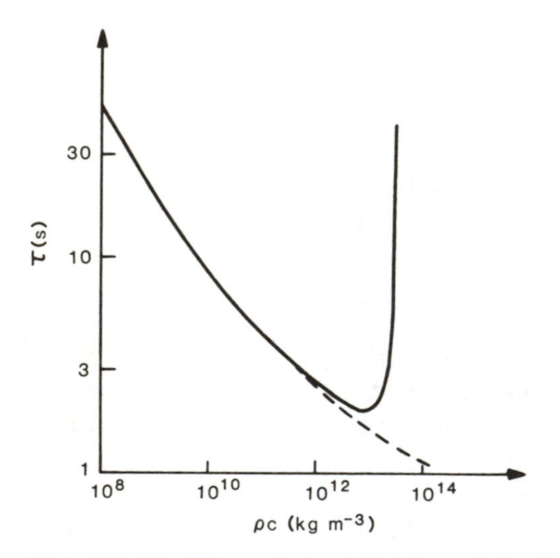

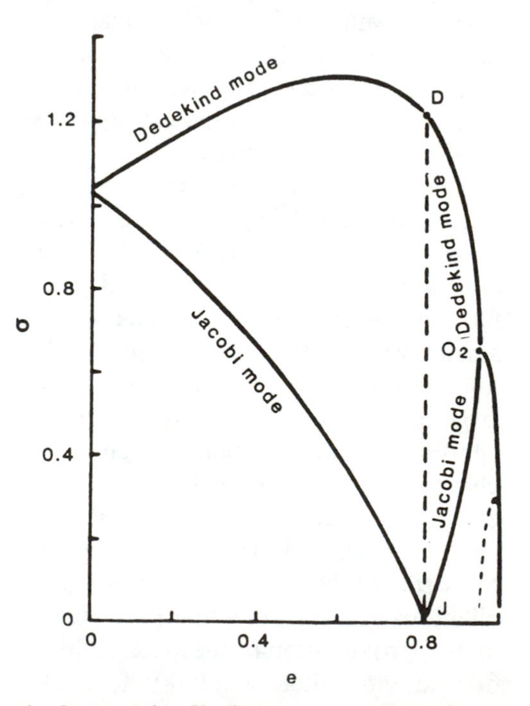

Fig. 6 (left). The variation of the period of radial oscillation along the completely degenerate configurations. Notice that the period tends to infinity for a mass close to the limiting mass. There is consequently a minimum period of oscillation along these configurations; and the minimum period is approximately 2 seconds. [J. Skilling, *Pulsating Stars* (Plenum, New York, 1968), reprinted with permission] Fig. 7 (right). The characteristic frequencies [in the unit $(\pi G\rho)^{1/2}$] of the two even modes of second-harmonic oscillation of the Maclaurin spheroid. The Jacobi sequence bifurcates from the Maclaurin sequence by the mode that is neutral ($\sigma = 0$) at $e = 0.813$; and the Dedekind sequence bifurcates by the alternative mode at D. At O_2 ($e = 0.9529$) the Maclaurin spheroid becomes dynamically unstable. The real and the imaginary parts of the frequency, beyond O_2, are shown by the full line and the dashed curves, respectively. Viscous dissipation induces instability in the branch of the Jacobi mode; and radiation-reaction induces instability in the branch DO_2 of the Dedekind mode.

near the limiting mass was ruled out; and this was one of the deciding factors in favor of the pulsars being neutron stars. (But by a strange irony, for reasons we have briefly explained in the preceding section, pulsars which have resulted from supernova explosions have masses close to 1.4 ⊙!)

Finally, we may note that the radial instability of relativistic origin is the underlying cause for the *existence* of a maximum mass for stability: it is a direct consequence of the equations governing hydrostatic equilibrium in general relativity. [For a complete investigation on the periods of radial oscillation of neutron stars for various admissible equations of state, see a recent paper by Detweiler and Lindblom (*15*).]

Instabilities of Relativistic Origin: Secular Instability of Rotating Stars

I now turn to a different type of instability which the general theory of relativity predicts for rotating configurations. This new type of instability (*16*) has its origin in the fact that the general theory of relativity builds into rotating masses a dissipative mechanism derived from the possibility of the emission of gravitational radiation by nonaxisymmetric modes of oscillation. It appears that this instability limits the periods of rotation of pulsars. But first, I shall explain the nature and the origin of this type of instability.

It is well known that a possible sequence of equilibrium figures of rotating homogeneous masses is the Maclaurin sequence of oblate spheroids (*17*). When one examines the second harmonic oscillations of the Maclaurin spheroid, in a frame of reference rotating with its angular velocity, one finds that for two of these modes, whose dependence on the azimuthal angle is given by $e^{2i\varphi}$, the characteristic frequencies of oscillation, σ, depend on the eccentricity e in the manner illustrated in Fig. 7. It will be observed that one of these modes becomes neutral (that is, $\sigma = 0$) when $e = 0.813$ and that the two modes coalesce when $e = 0.953$ and become complex conjugates of one another beyond this point. Accordingly, the Maclaurin spheroid becomes *dynamically unstable* at the latter point (first isolated by Riemann). On the other hand, the origin of the neutral mode at $e = 0.813$ is that at this point a new equilibrium sequence of triaxial ellipsoids—the ellipsoids of Jacobi—bifurcate. On this latter account, Lord Kelvin conjectured in 1883 that "if there be any viscosity, however slight . . . the equilibrium beyond $e = 0.81$ cannot be secularly stable." Kelvin's reasoning was this: viscosity dissipates energy but not angular momentum. And since for equal angular momenta the Jacobi ellipsoid has a lower energy content than the Maclaurin spheroid, one may expect that the action of viscosity will be to dissipate the excess energy of the Maclaurin spheroid and transform it into the Jacobi ellipsoid with the lower energy. A detailed calculation (*18*) of the effect of viscous dissipation on the two modes of oscillation, illustrated in Fig. 7, does confirm Lord Kelvin's conjecture. It is found that viscous dissipation makes the mode, which becomes neutral at $e = 0.813$, unstable beyond this point with an e-folding time which depends inversely on the magnitude of the kinematic viscosity and which further decreases monotonically to zero at the point, $e = 0.953$, where the dynamical instability sets in.

Since the emission of gravitational radiation dissipates *both* energy and angular momentum, it does *not* induce instability in the Jacobi mode; instead it induces instability in the *alternative* mode at the same eccentricity. In the first instance this may appear surprising; but the situation we encounter here clarifies some important issues.

If instead of analyzing the normal modes in the rotating frame we had analyzed them in the inertial frame, we should have found that the mode which becomes unstable by radiation reaction, at $e = 0.813$, is in fact neutral at this point. And the neutrality of *this* mode in the inertial frame corresponds to the fact that the neutral deformation at this point is associated with the bifurcation (at this point) of a new triaxial sequence—the sequence of the Dedekind ellipsoids. These Dedekind ellipsoids, while they are congruent to the Jacobi ellipsoids, differ from them in that they are at rest in the inertial frame and owe their triaxial figures to internal vortical motions. An important conclusion that would appear to follow from these facts is that in the framework of general relativity we can expect secular instability, derived from radiation-reaction to arise from a Dedekind mode of deformation (which is quasi-stationary in the inertial frame) rather than the Jacobi mode (which is quasi-stationary in the rotating frame).

A further fact concerning the secular instability induced by radiation-reaction, discovered subsequently by Friedman (*19*) and by Comins (*20*), is that the modes belonging to higher values of m (=3, 4, , ,) become unstable at smaller eccentricities though the e-folding times for the instability become rapidly longer. Nevertheless, it appears from some preliminary calculations of Friedman (*21*) that it is the secular instability derived from modes belonging to $m = 3$ (or 4) that limit the periods of rotation of the pulsars.

It is clear from the foregoing discussions that the two types of instabilities of relativistic origin we have considered are destined to play significant roles in the contexts we have considered.

The Mathematical Theory of Black Holes

So far, I have considered only the restrictions on the last stages of stellar evolution that follow from the existence of an upper limit to the mass of completely degenerate configurations and from the instabilities of relativistic origin. From these and related considerations, the conclusion is inescapable that black holes will form as one of the natural end products of stellar evolution of massive stars; and further that they must exist in large numbers in the present astronomical universe. In this last section I want to consider very briefly what the general theory of relativity has to say about them. But first, I must define precisely what a black hole is.

A black hole partitions the three-dimensional space into two regions: an inner region which is bounded by a smooth two-dimensional surface called the *event horizon*; and an outer region, external to the event horizon, which is asymptotically flat; and it is required (as a part of the definition) that no point in the inner region can communicate with any point of the outer region. This incommunicability is guaranteed by the impossibility of any light signal, originating in the inner region, crossing the event horizon. The requirement of asymptotic flatness of the outer region is equivalent to the requirement that the black hole is isolated in space and that far from the event horizon the space-time approaches the customary space-time of terrestrial physics.

In the general theory of relativity, we must seek solutions of Einstein's vacuum equations compatible with the two requirements I have stated. It is a startling fact that compatible with these very simple and necessary requirements, the general theory of relativity allows for stationary (that is, time-independent) black holes exactly a single, unique, two-parameter family of solutions. This is the Kerr family, in which the two parameters are the mass of the black hole and the angular momentum of the black hole. What is even more remarkable, the metric describing these solutions is simple and can be explicitly written down.

I do not know if the full import of what I have said is clear. Let me explain.

Black holes are macroscopic objects with masses varying from a few solar masses to millions of solar masses. To the extent they may be considered as stationary and isolated, to that extent, they are all, every single one of them, described *exactly* by the Kerr solution. This is the only instance we have of an exact description of a macroscopic object. Macroscopic objects, as we see them all around us, are governed by a variety of forces, derived from a variety of approximations to a variety of physical theories. In contrast, the only elements in the construction of black holes are our basic concepts of space and time. They are, thus, almost by definition, the most perfect macroscopic objects there are in the universe. And since the general theory of relativity provides a single unique two-parameter family of solutions for their descriptions, they are the simplest objects as well.

Turning to the physical properties of the black holes, we can study them best by examining their reaction to external perturbations such as the incidence of waves of different sorts. Such studies reveal an analytic richness of the Kerr space-time which one could hardly have expected. This is not the occasion to elaborate on these technical matters (*22*). Let it suffice to say that contrary to every prior expectation, all the standard equations of mathematical physics can be solved exactly in the Kerr space-time. And the solutions predict a variety and range of physical phenomena which black holes must exhibit in their interaction with the world outside.

The mathematical theory of black holes is a subject of immense complexity. But its study has convinced me of the basic truth of the ancient mottoes, "The simple is the seal of the true" and "Beauty is the splendour of truth."

References and Notes

1. A. S. Eddington, *The Internal Constitution of the Stars* (Cambridge Univ. Press, Cambridge, England, 1926), p. 16.
2. S. Chandrasekhar, *Mon. Not. R. Astron. Soc.* **96**, 644 (1936).
3. A. S. Eddington, *The Internal Constitution of the Stars* (Cambridge Univ. Press, Cambridge, England, 1926), p. 172.
4. R. H. Fowler, *Mon. Not. R. Astron. Soc.* **87**, 114 (1926).
5. S. Chandrasekhar, *Philos. Mag.* **11**, 592 (1931).
6. ———, *Astrophys. J.* **74**, 81 (1931).
7. ———, *Mon. Not. R. Astron. Soc.* **91**, 456 (1931).
8. ———, *Observatory* **57**, 373 (1934).
9. ———, *Mon. Not. R. Astron. Soc.* **95**, 207 (1935).
10. ———, *Z. Astrophys.* **5**, 321 (1932).
11. ———, *Observatory* **57**, 93 (1934).
12. ———, *Astrophys. J.* **140**, 417 (1964); see also *Phys. Rev. Lett.* **12**, 114 and 437 (1964).
13. ———, *Astrophys. J.* **142**, 1519 (1965).
14. ——— and R. F. Tooper, *ibid.* **139**, 1396 (1964).
15. S. Detweiler and L. Lindblom, *Astrophys. J. Suppl.* **53**, 93 (1983).
16. S. Chandrasekhar, *Astrophys. J.* **161**, 561 (1970); see also *Phys. Rev. Lett.* **24**, 611 and 762 (1970).
17. For an account of these matters pertaining to the classical ellipsoids, see S. Chandrasekhar, *Ellipsoidal Figures of Equilibrium* (Yale Univ. Press, New Haven, Conn., 1968).
18. S. Chandrasekhar, *ibid.*, chapter 5, section 37.

19. J. L. Friedman, *Commun. Math. Phys.* **62**, 247 (1978); see also J. L. Friedman and B. F. Schutz, *Astrophys. J.* **222**, 281 (1977).
20. N. Comins, *Mon. Not. R. Astron. Soc.* **189**, 233 and 255 (1979).
21. J. L. Friedman, *Phys. Rev. Lett.* **51**, 11 (1983).
22. My investigations on the mathematical theory of black holes, continued over the years 1974 to 1983, are summarized in my last book, *The Mathematical Theory of Black Holes* (International Series of Monographs on Physics, Clarendon Press, Oxford, 1983).
23. J. Skilling, *Pulsating Stars*, (Plenum, New York, 1968), p. 59.
24. The reader may wish to consult the following additional references: S. Chandrasekhar, "Edward Arthur Milne: His part in the development of modern astrophysics," *Q. J. R. Astron. Soc.* **21**, 93 (1980); *Eddington: The Most Distinguished Astrophysicist of His Time* (Cambridge Univ. Press, Cambridge, England, 1983).

This material originally appeared in *Science* **226**, 2 November 1984.

---

*I am grateful to Professor D. Arnett for guiding me through the recent literature and giving me advice.

Part III
Galaxies and Cosmology

14. The Most Distant Known Galaxies

Richard G. Kron

It is now possible to see astronomical objects at distances of many billions of light-years, with continuing gain in depth as new observational techniques are tried and old techniques are refined. Such distances are already appreciable fractions of the scale of the universe itself, which means that we may be able to see far enough to witness the formation of luminous objects (galaxies and quasars), the epoch that Sandage (*1*) has picturesquely called the edge of the world. In this chapter I review the situation at present, especially with regard to the discovery and spectroscopic study of distant galaxies.

The recessional velocities of galaxies derived from Doppler shifts are correlated with their distances, which can be estimated from their apparent brightnesses or angular sizes. This relationship is linear for small velocities and is more or less independent of direction in the sky: the hypothesis that we are not privileged in our vantage point leads to the concept of a uniformly expanding universe. These velocities, or redshifts, are measured by the dimensionless number $z = (\lambda_o - \lambda_e)/\lambda_e$, where λ_o is the observed wavelength of a spectral line and λ_e is the laboratory wavelength. In the following it will be assumed that the redshift is a strict function of distance and consequently of the travel time of light from the galaxy. The value of the proportionality factor between redshift and distance is controversial: a redshift of unity is believed to correspond to a distance somewhere between 5 billion and 9 billion light-years. This review will stress redshifts as the paramount method for determining relative distances. The enterprise of obtaining redshifts for distant galaxies is almost exclusively the story of optical techniques because of the concentration of strong spectral features in the optical band and the favorable signal-to-background ratio.

In the expanding universe picture sketched above, galaxies should evolve in various ways because, as time goes by, more and more gas is locked up in stars or stellar remnants (*2*) (unless there is counterbalancing replenishment by infalling gas). In addition, the abundance of heavy elements such as iron should be smaller in a young galaxy than in an older galaxy of the same total mass, since less time has elapsed for synthesis in the cores of stars. The supply of gas in a galaxy may change, through either star formation (which may result in a change in the optical appearance of the galaxy), rapid "stripping" of gas by an external agent (*3*), or infall of primordial gas. The internal stellar distribution may readjust on time scales of a rotation period or longer, due for instance to resonance in the field of a rotating nonaxisymmetric potential (*4*), such as the bar of a barred spiral galaxy. Another type of dynamical evolution is called galactic cannibalism, which refers to inelastic collisions and mergers between galaxies, occurring most favorably in places with a high density of slow-moving galaxies (*5*). Tidal forces during close encounters can lead to irreversible structural changes (*6*). One of the strongest motivations for finding the most remote possible objects is the expectation that such phenomena should be apparent from comparison of the properties of very distant galaxies (observed in a younger state because of the finite speed of light) with those of nearby galaxies.

Distant Galaxies and Cosmological Investigations

Historically, the most important motivation for studying faint galaxies was the task of mapping out the expansion of the universe by using galaxies to mark out the cosmological geometry. The idea is that by observing the detailed relationship between redshift and distance (deduced from the apparent brightness or angular size of a galaxy), it is possible to determine the deceleration of the expansion. Depending on the size of the deceleration, the universe may or may not continue to expand indefinitely, circumstances which correspond to an open, infinite geometry or a closed, finite geometry, respectively. In this fashion there is a straightforward theoretical connection between the large-scale dynamics and the large-scale geometry of the universe. But since galaxies can be expected to evolve in luminosity, the distance coordinate that one infers from the apparent brightness is accordingly uncertain. In practice it is extremely difficult to separate cosmological effects from evolutionary effects.

In any event, galaxies chosen to delineate the large-scale geometry should be luminous so that they can be seen to large distances. It was found early on that the brightest cluster galaxies have small dispersions in both luminosity and spectral characteristics. This latter property implies that these galaxies (called first-ranked galaxies, of either the giant elliptical, D, or the supergiant D class) are likely to evolve in similar ways. One disadvantage is that they have relatively little flux in the ultraviolet, which means that with redshift their optical apparent brightness becomes especially faint. If, say, the fourth-ranked galaxy in a cluster were a spiral with more ultraviolet light (due to a population of young, hot stars), it could appear at high redshift to be the first-ranked member (*7*). Near infrared broadband measurements are potentially very valuable in this regard, since such data are supposedly relatively insensitive to the young stellar population and are not as badly affected by the redshift dimming (*8*). Anyway, great care has to be taken to guarantee that the high-redshift galaxies in a sample are indeed of the same type as the low-redshift comparison galaxies. This is perhaps most efficiently done by "tuning" the bandpass in which the survey for clusters is done to longer wavelengths for higher expected redshifts, so that the same emitted wavelengths are sampled (*9*). Another problem is that of interloping field galaxies seen in projection against a cluster. These difficulties notwithstanding, the use of first-ranked cluster galaxies for testing cosmological models has continued to be very popular (*10*).

Limitations Imposed by Surface Brightness

Even the most distant galaxies usually have angular sizes somewhat larger than the blur circle imposed by the earth's atmosphere and the telescope optics. Extended images by definition have a sensible angular size, which means that the surface brightness for such images is another measurable quantity. The significance of this is that the observed surface brightness depends only on the intrinsic surface brightness and on the redshift, not on the cosmological model. Tolman (*11*) showed that the surface brightness, integrated over all wavelengths, is proportional to $(1 + z)^{-4}$, due to relativistic effects. The aberration of light increases angles by a factor of $(1 + z)$, thus accounting for two factors of $(1 + z)$ for the area of a receding disk. Also, the photon stream (rate of arrival) is diluted by the expansion by one factor of $(1 + z)$, and finally each photon appears with energy lower by $(1 + z)$ by virtue of the redshift.

To illustrate the observational limitations imposed by this strong dependence of surface brightness on redshift, consider that the net surface brightness of our galaxy at the position of the sun, as seen from the outside, would be about the same as the natural surface brightness of the night sky. Only the very brightest part of the nearest giant galaxy, the Andromeda Nebula, can be seen by eye. At redshift $z = 1$ the surface brightness would be lower by a factor of 16—actually more, since for almost all galaxies there is less energy at a wavelength of 250 nanometers than there is at 500 nanometers. This is the most important reason why galaxies have not been seen much beyond $z = 1$. For such galaxies to be detectable, there must have been a compensating increase at early epochs of the intrinsic surface brightness, as might be expected if the initial stellar population were formed rapidly (*12*). This idea has inspired a number of attempts to detect extended objects with very large redshifts (*13*), so far without success.

Remarks on Quasars

Quasar emission line redshifts are conventionally, but not universally, ascribed to the same cause as that of galaxy redshifts, namely the cosmological expansion. This view will be adopted here for the sake of the argument. Quasars can be seen to much greater redshifts because they are typically ~ 100 times more luminous than giant galaxies, and they are effectively point sources and so do not suffer any large decrease in surface brightness. Therefore, if quasars cluster with galaxies, they could be used as pointers to galaxies in their vicinity (*14*). Also, their high intensity compared with galaxies makes quasars practical background sources against which absorption lines in intervening gas clouds (in or outside galaxies) can be observed (*15*), although there is evidence that some of the absorption lines seen in quasar spectra are due to material physically associated with the quasar (*15*). In the former event, quasars can be used to study distant matter arguably associated with galaxies.

It has sometimes been suggested that very distant galaxies seen in an early phase might bear a superficial resemblance to the optical appearance of quasars (*16*); that is, they might be compact, blue, have strong emission lines from highly excited gas, and be sources of x-rays (*17*). The implication is that such "primeval galaxies" might already have been discovered, appearing in catalogs of quasars. The main difficulty with this line of thought is that quasars have spectra that do not look like an ensemble of stellar photospheres. Rather, the continuum spectra observed in the optical are consistent with radiation from a compact nonthermal source. Furthermore, since a large fraction of quasars vary in light with large amplitude on short time scales (*18*), in these cases the light cannot be dominated by starlight and the energy must be produced within a small volume.

Radio and X-ray Pointers to Distant Galaxies

Since about 1960, another popular way to find potential high-redshift galaxies has been the optical identification of powerful radio sources (*19*). Many of these sources are first-ranked cluster galaxies that have ordinary optical properties. Since there is a correlation between optical luminosity and the probability of being a strong radio source, it is not surprising that there is some evidence that radio galaxies are somewhat more luminous (*20*) than radio-quiet counterparts. Another feature of radio galaxies is their higher probability of displaying emission lines (the most important of which are due to O^{+} and O^{2+}) in their spectra, which allows a redshift to be determined much more easily than from spectra containing only absorption lines. While there is no doubt that radio identifications do produce samples of distant galaxies, they also tend to uncover peculiar galaxies, especially galaxies whose optical radiation is derived partially from nonthermal processes, as in quasars and N galaxies.

An analogous technique for finding distant objects is identification of extragalactic x-ray sources, which tend to be either quasars and N galaxies or clusters, much like radio sources. Scaling arguments indicated that the Einstein satellite would be able to detect clusters at $z = 1$ if their x-ray luminosities were similar to those of low-redshift clusters (*21*). Published results (*22*) have rather been detections of nearer clusters already known from other types of surveys, and the discovery rate of new, extremely distant clusters has been disappointing. This may be telling us something about the evolution of cluster x-ray luminosity (*23*), or it may be no more than a reflection of the extent of optical follow-up data.

Redshifts from Colors

Although redshifts are normally determined by measuring the shift in wavelength of individual spectral features, another technique involves measuring the slope of the spectral energy distribution for a galaxy between several fixed bands, as pioneered by Baum (*24*). These ratios of fluxes seen through two or more wave bands are called spectral indices, or color indices, or just colors. The idea is that since giant elliptical galaxies in clusters have integrated energy distributions that look quite similar, a plot of apparent spectral index against redshift should show small scatter (*25*). Empirically, this expected correlation is not only tight but also very steep (for suitably chosen bands), so that a rather accurate redshift ($\Delta z = \pm 0.02$) can be derived from reasonably good data. This technique is useful because obtaining broad- or medium-band fluxes is, in most circumstances, easier than obtaining a spectrum. Baum (*24*) was able to measure redshifts almost twice as large as those achieved with conventional spectroscopy. A modern example is a tentative and approximate redshift of $z = 1.5$ for 3C 65 (*26*), based on optical and near infrared colors. The main uncertainty in this procedure stems from the fact that there is naturally less information in the broadband fluxes than there is in a spectrum. Thus an object with a peculiar spectrum could well have spectral indices that are very misleading for redshift determination: Lebofsky (*27*) has illustrated this point with the strong emission line galaxies 3C 405 and 3C 133. Even so, redshifts derived from optical (*24, 28*) or infrared (*8, 26, 27, 29*) colors are likely to become a very powerful tool, at least in a statistical sense.

Practical Aspects of Faint Spectroscopy

The two principal motivations for studying distant galaxies are thus to test ideas concerning the evolution of galaxies and to use galaxies as tracers of the cosmological expansion. Both effects due to evolution and effects due to cosmology are amplified at higher redshifts, but the difficulty of obtaining high redshifts becomes accentuated because of the strong decrease of surface brightness discussed earlier. An observer is faced with the problem of detecting (and recognizing) spectral features seen together with the spectrum of all other sources that contribute to the light of the night sky in that direction. This "sky radiation" is partly due to zodiacal light (and so has the spectral features of the sun at zero redshift), partly due to airglow, which produces an emission line spectrum that is especially troublesome for wavelengths longer than 680 nm, and partly due to various sorts of scattered light, both within the earth's atmosphere and in interstellar space (*30*). If the sky spectrum is not subtracted from the galaxy spectrum, a redshift of 0.2 is the practical limit with photographic detectors for ordinary galaxies (*31*). Humason's (*32*) efforts in the 1930's often required several nights of successive exposures onto the same spectrogram. A brilliant exception to the limit $z = 0.2$ (established for the Hydra cluster in 1951, shortly after the completion of the Hale 5-meter telescope) was Minkowski's (*33*) redshift of 0.46 for 3C 295, based on a single emission line due to forbidden O^+. The line was identified by default: any other candidate identification for the observed line would have resulted in some contradiction, such as the prediction of another line at a position where none was observed. Also, Minkowski knew the radio flux density and the optical apparent brightness, plus some information about the surrounding cluster, which indicated that such a high redshift was not only plausible but likely.

Photographic plates used as spectroscopic detectors have two principal disadvantages with respect to modern electronic devices. In the first place, the quantum efficiency is very low, of the order of 1 percent. In the second place, there is a finite amount of information that can be impressed per unit area. This is a problem because this saturation can occur before sufficient statistics can be obtained at a given position on the photographic plate to allow sky subtraction to be effectively done from a very weak net signal. Electronic detectors, on the other hand, can be read out with sufficient frequency that saturation is not a problem. The detecting element itself is either a photoemissive surface or a solid-state diode array, such as a charge-coupled device (CCD). These devices have the important attributes of very high quantum efficiency (especially for red light), large dynamic range, linear response to light, and good stability. Spectrographs having CCD's as detectors can obtain redshifts in a fraction of the time formerly required (*34*); this is an important desideratum considering the high premium on large telescope time.

The highest redshift reported for any galaxy at a given time is, of course, partly a function of technological innovations, but it is also a function of persistence on the part of observers. Recent examples of the patience required are the redshifts for 3C 427.1 ($z = 1.175$) and 3C 13 ($z = 1.050$) reported by Spinrad *et al.* (*35*). Each of these galaxies required 30 to 40 hours of integration (with a photocathode as the detector) on 20 or so separate nights, all spectra being individually sky-subtracted and then added together. The observations were done in spite of the bright night sky above San Jose, California.

To give a rough idea of how the combination of persistence and technology has pushed the highest known redshift for galaxies to higher values, I have plotted in Fig. 1 the few highest known redshifts, derived from spectroscopic techniques, against the year of publication. The points come from (*9, 31, 33, 35–37*), with additional redshifts cited in (*27, 38*). The early points are the work of V. M. Slipher at Lowell Observatory, obtained at a time when the nature of galaxies was not well understood. Subsequently, the large reflectors at Mount Wilson produced results that dominated the field for three decades. The points from 1929 to 1954 are all due to Humason (*36*), who extended the observed depth by a factor

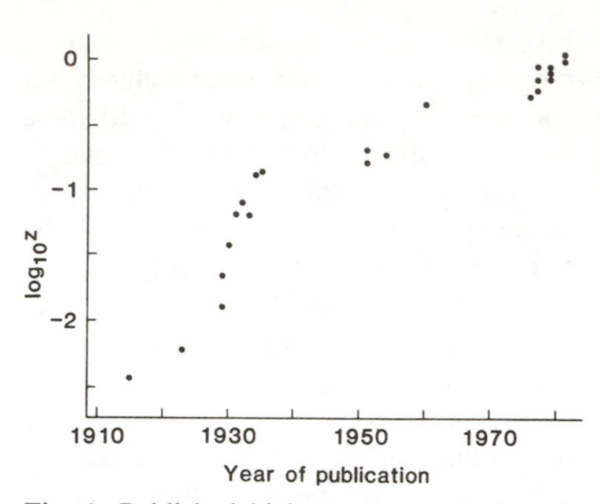

Fig. 1. Published highest spectroscopic redshifts for galaxies, illustrating the relative gains in observational sophistication. See text for sources of points plotted.

that has not been matched since, in spite of the new detectors. The most recent redshifts shown in Fig. 1 are especially uncertain: few have been confirmed by independent groups, and some are keyed to a single feature or are otherwise based on spectra with a low signal-to-noise ratio. Also, the most recent redshifts are likely to be dominated by peculiar objects (for instance, having strong emission lines or higher than normal luminosity), due to selection effects intrinsic to the search techniques discussed earlier.

Anecdotal Examples

Redshifts of order unity can be obtained with current instrumentation and selection procedures (*34, 35*). The claim for a high redshift is still often difficult to establish. A good example of some types of ambiguities and problems that occur is the case of the cluster 1305+2952 (*37*), originally reported to have a redshift of 0.947 on the basis of one absorption feature in the spectrum of the first-ranked galaxy and additional arguments based on the shape of the continuous spectrum. (The cluster was originally selected for study because it appeared to be qualitatively too blue to be at the redshift of ~ 0.5 that might have been guessed from its apparent brightness.) Alternatively, the redshift would be 0.285 if the absorption feature were instead identified with the "400-nm spectrum break" commonly seen in nearby galaxies. One of the arguments against this identification was that from early data the spectrum appeared to turn up in the far red, which would not be explainable on the low-redshift hypothesis but would have a natural explanation on the high-redshift hypothesis. However, CCD spectroscopy by Koo and Kron (*13*) failed to confirm this far-red upturn in any of the cluster members. In the meantime, Spinrad had been obtaining more data on the first-ranked cluster member, and at least one scan appeared to show an emission feature at the O^+ line position corresponding to $z = 0.28$. It thus seemed for a while that the case for the high redshift was weak, at best.

Subsequent events have, however, considerably strengthened the originally proposed high redshift. (i) As reported by Bruzual (*39*), a very red galaxy [cluster 8 (*40*)], presumed to be a true cluster member, shows a strong emission line (Fig. 2), which is most plausibly identified as O^+ at $z = 0.943$ by the same reasoning used by Minkowski to identify the single emission line in the spectrum of 3C 295. (ii) A very blue galaxy (cluster 3), also presumed to be a cluster member, does not show O^+ at $z = 0.285$, according to both Spinrad and Kron, although a galaxy as blue as this would be expected to show the line if the redshift were low. (iii) An infrared spectral index measured by Lebofsky (*27*) for the

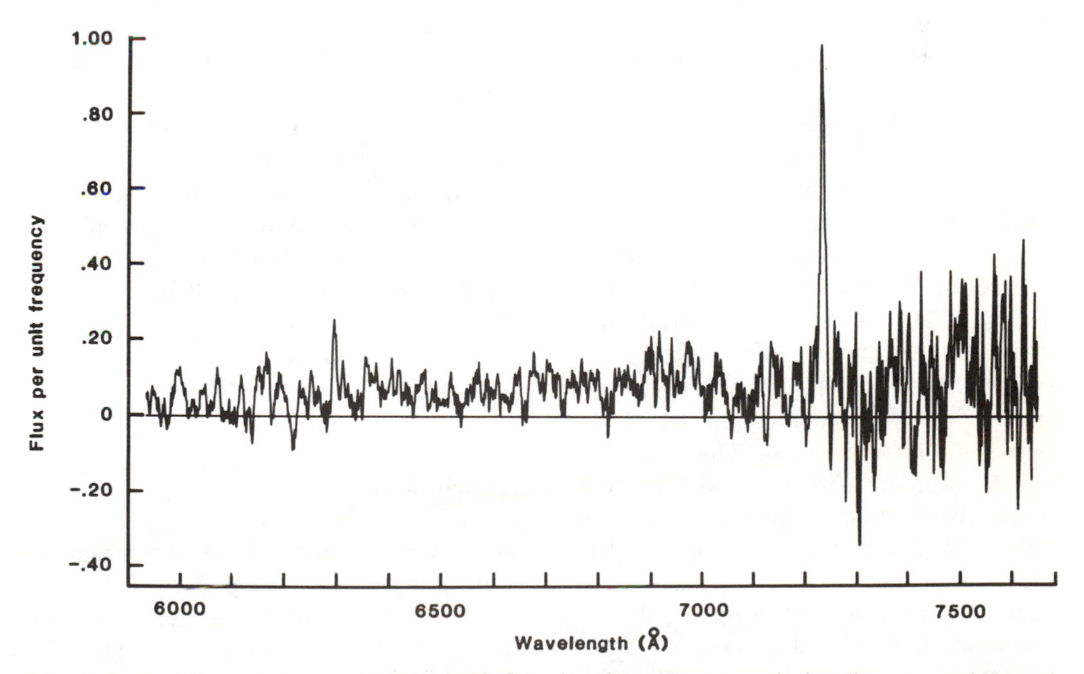

Fig. 2. Part of the spectrum of 1305+2952 G8, showing a strong emission line at an observed wavelength of 723 nm, identified as O^+ with a rest wavelength of 373 nm. This spectrum is the average of 11 separate scans by H. Spinrad with the Lick Observatory Shane 3-m telescope.

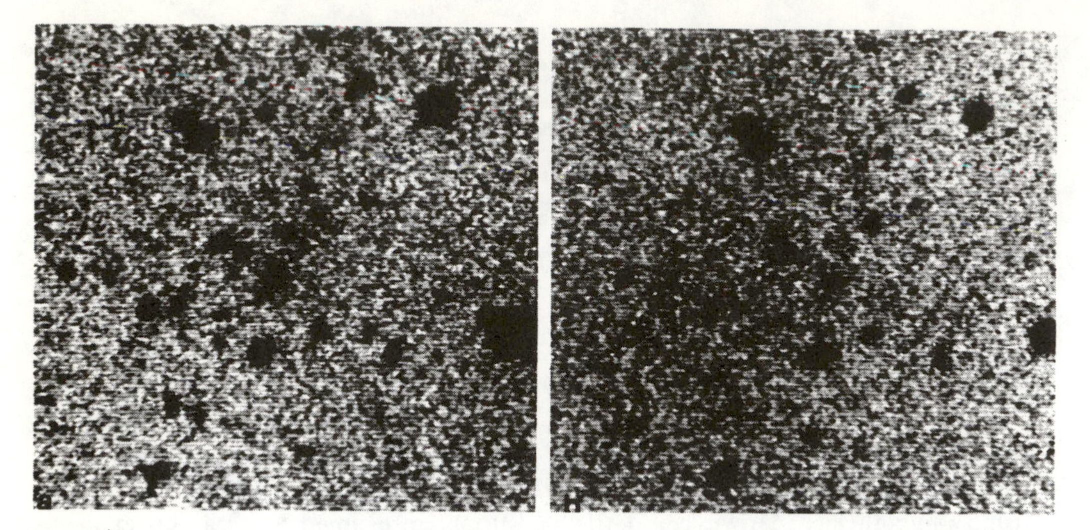

Fig. 3. The very faint, very blue cluster 0014.4+1551, depicting an area about 1.5 arc minutes on a side. (a) The digital sum of three photographic plates taken in ultraviolet (370-nm) light and two plates taken in blue light (obtained with the Kitt Peak Mayall 4-m telescope); (b) the sum of three plates taken in red light and three plates taken in far red (800-nm) light.

first-ranked galaxy indicates a color that would be normal for a redshift of 0.94 but would be too red for $z = 0.28$. According to Lebofsky, the apparent infrared flux of the galaxy is 2.6 sigma brighter than other first-ranked cluster galaxies, suitably scaled to a redshift of 0.94; this difference is not especially worrisome considering the large extrapolation. (iv) Another cluster in the same general direction, II Zw 1305.4+2941, has a redshift of 0.24 (measured by Spinrad) and appears to be much bigger, brighter, and redder than the 1305+2952 cluster, consistent with a large redshift difference between the two clusters. There are still some problems with the claim for the high redshift, namely the peculiar optical colors of the brighter galaxies of the cluster, as discussed by Bruzual (*39*); according to him, none of the galaxies has color indices that are expected for anything with redshift near 0.9. Also, the current version of the spectrum of the first-ranked galaxy (cluster 1) is just as plausibly at $z = 0.285$ as at $z = 0.94$; perhaps, ironically, this galaxy is not a member at all.

Another candidate for a very distant cluster is 0014.4+1551 (Fig. 3). This cluster is remarkable for its extreme blueness, quite unknown for low-redshift clusters but expected for very high redshifts (*39*). Unlike 1305+2952, there are no published spectroscopic data. The cluster is known not to be a radio source (*41*) or an x-ray source (*42*).

Conclusions

Recent progress in the identification and study of very distant galaxies has been rapid due to the development of new detectors and new techniques. In spite of this, no answers are yet available for the major motivating questions: what processes control the evolution of observable properties of galaxies, and what do observations of distant galaxies tell us about the large-scale structure of the universe? Regarding evolution, there are a number of examples of galaxies with colors that appear to be inconsistent with their redshifts, meaning that we do not understand the composition of these galaxies (assuming, of course, that the data are trustworthy). Regarding cosmology, different investigations (*10*) yield different values for the deceleration of the expansion, for reasons that are not at all clear. What is needed in both cases are strict selection criteria for the inclusion of any given galaxy in the statistical sample. There have been a number of noteworthy attempts to do this (*10*), but the very highest redshifts known are still those produced by some no-holds-barred technique, with inevitable biases.

It is likely that future efforts in this field will include the following developments. (i) Complete redshift surveys will be stressed, so that all visible galaxies in a patch of sky will be measured. Questions related to evolution and to cosmology would then address the distribution of observed redshifts, at a given apparent brightness. (ii) Spectroscopic techniques will exploit CCD's as detectors on a routine basis, especially for work in the range 600 to 1000 nm. (iii) Photometry in the near infrared will likely play a very important role, in particular with regard to measurements of color. (iv) The Space Telescope will allow spectroscopy of faint objects in the ultraviolet, as well as observations in the far red without interference from airglow. A combination of these and other techniques may reveal galaxies at $z \sim 2$, which may yield important insights into the origin of galaxies.

References and Notes

1. A. Sandage, *Astrophys. J.* **178**, 25 (1972).
2. W. W. Morgan, in *La Structure et l'Evolution de l'Universe* (Institut International de Physique Solvay, R. Stoop, Brussels, 1958), p. 297.
3. P. Biermann and S. L. Shapiro, *Astrophys. J. Lett.* **230**, L33 (1979).
4. J. Kormendy, *Astrophys. J.* **227**, 714 (1979); D. Lynden-Bell, *Mon. Not. R. Astron. Soc.* **187**, 101 (1979).
5. J. E. Gunn and B. M. Tinsley, *Astrophys. J.* **210**, 1 (1976); S. D. M. White, *Mon. Not. R. Astron. Soc.* **184**, 185 (1978); M. A. Hausman and J. P. Ostriker, *Astrophys. J.* **224**, 320 (1978); R. H. Miller and B. F. Smith, *ibid.* **235**, 421 (1980).
6. D. O. Richstone, *Astrophys. J.* **204**, 642 (1975); P. Hickson, D. O. Richstone, E. L. Turner, *ibid.* **213**, 323 (1977); J. Kormendy, *ibid.* **218**, 333 (1977).
7. J. L. Greenstein, *ibid.* **88**, 605 (1938); A. D. Code and G. A. Welch, *ibid.* **228**, 95 (1979).
8. G. L. Grasdalen, in *Objects of High Redshift, Int. Astron. Union Symp. 92* (1980), p. 269.
9. J. G. Hoessel, J. B. Oke, J. E. Gunn, *Carnegie Inst. Washington Yearb. 79* (1979–1980), p. 623.
10. J. Kristian, A. Sandage, J. A. Westphal, *Astrophys. J.* **221**, 383 (1978); J. G. Hoessel, J. E. Gunn, T. X. Thuan, *ibid.* **241**, 486 (1980).
11. R. C. Tolman, *Relativity, Thermodynamics, and Cosmology* (Clarendon, Oxford, 1934); E. Hubble and R. C. Tolman, *Astrophys. J.* **82**, 302 (1935); J. Kristian and R. K. Sachs, *ibid.* **143**, 379 (1965).
12. R. B. Partridge and P. J. E. Peebles, *Astrophys. J.* **147**, 868 (1967); D. L. Meier, *ibid.* **207**, 343 (1976).
13. R. B. Partridge, *ibid.* **192**, 241 (1974); M. Davis and D. T. Wilkinson, *ibid.*, p. 251; D. C. Koo and R. G. Kron, *Publ. Astron. Soc. Pac.* **92**, 537 (1980).
14. H. Spinrad, in *Objects of High Redshift, Int. Astron. Union Symp. 92* (1980), p. 39; A. Stockton, *Astrophys. J.* **223**, 747 (1978).
15. A. Boksenberg and W. L. W. Sargent, *Astrophys. J.* **220**, 42 (1978); R. J. Weymann, R. E. Williams, B. M. Peterson, D. A. Turnshek, *ibid.* **234**, 33 (1979); W. L. W. Sargent, P. J. Young, A. Boksenberg, D. Tytler, *Astrophys. J. Suppl.* **42**, 41 (1980).
16. D. L. Meier, *Astrophys. J. Lett.* **203**, L103 (1976).
17. J. Bookbinder, L. L. Cowie, J. H. Krolik, J. P. Ostriker, M. Rees, *Astrophys. J.* **237**, 647 (1980).
18. F. Bonoli, A. Braccesi, L. Federici, V. Zitelli, L. Formiggini, *Astron. Astrophys. Suppl.* **35**, 391 (1979).
19. H. Spinrad, *Publ. Astron. Soc. Pac.* **88**, 565 (1976).

20. J. E. Gunn, in *Decalages vers le Rouge et Expansion de l'Univers, Int. Astron. Union Colloq. 37* (1977), p. 183.
21. D. A. Schwartz, *Astrophys. J. Lett.* **206**, L95 (1976).
22. S. C. Perrenod and J. P. Henry, *ibid.* **247**, L1 (1981).
23. S. C. Perrenod, *Astrophys. J.* **226**, 566 (1978).
24. W. A. Baum, *Astron. J.* **62**, 6 (1958).
25. J. Stebbins and A. E. Whitford, *Astrophys. J.* **108**, 413 (1948); A. Sandage, *ibid.* **183**, 711 (1973).
26. J. J. Puschell, F. N. Owen, R. Laing, in *Extragalactic Radio Sources, Int. Astron. Union Symp. 97*, in press.
27. M. J. Lebofsky, *Astrophys. J. Lett.* **245**, L59 (1981).
28. D. C. Koo, thesis, University of California, Berkeley (1981).
29. S. J. Lilly and M. S. Longair, in *Extragalactic Radio Sources, Int. Astron. Union Symp. 97*, in press.
30. F. E. Roach and J. L. Gordon, *The Light of the Night Sky* (Reidel, Dordrecht, 1973).
31. A. Sandage, in *Galaxies and the Universe*, A. Sandage, M. Sandage, J. Kristian, Eds. (Univ. of Chicago Press, Chicago, 1975), p. 761.
32. M. L. Humason, *Proc. Natl. Acad. Sci. U.S.A.* **15**, 167 (1929).
33. R. Minkowski, *Astrophys. J.* **132**, 908 (1960).
34. *Carnegie Inst. Washington Yearb. 78* (1978–1979), p. 779.
35. H. Spinrad, J. Stauffer, H. Butcher, *Astrophys. J.* **244**, 382 (1981).
36. Separate announcements in the annual report of the director of the Mount Wilson Observatory from 1929 onward.
37. R. G. Kron, H. Spinrad, I. R. King, *Astrophys. J.* **217**, 951 (1977).
38. H. E. Smith, in *Radio Astronomy and Cosmology, Int. Astron. Union Symp. 74* (1977), p. 279.
39. G. Bruzual A., thesis, University of California, Berkeley (1981).
40. R. G. Kron, *Astrophys. J. Suppl.* **43**, 305 (1980).
41. From results of a new Westerbork survey by R. A. Windhorst, P. Katgert, and H. van der Laan.
42. Results of a collaborative effort with the Einstein Observatory, including P. Boynton, J. P. Henry, D. C. Koo, R. A. Schommer, and R. G. Kron.

43. I am indebted to H. Spinrad for permission to publish the spectrum of 1305+2952 G8 in Fig. 2; also I am grateful to A. Jankevics for assistance in the production of Fig. 3. This review was supported by grant AST-7920994 from the National Science Foundation.

44. Spectroscopic work on 1305+2952 by Spinrad subsequent to the original publication of this paper has shown no evidence for any redshift near Z = 0.95, Fig. 2 and other evidence notwithstanding. No two galaxies in the so-called cluster have been measured to have the same redshift.

This material originally appeared in *Science* 216, 16 April 1982.

15. Galactic Evolution: A Survey of Recent Progress

K. M. Strom and S. E. Strom

Since the 1920's, when a significant fraction of cataloged "nebulae" were recognized as "island universes" or galaxies external to our Milky Way, these enormous aggregates of stars, gas, and dust have been used as probes of the large-scale structure of the universe. For individual galaxies to be useful in such studies, it is necessary to understand their evolutional history so that we can estimate how their luminosity, size, and morphological appearance change with time. Lacking such information, it is impossible to use galaxies as standard candles or meter sticks in charting the shape and extent of the visible universe. It is not surprising therefore that considerable effort has been devoted to understanding the evolution of galaxies. Over the past 20 years, astronomers have gathered an impressive array of observational data bearing on this critical problem. At present, efforts to synthesize these data into a coherent picture of galactic evolution have been hindered by the unavoidable problem that the epoch of galaxy formation occurred between 10 billion and 20 billion years ago. Until recently, the study of galactic properties has been restricted to relatively nearby systems, many of which appear to be near the end of their active evolutionary phases. Our understanding of their past history is therefore to a large extent based on primitive models, many classes of which predict properties nearly identical to those found to characterize current-epoch galaxies. However, over the past few years, application of sensitive optical and infrared detectors has permitted astronomers to probe the properties of more distant, "younger" galaxies, with results which challenge evolutionary models and pose some serious new problems.

We provide here a compendium of basic properties that must be explained by any proposed picture of galaxy evolution. We attempt to sort those that appear to be "universal" properties of a particular galaxy type from those that appear to depend significantly on the environment in which a galaxy is located. Finally, we summarize the classes of evolutionary models currently believed to provide the best frameworks for interpreting these observed characteristics (*1*).

Galaxy Morphology

A necessary first step in the study of galactic evolution is to define a classification system that provides a basis for sorting galaxies into categories. By comparing the properties of galaxies of similar appearance as a function of time and environment, we hope to discern patterns of evolutionary development. Initially, such classification schemes are necessarily based partly on visual appearance. As our physical understanding of the factors that affect morphology deepens, classification criteria eventually become more refined.

The existence of two broad categories of galaxies—ellipticals and disk systems—is apparent from the most cursory examination of photographic surveys.

Elliptical or E galaxies are systems whose light distribution, deriving from the superposition of starlight along the line of sight, appears to fall off smoothly with radius and whose overall shape appears to represent the projection of a prolate or oblate spheroidal system on the plane of the sky. Typically, E galaxies are classified solely on the basis of their flattening (see Fig. 1). Detailed studies of surface brightness variations in ellipticals suggest that they can be well represented at large galactocentric distances by a variety of fitting functions; the most extensively used at present is the de Vaucouleurs law (*2*); $\log \mu = -3.33 [\log (r/r_e)^{1/4} - 1]$. Here μ is the observed surface brightness (magnitudes per arc second squared) and r and r_e are, respectively, the galactocentric distance and the effective radius within which half the luminosity of the galaxy is contained. This functional representation appears to fit well the light distribution of E galaxies in the luminosity range $-20 \lesssim M_V \lesssim -23$; the surface brightness distribution of fainter galaxies appears to fall off more rapidly at large galactocentric distances than does the de Vaucouleurs law.

Disk galaxies appear to be composed of two morphologically distinct parts—a spheroidal bulge and a flattened disk. The bulge components of disk galaxies are superficially similar to E galaxies. Disk components may exhibit a variety of appearances. Some are forming stars at the present epoch, either in relatively regular patterns (spirals; see Fig. 2) or in chaotic patterns (irregular systems; see Fig. 3); others appear to have ceased star formation (at the current epoch), either temporarily or permanently, and are characterized by relatively smooth disks (S0 systems). The relative size and luminosity of the bulge and disk components appear to vary continuously from pure bulge systems (E galaxies) to pure disk systems (late-type spirals and irregulars and their non–star-forming analogs). Hence a useful initial description of a disk galaxy is provided by (i) the bulge-to-disk ratio characteristic of the system, (ii) the presence or absence of current-epoch star formation in the disk, and (iii) the spatial regularity of the current-epoch star-forming episodes.

Classification schemes for disk galaxies generally follow precepts that are similar in spirit to those originally articulated by Hubble (*3*). In his system, as modified and extended by Sandage (*4*), the following categories are defined:

S0: galaxies with no or little evidence of star-forming activity in the disk; some may show evidence of dust lanes or dust patches.

Sa: galaxies with a low contrast of arm to disk light, a large bulge-to-disk ratio, and tightly wound spiral arms.

Sb: galaxies with a higher contrast of arm to disk light, a smaller bulge-to-disk ratio than those in class Sa, and a more open arm pattern.

Sc: galaxies with a high contrast of arm to disk light, a small bulge-to-disk ratio, and an open spiral pattern.

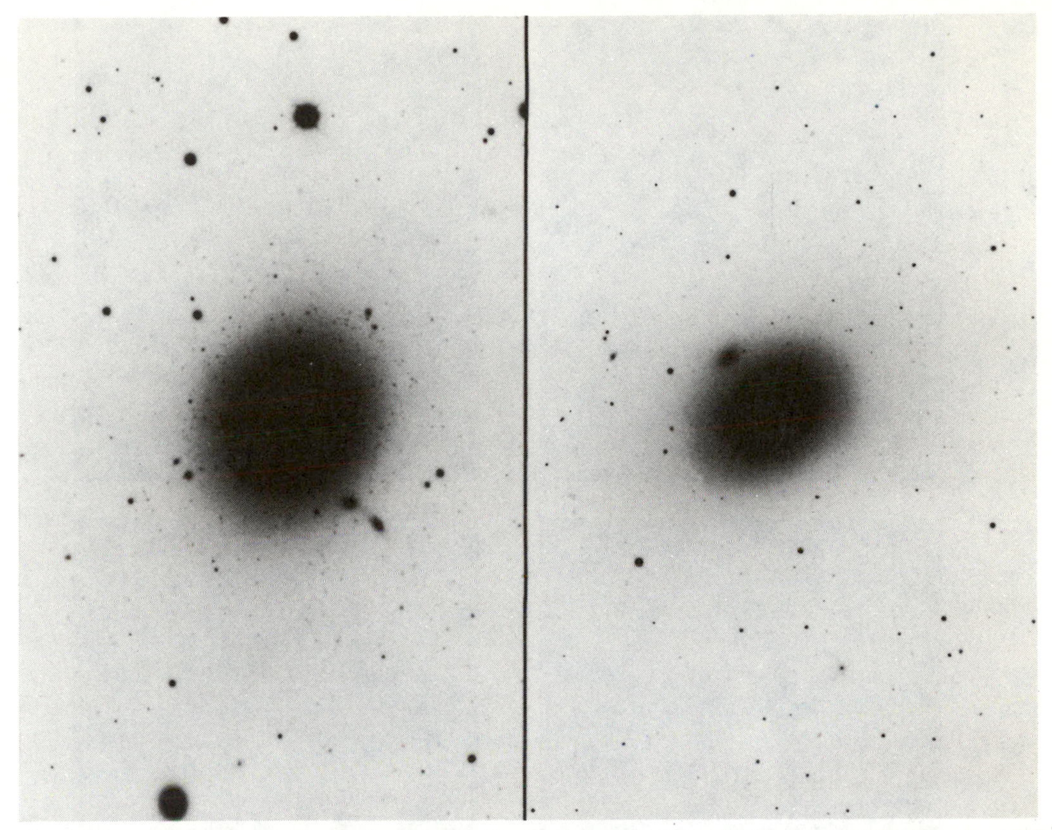

Fig. 1. Two examples of elliptical galaxies. On the left is a plate of Messier 87 (M87), a giant E0 galaxy in Virgo. Note the swarm of starlike images surrounding the galaxy; these are a population of globular clusters similar to those found in the halo of our Milky Way. On the right is a more flattened E galaxy (NGC 4406, E3).

Sd: galaxies with a small or absent bulge and an open, patchy, and ill-defined spiral pattern.

Sm: bulgeless systems characterized by open, poorly defined spiral patterns.

The Sa galaxies are often referred to as early Hubble types, while types Sc through Sm are referred to as late Hubble types.

As suggested above, the location of a spiral galaxy in the Hubble sequence depends primarily on the appearance ("texture") of the arms. However, other systems, such as those proposed by Morgan (*5*, *6*) and by van den Bergh (*7*), rely on the ratio of the bulge and disk sizes. In most cases the systems yield comparable classifications. In our view the systems based on bulge-to-disk ratio may eventually prove more powerful because of their more direct measure of the underlying dynamics of disk galaxies.

The light distribution of the bulge appears, at first glance, to be well fitted by a de Vaucouleurs law (*8*). More careful study suggests that this may not be the case (*9*) universally. The light distribution in the disk appears to follow an exponential law, although such a simple

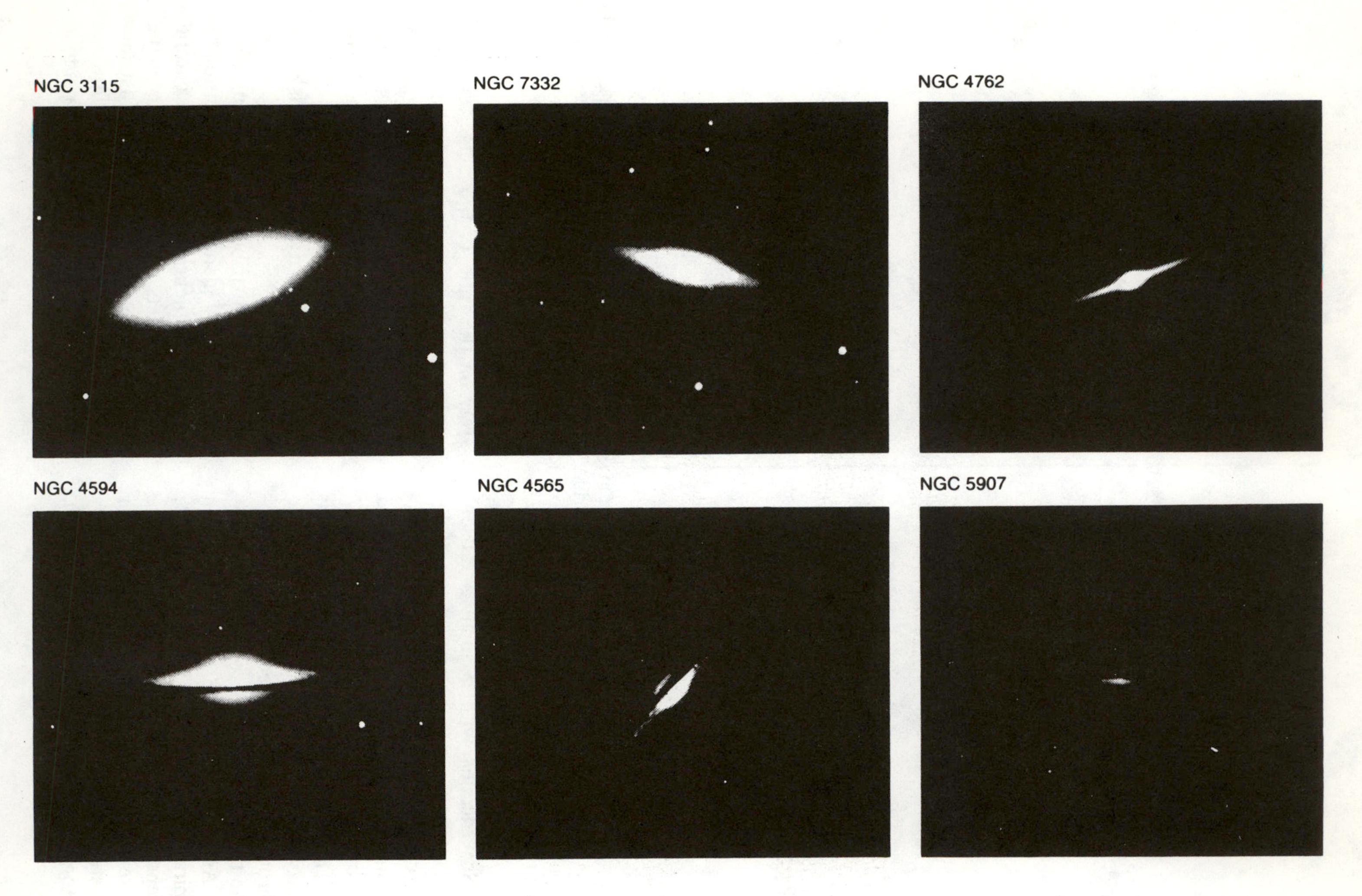
NGC 3115
NGC 7332
NGC 4762
NGC 4594
NGC 4565
NGC 5907

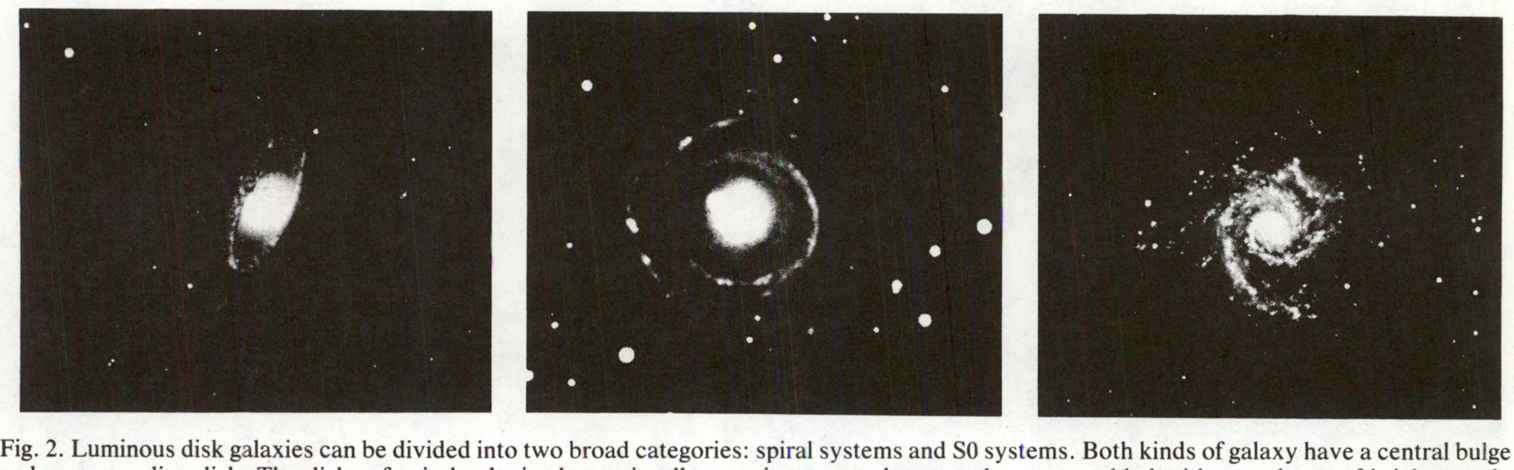

Fig. 2. Luminous disk galaxies can be divided into two broad categories: spiral systems and S0 systems. Both kinds of galaxy have a central bulge and a surrounding disk. The disks of spiral galaxies have visually prominent arms because they are studded with complexes of bright, newly formed stars. The disks of S0 galaxies, in contrast, are smooth, show no spiral structure, and are devoid of young stars. The three photographs in the top row depict S0 systems viewed nearly edge-on, arranged in order of the decreasing prominence of their bulge with respect to their disk. No evidence of recent star formation is visible. The photographs in the middle row show three galaxies of the spiral type, also seen edge-on and in order of decreasing bulge-to-disk ratio. The galaxies in the bottom row illustrate the probable face-on appearance of spiral galaxies in the middle row. The bright knots in the spiral arms of the galaxies represent newly formed stellar complexes.

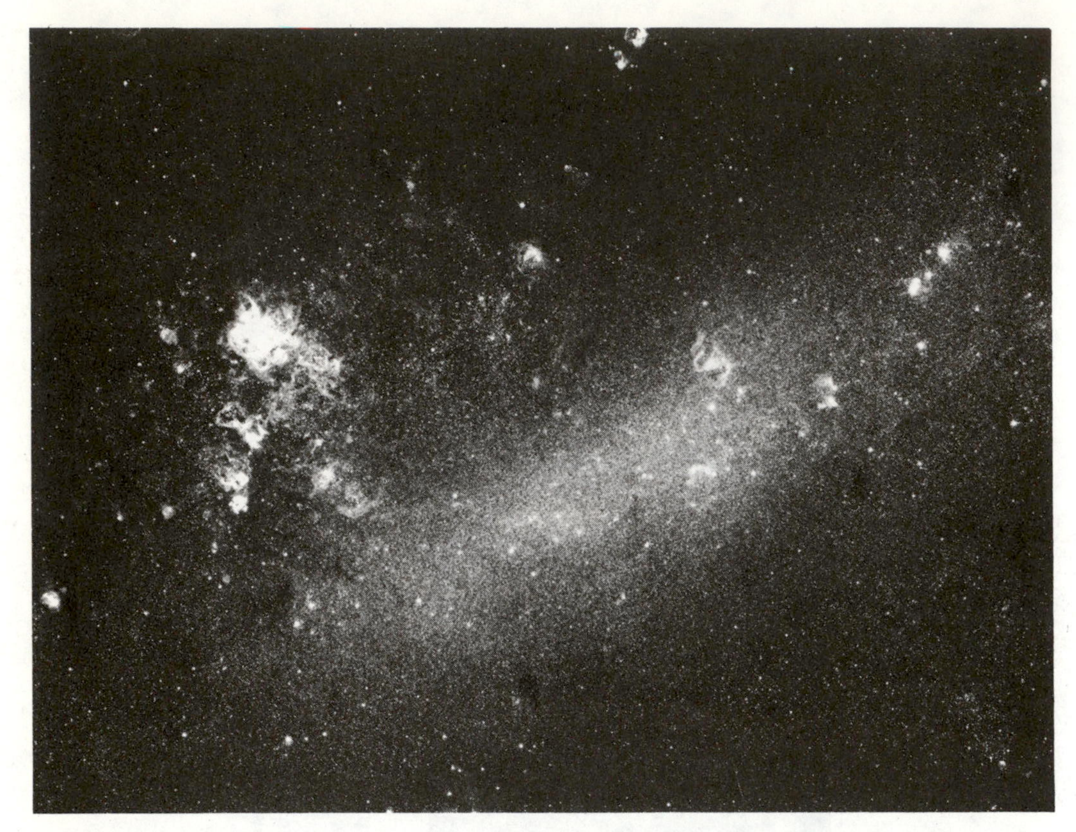

Fig. 3. An example of an irregular galaxy, the Large Magellanic Cloud. Note the bright patches of newly formed stars and associated nebulosity superposed randomly on the smooth disk population of relatively old stars. Irregular star-forming patterns are common among disk galaxies of low luminosity.

description for all systems has been questioned.

Properties of E Galaxies

Frequency distribution in differing environments. Five years ago, Oemler (*10*) presented the first modern study of galaxy morphological types in clusters. This work suggested the existence of three representative classes of galaxy clusters: (i) cD clusters—dense aggregates of galaxies characterized by a core-halo distribution of galaxies and dominated by a centrally located cD system, (ii) spiral-poor clusters—dense aggregates somewhat less centrally concentrated than cD clusters, and (iii) spiral-rich clusters—irregular, moderate- to low-density groups.

The three cluster classes exhibit significantly different proportions of galaxy types. The ratios of ellipticals to S0's to spirals (Sp's) for the three types are (i) 3:4:2, (ii) 1:2:1, and (iii) 1:2:3. Subsequent studies by Melnick and Sargent (*11*), Butcher and Oemler (*12*), and Dressler (*13*) confirm the dependence of galaxy content on cluster morphology.

Dressler notes that the Oemler classes are not discrete but rather representative of a continuous range of cluster morphologies. He also remarks that the E:S0:Sp ratio appears to vary continuously with current-epoch local density. Hence, within clusters of all types and among clusters of different types, the proportion of E and S0 galaxies is highest in regions of the highest local galaxy density. The cluster population studies of Melnick and Sargent are consistent with this result, although their sample is strongly biased toward dense, rich clusters of galaxies.

The best available modern data suggest that E (along with S0) galaxies are found most frequently in regions characterized by high galaxy density.

Sizes and luminosities. Over the past several years, Strom and Strom have published a survey of E-galaxy properties for a large sample of elliptical systems located in a variety of environmental settings [for example, see (*14*)]. They have presented ultraviolet and red surface photometry for nearly 600 galaxies, permitting quantitative discussion of the size, shape, and chemical composition of these galaxies. Two fundamental parameters that describe E galaxies are their size and luminosity. Strom and Strom used two measurements of size: the de Vaucouleurs effective radius r_e and the radius r_{26} at which the red surface brightness reaches a value of 26 mag (arc sec)$^{-2}$. The former quantity represents a metric size, whereas the latter is an isophotal size. As might be expected from cursory examination of galaxy photographs, they found that the more luminous galaxies were bigger. Their study permitted a quantitative comparison of the size-luminosity relationship (r,L) for E galaxies located in clusters of differing Oemler types. The (r,L) relationships for the spiral-rich clusters and for the low-density outer regions of all clusters are identical; however, galaxies located in the central region of rich clusters of galaxies appear to be smaller. Strom and Strom (*15*) attribute this difference to the effects of multiple gravitational encounters in the denser regions of the clusters included in their sample. Tidal stripping of stars from the outer regions of E galaxies appears to offer a reasonable explanation for the decrease in E-galaxy size in such regions.

The frequency distribution of E-galaxy luminosities appears to depend on environment. From Dressler's study (*16*), there appear to be several clusters, most of them characterized by high galaxy density, in which the number of galaxies as a function of luminosity differs significantly from the Schechter (*17*) relationship characteristic of the field. Whether this difference can be explained entirely in terms of postformation interactions [for example, tidal stripping and galaxy mergers (*18*)] is not yet clear.

Shape. The shape is another fundamental characteristic of E galaxies. Originally, it was believed that E galaxies were either prolate or oblate spheroids in which flattening was induced by rotation. Current results cast doubt on these assumptions (*19, 20*). It now appears that most ellipticals are triaxial (*20*) ellipsoids of revolution, although close in appearance to oblate spheroids (*21*). Their shape results primarily from anisotropies in the velocity field of their constituent stars rather than from rotation.

Strom and Strom (*14*) reported that the frequency distribution of ellipticities appears to depend on the shape of the cluster of galaxies in which E galaxies reside; there are more flattened ellipticals in flattened clusters. In some but not all cases the major axis of the E galaxies tends to be aligned along the major axis defined by the cluster galaxy distribution. Hence the environment in which an elliptical is located may determine its shape. The critical question we must face is whether this environmental difference results from relatively recent interactions among galaxies or was "built in" at the time of galaxy formation.

Gas content and star formation. Knapp *et al.* (*22*) attempted to measure the neutral hydrogen content of E galaxies. They concluded that the gas content of most ellipticals does not exceed 0.1 percent of their total mass.

Gisler (*23*) recently analyzed the results of all available observations of forbidden singly ionized oxygen, [O II], 3727-angstrom line emission in E and S0 galaxies located in a variety of environments. After correcting for observational selection effects and biases in the heterogeneous data sets, he found a higher percentage of [O II] emission among E galaxies located in low-density environments than in dense, rich clusters. Gisler suggested that the relative infrequency of observed ionized gas in clusters of E galaxies results from either thermal evaporation of intragalactic gas by electron conduction in the hot intracluster medium (*24*) or ablation by ram-pressure stripping (*25*).

Studies of the color distributions in E galaxies suggest that in most cases the galaxy becomes bluer at larger galactocentric distances (see the following section on chemical composition), suggesting a decrease in metal abundance in the outer parts of ellipticals. However, a small fraction of ellipticals exhibit blue colors in their nuclear region (*26*). Moreover, examination of the nearby E galaxy NGC 205 in the Local Group reveals a population of luminous blue stars near its nucleus. These observations suggest that some E galaxies are forming a limited number of stars at the current epoch. It is not clear whether such events occur episodically in all ellipticals or are restricted to a small subset of the class. Stauffer (*27*) and Gallagher (*28*) noticed the presence of dust clouds in ultraviolet and blue photographs of ellipticals. Their presence suggests that currently inactive systems contain material that could conceivably be assembled into stars. Faber (*29*) is attempting to map a sample of E galaxies of a variety of luminosities in a set of photometric indices sensitive to both metal abundance and the presence of a young stellar component. From her observations, it should be possible eventually to estimate the frequency and extent of star-forming events in elliptical galaxies.

Chemical composition. Perhaps the most solidly based result regarding the composition of E galaxies is the observed correlation between absorption-feature strength and luminosity; luminous E galaxies tend to have stronger absorption features (*30*) than do fainter systems. Although the exact relationship between absorption-feature strength and metal-to-hydrogen ratio Z may be questioned, it appears certain that the dominant factor influencing the observed variations in line strength is, in fact, the galaxian metallicity.

Faber (*30*) established that galaxian color varies directly with line strength (luminous E galaxies are redder), and subsequent studies (*31*) further refined the arguments which suggest that luminous E galaxies are more metal-rich. Integrated colors are available for a large number of E galaxies located in a variety of environments (*14, 26, 32*). In general, the relationship between color and luminosity is identical, independent of environment.

[*Note added in proof*: Caldwell (*32a*) notes that the dispersion in the color-luminosity relationship increases in regions of lower galaxy density. He attributes this difference to an increase of star-forming activity in E galaxies located in such environments.]

The variation of metallicity with position in a galaxy appears to provide a measure of star formation, element production, element dispersal, and gas flows that occurred during the early evolutionary history of an E galaxy. Unfortunately, the only extensive data available that measure $Z(r)$ are those of Strom and Strom (*14*). Their surface photometry in the ultraviolet and red permits estimates of $Z(r)$ for $r \gtrsim 1$ kiloparsec (that is, for regions well outside the nucleus). In their sample of cluster E galaxies only 20 percent exhibit observable halo ($r \gtrsim 1$ kiloparsec) color gradients (almost without exception, the galaxies that do are bluer and hence more metal-poor at larger galactocentric radii). The remainder show no observable variation of color with radius outside the nuclear regions. Whether this result applies to E galaxies in lower density environments is not known.

Globular Clusters Surrounding E Galaxies

High-quality photographs of E galaxies located within 20 megaparsecs reveal that they are surrounded by systems of globular clusters similar to the system populating the halo of the Milky Way. A photograph of the giant Virgo E galaxy M87 shows an excellent example of such a system (Fig. 1). Until recently, it was thought that the properties of the globular cluster system were identical to those of the stellar system that produces the smooth halo-light distribution in E galaxies. However, a recent study by Strom *et al.* (*33*) suggests otherwise. They find that at a given galactocentric distance the globular clusters are, on average, bluer and perhaps more metal-poor than are the stars composing the spheroid. Moreover, the spatial distribution of the globular clusters differs from that of the halo stars in the sense that the globular clusters are more distended. Hence the dynamical and chemical history of the globular clusters surrounding E galaxies appears different from that characterizing the halo population in such galaxies.

Mass-to-Luminosity Ratio in E Galaxies

It has recently become feasible to measure the velocity dispersion σ_V for the stellar component of E galaxies. For the most part, these studies have been restricted to the nuclear regions of ellipticals, although data concerning the radial dependence of σ_V are becoming more abundant. Faber and Jackson (*34*) established that over a range of 100 in L, $\sigma_V \approx L^{1/4}$. Photometry and measurements of σ_V lead to an estimated mean value of the ratio of mass to blue luminosity of $M/L_B \approx 8.5$ for E galaxies (*35*).

Sargent *et al.* (*36*) compared the run of σ_V and of surface brightness with position near the nucleus of M87. They argue that an object of mass $\approx 5 \times 10^9$ solar masses ($M_\odot$) and $M/L \approx 60$ occupies a region within 110 parsecs of the center of the elliptical, and they believe the object to be a massive black hole. However, more recent studies (*37, 38*) suggest that this interpretation does not provide a unique description of the data.

For regions well beyond the nucleus, little direct evidence is as yet available regarding mass-to-light ratios. A number of E galaxies have been observed to possess halos of hot ($\approx 10^7$ K) gas. If these hot halos are gravitationally bound to the galaxies and in hydrostatic equilibrium, the necessary mass predicts an M/L_B ratio of approximately several hundred (*39*).

Another estimate of M/L for E galaxies is provided by the virial-theorem masses for rich clusters such as Coma in which E galaxies are the dominant constituent. The distribution of Coma galaxies on the plane of the sky suggests that the inner regions of the cluster have reached virial equilibrium. If so, then the cluster mass estimated from the observed velocity dispersion for the galaxies provides a mean value of $M/L_B \approx 300$ for a typical E galaxy.

E-Galaxy Evolution

Empirical evidence. Recent optical and near-infrared measurements (*40–42*) have provided empirical evidence regarding color and luminosity evolution of E galaxies. For galaxies with redshifts $z > 1$, infrared observations are necessary in order to compare the observed spectral energy distribution of galaxies with the optical spectra of nearby galaxies. Moreover, such observations are far less sensitive to the presence of small admixtures of young stars than are measurements of distant galaxies made in the visible region of the spectrum. Grasdalen (*41*) was the first to exploit the possibility of using infrared measurements to assess E-galaxy evolution. His preliminary results suggest "mild" color evolution at $z \approx 0.5$. For a simple open-universe cosmology ($q_0 = 0$) the infrared measurements suggest measurable luminosity evolution. More complete surveys of E-galaxy colors and luminosities, particularly those based on infrared measurements, will place important constraints on evolution and on the "cosmic scatter" in galaxy evolutionary properties.

Evolutionary models. Theoretical models of E-galaxy evolution fall into three broad categories:

1) Merger models, in which E galaxies are assembled from slow collisions and the merger of disk galaxies (*43*).

2) Gravitational assembly models, in which an E galaxy is produced by assembly of preexisting stars or clusters of stars.

3) Gas dynamical models, in which E galaxies begin as primordial gas clouds in which star formation proceeds rapidly and accompanies the collapse of the cloud (*44*). The collapse is essentially dissipationless, and the E-galaxy structure is established within a few free-fall times.

All models successfully account for the luminosity profile of ellipticals. However, beyond this, each model encounters some degree of difficulty (*45, 46*). In general, gas dynamical models have proven most successful in predicting the overall properties of E galaxies. In such models the massive stars in initial stellar generations evolve and inject synthesized elements into the collapsing gas; successive generations of stars are thereby enriched in metals, thus producing a systemwide composition gradient. However, if the heating from supernovae and cloud-cloud collisions is sufficiently high to overcome the self-gravity of the protogalactic system, all of the enriched protogalactic gas can be ejected from the system in its later evolutionary stages, thus precluding the production of metal-rich stellar generations (*47*). Since lower mass systems are more susceptible to losing supernova-heated gas, their mean metallicity is expected to be smaller than that of more massive galaxies. Thus the observed metallicity-luminosity relationship can be explained in a natural way.

While more successful in confronting observations than other evolutionary models, the gas dynamical picture is at present heuristic at best. Its greatest uncertainties derive from our meager knowledge concerning the factors that influence star formation efficiency. Hence a critical factor in the evolutionary history of model E galaxies—the ratio of conversion of gas into stars as a function of gas density—is treated as an adjustable parameter. Moreover, the gas dynamical models computed to date suppose that rotation is the dominant factor controlling E-galaxy shape—in contradiction to the best current evidence. No models in which initial velocity anisotropies are assumed have been used to compute the chemical history of E galaxies.

Disk Galaxies

Frequency distribution in differing environments. Earlier we discussed the distribution of galaxy types in each of the Oemler cluster types; E and S0 galaxies dominate in the relatively regular and dense cD and spiral-poor clusters, while spirals dominate lower density irregular clusters and the field. Melnick and Sargent (*11*) and Bahcall (*48*) have studied the distribution of actively star-forming disk systems (spirals and irregulars) and non–star-forming disks (S0's) in several rich, dense system clusters known to be x-ray emitters. The ratio of Sp's to S0's increases monotonically with increasing distance from the center of the clusters in their sample. Melnick and Sargent also found that Sp:S0 is dependent on the velocity dispersion of the cluster. Dressler (*13*) argued that this systematic behavior is representative of a more general correlation between Sp:S0 and local galaxy density that applies not only to rich clusters but to all regions of space. He cites only one possible deviation from the smooth trend of Sp:S0; this ratio appears, at a marginal level, to be systematically smaller at given densities in clusters known to be x-ray emitters.

Dressler also finds the luminosity of the bulge component of disk galaxies to depend on local galaxy density; in regions of higher density, the bulge luminosity tends to be higher as well.

The frequency distribution of bulge sizes for actively star-forming and S0 disk systems has been found to differ by both Dressler (*13*) and Burstein (*49*). Systems with a large bulge-to-disk ratio are more likely to be S0's than spirals. S0 systems with small bulge-to-disk ratios are virtually absent.

Size. There are as yet no published comprehensive studies (analogous to those available for E galaxies) of disk system sizes located in a variety of environments. Peterson *et al*. (*50*) examined the size-luminosity relationship for spirals and S0 galaxies in the Hercules and Virgo clusters. For Hercules, they found that the isophotal diameters of disk systems exhibit a relationship to luminosity identical to that of the field galaxies studied by Holmberg (*51*). The Virgo disks appear to be approximately 20 percent smaller at constant luminosity. Strom and Strom (*52*) also studied the relationship between isophotal diameter and position for S0 galaxies in the Coma cluster. They found that the disk sizes for these galaxies tend to be smaller in the central, dense regions of this cluster, in analogy with their findings for E galaxies.

Shape and alignment. There are as yet no data that provide measurements of the frequency distribution of ellipticity for disk system bulges, although Kormendy (*53*) believes the average disk system bulge to be more flattened than the average elliptical. In contradistinction to E galaxies, Kormendy and Illingworth (*54*) find that rotation is sufficient to account for the flattening of disk system bulges.

Recent work by Adams *et al.* (*55*) suggests that the major axes of disk galaxies are aligned in two preferred directions: either along or perpendicular to the cluster major axis. Their results add to the list of clusters in which galaxy alignment has been noted (*16, 56*) for disk systems, ellipticals, or both. It appears as if galaxies in clusters may somehow reflect the effects of the cluster environment either at the time of galaxy formation or later, as a consequence of postformation interactions.

Gas content and star formation—spiral and irregular galaxies. All current models of disk galaxies presume that the bulge region forms within a few free-fall times of the initial collapses of a protogalactic cloud. In analogy to E galaxies, star formation in the bulge is supposed to occur rapidly, so that the subsequent evolution of this region is dominated by nondissipative, stellar-dynamical processes. Bulges are consequently presumed to have been inactive in the star-forming sense for nearly the entire lifetime of the disk galaxy. In discussing the evolutionary history of a disk galaxy, it is necessary to take into account the (possibly) large difference in characteristic time scale for star formation in the bulge and disk subsystems; the evolutionary behavior of each component must be evaluated separately.

Assessment of star formation in disk galaxies at present depends on interpretation of integrated colors. Qualitatively, the bluest galaxies are believed to contain the largest fraction of newly formed stars. However, the observed colors include a nonnegligible contribution from the bulges; the relative contribution of the bulge is, of course, larger for earlier Hubble types. Hence if one sets out to compare the relative numbers of new and old stars in the actively star-forming region of the galaxy—the disk—the use of integrated colors will result in underestimating the young star contribution in systems of early Hubble type.

Taken at face value, the observed colors of spirals (*57, 58*) suggest that the ratio of new ($t \lesssim 10^8$ years) to old ($t \gtrsim 10^9$ years) stellar populations increases toward later Hubble types; the largest fraction of new stars is found in irregular galaxies. It is therefore believed that star formation was extremely efficient at early epochs in early Hubble types but is relatively inefficient at present; the opposite is believed true for late Hubble types. However, this conclusion rests heavily on the assumption that the metallicity of disk stars is constant with type.

A further difficulty in the discussion of disk system star-forming histories is presented by the one-dimensional classification scheme currently in use by most astronomers. For E galaxies, we now recognize that the observed colors depend on luminosity, presumably because they reflect a systematic variation of metallicity with galaxy mass (see earlier section on chemical composition). Such a luminosity-dependent phenomenon should serve as a warning to compare galaxy properties at constant luminosity in disk galaxies as well; physical arguments presented here will reinforce this intuitive belief. In this context Strom (*59*) has emphasized [see also van den Bergh (*7*)] that the Hubble sequence is also, in part, a luminosity sequence. Late-type galaxies are found to persist at lower luminosities than do the earlier types. Moreover, there are no high-luminosity representatives of types Sd and Sm. Consequently, a statement such as "galaxy colors become bluer at later Hubble types" contains information regarding the star-forming history of galaxies not only as a function of type but of luminosity as well. It will be critical in future discussions of the evolution of disk galaxies to establish an objective two-dimensional galaxy classification system.

Combined with an assessment of new and old population ratios, the relative mass in gas and stars provides another indication of the evolutionary status of a galaxy. The ratio of hydrogen mass to luminosity (M_H/L_B) appears to increase toward later Hubble types (*60*), suggesting that such galaxies are less advanced in an evolutionary sense than are earlier types. However, the total hydrogen mass rarely exceeds 20 percent of the total stellar mass. Hence in most disk galaxies a large fraction of the initial gas appears to have been converted into stars. Maps of the neutral hydrogen column density and the surface brightness of the stellar disk suggest that within a galaxy the ratio of gas to stellar mass increases outward. This suggests that the conversion of gas to stars is slower in the outer regions of disk galaxies, a result consistent with the radial distribution of chemical abundance described in the forthcoming section on chemical composition.

Some attempts (*61*) to study M_H/L_B as a function of luminosity suggest that the ratio increases toward lower luminosities within a given Hubble type. The highest values of M_H/L_B are found for irregular galaxies. Following the reasoning in our discussion of galaxy colors, it would be instructive in comparing the relative hydrogen content of galaxy types to compute the ratio of hydrogen mass to (red) disk luminosity and to compare systems of comparable disk luminosity.

Another difficulty in quantitatively assessing these M_H/L_B ratios is the possibility that a significant fraction of the gas content of a galaxy may be in the form of molecular as opposed to atomic hydrogen. Maps of some external galaxies (*62*) suggest that $M_{H_2}/M_H \gtrsim 1$ over a significant fraction of the galactic disks in late-type galaxies. Until we are able to estimate this quantity for a wide range of galaxy types and luminosities, it will be difficult to provide a definitive discussion of the relative degree of gas consumption in disk galaxies.

A further question regarding the gas content of spirals is the mass of gas contained in galactic halos. Larson *et al.* (*63*) suggested that halo gas "reservoirs" play a significant role in the evolution of disk galaxies. They proposed that removal of such halos in dense regions through galaxy-galaxy collisions may be responsible for the dominance of S0 galaxies in such regions. Neutral hydrogen (H I) maps of edge-on, isolated disk galaxies and careful observations of H I velocity dispersion in face-on systems should be made in order to assess the importance of gaseous halos.

A final concern regarding the star-forming histories of disk galaxies arises from recent observations (*64, 65*) which suggest that the oldest stars in the disk of the Milky Way may be no older than 6×10^9 years and in the Magellanic Clouds no older than 4×10^9 years. If true, these results suggest that the bulk of disk star formation may have begun more than 5×10^9 years after the formation of globular clusters and halo stars. Consequently, disk systems should exhibit significant changes in luminosity and color at relatively modest look-back times. Detailed population surveys of nearby galaxies with Space Telescope should provide definitive information regarding the age range of the constituents of the "old" disk population.

Gas content and star formation—S0 galaxies. Biermann *et al.* (*66*) have made searches for neutral hydrogen in S0 galaxies. In most cases the results have been negative, although some systems do contain small amounts of neutral hydrogen (1 percent of the total mass). In most cases detailed examination of the gas-bearing S0 galaxies reveals some evidence of recent star formation.

Searches for [O II] emission analogous to those carried out for E galaxies have also been conducted for S0 galaxies located in a variety of environments. The surveys to date provide primarily an indication of whether ionized gas is present in the nuclear bulge region. As for E galaxies, the frequency of [O II] emission is lower for the S0's located in rich clusters than for those in the field (*23*).

Wilkerson *et al.* (*67*) discussed an unusual class of spiral galaxies that appear not to be forming stars at the current epoch. They have smooth arms and no evidence of ionized hydrogen (H II) regions, OB complexes (associations of spectral type O and B stars), or dust and are found primarily, although not exclusively, in dense, rich clusters. Wilkerson *et al.* observed a sample of these galaxies and found that M_H/L_B is deficient, compared with normal spirals of similar bulge-to-disk ratio and arm winding, by more than a factor of 5 in all cases studied to date. They concluded that these systems are hydrogen-poor and suggested that some mechanism has accelerated the consumption of gas or removed most of the disk hydrogen.

A question related to the gas content of disk galaxies is whether the star-forming rates in actively star-forming galaxies vary as a function of environment. To answer this question, it will be necessary to observe the disk colors of spiral galaxies and compare the colors of systems of similar Hubble type located in a variety of settings. No such data are available at present.

Data are available that permit comparison of the integrated colors of S0 galaxies located within and outside rich clusters. From a sample of over 400 galaxies, Visvanathan and Sandage (*32*) argued that the colors of S0 galaxies are independent of environment. It should be noted that their result applies to the combined light of the bulge and disk. If the bulge light dominates the observed colors, then the test suggests that the bulge colors are environment-independent; the disk colors could be different. That this may be the case was suggested by Strom and Strom (*52*), who found the disk colors of S0 galaxies in the outer parts of the Coma cluster to be bluer than those in dense central regions. Their result suggests that star formation ceased more recently for galaxies located in the outer parts of the cluster.

Another indication of the star-forming history of disk systems located in rich clusters comes from recent work of Butcher and Oemler (*12*). From an examination of the frequency distribution of galaxy colors in two clusters similar to Coma but located at $z = 0.4$, they infer that a much larger fraction (compared with Coma) of the galaxies in the two distant clusters are actively star-forming systems (presumably spirals). Their result has, however, been challenged recently by Eastwood and Grasdalen (*68*), for example. If the Butcher-Oemler effect is real, it suggests that some mechanism acts to truncate star formation in such systems on a time scale comparable to 5 billion years. Alternatively, disks may be formed relatively recently.

Chemical composition—bulges. Burstein (*69*) recently completed an initial survey of S0 bulges and believes them to differ significantly from E galaxies in the sense that the nuclear region metallicity of S0's is greater at a fixed luminosity. Boroson (*70*) has challenged these results and attempted to extend the study of the bulge luminosity-composition relationship to actively star-forming disk systems. At present, the samples are too small to provide a definitive comparison of E galaxies and S0 and spiral bulges.

Wirth (*71*) has reported the first results of a survey of composition gradients in disk system bulges. He finds that the frequency and magnitude of detectable composition gradients are significantly higher among disk system bulges than in ellipticals. Combined with the Kormendy and Illingworth (*54*) results suggesting higher rotation for disk system bulges as compared with ellipticals, Wirth's observations seem to suggest that the evolutionary history of a disk system bulge may differ significantly from that of an E galaxy.

Chemical composition—disks. Considerable effort has been invested in the study of metal abundances in galactic disks [for example, see (*72*)], primarily through analysis of H II regions. All investigators agree that the metal-to-hydrogen ratio is greatest near the galactic center and decreases outward; disagreements regarding the magnitude of the composition gradients persist. Such gradients are predicted naturally in the context of gas dynamical models (*73*). Jensen *et al.* (*72*) attributed the gradients and the differences in mean metal abundances among galaxies of a given type to differences in star formation, element production, and gas depletion rates induced by galactic shocks.

Infrared properties of disk galaxies. The integrated light from actively star-forming disk galaxies derives both from newly formed stars and from an "old disk" population. Most discussions of disk properties such as scale length and total luminosity have been based on blue-light photometry. This is in hindsight an unfortunate choice, since the reported properties depend on the fractional contribution of young and old stellar populations. At wavelengths of 1.6 and 2.2 micrometers, however, the observed luminosity is normally dominated by the old disk population. Hence observations at these wavelengths provide a direct measurement of the underlying structure of the stellar component of disk galaxies. Aaronson (*74*) has provided the most extensive survey of disk galaxies to date. His work clearly demonstrates that light arising from old disk K giants dominates the luminosity observed in the near-infrared. Potentially, a comparison of near-ultraviolet and infrared maps of disk galaxies offers the possibility of separating the contributions of new and old populations and of leading thereby to an assessment of the ratio of present to past star-forming rates as a function of position. As yet this potential for explaining the star-forming behavior of disk galaxies has not been exploited fully.

Infrared observations at long wavelengths ($\approx$ 10 micrometers) have thus far been restricted primarily to the nuclear regions of disk galaxies. The results suggest the presence in some galaxies of large complexes of newly formed stars (of mass 10^7 to 10^8 $M_\odot$) still surrounded by molecular cloud complexes (*42*). The cause of such extremely vigorous star formation in the nuclear region is at present unknown. However, the existence of such stars is an important consideration in attempts to model active disk galaxy nuclei.

Efforts to map the disks of actively star-forming galaxies in the 10-micrometer region have just begun (*75*). Such observations will clearly be of importance in assessing the contribution of newly formed stars that are presently obscured from optical detection by optically thick dust clouds. Until such a census is available, the range of star-forming activity in galaxies in the current epoch will not be known with certainty.

Disk Galaxies as Laboratories for Understanding Star Formation

The apparent difference in the rates at which disk systems convert gas to stars suggests that by identifying systems of high M_H/L_B we may be able to gain some insight into the star-forming process as it operated at earlier epochs. Searches for such systems appear most profitably directed at lower luminosity, irregular systems, although in some cases relatively luminous galaxies of high M_H/L_B have been found (*76, 77*).

From studies of the relationship between observed metallicity and relative gas content M_H/M_{total} in such systems, it may be possible to estimate the yield of heavy elements per stellar generation. For closed systems with a constant yield per stellar generation, metallicity ~ $\ln(M_H/M_{total})$ (*78*). In more complex models the functional relationship will differ. The metal yield and its variation with M_H/M_{total} provides insight into the number of stars born as a function of mass—the initial mass function (IMF)—at least at the high-mass end. Attempts to evaluate the yield and place limits on the variation of IMF with age have been made by Lequeux *et al.* (*79*) and others (*80–82*). Thus far, the number of systems analyzed and the precision of metal abundance determination have not been sufficient for a definitive statement regarding yield (and hence IMF behavior) as a function of gas consumption fraction.

Observations of the *range* in observed current-epoch star formation as a function of M_H/M_{total} and of metallicity may yield important information about the frequency and efficiency of star-forming events in gas-dominated systems. Such a statistical study is rendered difficult because our census of gas-rich, low-luminosity systems is woefully incomplete and naturally biased toward systems that at present are bright. If "bursts" of star formation followed by long nascent periods are the rule, then large numbers of systems will be faint most of the time [for example, see (*58*)]. Nevertheless, studies of a complete sample of low-luminosity galaxies may prove to be critical in understanding the nature of star-forming events in the earlier evolutionary phases of more luminous galaxies.

Mass-to-Luminosity Ratios

Evidence that the stellar population producing the visible light observed for disk galaxies represents only a fraction of the total system mass has been growing over the past decade [for a review, see (*35*)]. The primary evidence supporting this view derives from rotation curves that, after an initial rise near the nucleus, reach a maximum and remain flat to very large galactocentric distances (*83*). If such rotation curves measure the true circular velocity of the stars and gas, the observations suggest a linear increase of mass with radius. As noted previously, the light distribution in disk galaxies decreased exponentially. Hence the mass-to-luminosity ratio in the outer parts of disk galaxies increases outward; local values of M/L exceeding 100 have been reported. Typical systemwide mean values for M/L_B range from 2 to 10. In computing the average M, only mass interior to isophotes of blue surface brightness $B = 26.5$ mag (arc sec)$^{-2}$ is included (*35*).

Average mass-to-luminosity ratios for disk galaxies have been derived from studies of binary pairs of disk galaxies (*84, 85*). However, it is not yet possible to compare with confidence the M/L ratios derived from binary galaxies with those estimated from rotation curves (*86*).

Whitmore *et al.* (*87*) compared the velocity dispersion of the nuclear bulge of disk galaxies with the peak rotational velocity. From such a comparison, they deduced that the unseen material constitutes a third, "hot" component independent of bulge and disk.

Attempts to isolate the agent responsible for the unseen mass have, to date, proved unsuccessful. Stringent limits on contributions from very red, low-mass stars have been derived from near-infrared measurements. Such objects appear incapable of supplying the missing mass. However, attempts to uncover the nature of the remaining mass continue to be a major challenge to students of galactic evolution, since the constituents of the unseen halo may play a critical role in the chemical and dynamical history of disk galaxies.

Interpretation

To understand these results, we must first attempt to understand the factors that affect the evolution of isolated disk galaxies. Most models (*73*) assume that disk galaxies begin as rotating protogalactic clouds containing large numbers of gaseous clumps. If the mass of cloud is large enough, its self-gravity "wins" over the Hubble expansion and the cloud begins to collapse. The collapse proceeds nonhomologously. In the denser central regions of the cloud, star formation proceeds vigorously, rapidly consuming the gas. This region evolves into the stellar-dominated nuclear bulge on a time scale comparable with a free-fall time ($\sim 10^8$ years for a Milky Way system). Star formation in the lower density outer regions proceeds far less efficiently. In collisions between gas clumps, the kinetic energy of motion is converted into heat, which is then radiatively lost to the system. Such collisions occur most easily along the rotational axis of the system. By dissipating the energy in random motions through gas-cloud collisions, the outer regions eventually settle into a cold, rotating disk. The subsequent changes in disk system appearance will depend primarily on the star-forming history of the disk component. Perhaps the major factor that determines the ability of a disk to form stars is the availability of gaseous material. In turn, the amount of gas available at any time depends on the initial gas content of the disk, the rate of injection from dying stars in the disk or possibly in a halo, and the rate of depletion through star-forming events. Self-sweeping mechanisms such as winds emanating from the nuclear bulge region can also deplete disk gas. Suppose that initially disk galaxies are born with a wide range of bulge-to-disk ratios—from the small values characterizing late-type spirals to the large values found today for only the most extreme Sa galaxies and for S0 systems. At birth, each Hubble type starts out with comparable ranges of initial disk gas content.

Two models for the subsequent star-forming history of the disk have been suggested in recent years. One proposes that astration in the disk is driven by galactic shocks, while the other posits that star formation occurs stochastically.

The galactic shock model (*88*) presupposes the existence of a spiral density wave in the old disk stars. The wave pattern propagates through the disk with a characteristic angular speed Ω_p, the pattern frequency. The disk gas is supposed to be in circular orbit about the galaxy center with an angular speed $\Omega(r)$ at galactocentric distance r. At any r, the spiral arm intersects a circular path at an angle i called the pitch angle. The velocity of gas normal to the spiral pattern is

$$w_\perp = [\Omega(r) - \Omega_p]\, r \sin i$$

The galactic shock model then presumes that if $w_\perp$ exceeds the acoustic speed a in the gas, the gas will be shocked and compressed behind the shock; the compression is supposed sufficient to initiate the collapse of gas clouds and the formation of stars. Hence the model predicts that star-forming events will occur each time the disk gas "encounters" the spiral wave pattern, provided the circular velocity of the gas is sufficiently high to produce a shock. Moreover, the star-forming regions are predicted to lie adjacent to the density wave crests. This picture appears to be highly successful in predicting the star-forming patterns and gas flows in well-defined, luminous spiral galaxies [for example, see (*89*)]. However, there are classes of galaxies at all luminosity ranges in which no obvious underlying density wave pattern is found and in which the star-forming pattern is less obviously spiral.

Seiden and Gerola (*90*) posit that star formation initially begins at random in a differentially rotating disk. Each initial star-forming event has a probability of "stimulating" star formation in a nearby region (for example, through a supernova explosion and subsequent compression of "nearby" gas behind the supernova shock). They find that for a narrow range of random and stimulated star-forming probabilities the differential rotation of the galaxy naturally produces patterns quite similar to those observed in nature. In one sense their model is purely mathematical. However, if the values derived for random and stimulated probabilities in a wide variety of types occupy relatively small ranges, we may be able to deduce important constraints on the factors influencing the star-forming history of galaxies, although at present we lack even a crude outline of an adequate model of star-forming physics.

Both models predict that the mass (luminosity) of the galaxy should play a major role in determining star-forming patterns. The galactic shock model demands that $w_\perp \gtrsim a$ in order that star-forming events may be induced. Let us define a quantity

$$w_\perp(\max) = V_{\max} \sin i$$

where $V_{\max}$ is the maximum value of the quantity $(\Omega - \Omega_p)r$. For a galaxy of mass m and characteristic radius R

$$V_{\max} \approx (M/R)^{1/2}$$

Assuming that the luminosity L of a galaxy is proportioned to M and that, from observation (*50*), $R \sim L^{0.4}$,

$$w_\perp(\max) \approx L^{0.3} \sin i$$

For a fixed pitch angle $w_\perp$ therefore decreases with decreasing L until, at some critical luminosity $L_{\min}$, $w_\perp = a$.

Below L_{min}, $w_{\perp}$ is too small to initiate star-forming events in galactic shocks, and hence star formation will no longer take place in spiral patterns unless the amplitude of the wave pattern is extremely large. The pitch angle of the arms i is larger for open-armed, late Hubble-type systems (Sc and later) than for tightly wound spirals of early Hubble type (Sb and earlier). Hence $L_{min}(Sa) > L_{min}(Sc)$, and we expect to see spiral star-forming patterns begin to die out at higher luminosities for Sa galaxies than for Sc's. Below L_{min}, only irregular star-forming patterns are expected.

Indeed, observations show that spiral patterns are present in late-type systems at much lower luminosities than in early-type galaxies; irregular galaxies dominate low-luminosity systems. This picture also suggests that for systems of luminosity below L_{min} star formation can no longer be driven. As a result, we expect the formation of stars and the consumption of gas to proceed more slowly, at least if shock-driven star formation plays an important role in the star-forming history of spiral galaxies. Hence the ratio of new to old stellar population and gas content should be a function of galaxy luminosity—a result in qualitative agreement with observations.

The stochastic model attributes the decrease in average star-forming activity and the tendency toward irregular patterns in low-luminosity galaxies to (i) a decrease in system scale (and the consequent low-average rotational velocity) and (ii) the apparent rigid (as opposed to differential) rotation of low-luminosity systems. The scale determines the number of star-forming "cells" that can be active at any given time—the smaller the number of cells, the smaller the average star-forming rate. The lack of differential rotation prevents the development of a spiral pattern, since all star-forming cells move at the same rate relative to one another. Hence a "snapshot" at any time reveals only a random pattern.

Observers are faced with two major tasks at present. The first is to determine the relationship between star-forming patterns and the underlying density wave pattern, if any. Strong coupling between the two would strongly suggest a causal relationship of astration events and the density wave pattern. Such studies should be carried out for all Hubble types at constant system luminosity. Quantitative measurements of spiral pattern shape and the relationship of the pattern to the distribution of matter in the galaxy will be especially valuable in providing insight into the physical basis for the underlying spiral patterns.

The second task is to determine quantitatively the ratio of new to old stellar populations and the fractional gas content as a function of luminosity for fixed Hubble types. Overall galaxy dynamics may play an essential role in determining the evolutionary history of disk systems. If so, such effects will become apparent through a classification scheme that provides an appropriate measure of disk system dynamics.

At some point in the history of a system, a significant fraction of disk gas must be depleted in star-forming events. As noted previously, in most luminous galaxies the neutral hydrogen mass constitutes less than 10 percent of the total system mass. Suppose, for the sake of argument, that at some point in a system's evolutionary history the disk gas content drops below the point critical to sustain star formation. What happens when the disk gas is depleted? It seems logical to assume that a gas-free spiral should appear similar to the smooth-arm systems discussed by Wilkerson *et al.* (*67*). In such systems the underlying arm pattern is readily visible; however, there appear to be no accompanying star-forming events, and moreover the gas content is unusually low. It is tempting to speculate that such systems evolve into S0's. However, no convincing models as yet predict whether smooth-arm systems evolve naturally into featureless S0 disks.

Dressler (*13*) notes that S0's tend to have significantly larger bulge-to-disk ratios than do actively star-forming spirals. Suppose rapid astration is the major cause of disk gas depletion. If so, the currently available data suggest that among still star-forming systems, types with large bulge-to-disk ratios (Sa and Sab) have depleted their disk gas mass most rapidly. It seems natural to extrapolate this behavior to systems with even higher bulge-to-disk ratios and to posit that such systems have already consumed their available disk gas. A satisfactory explanation of why large bulge/disk systems should be more likely to form stars more rapidly is, however, not available.

The evolutionary fate of disk systems located in a rich cluster may be affected by interaction both with other galaxies and with the intracluster medium. Gravitational encounters may tidally strip the gaseous halos and the outer disks of individual galaxies, perhaps resulting in the observed decrease of disk sizes in the central region of the Coma cluster (note that such a reduction will increase the apparent bulge-to-disk ratios for such systems). As galaxies traverse clusters containing intracluster gas of sufficient density, ram-pressure stripping may remove their disk or halo gas (*25*); partially stripped galaxies may continue active star formation although they possess less disk gas. In some cases, for galaxies bathed in hot intracluster gas, thermal evaporation may also remove disk gas. If gas is removed by either process, we must again ask, what is the fate of an actively star-forming system? It seems logical that the evolutionary path followed by such galaxies should be similar to that followed by isolated disk galaxies that deplete their disk gas; the spiral will at first appear as a smooth-arm system and later as an S0. If gas removal by evaporation or ram-pressure stripping is important in driving evolution from spirals to S0's, two major effects should be observed: (i) the ratio of Sp's to S0's should be smaller in x-ray clusters, and (ii) the fraction of S0's with small bulge-to-disk ratios may be larger in such clusters. Dressler's data (*13*) are marginally supportive of the first effect and say little about the second. However, it is noteworthy (*13*) that the median bulge luminosity in dense regions (although not necessarily those pervaded by hot gas) is actually larger, suggesting that the bulge size may reflect initial conditions at the epoch of galaxy formation that may, in turn, be somehow related to the current-epoch environment.

Gisler (*91*) noted that the gas densities in disk galaxies were probably higher at earlier epochs. Consequently, they would have been more resistant to ram-pressure stripping. Hence, even if the intracluster gas density at earlier epochs was as high as it is at present, many actively star-forming systems (particularly of later Hubble type) could have resisted stripping. Therefore the presence of a large admixture of blue galaxies in the Butcher-Oemler clusters is not surprising. However, we must keep in mind that the Butcher-Oemler data may also be telling us that star formation in galactic disks may be a more recent phenomenon than currently popular thought would have it (*64, 65*).

Finally, Dressler's results on the dependence of bulge luminosity on environment and the variety of work suggesting that the shape and orientation of both disk and elliptical galaxies reflect the properties of their host clusters both argue strongly that we heed the possibility that environment-dependent initial conditions greatly influence the course of galaxy evolution.

References and Notes

1. Because of space limitations we have not been able to adequately discuss several subjects of active research, including the role of active nuclei and "heavy halos" in galactic evolution and the role of bars in disk system evolution. Moreover, in a limited bibliography we have almost certainly been unable to do justice either to the pioneers or the current active practitioners in the field.
2. G. de Vaucouleurs, *Mon. Not. R. Astron. Soc.* **113**, 134 (1953).
3. E. Hubble, *Astrophys J.* **64**, 321 (1926).
4. A. R. Sandage, *The Hubble Atlas of Galaxies* (Carnegie Institution of Washington, Washington, D.C., 1961).
5. W. W. Morgan, *Publ. Astron. Soc. Pac.* **70**, 372 (1958).
6. ———, *ibid.* **71**, 394 (1959).
7. S. van den Bergh, *Publ. David Dunlap Obs.* **2** (No. 6) (1960).
8. J. Kormendy, *Astrophys. J.* **217**, 406 (1977).
9. H. Spinrad, J. P. Ostriker, R. P. S. Stone, L.-T. G. Chiu, G. Bruzual A., *ibid.* **225**, 56 (1978).
10. A. Oemler, Jr., *ibid.* **209**, 693 (1976).
11. J. Melnick and W. L. W. Sargent, *ibid.* **215**, 401 (1977).
12. H. Butcher and A. Oemler, Jr., *ibid.* **219**, 18 (1978).
13. A. Dressler, *ibid.* **236**, 351 (1980).
14. S. E. Strom and K. M. Strom, *Astron. J.* **84**, 1091 (1979).
15. ———, *Astrophys. J. Lett.* **225**, L93 (1978).
16. A. Dressler, thesis, University of California, Santa Cruz (1976).
17. P. Schechter, *Astrophys. J.* **203**, 297 (1976).
18. M. A. Hausman and J. P. Ostriker, *ibid.* **224**, 320 (1978).
19. P. L. Schechter and J. E. Gunn, *ibid.* **229**, 472 (1979).
20. T. B. Williams and M. Schwarzschild, *ibid.* **227**, 56 (1979).
21. A. B. Marchant and D. W. Olsen, *Astrophys. J. Lett.* **230**, L157 (1979).
22. G. R. Knapp, F. J. Kerr, B. A. Williams, *Astrophys. J.* **222**, 800 (1978).
23. G. R. Gisler, *Mon. Not. R. Astron. Soc.* **183**, 633 (1978).
24. L. L. Cowie and A. Songaila, *Nature (London)* **266**, 501 (1977).
25. J. E. Gunn and J. R. Gott, III, *Astrophys. J.* **176**, 1 (1972).
26. W. G. Tifft, *Astron. J.* **74**, 354 (1969).
27. J. Stauffer, personal communication.
28. J. S. Gallagher, personal communication.
29. S. M. Faber, paper presented at the Aspen Physics Workshop (1981).
30. ———, *Astrophys. J.* **179**, 731 (1973).
31. J. A. Frogel, S. E. Persson, M. Aaronson, K. Matthews, *ibid.* **220**, 75 (1978).
32. N. Visvanathan and A. Sandage, *ibid.* **216**, 214 (1977).

32a. E. N. Caldwell, thesis, Yale University (1982).

33. S. E. Strom, J. C. Forte, W. E. Harris, K. M. Strom, D. C. Wells, M. G. Smith, *Astrophys. J.* **245**, 416 (1981).
34. S. M. Faber and R. E. Jackson, *ibid.* **204**, 668 (1976).
35. S. M. Faber and J. S. Gallagher, *Annu. Rev. Astron. Astrophys.* **17**, 135 (1979).
36. W. L. W. Sargent, P. J. Young, A. Boksenberg, K. Shortridge, C. R. Lynds, F. D. A. Hartwick, *Astrophys. J.* **221**, 731 (1978).
37. A. Dressler, *Astrophys. J. Lett.* **240**, L11 (1980).
38. M. Duncan and C. Wheeler, *ibid.* **237**, L27 (1980).
39. J. N. Bahcall and C. L. Sarazin, *ibid.* **213**, L99 (1977).
40. J. Kristian, A. Sandage, J. A. Westphal, *Astrophys. J.* **221**, 383 (1978).
41. G. L. Grasdalen, in *Proceedings, IAU Symposium 92, Objects of High Redshift*, G. O. Abell and P. J. E. Peebles, Eds. (Reidel, Boston, 1980), p. 269.
42. G. H. Rieke and M. J. Lebofsky, *Annu. Rev. Astron. Astrophys.* **17**, 477 (1979).
43. A. Toomre, in *The Evolution of Galaxies and Stellar Populations*, B. M. Tinsley and R. B. Larson, Eds. (Yale University Observatory, New Haven, Conn., 1977).
44. R. B. Larson, *Mon. Not. R. Astron. Soc.* **166**, 585 (1974).
45. J. P. Ostriker, *Comments Astrophys.* **8**, 177 (1980).
46. J. R. Gott, III, and T. X. Thuan, *Astrophys. J.* **204**, 649 (1976).
47. R. B. Larson, *Mon. Not. R. Astron. Soc.* **169**, 229 (1974).
48. N. A. Bahcall, *Astrophys. J. Lett.* **218**, L93 (1977).
49. D. Burstein, *Astrophys. J.* **234**, 435 (1979).
50. B. M. Peterson, S. E. Strom, K. M. Strom, *Astron. J.* **84**, 735 (1979).
51. E. Holmberg, *Medd. Lunds Astr. Obs. Ser. 2 No. 136* (1958).
52. S. E. Strom and K. M. Strom, in *Proceedings, IAU Symposium 77, Structure and Properties of Nearby Galaxies*, E. M. Berkhuijsen and R. Wielebinski, Eds. (Reidel, Dordrecht, Netherlands, 1978).
53. J. Kormendy, personal communication.
54. ——— and G. Illingworth, in *Photometry, Kinematics and Dynamics of Galaxies*, D. S. Evans, Ed. (Department of Astronomy, University of Texas, Austin, 1979), p. 195.
55. M. T. Adams, K. M. Strom. S. E. Strom. *Astrophys. J.* **238**, 445 (1980).
56. L. A. Thompson, *ibid.* **209**, 22 (1976).
57. G. de Vaucouleurs, A. de Vaucouleurs, H. G. Corwin, *Second Reference Catalogue of*

Bright Galaxies (Univ. of Texas Press, Austin, 1976).
58. L. Searle, W. L. W. Sargent, W. G. Bagnuolo, *Astrophys. J.* **179**, 427 (1973).
59. S. E. Strom, *ibid.* **237**, 686 (1980).
60. M. Roberts, in *Galaxies and the Universe*, A. Sandage, M. Sandage, J. Kristian, Eds. (Univ. of Chicago Press, Chicago, 1975), vol. 9. p. 309.
61. C. Balkowski, *Astron. Astrophys.* **29**, 43 (1973).
62. N. Scoville and J. R. Young, *Astrophys. J.*, in press.
63. R. B. Larson, B. M. Tinsley, C. N. Caldwell, *ibid.* **237**, 692 (1980).
64. P. DeMarque and R. D. McClure, in *The Evolution of Galaxies and Stellar Populations*, B. M. Tinsley and R. B. Larson, Eds. (Yale University Observatory, New Haven, Conn., 1977).
65. H. Butcher, *Astrophys. J.* **216**, 372 (1977).
66. P. Biermann, J. N. Clarke, K. J. Fricke, *Astron. Astrophys.* **75**, 7 (1979).
67. M. S. Wilkerson, S. E. Strom, K. M. Strom, *Bull. Am. Astron. Soc.* **9**, 649 (1977).
68. K. Eastwood and G. L. Grasdalen, *Astrophys. J. Lett.* **239**, L1 (1980).
69. D. Burstein, *Astrophys. J.* **232**, 74 (1979).
70. T. A. Boroson, thesis, University of Arizona (1980).
71. A. Wirth, *Astrophys. J.*, in press.
72. E. B. Jensen, K. M. Strom, S. E. Strom, *ibid.* **209**, 748 (1976).
73. R. B. Larson, *Mon. Not. R. Astron. Soc.* **176**, 31 (1976).
74. M. Aaronson, thesis. Harvard University (1977).
75. G. Wynn-Williams, personal communication.
76. W. Romanishin, thesis, University of Arizona, Tucson (1980).
77. B. L. Webster, W. M. Goss, T. G. Hawarden, A. J. Longmore, U. Mebold, *Mon. Not. R. Astron. Soc.* **186**, 31 (1979).
78. L. Searle and W. L. W. Sargent, *Astrophys. J.* **173**, 25 (1972).
79. J. Lequeux, M. Peimbert, J. F. Rayo, A. Serrano, S. Torres-Peimbert, *Astron. Astrophys.* **80**, 155 (1979).
80. D. Talent, thesis, Rice University (1980).
81. T. D. Kinman and K. Davidson, *Astrophys. J.* **243**, 127 (1981).
82. H. B. French, thesis, University of California, Santa Cruz (1979).
83. V. C. Rubin, W. K. Ford, Jr., N. Thonnard, *Astrophys. J. Lett.* **225**, 107 (1978).
84. E. L. Turner, *Astrophys. J.* **208**, 304 (1976).
85. P. D. Noerdlinger, *ibid.* **229**, 877 (1979).
86. H. Rood, *ibid.*, in press.
87. B. C. Whitmore, R. P. Kirshner, P. L. Schechter, *ibid.* **234**, 68 (1979).
88. W. W. Roberts, Jr., M. S. Roberts, F. H. Shu, *ibid.* **196**, 381 (1975).
89. H. C. D. Visser, *Astron. Astrophys.* **88**, 159 (1980).
90. P. E. Seiden and H. Gerola, *Astrophys. J.* **233**, 56 (1979).
91. G. R. Gisler, *ibid.* **228**, 385 (1979).
92. We express our deepest gratitude to the members of the Physics and Astronomy Department of the University of Wyoming for providing an environment conducive to the completion of this work. Kitt Peak National Observatory is operated by the Association of Universities for Research in Astronomy, Inc., under contract with the National Science Foundation.

This material originally appeared in *Science* 216, 7 May 1982.

Color Plates

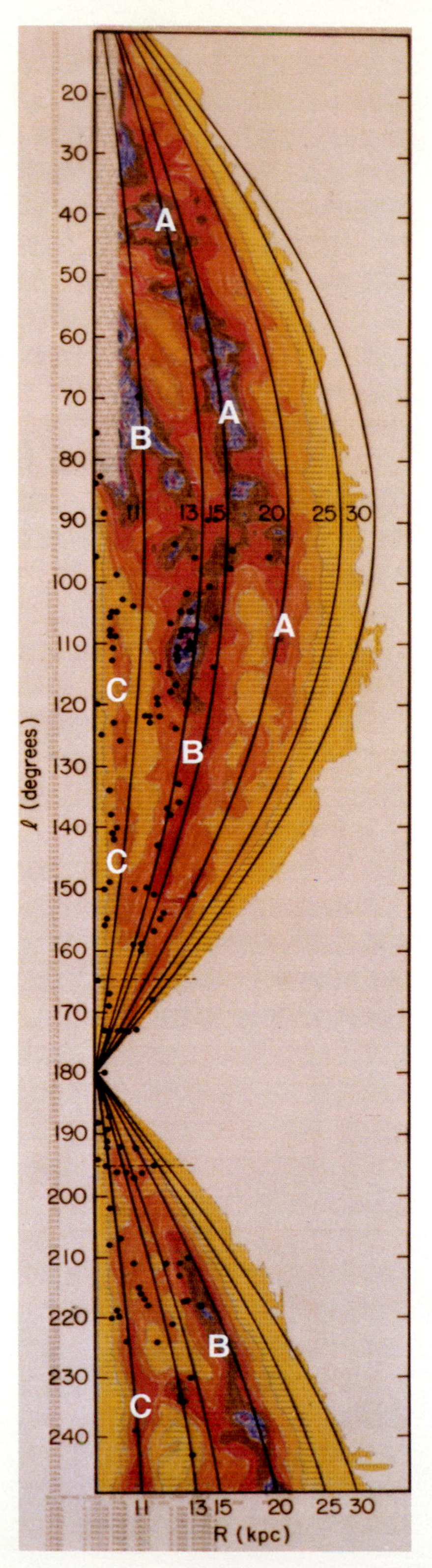

Plate I, Chapter 5

Surface density of atomic hydrogen beyond the solar distance, showing the large-scale ordered spiral structure. The ordinate is galactic longitude (see Fig. 2 on page 81), and the solid lines are lines of constant distance from the center of the Galaxy. The contours are in units of 1.0 solar mass per square parsec. Features *A*, *B*, and *C* are discussed in the text. The black dots represent the positions of starforming molecular clouds, which are seen to be well correlated with the arms defined by the atomic hydrogen.

Plate II, Chapter 5 (facing page)

Deviations of the mean plane of the atomic hydrogen emission from the galactic equator: (a) positive deviations and (b) negative deviations. Contours are in units of 200 pc. The large-scale warp between longitudes 10° and 180° is fairly gentle to a distance of ~ 18 kpc from the center, where it appears to increase sharply. Scalloping is indicated by arrows at positions where deviations at the edge of the disk are opposite to the large-scale warp.

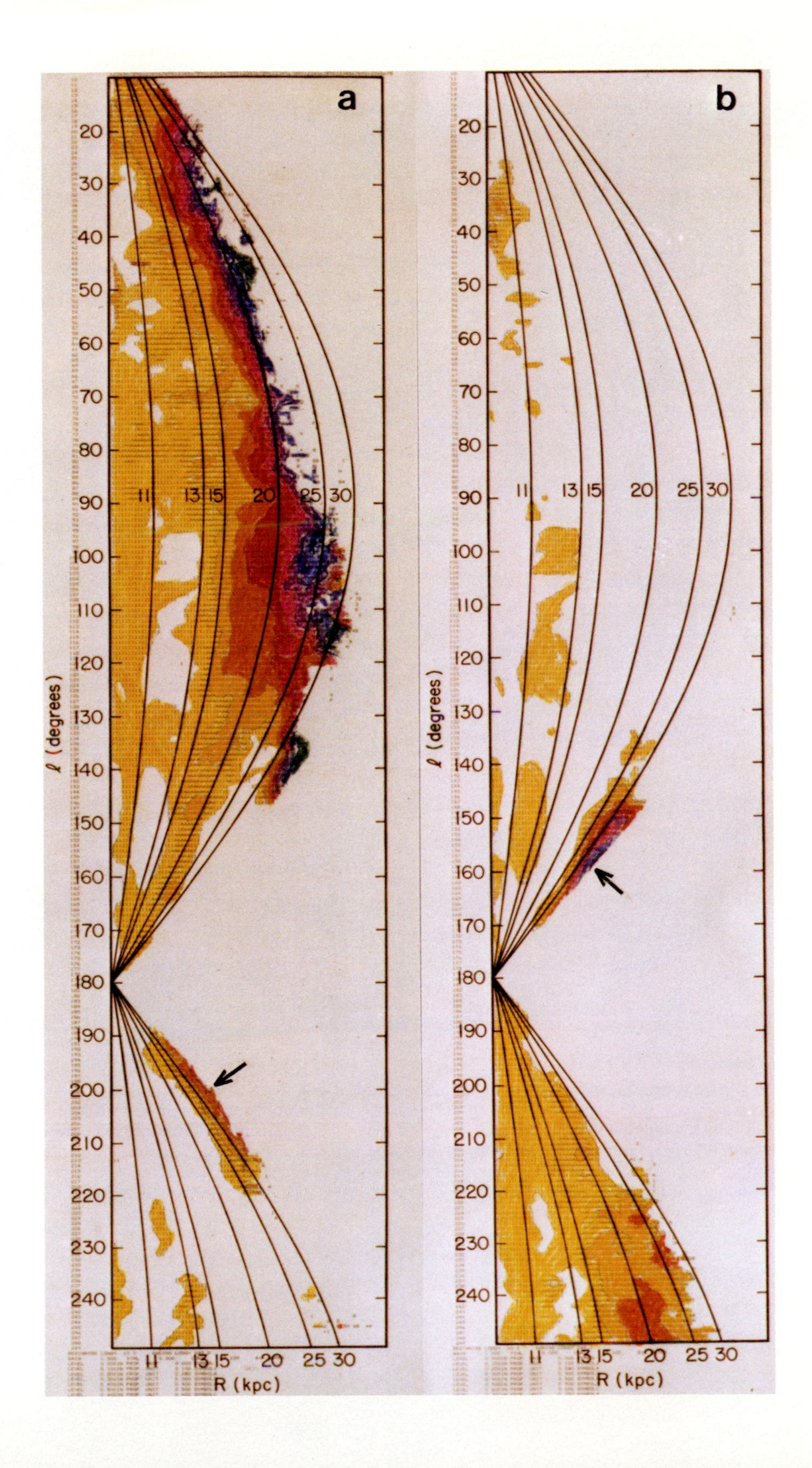
a
b
ℓ (degrees)
R (kpc)
11
13
15
20
25
30
20
30
40
50
60
70
80
90
100
110
120
130
140
150
160
170
180
190
200
210
220
230
240

Plate III, Chapter 6

Massive star Eta Carinae (center) and its surrounding cloud of dusty gas. The ejection of this gas from the central star was observed about 140 years ago. This is a logarithmic isophote map in red light; the outer contours are fainter than the central spot by a factor of roughly 2000 in surface brightness. The width of the picture is about 0.5 arc minute or 1.3 light years at the 9000-light-year distance of Eta Carinae. Recently it was discovered that the large outer condensation on the right side is nitrogen-rich. [Data for image were obtained with SIT vidicon and 60-inch telescope at Cerro Tololo Inter-American Observatory, La Serena, Chile]

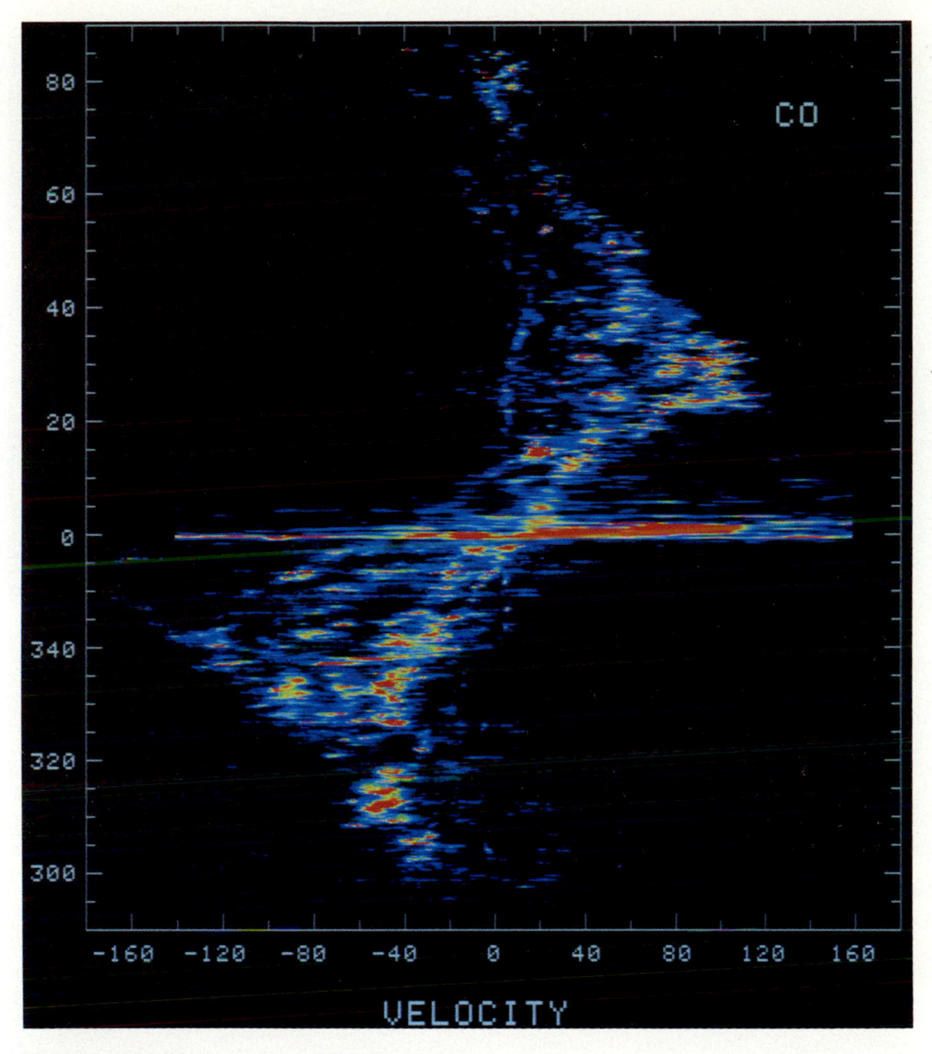

Plate IV, Chapter 9

A representation of the intensity of millimeter-wave emission from CO at a wavelength of 2.6 mm along the inner galactic plane between 290° and +90° longitude (*L*). The velocity of the emission indicates the distance from the galactic center, with the highest velocity at each longitude (or lowest velocity at negative longitude) corresponding to the closest approach of that line of sight to the galactic center. The strongest emission arising from giant molecular clouds (GMC's) and clusters of GMC's is apparent in the yellow and red structures. For example, at longitude = 15°, velocity = 15 km/sec (M17) and longitude 30.5°, velocity = 90 km/sec (W43), two well known GMC's associated with active star formation and ionized gas (from hot stars) are very prominent. The strong emission near longitude 0 is from the galactic center region. This picture is a composite of data from two surveys. The southern data are from the Australian survey by B. J. Robinson, R. N. Manchester, J. B. Whiteoak, and W. H. McCutcheon. The northern data are from the Massachusetts-Stony Brook survey by D. B. Sanders *et al.* (*53*).

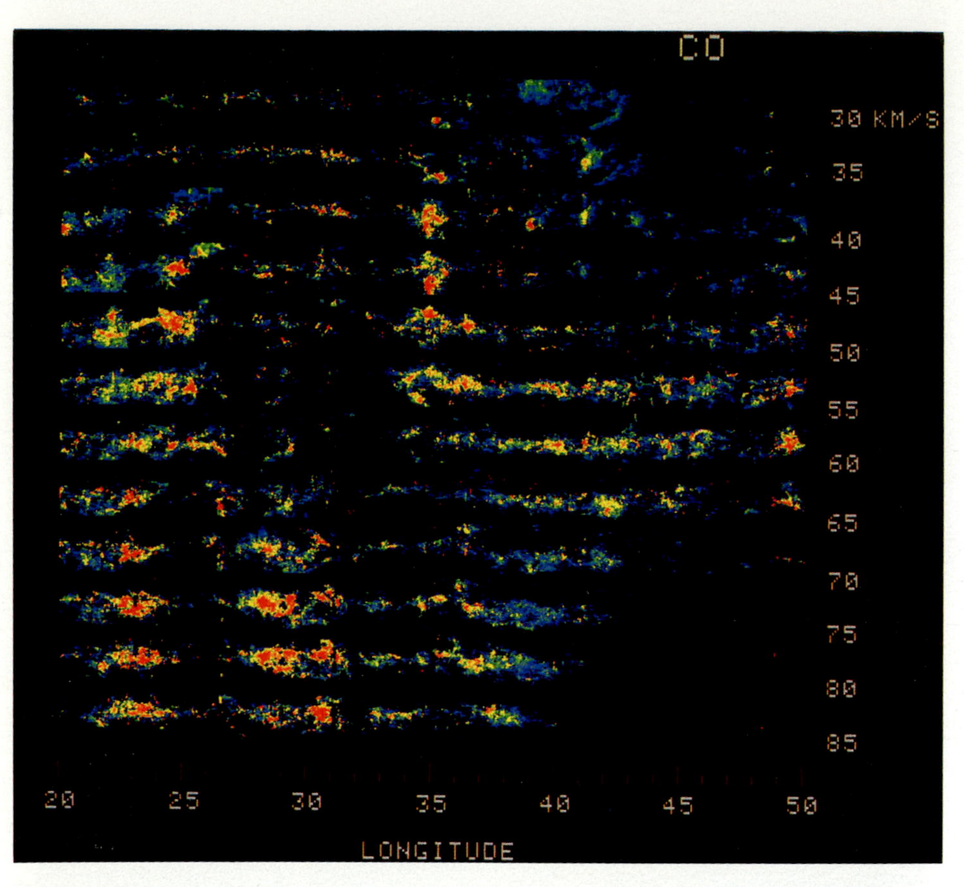

Plate V, Chapter 9

A map showing the location and intensity of CO emission from interstellar molecular clouds. Each strip is a view of the galactic plane from longitude 20° to 50° and latitude − 1° to + 1.0°, at a fixed velocity (Doppler shift); the map consists of a composite of 24,600 observations at 0.05° intervals. Individual GMC's as well as clusters of molecular clouds, which are the most massive objects in the galaxy, dominate the emission. The clouds in this picture are at a distance of from 6,000 to 36,000 light-years from the sun. The data are from the Massachusetts-Stony Brook survey of the galaxy.

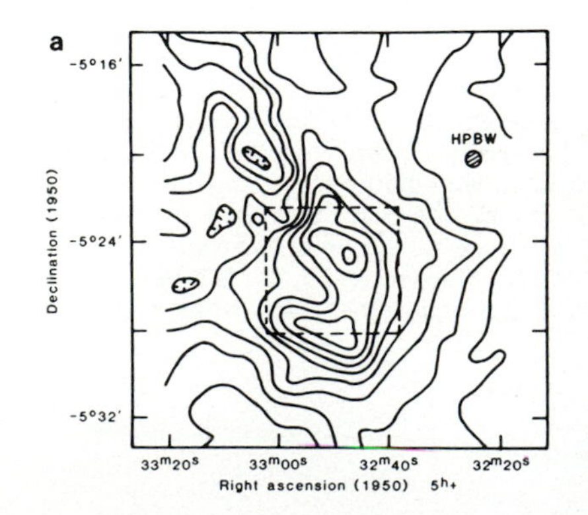

Plate VI, Chapter 9

(a) The brightness temperature distribution of CO ($J = 1 \rightarrow 0$) emission from the central 0.3° of the Orion Molecular Cloud (OMC) (*49*). The black dashed box indicates the area covered by (b); *HPBW*, half-power beam width. North is at the top, east is to the left. (b) A 5′ square optical photograph of M42, the active star formation region in the OMC, with 10-μm dust emission contours superimposed (*50*). Stars of several ages are represent-

Plate VI, Chapter 9 (continued)

ed. The white dashed box indicates the area covered by (c). North is at the top, east is to the left. (*c*) The core region (1′ square) of M42 showing infrared emission (color) from the BNKL complex and H_2 emission (white contours) from a shock front in the vicinity (*51*). BN is possibly the youngest stellar object to have been observed to date. Its formation may be associated with the shock front. The infrared image was produced by displaying 11-, 12-, and 20-μm images on the blue, green, and red guns of a cathode-ray tube (*52*). North is at the top, east is to the left.

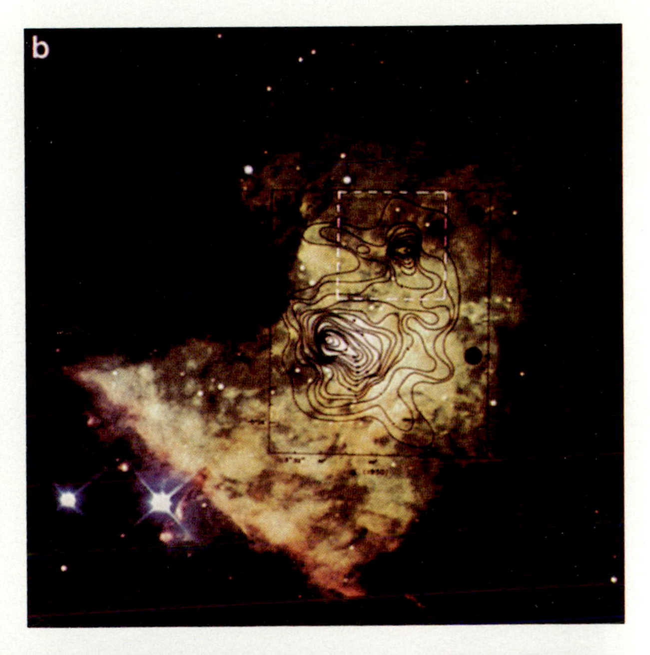

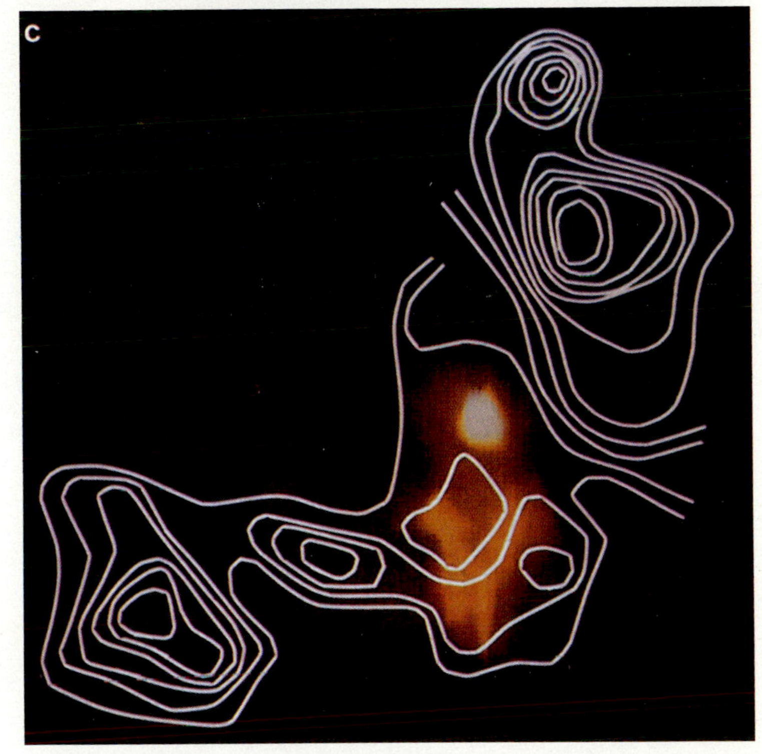

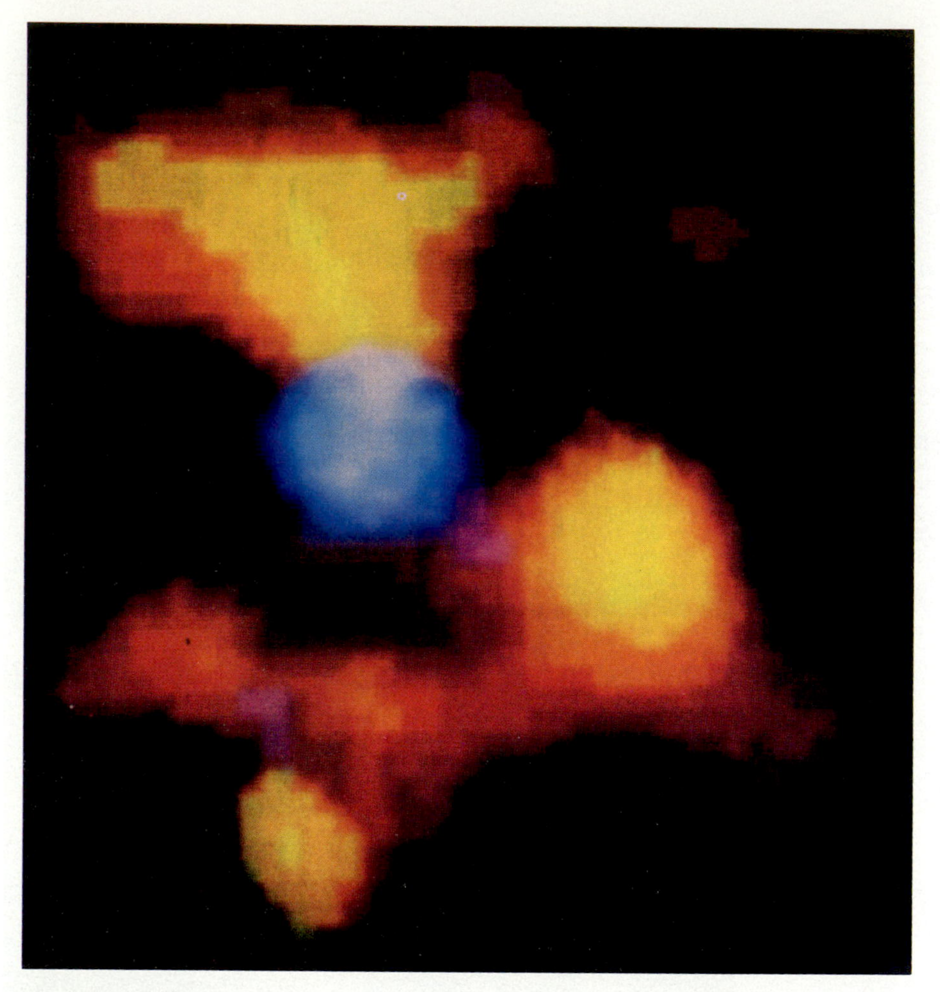

Plate VII, Chapter 9

False-color image of Sharpless 106 produced by displaying 3-, 10-, and 20-μm images on the blue, green, and red guns of a cathode-ray tube (*38*). The young star is the blue spot in the center. An outflow is observed in the orange biconical lobes. North is at the top, east is to the left.

Plate VIII, Chapter 18

The distribution of galaxies observed on a limiting exposure plate taken with the 4-meter telescope at Kitt Peak National Observatory. The cluster of galaxies clearly visible spans several million light-years in radius and is approximately 2×10^9 light-years away. [Courtesy of J. A. Tyson and J. F. Jarvis]

Plate IX, Chapter 18

Some of the windows on cosmology. (a) An underwater photograph of the Irvine-Michigan-Brookhaven detector "pool," which is part of the proton decay apparatus in the Morton Salt Mine outside of Cleveland [courtesy of K. Luttrell].

(b) The neutron electric dipole moment apparatus at the Grenoble HFR-ILL [courtesy of Institut Lave-Langevin].

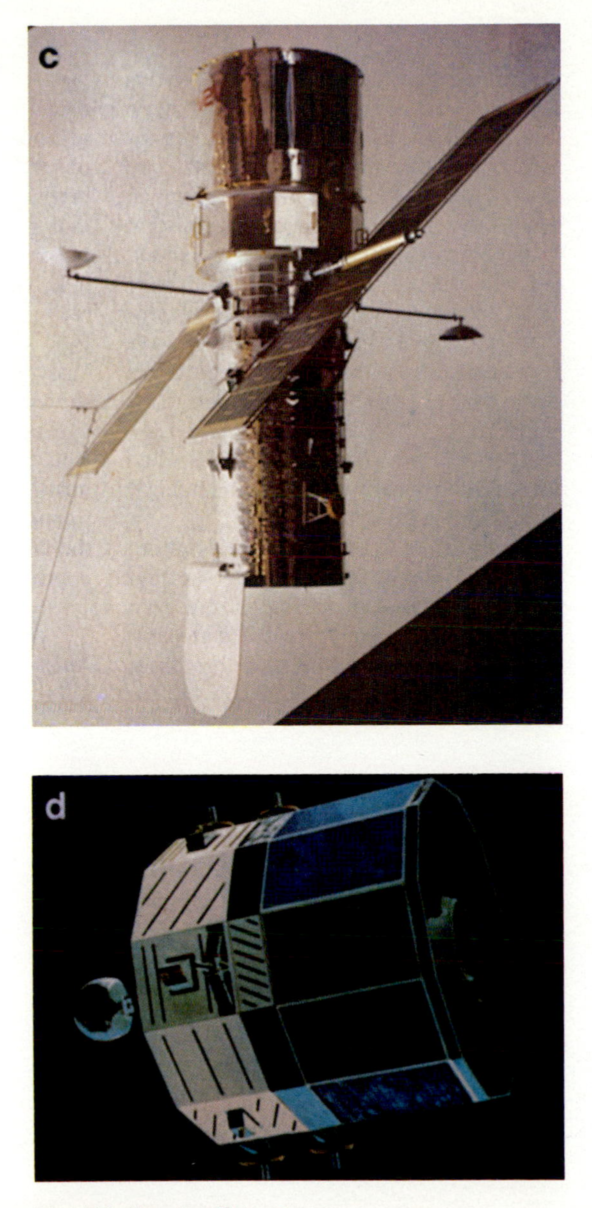

Plate IX, Chapter 18 (continued)

(c) A model of the Space Telescope. (d) An artist's rendering of the Cosmic Background Explorer satellite, which will probe the spectral characteristics and large-scale homogeneity of the 2.7-K background radiation.

16. The Rotation of Spiral Galaxies

Vera C. Rubin

Historians of astronomy may some day call the mid-20th century the era of galaxies. During those years, astronomers made significant progress in understanding the structure, formation, and evolution of galaxies. In this chapter, I will discuss a few selected early steps in determining the internal dynamics of galaxies, as well as current research concerning the rotation and mass distribution within galaxies. These studies contribute to the present view that much of the mass of the universe is dark.

In 1845, Lord Rosse (*1*) constructed a mammoth 72-inch reflecting telescope, which remained the largest telescope in the world until the construction of the "modern" 100-inch telescope on Mount Wilson in 1917. Unfortunately, the Rosse telescope did not automatically compensate for the rotation of the earth, so an object set in the field of the telescope would race across the field. Nevertheless, Lord Rosse made the major discovery that some nebulous objects show a spiral structure (*2*).

Insight into the nature of these enigmatic objects was slow in coming, but by 1899 the combination of good tracking telescopes and photographic recording made it possible for Scheiner (*3*) to obtain at the Potsdam Observatory a spectrum of the nucleus of the giant spiral in Andromeda, M31. From the prominence of the *H* and *K* absorption lines of calcium, Scheiner correctly concluded that the nucleus of M31 was a stellar system composed of stars like the sun, rather than a gaseous object.

Even earlier, Lord and Lady Huggins (*4*) had attempted at their private observatory to obtain a spectrum of M31, but their labors were fraught with disaster. Exposure times were so long that each fall when M31 was overhead they would expose consecutively on one plate during the early evening hours, until they could build up the requisite exposure (perhaps 100 hours). They would then superpose a solar spectrum adjacent to the galaxy spectrum to calibrate the wavelength scale. Year after year, their single exposure was a failure, sometimes showing no galaxy spectrum, sometimes contaminated by the solar spectrum; once the plate was accidentally placed emulsion side down, where it dried stuck to the laboratory table. A copy of their 1888 failure is published along with their remarkable stellar spectra (*4*).

Slipher (*5*), working with the 24-inch reflector at Lowell Observatory, initiated in 1912 systematic spectral observations of the bright inner regions of the nearest galaxies. For NGC 4594, the Sombrero galaxy, he detected inclined lines [(*6*); see also (*7*)], which he correctly attributed to stars collectively orbiting about the center of that galaxy. On the side of the nucleus where the rotation of the galaxy carries the stars toward the observer, spectral lines are shifted toward the blue region of the spectrum with respect to the central velocity. On the opposite side, where rotation carries the stars away from the observer, lines are shifted toward the red spectral region.

Several years later, Pease (*8*) convincingly illustrated that rotation *was* responsible for the inclined lines. Using the 60-inch telescope on Mount Wilson during the months of August, September, and October 1917, he patiently acquired a 79-hour exposure of the spectrum of the nuclear region of M31 with the slit aligned along the apparent major axis of the galaxy. Inclined lines appeared. A second exposure made during these 3 months in 1918, with the slit placed perpendicular to the major axis along the apparent minor axis, showed no inclination of the lines. Along the minor axis, stars move at right angles to the line-of-sight of the observer, so no Doppler shift appears. The rotation of galaxies was established even before astronomers understood what a galaxy was.

By the mid-1920's, astronomers knew that we lived in a Galaxy composed of billions of stars distributed principally in a disk, all rotating about a distant center. Each spiral viewed in a telescope, located at an enormous distance beyond our Galaxy, is itself a gigantic gravitationally bound system of billions of stars all orbiting in concert about a common center. This knowledge came from a variety of observational discoveries: Henrietta Leavitt's (*9*) discovery of Cepheid variables, stars whose periods of light fluctuation reveal their true brightnesses and hence distances; Shapley's (*10*) demonstration that the globular clusters surround our own Galaxy like a halo, with a geometric center located not at the sun but at a great distance in the dense star clouds of Sagittarius; and Hubble's (*11*) discovery of Cepheids in M31 and M33.

In a brilliant study, Öpik (*12*) used the rotational velocities in M31 to estimate its distance. His result, 450,000 parsecs (1 pc = 3×10^{13} km), comes closer to the distance in use today, 750,000 pc, than the distance of 230,000 pc which Hubble derived from the Cepheids and which he was still using in 1950 (*13*). Öpik's distance was based on the procedure used currently to determine the masses of galaxies, and I will discuss it in some detail. For a particle of mass m moving in a circular orbit with velocity V about a spherical distribution of mass M at distance r from the center, the equality of gravitational and centrifugal forces gives

$$\frac{GM(r)m}{r^2} = \frac{mV^2(r)}{r} \quad (1)$$

where G is the constant of gravitation and $M(r)$ is the total mass contained out to a distance r from the center. It follows that the mass interior to r is given by

$$M(r) = rV^2(r)/G \quad (2)$$

or

$$V(r) = [GM(r)/r]^{1/2} \quad (3)$$

Today we adopt a distance to a galaxy (generally from its Hubble velocity), determine the rotational velocity V at each r, and then calculate the variation of M with r. Öpik assumed that the ratio of mass to luminosity was the same for M31 as for our Galaxy, and used M and V to determine r. Both the dynamics and the astrophysics were sound, and the result convinced many astronomers that spiral galaxies were external to our own Milky Way system.

Modern Optical Observations of Spiral Galaxy Rotation

Several observational procedures are available today to study the rotation (that is, orbiting) of stars and gas in a spiral galaxy. A spectrograph with a long slit will record the light arising from all of the stars in the galaxy along each line-of-sight. The resulting spectrum will be composite, the sum of individual stellar spectra. A stellar motion toward the observer will displace each spectral line toward the blue region of the spectrum; a motion away from the observer will displace each line toward the red. Measurement of the successive displacements along a spectral line will give the mean velocity of the stars corresponding to that location in the galaxy. The shape and width of the line contain information concerning the random motions along the line-of-sight.

Starlight from external galaxies is generally too faint to permit a dense spectral exposure on which accurate positional measurements may be made from stellar absorption lines. One way around this difficulty is to measure velocities from the emission lines arising in the ionized gas clouds surrounding the hot young blue stars which delineate the spiral structure. Because the light from these clouds is emitted principally in a few lines of abundant elements (hydrogen, ionized oxygen, ionized nitrogen, and ionized sulfur), a measurable exposure is obtained in a fraction of the time required for the stellar exposure. This is the method we employ in our optical observations. A third procedure is to observe in the radio spectral region, at the wavelength of 21 cm emitted by the hydrogen atom. Hydrogen gas, a major constituent of the prominent dust lanes seen in spiral galaxies and hence opaque for an optical astronomer, becomes the object of observation for the radio astronomer.

Following the pioneering work discussed above, observations of the rotation of galaxies proceeded very slowly. A major addition to our knowledge came from the extensive work of Margaret and Geoffrey Burbidge (*14*). Due to the long observing times, velocities were obtained only for the brightest inner regions. Only for a few of the nearest galaxies was it possible to observe individual regions and map the rotation to large nuclear distance (*15*). Until recently, radio observations had limited spatial resolution and hence poor velocity accuracy. Astronomers attempting to examine the dynamical properties of galaxies (*16*) had few measured velocities well beyond the bright nuclear regions.

For the past several years, W. K. Ford, Jr., N. Thonnard, D. Burstein, B. Whitmore, and I (*17–19*) have been using modern detectors to study the dynamical properties of isolated spiral galaxies. We attempt to measure the rotational velocities across the entire optical galaxy. We have observed galaxies of Hubble types Sa (disk galaxies with large central bulges, and tightly wound weak arms with few emission regions), Sb, and Sc (disk galaxies with small central bulges, and open arms with prominent emission regions), emphasizing galaxies of high and low luminosity within each class. Our aim is to learn how rotational properties vary along the Hubble sequence of galaxies, and within a Hubble class, and to relate the dynamical properties to other galaxy parameters.

For 60 Sa, Sb, and Sc program galaxies, we have obtained spectra with the 4-m telescopes at Kitt Peak National Observatory (near Tucson, Arizona) and Cerro Tololo Inter-American Observatory (near La Serena, Chile). A few spectra come from the 2.5-m du Pont Telescope of the Carnegie Institution of Washington at Las Campanas (also near La Serena). Optical observations are made with spectrographs which incorporate an RCA C33063 image tube (*20*). The image is photographed from the final phosphor of this tube. Use of this electronic enhancement device makes it possible to obtain spectra at a high spatial scale and of high velocity accuracy with exposure times of about 3 hours, on Kodak IIIa-J plates which have been baked in forming gas and preflashed to enhance their speed.

We choose galaxies that are relatively isolated, that are not strongly barred, that subtend an angular size at the telescope which approximately matches the length of the spectrograph slit, and that are viewed at relatively high inclination to minimize uncertainties in transforming line-of-sight velocities to orbital velocities in the galaxy. Distances are calculated from velocities arising from the cosmological expansion, adopting a Hubble constant of 50 km sec^{-1} Mpc^{-1}.

A sample of Sc program galaxies and spectra is shown in Fig. 1; the spectra were taken with the spectrograph slit aligned along the galaxy major axis. The galaxies are arranged by increasing luminosity. Velocities are determined by measuring the displacement of the emission lines with respect to the night sky lines to an accuracy of 1 μm. We measure with a microscope which moves in two dimensions, a rather old-fashioned but still very accurate technique. Many astronomers now trace plates with a microdensitometer or obtain digital data directly at the telescope. From the measured velocities on each side of the nucleus a mean curve is formed. It describes the circular velocity in the plane of the galaxy as a function of distance from the nucleus, as shown on the right in Fig. 1.

Within a Hubble type, rotation curves vary systematically with luminosity. For a low-luminosity galaxy (NGC 2742), velocities rise gradually from the nucleus and reach a low maximum velocity only in the outer regions. For a high-luminosity galaxy (UGC 2885), rotational velocities rise steeply from the nucleus and reach a "nearly flat" portion in a small fraction of the galaxy radius. An Sc of low luminosity (6×10^9 solar luminosities) has a maximum rotational velocity V_{max} near 100 km sec^{-1}, compared with V_{max} near 225 km sec^{-1} for an Sc of high luminosity (2×10^{11} solar luminosities). Dynamical variations from one Hubble class to another are equally systematic. What differentiates Sa and Sc galaxies of equal luminosity are the amplitudes of their rotational velocities. At equal luminosity, an Sa has a higher rotational velocity than an Sb, which in turn has a higher velocity than that of an Sc. A low luminosity Sa has V_{max} = 175 km sec^{-1}, a high-luminosity Sa has V_{max} = 375 km sec^{-1}. Each value is higher than that of the corresponding Sc galaxy.

Mass Distribution Within Spiral Galaxies

What do these flat rotation curves tell us about the mass distribution within a disk galaxy? Unfortunately, we cannot determine a unique mass distribution from an observed rotation curve; only the mass distribution for an assumed model can be deduced. In practice, a spheroidal or disk model is adopted, and an integral equation is solved for the density distribution as a function of r (*21*).

In several cases of special interest, the velocity or mass distribution follows directly from Eq. 2 or Eq. 3, for a simple spheroidal model. These cases are now briefly discussed.

1) *Central mass*. In the solar system, where essentially all of the mass is in the sun, $M(r)$ is constant for all distances beyond the sun. Velocities of all planets decrease as $r^{-1/2}$, that is, a Keplerian decrease, with increasing distance from the sun (Eq. 3). Mercury (solar distance = 0.39 of the earth's distance) orbits with a velocity of 47.9 km sec^{-1}. Pluto, 100 times farther (39 earth distances), moves with a velocity one-tenth that of Mercury, V = 4.7 km sec^{-1}. For galaxies, the lack of a Keplerian falloff in velocities indicates that most of the mass is not located in the nuclear regions.

2) *Solid body*. For a body of uniform density, $M(r) \propto r^3$. Thus $V(r)$ increases linearly with r; the object will exhibit a constant angular velocity which produces solid body rotation. Some galaxies show linear velocity curves near the nucleus, although the spatial resolution is generally poor. Solid body rotation is rare in the outer regions of disk galaxies.

3) *Constant velocity*. For $V(r)$ constant, $M(r)$ increases with r. This variation describes the velocities we observe in galaxies. Density, $M(r)/r^3$, decreases with increasing r as $1/r^2$ for spherical models. Although the density is decreasing outward, the mass in every concentric shell r is the same for all r. Hence a rotation curve which is flat (or slightly rising) out to the final measurable point indicates that the mass is not converging to a limiting mass at the present observational limit of the optical galaxy for a galaxy modeled as a single spheroid.

The mass interior to r is plotted against r in Fig. 2 for four spirals; the mass is calculated from the observed velocity and Eq. 2. Masses range up to several times 10^{12} solar masses for the most luminous spirals. Because rotational velocity (at a fixed luminosity) is higher in an Sa than in an Sb, and higher in an Sb than in an Sc, big-bulge spirals (Sa's) have a higher mass density (V^2/r^2) at every r than Sb's and Sc's of the same luminosity. Moreover, the distribution of mass with radius is similar, within a single scale factor, for all of the Sc's, most of the Sb's, and a few of the Sa's.

In the disk of a normal spiral galaxy, surface brightness is observed to decrease exponentially with increasing radial distance, while the flat rotation curves imply that mass density falls slower, as $1/r^2$. Hence locally the ratio of mass to luminosity M/L increases with increasing distance from the nucleus. If we average the mass and the luminosity across the entire visible galaxy, then the ratio of mass to luminosity is a function of Hubble type, but is independent of luminosity within a type. This statement holds for observed ranges of (blue) luminosity L_B up to a factor of 100. Consequently, the average mass per unit (blue) luminosity in a galaxy, a measure of the stellar population in the galaxy, is a good indicator of Hubble type.

These observational results are summarized in Fig. 3 and as follows. Within a Hubble type, high-luminosity galaxies are larger, have higher mass and density, have stellar orbital velocities which are larger, but have the same value of M/L_B within the isophotal radius as do low-luminosity spirals. At equal luminosity, Sa, Sb, and Sc galaxies are the same size, but the Sa has a higher density, larger mass, higher rotational velocity, and larger value of M/L_B. This interplay

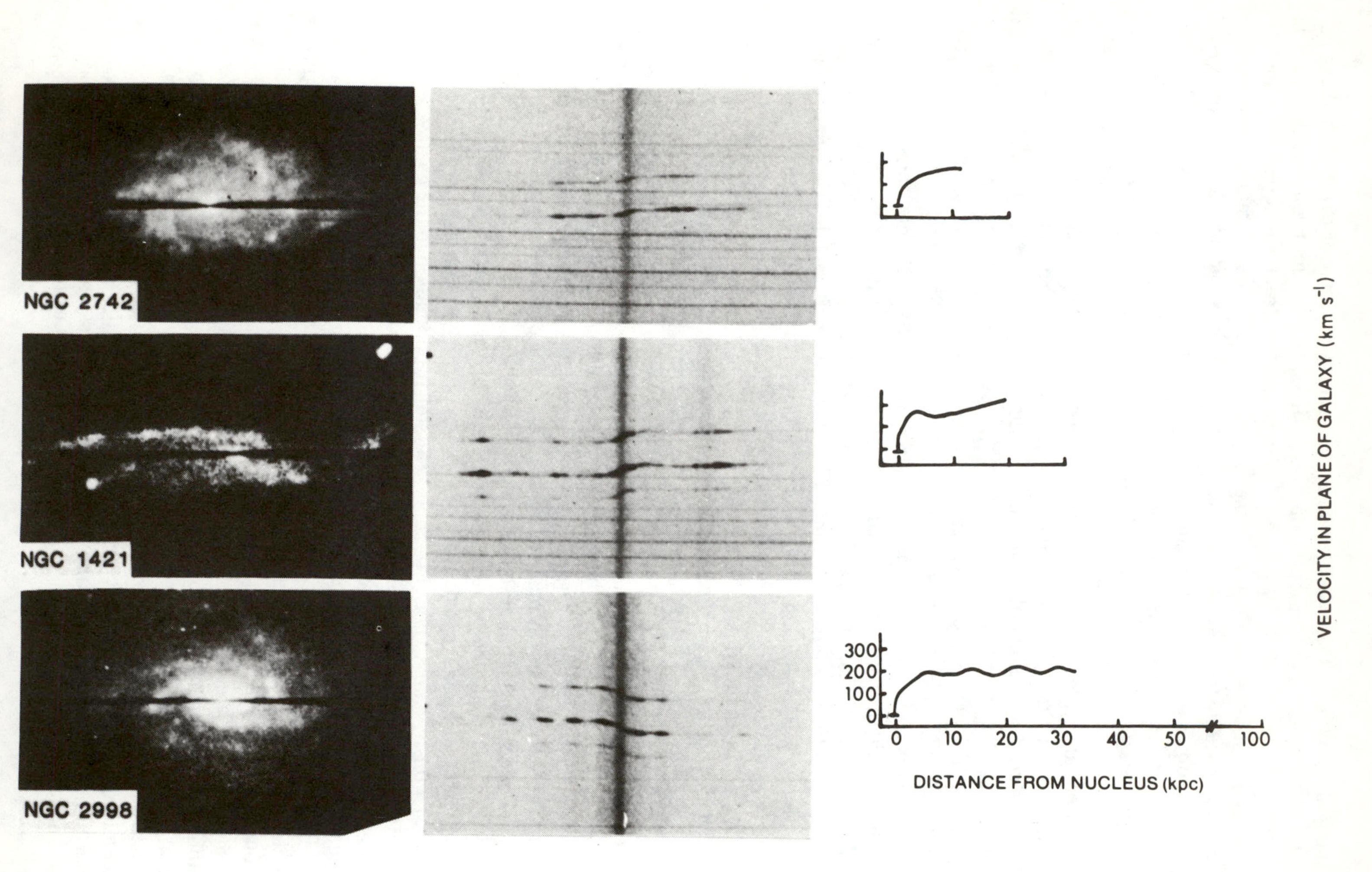

NGC 2742
NGC 1421
NGC 2998
VELOCITY IN PLANE OF GALAXY (km s-1)
300
200
100
0
0
10
20
30
40
50
100
DISTANCE FROM NUCLEUS (kpc)

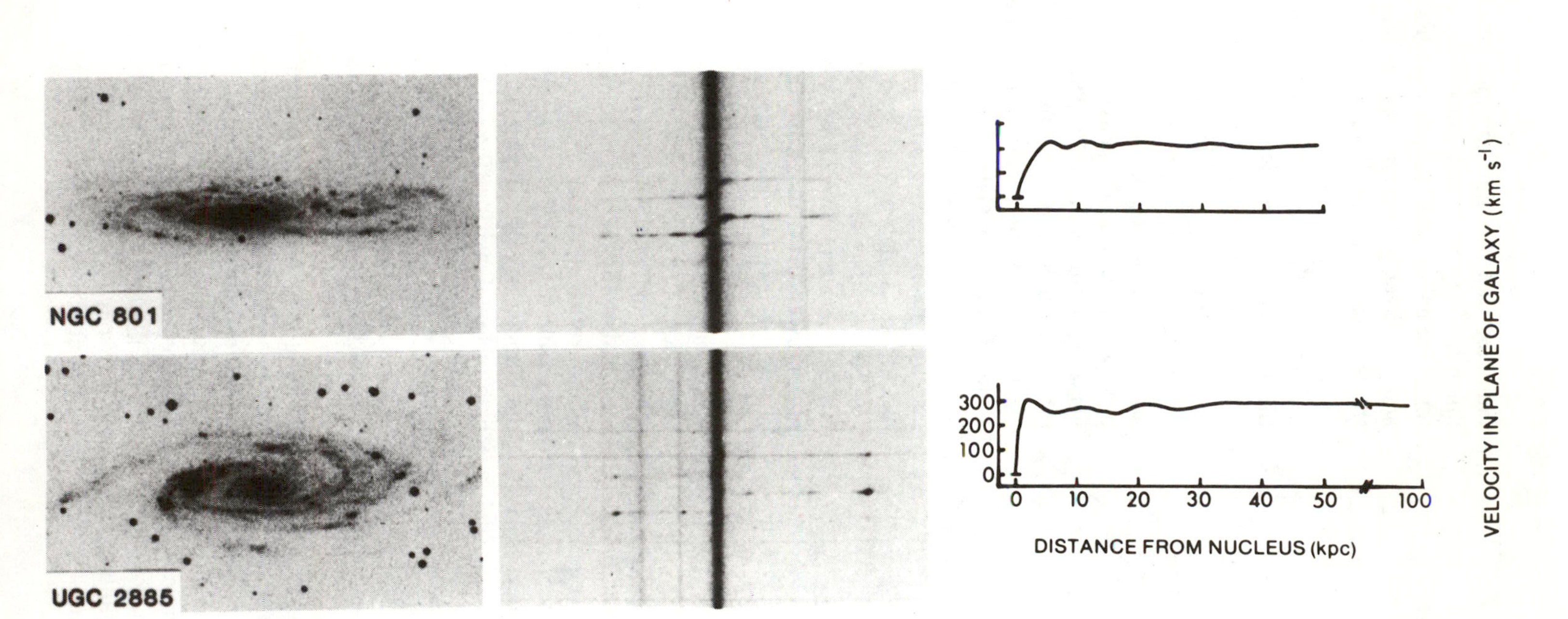

Fig. 1. Spectra and rotation curves for five Sc galaxies, arranged according to increasing luminosity. Photographs for NGC 2742, 1421, and 2998 are copies of the television screen which displays the image reflected off the spectrograph slit jaws. The dark line crossing the galaxy is the spectrograph slit. NGC 801 and UGC 2885 are reproduced from plates taken at the prime focus of the 4-m telescope at Kitt Peak National Observatory by B. Carney. The corresponding spectra are arranged with wavelength increasing from the bottom to the top. The strongest step-shaped line in each spectrum is from hydrogen in the galaxy, and is flanked by weaker lines of forbidden ionized nitrogen. The strong vertical line in each spectrum is the continuum emission from stars in the nucleus. The undistorted horizontal lines are emission from the earth's atmosphere, principally OH. The curves at the right show the rotational velocities as a function of nuclear distance, measured from the emission lines in the spectra.

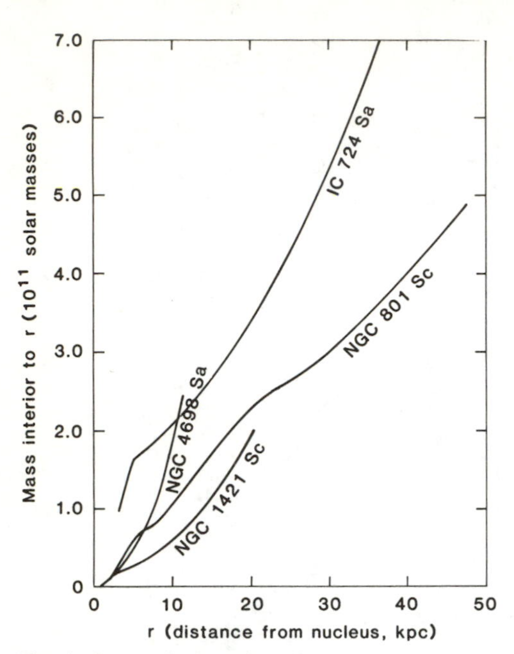

Fig. 2. Integral mass interior to r, as a function of r, for Sc and Sa galaxies of high (IC 724, NGC 801) and intermediate (NGC 4698, 1421) luminosity. Mass is increasing approximately linearly with r and is not converging to a limiting mass at the edge of the optical galaxy.

of luminosity and Hubble type has convinced us that galaxies cannot be described by a single parameter sequence, be it Hubble type or luminosity or mass. At least two parameters are necessary; Hubble type and luminosity are one such pair.

Nonluminous Mass in Galaxies

Observations of rotation curves which do not fall support the inference that massive nonluminous halos surround spiral galaxies. The gravitational attraction of this unseen mass, much of it located beyond the optical image, keeps the rotational velocities from falling. The increase in M/L_B with nuclear distance indicates that this nonluminous mass is much less concentrated toward the center of the galaxy than are the visible stars and gas. Even at large nuclear distances, the average density of the nonluminous matter is several orders of magnitude greater than the mean density of mass in the universe. Hence it is clumped around galaxies and is not just an extension of the overall background density.

During the past decade, there has been growing acceptance of the idea that perhaps 90 percent of the mass in the universe is nonluminous (*22*). The requirements for such mass come from a variety of observations on a variety of distance scales. In our own Galaxy, we can calculate the disk mass at the position of the sun by observing the attraction of the disk on stars high out of the plane, and comparing this mass with that which we can enumerate in stars and gas. Oort (*23*) first showed that the counted density at the sun is less by about one-third than that implied by the dynamics, 0.15 solar mass per parsec cubed. But self-gravitating disks of stars are unstable against formation of barlike distortions. Hence it is appealing to place the unseen matter in a halo, rather than a disk, and solve both the problem of the Oort limit and that of the stability of spiral disks. Moreover, a halo of low density at the solar position is consistent both with the Oort limit and with a halo which becomes dominant beyond the optical galaxy.

Additional evidence for a massive halo surrounding our Galaxy comes from individual stars and gas at greater nuclear distance than the sun, whose rotational velocities continue to rise (*24*). The galaxy rotational velocity is high, befitting the enormous distances. The sun, carrying the planets with it, orbits the galaxy with a speed of 220 km sec^{-1} (500,000 miles per hour); even so, it takes us 225

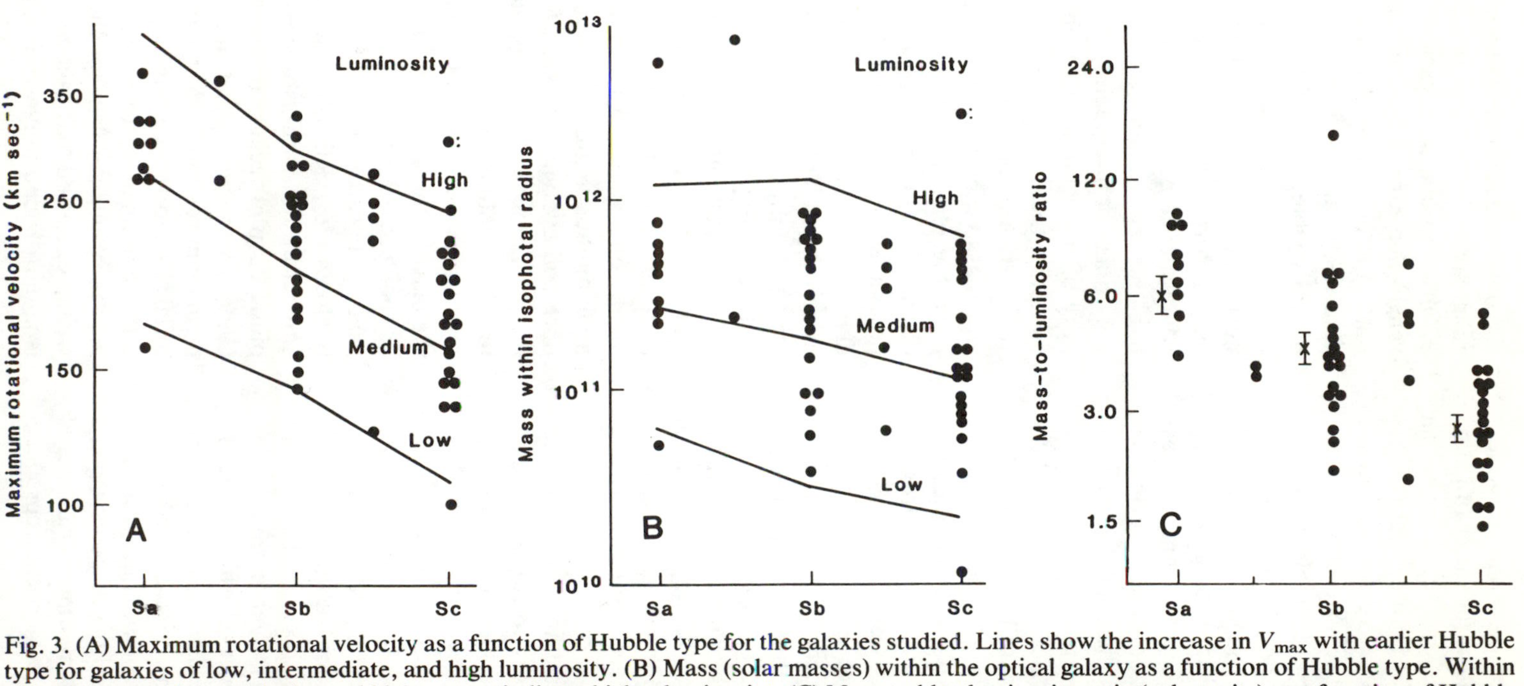

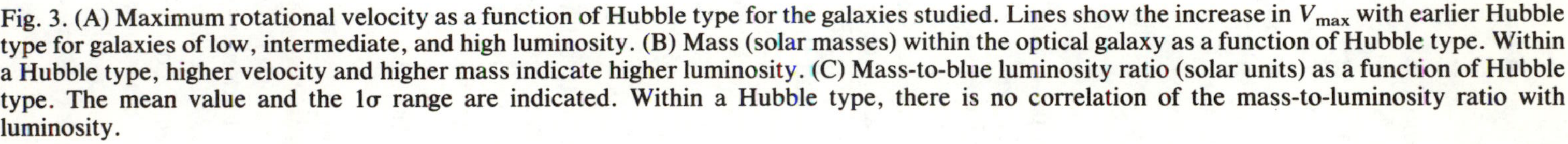

Fig. 3. (A) Maximum rotational velocity as a function of Hubble type for the galaxies studied. Lines show the increase in V_{max} with earlier Hubble type for galaxies of low, intermediate, and high luminosity. (B) Mass (solar masses) within the optical galaxy as a function of Hubble type. Within a Hubble type, higher velocity and higher mass indicate higher luminosity. (C) Mass-to-blue luminosity ratio (solar units) as a function of Hubble type. The mean value and the 1σ range are indicated. Within a Hubble type, there is no correlation of the mass-to-luminosity ratio with luminosity.

million years to complete one revolution.

On scale lengths of the order of galaxies, $20 < R < 100$ kpc, the flat rotation curves observed at both optical (*17, 19, 25*) and radio (*26*) wavelengths and the consequent increase with radius of the mass-to-luminosity ratio offer supporting evidence for heavy halos. Moreover, the dynamics of the globular clusters (*27*) and of the dwarf satellite galaxies (*28*) which orbit our Galaxy imply a mass which increases approximately linearly with increasing radius, to distances as great as 75 or 100 kpc.

On distance scales as great as hundreds to thousands of kiloparsecs, there is equally impressive evidence that the gravitational mass far exceeds the luminous mass. The evidence comes from the orbits of binary galaxies (*29*) (our Galaxy and the Andromeda galaxy form one such pair), random motions of galaxies in clusters (*30*), and the distribution of hot gas in clusters of galaxies (*31*) as observed by the x-ray emission. Such observations compose a body of evidence which lends support to the concept of heavy halos clumped around individual galaxies.

Increased interest in rotation curves has led astronomers to related studies. Radio astronomers have determined rotational velocities beyond the optical image for a few galaxies in which the neutral hydrogen distribution extends farther than the optical galaxy. Sancisi (*32*) suggests that rotation curves do fall beyond the optical galaxy, perhaps by 10 percent in velocity over a small radial distance, but then level off once again. Casertano (*33*), Bahcall and collaborators (*34*), and Caldwell and Ostriker (*35*) have constructed multicomponent mass distributions, consisting of central point masses, disks (sometimes truncated), bulges, and halos, which can reproduce the observed rotational properties of spiral galaxies.

Observations in a variety of spectral ranges have been unable to detect the dark matter. Massive halos do not appear to radiate significantly in the ultraviolet, visible, infrared, or x-ray regions; they are not composed of gas or of normal low-luminosity stars. Other possible forms include fragments of matter which never became luminous (Jupiter-like planets, black holes, neutrinos, gravitinos, or monopoles) (*36*). The enormity of our ignorance can be measured by noting that there is a range in mass of 10^{70} between non-zero-mass neutrinos and massive black holes. Not until we learn the characteristics and the spatial distribution of the dark matter can we predict whether the universe is of high density, so that the expansion will ultimately be halted and the universe will start to contract, or of low density, and so that the expansion will go on forever.

Theoretical models of the formation and evolution of galaxies currently attempt to take into account the environmental effects of enormous quantities of nonluminous matter surrounding newborn and evolving galaxies. Most likely, the formation processes for galaxies and for the nonluminous matter were distinct; one attractive idea is that much of the mass of the universe was arranged in its nongaseous dark guise before the galaxies began to form. Recent reviews by Rees (*36*), Silk (*37*), and White (*38*) enumerate the various interesting possibilities. An especially pleasing model by Gunn (*39*) suggests that disks of spirals form from relatively slowly infalling material. Regardless of which evolutionary scheme ultimately explains the current phases of galaxies and of the universe, we have learned that galaxies are not the isolated ''island universes'' imagined by

Hubble (*40*), but that environmental effects play a prominent role in directing their evolution.

For nonbelievers in heavy halos and invisible mass, an alternative explanation (*41*) modifies the $1/r^2$ dependence (Eq. 1) in Newton's law of gravitation for large r. Under this circumstance, the distribution of mass follows the distribution of light, but the velocities resulting from the mass distribution remain high due to a modified law of gravitation. For the present, this possibility must remain as a last resort.

Why Was It Thought That Rotational Velocities Decreased?

Most present-day astronomers grew up believing that disk galaxies had Keplerian velocities at moderate distances. The reasons for this belief are easy to identify. In the early 1900's, astronomers were more at home with the planets than with the galaxies. Slipher's early observations concerned the study of the planets, Percival Lowell's major interest. Galaxies were studied only to learn whether these nebulous disks in the sky were the stuff from which planets were formed. Slipher used Saturn as a radial velocity standard for his M31 spectra; he characterized the spectrum of the Sombrero galaxy as "planetary." Astronomers were predisposed to draw an analogy between the distribution of luminosity in a galaxy and the distribution of mass in the solar system.

Early spectral observations of galaxies generally did not extend beyond the bright nuclear bulges. Measured velocities usually increased with increasing nuclear distances and mass distributions were in the solid body domain. By 1950, only a handful of rotation curves had been determined, and astronomers were acting on their expectations when they "saw" a region of Keplerian falling velocities. Mayall (*42*) described his newly determined velocities in M33 as follows: "As in the Andromeda nebula, the results suggested more or less constant angular velocity . . . for the main body of the spiral. Beyond the main body these outer parts rotate more slowly . . . as in a planetary system where most of the mass lies inside the orbit concerned." De Vaucouleurs (*43*) reviewed the eight available rotation curves in 1959 and concluded, "In all cases the rotation curve consists of a straight inner part in the region of constant angular velocity up to a maximum at $R = R_m$ beyond which the rotational velocity decreases with increasing distance to the center and tends asymptotically toward Kepler's third law (Figure 27)." But with the 20/20 vision of hindsight, plots of the data reveal only a scatter of points, from which no certain conclusion can be drawn.

Yet not all astronomers were blind to the contradication posed by rapidly falling light distributions and nonfalling rotational velocities. In a classic paper discussing the structure and dynamics of NGC 3115, an almost featureless disk S0 galaxy, the ever-wise Oort (*44*) wrote, "It may be concluded that the distribution of mass in the system must be considerably different from the distribution of light." And he concluded, "The strongly condensed luminous system appears imbedded in a large and more or less homogeneous mass of great density." Schwarzschild (*45*), too, looked at the rotation curve of M31 and noted that "Contrary to earlier indication, the five normal points in Figure 1 do not suggest solid body rotation. . . . Rather, they suggest fairly constant circular velocity over the whole interval from 25′ to 115′ (5–25 kpc with the current distance)."

Indeed, the literature is replete with isolated comments stressing the same point. When in 1962 Rubin *et al.* (*46*) examined the kinematics of 888 early type stars in our Galaxy, they concluded that beyond the sun "the rotation curve is approximately flat. The decrease in rotational velocity expected for Keplerian orbits is not found." Shostak (*47*) noted that "the overwhelming characteristic of the velocity field of NGC 2403 is the practically constant circular rotation seen over much of the object."

Now that systematic studies indicate that virtually all spiral galaxies have rotational velocities which remain high to the limits of the optical galaxy, we can recognize the circuitous route which brought us to this knowledge. Astronomers can approach their tasks with some amusement, recognizing that they study only the 5 or 10 percent of the universe which is luminous. Future astronomers will have to be clever in devising detectors which can map and study this ubiquitous matter which does not reveal itself to us by its light.

References and Notes

1. Third Earl of Rosse, *Br. Assoc. Adv. Sci. Rep.* (1884), p. 79; P. Moore, *The Astronomy of Birr Castle* (Mitchell Beazley, London, 1971).
2. Third Earl of Rosse, *Philos. Trans. R. Soc. London* (1850), p. 110.
3. J. Scheiner, *Astrophys. J.* **9**, 149 (1899).
4. Sir William Huggins and Lady Huggins, *An Atlas of Representative Spectra* (Wm. Wesley and Son, London, 1899), plate II.
5. V. M. Slipher, *Lowell Obs. Bull. No. 58* (1914).
6. ——, *ibid. No. 62* (1914).
7. M. Wolf, *Vierteljahresschr. Astron. Ges.* **49**, 162 (1914).
8. F. G. Pease, *Proc. Natl. Acad. Sci. U.S.A.* **4**, 21 (1918).
9. H. Leavitt, *Harvard Coll. Obs. Ann.* **60**, 87 (1908); *Harvard Coll. Obs. Circ. No. 179* (1912).
10. H. Shapley, *Proc. Astron. Soc. Pacific* **30**, 42 (1918); *Astrophys. J.* **48**, 154 (1918).
11. E. Hubble, *Observatory* **43**, 139 (1925).
12. E. Öpik, *Astrophys. J.* **55**, 406 (1922).
13. E. Hubble, *The Realm of the Nebulae* (Yale Univ. Press, New Haven, Conn., 1936), p. 134; E. Holmberg, *Medd. Lunds Astron. Obs. Ser. 2* (No. 128) (1950).
14. E. M. Burbidge and G. R. Burbidge, in *Galaxies and the Universe*, A. Sandage, M. Sandage, J. Kristian, Eds. (Univ. of Chicago Press, Chicago, 1975), p. 81.
15. V. C. Rubin and W. K. Ford, Jr., *Astrophys. J.* **159**, 379 (1970).
16. P. Brosche. *Astron. Astrophys.* **23**, 259 (1973).
17. V. C. Rubin, W. K. Ford, Jr., N. Thonnard, *Astrophys. J. Lett.* **225**, L107 (1978); V. C. Rubin and W. K. Ford, Jr., *Astrophys. J.* **238**, 471 (1980).
18. D. Burstein, V. C. Rubin, N. Thonnard, W. K. Ford, Jr., *Astrophys. J.* **253**, 70 (1982).
19. V. C. Rubin, W. K. Ford, Jr., N. Thonnard, *ibid.* **261**, 439 (1982).
20. This tube is known throughout the astronomical community as the "Carnegie" image tube, for its development was the product of a cooperative effort by RCA and a National Image Tube Committee, under the leadership of M. Tuve and W. K. Ford, Jr., of the Carnegie Institution of Washington, funded by the National Science Foundation.
21. L. Perek, *Adv. Astron. Astrophys.* **1**, 165 (1962); A. Toomre, *Astrophys. J.* **138**, 385 (1963). See also Burbidge and Burbidge (*14*).
22. J. P. Ostriker, P. J. E. Peebles, A. Yahil, *Astrophys. J. Lett.* **193**, L1 (1974); J. Einasto, A. Kaasik, E. Saar, *Nature (London)* **250**, 309 (1974). For a dissenting view, see G. R. Burbidge, *Astrophys. J. Lett.* **196**, L7 (1975).
23. J. Oort, *Bull. Astron. Inst. Neth.* **15**, 45 (1960); in *Galaxies and the Universe*, A. Sandage, M. Sandage, J. Kristian, Eds. (Univ. of Chicago Press, Chicago, 1975), p. 455.
24. S. Kulkarni, L. Blitz, C. Heiles, *Astrophys. J. Lett.* **259**, L63 (1982).
25. S. M. Faber and J. S. Gallagher, *Annu. Rev. Astron. Astrophys.* **17**, 135 (1979).
26. M. S. Roberts and A. H. Rots, *Astron. Astrophys.* **26**, 483 (1973); A. Bosma, thesis, Rijksuniversiteit te Groningen (1978); *Astron. J.* **86**, 1791 (1981); *ibid.*, p. 1825.
27. F. D. A. Hartwick and W. L. W. Sargent, *Astrophys. J.* **221**, 512 (1978).
28. J. Einasto, in (*22*); D. Lynden-Bell, in *Astrophysical Cosmology*, H. A. Bruck, G. V. Coyne, M. S. Longair, Eds. (Pontifical Academy, Vatican City State, 1982), p. 85.
29. S. D. Peterson, *Astrophys. J.* **232**, 20 (1979).
30. F. Zwicky, *Helv. Phys. Acta* **6**, 110 (1933); H. J. Rood, *Astrophys. J. Suppl.* **49**, 111 (1982); W. Press and M. Davis, *Astrophys. J.*, in press.
31. D. Fabricant, M. Lecar, P. Gorenstein, *Astrophys. J.* **241**, 552 (1980); W. Forman and C. Jones, *Annu. Rev. Astron. Astrophys.* **20**, 547 (1982).
32. R. Sancisi, private communication; —— and R. J. Allen, *Astron. Astrophys.* **74**, 73 (1979).
33. S. Casertano, *Mon. Not. R. Astron. Soc.*, in press.
34. J. N. Bahcall, M. Schmidt, R. M. Soneira, *Astrophys. J. Lett.* **258**, L23 (1982).
35. J. A. R. Caldwell and J. P. Ostriker, *Astrophys. J.* **251**, 61 (1982).

36. M. J. Rees, in *Observational Cosmology* (Geneva Observatory, Geneva, Switzerland, 1978), p. 259; in *Astrophysical Cosmology*, H. A. Bruck, G. V. Coyne, M. S. Longair, Eds. (Pontifical Academy, Vatican City State, 1982), p. 3.
37. J. Silk, in *Astrophysical Cosmology*, H. A. Bruck, G. V. Coyne, M. S. Longair, Eds. (Pontifical Academy, Vatican City State, 1982) p. 427; see also S. M. Faher, in *ibid.*, p. 191; J. P. Ostriker, in *ibid.*, p. 473.
38. S. White, in *Morphology and Dynamics of Galaxies* (Geneva Observatory, Geneva, Switzerland, 1982), p. 291.
39. J. E. Gunn, in *Astrophysical Cosmology*, H. A. Bruck, G. V. Coyne, M. S. Longair, Eds. (Pontifical Academy, Vatican City State, 1982), p. 233.
40. E. Hubble, in *The Realm of the Nebulae* (Yale Univ. Press, New Haven, Conn., 1936), p. 98.
41. J. Bekenstein, *Int. Astron. Union Symp.*, in press; J. Bekenstein and M. Milgrom, preprint.
42. N. U. Mayall [*Publ. Obs. Univ. Mich. No. 10* (1950), p. 9] notes that the form of the rotation curve of M31, compared with that of our Galaxy, suggests that the distance of M31 (then adopted as 230,000 pc) is too small. The use of rotational properties to establish distances is only now coming into wide use.
43. G. de Vaucouleurs, in *Handbuch der Physik*, S. Flugge, Ed. (Springer, Berlin, 1959), vol. 20, p. 311.
44. J. H. Oort, *Astrophys. J.* **91**, 273 (1940).
45. M. Schwarzschild, *Astron. J.* **59**, 273 (1954).
46. V. C. Rubin, J. Burley, A. Kiasaipoor, B. Klock, G. Pease, E. Rutscheidt, C. Smith, *ibid.* **67**, 527 (1962).
47. G. S. Shostak, *Astron. Astrophys.* **24**, 411 (1973).
48. More complete references are contained in the literature cited. I thank my colleagues Kent Ford and Norbert Thonnard for their continued assistance and support in all the phases of this work and Drs. Robert J. Rubin, Robert Herman, Francois Schweizer, and W. Kent Ford, Jr., for wise comments on the manuscript.

This material originally appeared in *Science* **220**, 24 June 1983.

17. Quasars and Gravitational Lenses

Edwin L. Turner

Even though the discovery of quasars (objects with the highest red shift and probably the most distant and luminous sources in the Universe) was one of the most exciting events in the recent history of astronomy (*1*), in many ways their study has proven to be one of the least successful enterprises in astronomy during the past two decades. Despite intensive and systematic observations in every available wavelength range, few if any of even the simplest and most basic questions about quasars can be answered with certainty. For the key question of the nature and structure of the central energy source, progress during the past 15 years is barely discernible. Similarly, in spite of their large red shifts, quasars have shed little light on cosmological problems. The enormous investment of observational resources in quasar studies has led to the discovery of ~ 3000 quasars and has generated a large and complex body of empirical and phenomenological information, but even in this research effort the situation is far from satisfactory. Selection biases may still be seriously distorting our empirical assessment of quasar properties. No well-defined and generally accepted morphological classification system has appeared. Very few clear correlations or regularities in quasar properties have been observed.

Notwithstanding these difficulties and disappointments, the enthusiasm of astronomers for the study of quasars has not declined. Observational work continues at an undiminished rate, and theoretical attention is increasingly focused on a particular "best guess" model (*2*) for the nature of the central energy source. Moreover, recently there has been important progress in some areas of quasar research. On the observational side, x-ray observations of significant samples of quasars are now available (*3*), an all-sky bright quasar survey has been carried out (*4*), and new slitless spectral techniques (*5*) have greatly increased the number of known radio-quiet quasars. Our knowledge of the population statistics and statistical evolution of quasars (*4, 6*) has attained a new level of sophistication and reliability, subject to the highly likely, but not quite certain, hypothesis of cosmological red shifts. Perhaps the most dramatic results of quasars studies do not bear on quasars at all but rather on the intergalactic gas clouds and gravitational lenses through which we view some quasars. In these studies, the quasar simply acts as a convenient background light source which allows observations of absorption lines or gravitational deflections due to intervening objects. Over all, despite the slow growth in our understanding of their physical natures, quasars still provide the best available probe of the most distant observable regions of the universe.

In this chapter I present a selective review of the current state and apparent prospects of various areas of quasar research. The choice of topics and emphasis reflect my personal research interests to some extent.

Empirical Properties

Quasars have been detected in essentially every wavelength band used by astronomers from radio to γ-rays. Extensive radio, infrared, optical, and x-ray surveys have been carried out. Thus the purely empirical properties of quasars are fairly well known.

Quasars are essentially defined by their optical properties to be unresolved or nearly unresolved (that is, having nearly all of their flux in a component unresolved at the ~ 1″ limit imposed by atmospheric blurring) objects with spectra containing highly red-shifted emission lines (*7*). These red shifts ($= \Delta\lambda/\lambda$) generally lie in the range from 0.1 to nearly 3.8 with many values near 2.0. If not completely stellar (unresolved) in appearance, quasars generally show faint and frequently asymmetric halos with sizes of the order of a few arcseconds. In addition to these defining optical features, many quasars share several other optical characteristics. The observed emission lines may generally be identified with resonance lines of hydrogen or of ions of heavy elements present in high cosmic abundance. The two most common strong quasar emission lines are Lyman-α (hydrogen) and C IV (triply ionized carbon) with rest wavelengths in the vacuum ultraviolet. In addition to exhibiting large red shifts, quasar emission lines are generally very broad with widths up to several percent of their observed wavelengths. The continuum emission from quasars is usually a nonthermal spectrum with strong ultraviolet and infrared excesses relative to a thermal Planck spectrum; in many cases the flux per unit frequency may be approximated by a ν^{-1} power law, although there are wide variations about this mean slope. Almost all quasars show slow (over years) continuum flux variations of a few tens of percent, and a few vary rapidly (in as little as a few days) by factors up to 100 or more. Large and variable polarization has also been seen in some of the optically active quasars. Finally, most quasars of high red shift show a multitude of narrow and weak absorption lines when observed at high spectral resolution with good signal-to-noise ratio. It is only possible to identify these lines by associating them with a number of discrete objects each having a different red shift (in almost all cases smaller than the emission-line red shift) through which the quasar continuum source is being seen.

Only a small fraction (~ 3 percent) of quasars are radio sources; however, radio quasars are strongly overrepresented among known (catalogued) and well-studied quasars because, until fairly recently, the optical identification of radio sources was by far the easiest way to locate quasars seen in projection among the ordinary stars in our Galaxy. Those quasars that are radio sources are frequently very powerful and, in fact, number among the brightest radio objects in the sky. Radio quasars generally contain a compact central source with structure frequently extending down to scales of milliarc-seconds, emitting a flat (spectral index ~ −0.25) power-law spectrum over a wide range of frequencies. Strong variability both in flux and in milliarc-second image structure on a time scale of months to years is common. In many cases the central source will be flanked by much larger nonvariable double radio lobes with a steep spectrum, reminiscent of typical radio galaxy structure.

X-ray surveys of quasars made with the Einstein satellite (*3*) have established that many quasars are strong x-ray sources and that their combined emission must account for at least a modest fraction of the cosmic x-ray background. There are indications in the data that quasar x-ray emission may be well correlated with optical emission, although selection effects and other difficulties with the x-ray sample cast some doubt on this and other statistical inferences about x-ray emission (*8*).

Classifications

The complex phenomenology summarized above has given rise to a rather confused system of subclassifications for the sources generically known as quasi-stellar objects (quasar or QSO) and the apparently related phenomena of active galactic nuclei (AGN). Quasi-stellar radio sources (QSS) are quasars with detectable radio emission. BL Lacerta objects (BL Lacs) are quasar-like objects that show no or only extremely weak emission lines. Optically active quasars are those that show unusually rapid or large-amplitude light variations and strong optical polarization. N galaxies are essentially ordinary galaxies containing a quasar-like object, which dominates their total luminosity. Seyfert galaxies of type I are galaxies (usually spirals) containing a central object with optical properties much like those of a quasar though of sufficiently low luminosity that they do not dominate the total galaxy luminosity. Radio galaxies are galaxies (usually ellipticals) with radio emission properties like those of quasars although frequently without the central flat spectrum component. There are also various classifications based on details of the emission- or absorption-line spectra of quasars.

Unfortunately, these classifications are not mutually exclusive, distance-independent, or universally applied in any consistent way. Our understanding of the morphology of quasars and AGN has not yet reached a state analogous to the invention of spectral types for stars much less to the discovery of a "main sequence." The explanation for this unsatisfactory state of affairs is the absence of any strong or universal correlations or regularities in the empirical properties of quasars. There are a few intriguing hints such as the absence of radio lobes in BL Lacs and the correlation of variability with spectral steepness in the optical band but no sign of anything comparable to the Hubble sequence for galaxies. Of course, it is not necessary that quasars have any such regular morphology; nevertheless, its absence and the empirically rather meager definition of a quasar leave room for worry that a rather diverse set of physical phenomena might all have been lumped together (*9*).

Red-shift Distances and Population Statistics

Fortunately, it is not necessary to understand the nature of a class of astronomical objects in order to study its population statistics, that is, the mean number density, spatial distribution, and luminosity function (distribution function of luminosities in one or more bands) of the objects. What is necessary is to be able to measure the objects' distances. For quasars this is straightforward if one assumes that their red shifts are cosmological and give their distances via the Hubble law. On the other hand, if this assumption is not granted as a given, it has proven very difficult to establish by independent means. The study of quasars has been accompanied by a vigorous controversy over this point ever since it began (*10*). At present, there is no independent distance indicator for the vast majority of known quasars. Most quasars of low red shift are associated with faint objects which could be (and in some cases demonstrably are) galaxies of the same red shift (*11*), thus associating them with the well-established galaxy red shift–distance relation. A very few quasars of high red shift are constrained to be at large cosmological distances by gravitational lensing or spectroscopic line ratio arguments (*12*). On the other hand, a small but at least intuitively surprising number of quasars appear in the sky in the company of nearby galaxies of much smaller red shift (*13*). In short, although the controversy has not yet been decisively resolved, the cosmological interpretation of quasar red shifts remains observationally viable and scientifically productive. It will be assumed for the remainder of this chapter.

The most striking feature of quasar population statistics is the extremely rapid cosmic evolution apparent in the quasar luminosity function. Both the number density and typical luminosity of quasars appear to increase rapidly with increasing red shift, or, equivalent, light-travel look-back time (*4*). This evolution is most extreme for quasars of high optical luminosity; for instance, quasars with absolute blue magnitudes $M_B < -29$ (that is, at least 2000 times brighter than a typical normal galaxy) have a number density that changes by a factor of 2 in only about 3 percent of the age of the Universe. For more typical quasars with luminosities a factor of 10 to 20 times lower, this population half-life is up to over 5 percent. For AGN or quasars with luminosities comparable to that of a normal galaxy, this half-life is not reliably known but is probably at least a substantial fraction ($\gtrsim 1/4$) of the Universe's age; the data may even be compatible with no evolution for these objects.

The quasar luminosity function (the number of quasars per unit volume per unit luminosity interval) varies roughly as L^{-1} for luminosities corresponding to $M_B \gtrsim -26$ (less than roughly 100 typical galaxy luminosities) and then breaks more steeply downward for greater luminosities. At low red shifts the high-luminosity tail of the luminosity function is somewhat steeper than L^{-6}, but this slope rapidly flattens at higher red shifts as a result of the differential density evolution described above.

Another striking feature of quasar population statistics is the absence of significant numbers of quasars with red shifts above ~ 3.5 (*14*). Only a handful of objects with greater red shifts are known at all, and the upper limit on the comoving density of quasars in the red-shift range 3.7 to 4.7 is below the measured density for the range 3.0 to 3.5 by at least a factor of several. The combination of this red-shift cutoff with the steep density evolution described above implies a rather special "quasar epoch" in the evolution of the Universe. Taken literally, it implies that number density versus time for luminous quasars consists primarily of a very narrow spike with a full width at half maximum well less than 10 percent of the Universe's age. If the large red-shift cutoff cannot be explained as a result of dust obscuration (*15*) or some as yet undiscovered selection bias, the explanation of this extraordinary evolutionary behavior will stand as a major challenge to any comprehensive theory of quasars. Some would even argue that it mitigates against the cosmological interpretation of the quasar red shifts.

Rather less is known of quasar luminosity functions and evolution as measured in bands other than the optical. The radio luminosity function appears to be a featureless power law of index approximately −3 for QSS (*16*). There are also indications of strong density evolution for QSS (*17*). There are somewhat conflicting indications that x-ray luminosity is well correlated with optical luminosity and that the average x-ray–to–optical luminosity may be decreasing with increasing red shift (*3, 18*).

Nature of the Energy Source

On the basis of the preceding comments, the reader might expect that astronomers have no idea of the fundamental nature of quasars. This is not the case; in fact, there is a widely agreed upon "best guess" model. In this model a quasar is thought to be a massive ($\sim 10^7$ to 10^9 $M_\odot$, where $M_\odot$ is the mass of the sun) black hole surrounded by a disk of accreting gas and located in the nucleus of a large galaxy (*2*). The basic power source is the gravitational potential energy released by the infalling gas. The primary sources of radiation are thermal radiation from the disk, coherent high-energy phenomena very near the black hole, and perhaps magnetic phenomena (that is, flares) in the differentially rotating accretion disk. Collimated beams of magnetized relativistic plasma are thought to be produced by the confining effects of the inner edge of the accretion disk on the radiation emitted near the black hole. In this model the radio emission is identified with synchrotron radiation in this plasma; the optical and near-optical continuum is the thermal disk radiation (with its spectral shape masked by the strong radial variation of temperature) or possibly also synchrotron emission; x-rays are generated by inverse Compton scattering in the plasma; and emission lines are produced in a large surrounding cloud of photo-ionized gas. The observed narrow absorption lines are generally attributed to intervening clouds of intergalactic gas that are not directly associated with the quasar. The required accretion rate for this model is ~ 1 $M_\odot$ per year for a typical quasar and perhaps 100 times as much for the most extreme objects. The source of this gas is uncertain but might include primordial gas in the galactic nucleus, mass lost by stars in the course of their normal evolution, stars tidally disrupted by the black hole, and clouds of gas accreted by the galaxy through collisions with other galaxies or directly from the intergalactic medium. Variability is the result of changes or instabilities in the accretion flow in this picture. Evolution in the number or luminosity of quasars with cosmic epoch is explained (poorly?) by changes in the mean availability of gas (fuel) in galactic nuclei with time.

This standard model of quasars owes its popularity to several strengths. First, it explains at least qualitatively most of the general empirical properties of quasars. Second, the discovery of SS433, a stellar system (~ 1 $M_\odot$) in our Galaxy, which displays in miniature many of the properties of quasars and which is almost certainly explained by a scaled-down accretion disk model, indicates that such systems can exist and that most of the relevant physical mechanisms can operate as required by the model (*19*). Third, essentially all models for quasars are forced to concentrate enough matter into a small enough volume that the eventual formation of a large black hole seems likely. Fourth, the standard model is perhaps more popular than is strictly scientifically warranted because it provides a convenient way of focusing and organizing the theoretical discussion of quasars between various investigators.

The standard model also has weaknesses. Chief among these is that it predicts very little and is of no real assistance in organizing or classifying the observed phenomenology of quasars. Another important point is that very few of the competing models for quasars can be eliminated by the data even though most of them were proposed soon after the discovery of quasars. The competition includes models for various types of discrete supermassive stars or pulsars and models involving dense clusters of objects of low (stellar) mass (*20*). Essentially any model that can hope to produce relativistic plasma and high-energy density radiation fields with reasonable efficiency remains in the running.

Absorption Lines and the Intergalactic Medium

Perhaps the most scientifically productive result of quasar studies to date has nothing directly to do with quasars nor does it depend on a knowledge of their physical natures although it does (again) depend on the association of their red shifts with cosmological distances. This is the study of clouds in the intergalactic medium through observation of the absorption lines they produce in the spectra of quasars of high red shift. This technique effectively allowed the "discovery" of the intergalactic medium, giving rise to a large and complex field. Only a bare summary of the main results can be given here.

At least two basic types of absorbing clouds have been discovered (*21*). The two types differ in their content of heavy elements, in their total number, and in their red-shift clustering properties. The rarer of the two types contains heavy elements at abundances that may approach solar values. The absorption red shifts of these clouds often occur in clumps with spreads of a few hundred to a thousand kilometers per second. Clouds of the second type are much more numerous but reveal themselves only by hydrogen Lyman lines, primarily Lyman-α. Typically they appear as a forest of narrow absorption features blueward of a quasar's Lyman-α emission line. They apparently are low in heavy-element abundance; the upper limit is uncertain, in that it depends somewhat on the clouds' ionization state, but is probably in the vicinity of a few percent of solar abundances. The red shifts of these Lyman-α clouds are distributed randomly and independently in any particular red-shift interval. The number density of Lyman-α clouds per unit comoving volume appears to increase slowly with increasing red shift but appears to be relatively constant from one quasar to another at any given red shift.

These two cloud populations are believed to be due to the interstellar medium of normal galaxies on the one hand and to perhaps primordial, isolated intergalactic clouds on the other (*21*). The number, high metal content, and redshift clustering properties of clouds of the first type may be roughly explained if the interstellar medium in the outer regions of normal spiral galaxies is similar to that in our own Galaxy and extends to an effective radius a few times larger than that of the visible stars; there is some independent evidence that this may be the case (*22*). The absence of redshift clustering in the Lyman-α clouds makes it impossible to associate them with ordinary galaxies for most (but not all) theories of the evolution of galaxy clustering. This fact, their low heavy-element abundance, and the change in their number density with cosmic time suggest that they are primitive and perhaps primordial objects, which may therefore provide information on the course of cosmic evolution.

Gravitational Lenses

The recent, though long predicted, discovery and the ongoing detailed study of gravitational lens systems among the quasars open the door for a wide variety of qualitatively new types of observations and may provide a major new tool for extragalactic astronomy and cosmology (*23*). In such gravitational lens systems, the line of sight to a distant quasar passes very near an intervening galaxy, cluster of galaxies, or other object. This results in a strong magnification, displacement, distortion, and in some cases fragmentation of the quasar's image due to the gravitational deflection of photons from the quasar by the intervening object. As in absorption-line studies, the physical nature of quasars is unimportant for gravitational lens studies; the quasar simply provides a convenient distant but bright light source.

In addition to the uses of gravitational lenses discussed below, it is well to recall that they do represent a fundamental physical phenomenon of some importance. Their existence was predicted (*24*) and their astronomical significance discussed (*25*) during the same period when the advent of modern physics led to the widely celebrated predictions of degenerate dwarfs, neutron stars, black holes, interstellar clouds, the microwave background radiation, and so forth. It is not impossible that gravitational lensing will someday also join this list in terms of its fundamental importance for astronomy.

The first major use for gravitational lensing is as a probe of the distribution of matter, particularly dark matter, both in the lensing objects themselves and along the line of sight out to the lenses. During recent years it has become increasingly clear that most of the mass density of the Universe and even much of the matter in individual galaxies must reside in some form that does not emit strongly (at least at any accessible wavelength) (*26*). The factor by which this dark material exceeds the mass density in ordinary stars, interstellar matter, and other detectable objects is not known but could be ~ 10 to 100. Discovery of the nature and detailed properties of this component of the Universe is generally recognized to be one of the most important problems in astronomy today. Gravitational lensing offers hope of addressing this problem in several new ways because it depends entirely on the space distribution of gravitating material between the source and the observer and not at all on any of its other properties. Specifically, observations of individual lens systems and the statistical properties of samples of lenses can be used in principle to determine the total masses of lensing galaxies, to characterize the distribution of dark matter in rich clusters, to discriminate between very different models for the nature of the dark material (for example, neutrinos versus very-low-mass, nonburning stars versus massive black holes), to detect a uniformly distributed component of the Universe's mass density, and even to identify specific condensed dark objects if such things exist.

A second major application for gravitational lens observations will be the determination of cosmological parameters. This possibility arises because the background cosmology in which lensing occurs determines the geometrical optics of the situation. Measurements of differential time delays between images in lens systems and the statistics of the properties of samples of lenses can be used to determine all the classical cosmological parameters (H_0, q_0, and Λ) (*27*) in principle. Moreover, the internal consistency of such determinations can be used to check the validity of the standard cosmological models. The possibility of measuring or at least placing limits on Λ is particularly good because lensing is much more sensitive to this parameter than other available astronomical observables.

A third possible application for gravitational lenses is to use them directly as lenses in a sort of natural telescope that will allow us to look much more deeply into space (in a few small patches of the sky) than would otherwise be possible. This possibility, which was first pointed out in 1937 (*25*), may in a few special situations improve the effective performance of our telescopes by a factor of several (at no additional cost!).

The reader will have noted the phrase "in principle" or the word "possible" in most of the critical sentences above and will realize that many of these potential applications of lensing will be quite difficult to realize. Indeed, some may be completely impractical even in the long term. Gravitational lenses are rare and hard to locate; so far, only five cases have been found among the ~ 3000 known quasars (though many could have been missed). Many of the idealized lensing tests such as the determinations of cosmological parameters may be confused by other astrophysical effects in reality. Many of the required observations will be practically difficult since they will require that faint objects be monitored with high resolution and good time coverage. Nevertheless, the questions that lenses might enable us to answer are so important and so difficult to attack by other means that the incentive for at least trying to overcome the problems will be great. One probably should not be too discouraged yet that work on the five known examples of lensing have not produced any breakthroughs [in fact, it can be demonstrated from the properties of 0957 + 561 that the matter in the rich cluster that forms part of that lens cannot be distributed like the galaxies (*28*)] for two reasons. First, ground-based observations do not have the resolution or sensitivity to objects of low surface brightness needed to clearly disentangle the quasar images from the lensing galaxy and to determine the positional and structural parameters of the latter accurately; Space Telescope will decisively relieve this problem in a few years. Second, only five cases have been studied so far (only two extensively); the first five pulsars discovered did not teach us much about neutron stars or general relativity, although larger samples, the Crab pulsar, and the binary pulsar eventually did. It may well be the same for gravitational lenses; the major discoveries may have to await statistical samples and particularly simple or otherwise special objects.

References and Notes

1. T. A. Matthews and A. R. Sandage, *Astrophys. J.* **138**, 30 (1963); M. Schmidt, *Nature (London)* **197**, 1040 (1963).
2. M. J. Rees, *Ann. N.Y. Acad. Sci.* **302**, 613 (1977), and references therein.
3. J. P. Henry, in *X-Ray Astronomy with the Einstein Satellite*, R. Giacconi, Ed. (Reidel, Boston, 1981), p. 261.
4. M. Schmidt and R. F. Green, *Astrophys. J.* **269**, 352 (1983).
5. P. S. Osmer, in *Objects of High Redshift*, G. O. Abell and P. J. E. Peebles, Eds. (Reidel, Boston, 1980), p. 77.
6. E. L. Turner, *Astrophys. J.* **228**, L51 (1979).
7. M. Schmidt, in *Galaxies and the Universe*, A. Sandage, M. Sandage, J. Kristian, Eds. (University of Chicago, Chicago, 1975), p. 283.
8. B. Margon, G. A. Chanan, R. A. Downes, *Astrophys. J.* **253**, L7 (1982).
9. Before the introduction of photography and the spectrograph into astronomical research, a wide variety of objects including gaseous ejecta from stars, clouds of interstellar dust and gas, clusters of Galactic stars, and distant external galaxies were classified together as "nebulae" (clouds) because they shared a diffuse fuzzy appearance as seen through telescopes of the day. The last of these confusions continued into the 1920's, and the confusing nomenclature persists in modern astronomical usage. Is there an analogy to quasars?
10. G. Burbidge, in *Objects of High Redshift*, G. O. Abell and P. J. E. Peebles, Eds. (Reidel, Boston, 1980), p. 99; H. C. Arp, *Science* **151**, 1214 (1966).
11. A. Stockton, *Astrophys. J.* **251**, 33 (1982), and references therein.
12. P. Young, J. E. Gunn, J. Kristian, J. B. Oke, J. A. Westphal, *ibid.* **241**, 507 (1980); C. L. Sarazin and E. J. Wadliak, *Astron. Astrophys.* **123**, L1 (1983).
13. H. C. Arp, *Astrophys. J.* **271**, 479 (1983), and references therein.
14. P. S. Osmer, *ibid.* **253**, 28 (1982).
15. J. P. Ostriker and L. L. Cowie, *ibid.* **243**, L127 (1981); J. P. Ostriker and J. Heisler, *ibid.*, in press.
16. E. L. Turner, *ibid.* **231**, 645 (1979), and references therein.
17. M. Schmidt, *ibid.* **151**, 393 (1968).
18. G. Zamorani, *ibid.* **260**, L31 (1982).
19. B. Margon, S. A. Grandi, R. A. Downes, *ibid.* **241**, 306 (1980), and references therein.
20. D. Lynden-Bell, *Phys. Scr.* **17**, 185 (1978), and references therein; M. Rees, *ibid.*, p. 193, and references therein.
21. W. L. W. Sargent, *Philos. Trans. R. Soc. London Ser. A* **307**, 87 (1982), and references therein.
22. A. Boksenberg and W. L. W. Sargent, *Astrophys. J.* **220**, 42 (1978).
23. D. Walsh, R. F. Carswell, R. J. Weymann, *Nature (London)* **279**, 381 (1979); J. E. Gunn, in *Tenth Texas Symposium on Relativistic Astrophysics* (New York Academy of Science, New York, 1981), p. 287, and references therein.
24. A. Einstein, *Science* **84**, 506 (1936).
25. F. Zwicky, *Phys. Rev.* **51**, 290 (1937); H. N. Russell, *Sci. Am.* No. 2 (1937), p. 76.
26. S. M. Faber and J. S. Gallagher, *Annu. Rev. Astron. Astrophys.* **17**, 135 (1979), and references therein.
27. The value of H_0 gives the expansion rate and hence age of the Universe; q_0 measures the deceleration of the expansion and determines whether the Universe is open or closed. The cosmological constant Λ may be regarded as a measure of the energy density of the vacuum.
28. I. I. Shapiro, paper presented at the MIT Workshop on Gravitational Lensing, Cambridge, Mass., 1982.

29. I thank J. P. Ostriker, who provided several useful suggestions for improvements in this review. Supported in part by NSF grant AST82-16717 and by the Alfred P. Sloan Foundation.

This material originally appeared in *Science* **223**, 23 March 1984.

18. Windows on a New Cosmology

George Lake

At first sight, there seem to be remarkably few probes of the early Universe. The expansion of the Universe was first discovered by Edwin Hubble. Together with the 2.7-K blackbody radiation discovered by Penzias and Wilson (*1*), this expansion implies that the Universe was once very hot and dense (the Universe expands adiabatically, cooling in the process). There are other signatures of this hot, dense state of the Universe. Nucleosynthesis of the lighter elements (*2*) occurred when the Universe was at temperatures of 10^8 to 10^9 K (a thermal energy of 0.1 to 1 MeV; hereafter such energy units will often be used to characterize the temperature), producing the currently observed abundances of ^{2}H, ^{3}He, ^{4}He, and ^{7}Li.

The measurements of these elemental abundances are in such good accord with the calculations of primeval nucleosynthesis that one is encouraged to consider the state of the Universe at still higher temperatures and densities, ones that will never be available in the laboratory. Consideration of these conditions places cosmology in a new position with respect to fundamental physics. It can provide a unique laboratory for the understanding of new physical laws, perhaps in the same way that Galileo's observations of Jupiter led to the formulation of the laws of gravity or in the way that nuclear physics was stimulated by the mystery of energy generation in stars. A key to further discovery may well be the tension of an ill fit between particle physics and cosmology. At present they are sewn together in an awkward manner, similar to the patchwork-quilt nature of the fundamental forces before the electroweak unification (*3*, *4*). I will make no attempt to hide the ugly stitches of the seam, as they are likely to be the source of the most lively research in the coming years.

Old Global Symmetries and New Gauge Symmetries of Grand Unified Theories

The use of symmetries has a long history in the formulation of physical laws. In the *Phaedo*, Plato explains that objects fall toward the center of the earth as there is no other direction for them to go. Even in modern physical theory, all things are permitted unless specifically excluded by symmetry laws. [In this regard, it is interesting that, only 40 years ago, baryon number was invented as a symmetry to exclude proton decay (*5*). For the purpose of our discussion here, baryon number may be considered to be the number of protons and neutrons minus the number of their antiparticles.] Until recently, the types of symmetry considered were nearly all "global" symmetries: C or charge symmetry—the replacement of particle by antiparticle; P or parity—spatial inversion; T—time reversal, isospin, and strangeness. All these symmetries are now known to be only approximate (*3*). There are also "local" or "gauge" symmetries. In these there are continuous transformations which differ from place to place. One example is the phase changes that can be put into the electric field when accompanied by a corresponding transformation of the magnetic vector potential.

At present, all global symmetries are observed to be violated with the exception of baryon and lepton number. It is hoped that experiments on proton decay and neutrino mass will reveal that these symmetries are also violated, which would leave only gauge symmetries as the true symmetries of nature. If baryon number is not conserved, this opens the possibility of generating the matter content of the Universe, as will be discussed later.

A symmetry may be either violated (which is to say it is not really a symmetry) or "broken" (*6*). A macroscopic example of a broken symmetry is the absence of translational and rotational invariance in crystals despite the existence of these symmetries in the electromagnetic forces that determine the structure of the crystal. This occurs because the physical states of a system can have a lower energy by breaking the symmetries of the Hamiltonian. This can even occur in the vacuum state of a field theory. In this way the properties of the space between particles can become dynamically important to cosmology, as will be explained later.

Symmetry is also important in cosmology. In the next section, I use the assumptions of homogeneity and isotropy to construct a tractable description of universal expansion. In the following sections, I describe the exquisite tuning required for these symmetries and attempt to relate them to physics at very high energies, producing a model known as the inflationary Universe.

Dynamical Equations for the Expansion

To get a good handle on the conundrums of cosmology, we need to start by describing the dynamical history of the Universe. The gross features should be clear at the outset. The gravitational force of matter pulls everything in the Universe toward every other thing. It halts expansion and speeds contraction. A pressure affects the expansion in two ways: (i) its energy density has an equivalent mass density, which contributes to the gravitational forces, and (ii) it does work (PdV, where P is the pressure and V is the volume) as the Universe expands.

Gravity determines an expansion law, where the declaration or acceleration is determined by the density and pressure. The evolution of density is determined by the equation of state, which relates the density to the pressure (which is different for radiation, matter, and the vacuum) and the conservation of energy, which balances the books for the dilution of density by the expansion and the work done by the pressure. With these features in mind, we can easily construct a mathematical description of the expansion.

In order to describe the expansion of the Universe, we start with a Newtonian calculation of the expansion of a small spherical volume. The reason this suffices is that the matter outside the volume exerts no force (by spherical symmetry), and in this small volume the forces are small and the expansion velocities are nonrelativistic. As the expansion is universal, the laws derived for any small volume must apply to the Universe as a whole (*7*). There are some features that must be added to this approach. The equivalent "mass" of energy or radiation, E/c^2, is important in the early history of the Universe. Another problem is that over large enough distances, the Universal expansion approaches and even exceeds the speed of light, c. The full treatment of these effects requires general relativity; but the most important aspect is that, when the Newtonian calculation indicates that two regions are expanding faster than light, it really means that they are not in communication.

Consider a particle at a distance $L(t)$ from the center of the small spherical volume. First, we introduce a coordinate r, which is a "comoving coordinate" and does not change as the expansion occurs. A scale function $a(t)$ describes the expansion in that

$$L(t) = a(t)r \tag{1}$$

The acceleration of the particle at $L(t)$ is given by the gravitational attraction of the matter inside a shell of radius $L(t)$, that is,

$$\ddot{L}(t) = -\frac{GM(\text{int})}{L(t)^2} = -\frac{4\pi}{3}G\rho\frac{L^3(t)}{L^2(t)} = -\frac{4\pi}{3}G\rho L(t) \tag{2}$$

where G is Newton's gravitational constant. Here the mass interior to L, M(int), and the density, ρ, include the normal matter density as well as the "mass" of energy or radiation, E/c^2. If we divide by r, we find the dynamical equation for the scale factor:

$$\ddot{a}(t) = -\frac{4\pi}{3}G\rho a \tag{3}$$

Pressure has been neglected here. When it is included, the correct equation reads

$$\ddot{a}(t) = -\frac{4\pi}{3}G(\rho + 3P)a \tag{4}$$

where P is measured in units that are somewhat unfamiliar; this pressure is the familiar one divided by c^2. Normally one is used to only gradients in the pressure having significance. The pressure here is defined as $-(dE/dV)$ where E is the energy and V the volume. Its absolute value is important because of its equivalent mass density. This effect only arises in the full general relativistic treatment, which is why it has to be inserted here "by hand." We add to this the first law of thermodynamics (which is nearly the conservation of energy, except for the PdV work done in the expansion):

$$\frac{d\rho}{da} = -\frac{3(P + \rho)}{a} \tag{5}$$

To complete the description of the evolution of the density and the scale factor, an equation of state relating the density and the pressure has to be specified. There are three cases normally considered: (i) "dust": $P = 0$; (ii) radiation: $P = \rho/3$; and (iii) vacuum energy: $P = -\rho$. We can now look at the evolution of the density and scale factor in these three cases.

"Dust," $P = 0$: The solution to Eq. 5 in this case is $\rho_{\text{dust}} = \rho_0(a_0/a)^3$, where the subscript zero denotes the current epoch. Integrating Eqs. 4 and 5, we find

$$\dot{a}^2 = \frac{8\pi}{3}G\rho_{\text{dust}}\, a^2 - \epsilon A^2 \tag{6}$$

where ϵA^2 is an integration constant with $\epsilon = -1, 0, +1$ and $A > 0$. If we redefine $a \rightarrow Aa$ and $r \rightarrow A^{-1}r$, we have

$$\dot{a}^2 = \frac{8\pi}{3}G\rho_{\text{dust}}\, a^2 - \epsilon \tag{7}$$

If $\epsilon = +1$, the scale factor expands to a maximum value and then the Universe recontracts. If $\epsilon = 0$ or -1, $a(t)$ can increase monotonically from zero to infinity. The parameter ϵ is called the sign of the spatial curvature. To show the physical significances of this number, we rearrange Eq. 7 and evaluate it at our current epoch to find

$$\frac{-\epsilon}{a_0^2} = \left(\frac{\dot{a}_0}{a_0}\right)^2 - \frac{8\pi}{3}G\rho_0 \equiv H_0^2 - \frac{8\pi}{3}G\rho_0 \qquad (8)$$

Here (ϵ/a_0^2) is the present spatial curvature and H_0 is the current expansion rate (the Hubble constant), given by $\dot{a}/a$ at the current epoch. The current value of the Hubble constant is 50 km sec^{-1} Mpc^{-1} (1 megaparsec is 3.2×10^6 light-years). This number is uncertain, with some techniques indicating a value nearly twice as large. The spatial curvature is then positive, negative, or zero depending on whether the density ρ_0 is greater than, equal to, or less than

$$\rho_{crit} = \frac{3H_0^2}{8\pi G} = 4.9 \times 10^{-30}\left(\frac{H_0}{50 \text{ km sec}^{-1} \text{ Mpc}^{-1}}\right)^2 \text{ g cm}^{-3}$$

This shows that the Newtonian distinction of a system being bound, unbound, or marginally bound is equivalent to the general relativistic "sign of the spatial curvature." An unbound Universe ($\rho < \rho_{crit}$) is usually called "open" and expands forever. A bound Universe ($\rho > \rho_{crit}$) is "closed"; the force of gravity is sufficient to reverse the expansion and cause collapse. A marginally bound Universe ($\rho = \rho_{crit}$) is "flat," as the spatial curvature is zero.

The evolution of the scale factor in a matter-dominated Universe is shown in Fig. 1. This plot shows the infinite expansion of an open ($\epsilon = -1$) or flat ($\epsilon = 0$) Universe and the recollapse of a closed ($\epsilon = +1$) Universe.

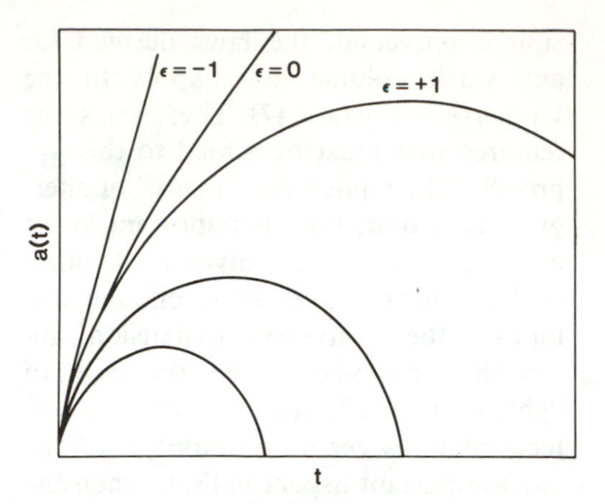

Fig. 1. The evolution of the scale factor $a(t)$ with time in a matter-dominated Universe. The $\epsilon = -1$ curves correspond to open, infinitely expanding Universes. A Universe with positive curvature ($\epsilon = +1$) is closed; the force of gravity is sufficient to halt the expansion and cause collapse. The $\epsilon = 0$ Universe is "flat"; small perturbations will cause it to behave as either the open or the closed case.

"Radiation," $P = \rho/3$: In this case, Eq. 5 becomes $\rho_{rad} = \rho_0(a_0/a)^4$. Once again, integrating Eqs. 4 and 5 and transforming as before, we find

$$\dot{a}^2 + \epsilon = \frac{4\pi}{3}G\rho_{rad}\, a^2 \qquad (9)$$

I will say more on this case later.

"Vacuum energy," $P = -\rho$: Equation 5 now yields the particularly simple form; $\rho_{vac} = \rho_0$, a constant. As the Universe expands, the energy density of the vacuum remains unchanged. For $\epsilon = 0$, this case yields a qualitatively different equation:

$$\ddot{a} = \frac{8\pi G\rho_{vac}}{3}\, a \qquad (10)$$

This is the form of Einstein's equations with a cosmological constant, where $8\pi G\rho_{vac}$ is normally referred to as Λ.

This form was first introduced to satisfy a "perfect cosmological principle": the Universe is the same at any time and place. The particular form chosen by Einstein was a combination of matter density and vacuum energy that gave $\rho = -3P$, making the scale factor constant in time. It was not treated as just another equation of state because of the unphysical feature of either negative pressure or density. After the discovery of the universal expansion by Hubble, the cosmological constant was later denounced by Einstein as the biggest blunder of his life. That Einstein would regard this as such a colossal blunder is easy to see. Since gravity is a Universal force, Newtonian physics and general relativity automatically lead to the picture of a dynamic Universe, except in the very special case of a purely homogeneous, isotropic Universe with $\rho = -3P$. Even in this idealized case, small perturbations will immediately set a Universe in motion. To come so close to a prediction of an expanding, dynamic Universe and yet abandon the idea because of preconceived belief must surely have seemed to Einstein a grave mistake. The backlash led to a veritable banishment of the cosmological constant in physical cosmology. The revival of this term is due to its modern use as the energy density of the vacuum (which can have a negative pressure) which arises in field theories.

When the vacuum energy density is large, the curvature, ϵ, is unimportant and the evolution of the scale factor is given by

$$a = a_0 e^{\left(\frac{8\pi}{3} G\rho_{\text{vac}}\right)^{1/2} t} \tag{11}$$

The scale factor evolves exponentially with time, t. This will be a critical feature later.

A Matter and Radiation Universe

The observed Universe expands at a rate of 50 km sec^{-1} Mpc^{-1}, contains a matter density equal to $\sim 10^{-30}$ g cm^{-3}, and has a radiation density equal to 10^{-33} g cm^{-3}. Matter now dominates radiation by a factor of approximately 10^3. The expansion of the Universe diminishes the matter density by $(\text{volume})^{-1}$ or $[a(t)]^{-3}$, while the radiation density evolves as $[a(t)]^{-4}$. This can be viewed in terms of the PdV work done by the radiation or, equivalently, the red-shifting of individual quanta by the expansion. As a consequence, when the scale factor was one-thousandth of its current value, the density of matter and radiation were equal. Before this time, the dynamics of the Universe was dominated by the radiation.

The Bang and Causal Structure

If we integrate the equations for our expanding Universe backward, we find a time when the scale factor is zero. As this is the natural starting point for the Universe, we call this $t = 0$. At $t = 0$, the scale factor in the radiation case is proportional to $t^{1/2}$ (at these early times, the curvature term is unimportant and the Universe is radiation-dominated). The expansion rate at $t = 0$ is singular, the Universe starts with a Bang.

In the case of a Universe comprised of radiation and matter, the singular expansion has an important consequence for the causal structure of the Universe. Although starting from an initially small volume, the region of the Universe in causal contact (defined as having the ability to have exchanged a signal at the speed of light) is initially zero and grows as the rate of expansion decreases and time elapses for signals to propagate. This leads to the concept of a particle horizon, that region of space that it is possible to have seen. Note that once seen, always seen.

The case of a "vacuum energy" is different in that the expansion rate is uniform, and, looking to the past, one sees a Universe that is forever shrinking, but the scale factor only approaches zero as time goes to $-\infty$. A forward evolution shows a fundamental difference in the causal structure. As the particles are carried apart by expansion, the recessional velocity between them increases. Eventually the inferred speed approaches and even exceeds the speed of light. At this point the particles drop out of communication. In this type of Universe, the volume of communication remains fixed and expansion carries particles out of this volume through an event horizon (just as particles falling into a black hole pass through a point of last communication). Whereas particles say hello and never farewell in a radiation-matter Universe with particle horizons, there are only farewells in a Universe with a vacuum energy density and event horizons.

Starting Up the Universe

What constitutes "initial data" in cosmology? (*8*). In the standard model of a radiation-dominated Universe, $t = 0$ is a singular point and as such it is a poor place to specify any parameters. The fundamental constants, Newton's gravitational constant, G; Planck's constant characterizing the quantum aspects of matter, $\hbar$; and the speed of light, c, can be used to define units of mass (or equivalently energy or temperature), time, and length and thus yield a starting point for a discussion of the early Universe. These are known as Planck units and are given by

$$M_{PL} = \left(\frac{\hbar c}{G}\right)^{1/2} = 2.2 \times 10^{-5}\text{g} = 1.2 \times 10^{19} \frac{\text{GeV}}{c^2}$$

$$t_{PL} = \left(\frac{\hbar G}{c^5}\right)^{1/2} = 5.3 \times 10^{-44} \text{ second}$$

$$\ell_{PL} = \left(\frac{\hbar G}{c^3}\right)^{1/2} = 1.6 \times 10^{-33} \text{ cm}$$

A particle of Planck energy is one whose Compton wavelength (the radius where its quantum behavior becomes evident, $\hbar/mc$) is the same as its Schwarzschild radius (the radius inside which it becomes a black hole, GM/c^2), a ghastly beast beyond the realm of known physical law. We will restrict our discussion to temperatures of less than 1 percent of the Planck temperature, that is, $T < 10^{17}$ GeV. This regime is interesting as the symmetry breaking which separates the strong and electroweak interactions is believed to occur at 10^{15} to 10^{16} GeV.

In the standard scenario, the Universe cools to a temperature of 10^{17} GeV just 10^{-39} second after the initial singularity. The Hubble constant has the dimensions of (velocity/distance) or (1/time) and defines the current age of the Universe of roughly 10^{18} seconds. Comparing this age to the Planck time, we find that the Universe has clearly persisted for a very large number (10^{61}) of fundamental time units ($t_{PL} = 10^{-43}$ second). To do so requires an exquisite tuning of the density and the expansion rate at early times. If we rearrange Eq. 8 into

$$[H(t)]^2 = \frac{8\pi}{3}G\rho(t) - \frac{\epsilon}{a^2(t)}$$

we can consider the relative importance

of the mass density and curvature term in determining the expansion rate. Currently ρ is poorly known, but the first term is in the range of a tenth to ten times the second. If we trace both in time $\rho \propto a^{-3}$ in a matter-dominated Universe (or a^{-4} for a radiation-dominated Universe). At a time $t \sim 1$ second, when the light elements were formed, the first term is larger than the second term by some 14 orders of magnitude. At a temperature of 10^{17} GeV ($t = 10^{-39}$ second), the ratio of the two numbers is 10^{50}. This is seen in Fig. 1 as the rapid divergence of the curves that are nearly identical close to the origin. What we find from this is that the initial data (expansion rate and density), specified at a time 10^{-39} second after the Bang, have to be tuned to an accuracy of one part in 10^{50}, even though ρ is so poorly known today. This tuning would have to be done to 1 part in 10^{14} even if we imagined starting the Universe just before the light elements were formed.

To describe this in more physical terms, we note that the basic time scale for gravitational instability is the free-fall time $[(4\pi/3)G\rho]^{-1/2}$. The expansion of the Universe can be described as a time-reversed free fall, so the age of the Universe is roughly the free-fall time. This implies that the time scale for deviations from a flat Universe to grow and the age of the Universe are the same at any time. Any irregularities at $t = 10^{-39}$ second would have rapidly become black holes or regions of space that expand so rapidly as to now appear empty and devoid of structure. Clearly this does not describe the Universe around us. This difficulty is known as the "flatness problem."

The flatness problem is compounded by what is known as the "horizon problem." In the standard model of a Universe with only matter and radiation, the volume in causal contact is constantly growing. The Universe we see today was 10^{83} separate horizon volumes when the temperature was 10^{17} GeV. Not only do we have to specify expansion parameters to 1 part in 10^{50}, it has to be done 10^{83} times in each of the tiny causally disconnected regions!

Are these really problems? Isn't the most simple specification of a Universe just the statement that it is described by a homogeneous and isotropic flat cosmology? This is a statement whose simplicity and elegance might be overpowering, if only it were true. The problem is that it is only approximately true (approximate simplicity and elegance are not powerful tools), as the Universe is known to exhibit structure on scales up to the largest explored by examining the distributions of galaxies. Plate VIII shows the distribution of galaxies seen on a photographic plate at the limit of the observing power of large telescopes. It is only on still larger scales, probed by fluctuations in the 2.7-K background, that the Universe appears structureless. There are clearly small fluctuations; the puzzle is why they are so very small, yet there.

Both the horizon and flatness problems may be solved by a model first proposed independently by Guth (*9*) and by Brout, Englert, Gunzig, and Spindel (*10*) and termed the inflationary Universe by Guth (*9*). In this model there is an epoch of exponential expansion derived by the energy of the vacuum. During this time an initially small volume is expanded by a large factor. In this scenario, two particles that are initially in causal contact (contained in one another's particle horizon) are carried through event horizons. At the end of the period of exponential expansion, the Universe reverts to one with particle horizons. As time passes, the two particles eventually are again contained in each other's particle horizons. This time they are not strangers, having met before the inflation. If at a temperature of 10^{17} GeV the scale factor of the Universe is increased by 28 orders of magnitude, what was one small horizon volume at the beginning becomes the entire volume of the Universe now seen by us.

Can this scheme be realized from the microphysics of grand unified theories? It is allowed by the theory and is vaguely plausible but certainly not required, and it has some "tuning" problems of its own. In order to understand this microphysics, it is useful to first describe Heisenberg's model of ferromagnetism as an example of the type of symmetry breaking that is essential to theories of unification.

Ferromagnets display a spontaneous magnetic moment, one that persists in the absence of applied fields. This orientational order is promoted by the magnetic and exchange forces between the magnetic moments produced by the electron spins. Although the laws of motion that govern these materials are isotropic, that is, display no preference for direction in space, it is energetically favorable for neighboring spins to be aligned (there is no preference for an absolute direction, only a relative one). At high temperatures, when the thermal energy is large as compared to the interaction energy, no such alignment is seen; the system displays the full symmetry of the Hamiltonian at high temperatures. Inside the ordered material there may be many ferromagnetic domains, wherein the spins are locally aligned but domain walls separate regions that have selected different directions of alignment. This occurs to minimize the value of the energy contained in the magnetic field, while allowing most neighboring spins to be aligned.

The vacuum in grand unified theories has many properties analogous to those of a ferromagnet. The spin orientation of electrons is replaced by an order parameter for the orientation of a set of "Higgs fields," ϕ, entities that lead to the symmetry breaking. Interactions lead to local alignments and a lowering of the total energy by the correlations. Domains form, but the defects between them are not walls but point magnetic monopoles. As causality constrains each domain to be no larger than a horizon volume, if the Universe doesn't inflate there would be as many monopoles as there are protons. Detection of a monopole would be a sensational discovery, as it would provide direct evidence that the Universe was once at a temperature of 10^{17} GeV and would also provide some information on the correlations in the Higgs fields and the amount of inflation. Completing the analogy, at high temperatures the symmetry is restored and the expectation value of the field reverts to zero.

The energy $V_T(\phi)$, as a function of $<\phi>$ is shown for several temperatures in Fig. 2. Similar diagrams abound in studies of solids. From the standpoint of describing the dynamics of the fields only relative energies are important in Fig. 2, but the absolute value is important cosmologically as it contributes to the mass-energy density. Where is one to place the zero point? Since we know that the cosmological constant is now small, we will assume that Fig. 2 should have been drawn such that our T-nearly-zero state is the one at zero energy. We have to change Fig. 2, moving all the curves up by the quantity β shown there. This implies that at an early time, when the temperature was large, the vacuum had

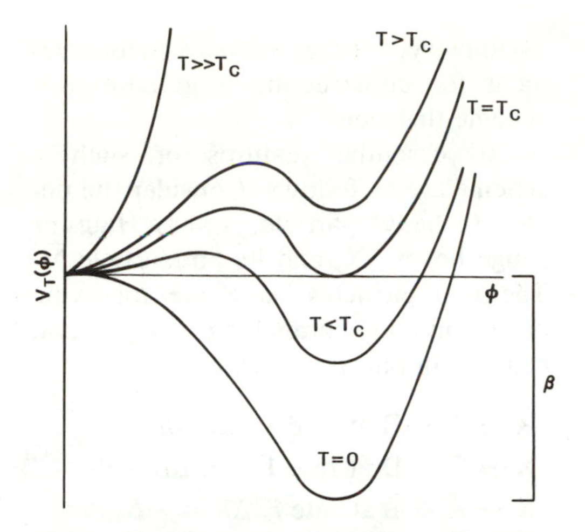

Fig. 2. The free energy density as a function of the order parameter in the Higgs field (ϕ) at various temperatures. The critical temperature, T_c, is the point where the two possible vacuum states are degenerate in energy; T_c is approximately 10^{15} GeV in current grand unified theories.

an enormous energy density which might trigger the inflation.

This scheme is an exciting one but not without its share of problems. The first one is the cavalier setting of the zero point which transpired in the last paragraph. The change in the zero point from that shown in Fig. 2 relative to what we require from current observations is in the ratio of 10^{110}, a problem of "tuning" that is orders of magnitude larger than any inherent in a Universe containing only matter and radiation. Even when one considers only the symmetry breaking in the electroweak interactions, the tuning of the final vacuum energy density must be "done by hand" to 1 part in 10^{60}. There is a further problem in getting the exponential phase to last long enough to change the scale factor by 10^{28}. As the Universe expands, it "supercools" to very low temperatures. Further parameter tunings are required to keep the Universe from rolling off the high-temperature state too soon or having various regions change at different times, which would yield "bubbles" of the standard model Universe inside an exponentially expanding vacuum.

The solution to some of these problems may be within sight. Gott (*11*) has proposed that, if "bubbles" form slowly, just one of them might contain the entire observable Universe. Linde (*12*) and Albrecht and Steinhardt (*13*) have shown that the Coleman-Weinberg (*14*) potential introduces a small potential barrier near the origin with a steep drop to a relative minimum at the origin of the $V_{T=0}(\phi)$ potential. This turns the obviously unstable state at $\phi = 0$, $T = 0$, shown in Fig. 2 into a metastable state and helps prolong the exponential expansion in what is known as the "new inflationary" model. The period of inflation is still too short and the amplitude of the fluctuations is too high (*15*), but this is progress in a difficult problem.

Another Problem: The Matter-Antimatter Asymmetry

There is a second problem inherent in starting up the Universe: where did the matter come from? In our Galaxy, the absence of antimatter in low-energy cosmic rays leads us to conclude that the Galaxy is all matter. At present we have no way to refute the hypothesis that the nearest Galaxy, the Andromeda nebula, is pure antimatter (*16*). To arrange for the degree of homogeneity that must have been present in the early Universe and yet separate matter and antimatter on scales of galaxies or clusters of galaxies seems an impossible task after nearly two decades of effort (*16*). Why then is there a pronounced matter-antimatter asymmetry in the Universe?

In order for an excess of matter to arise, charge-parity (CP) must be an imperfect symmetry (*17*). Somehow matter must be created in preference to antimatter. There is a second requirement that the process occur out of equilibrium. If the system were in equilibrium, CP violation would introduce a reaction channel between matter and antimatter and prevent any imbalance. Generally, it has been assumed that, at extremely high temperatures, gravitational interactions succeed in establishing an equilibrium where the initial baryon number (B, defined as the number of particles minus antiparticles) is zero. At a later time the expansion of the Universe breaks the equilibrium.

Before describing a scenario of net baryon generation, it is useful to enumerate the reasons that have convinced the physics community at large that baryon number is not conserved:

1) Gravity violates baryon number. A black hole made of particles is indistinguishable from one made of antiparticles, even if it evaporates by quantum processes (*18*).

2) Baryon nonconservation arises as a natural consequence of the standard Weinberg-Salam model of the electroweak interaction (*15*).

3) In unifying the strong and electroweak interactions, grand unified theories break the global symmetry of baryon number leaving only local gauge symmetries (*4*).

4) The Universe contains an excess of baryons over antibaryons.

Although items 1 and 2 have not been experimentally verified, they are such natural consequences of well-established theories that there is little doubt as to their existence. These first two violations of baryon number are far too small to account for the matter-antimatter asymmetry, but they serve as encouragement for constructing a grand-unified scheme that does.

The essential features of such a scheme are as follows. Consider the decay of a heavy particle, either a Higgs or gauge boson, X, and its antiparticle $\overline{\mathrm{X}}$. The two particles have the following decay modes, branching ratios, and changes in baryon number:

$$\mathrm{X} \rightarrow \mathrm{A} + \mathrm{B} \text{ at rate } r,\ \Delta B = B_1$$
$$\mathrm{X} \rightarrow \mathrm{C} + \mathrm{D} \text{ at rate } 1 - r,\ \Delta B = B_2$$
$$\overline{\mathrm{X}} \rightarrow \overline{\mathrm{A}} + \overline{\mathrm{B}} \text{ at rate } \bar{r},\ \Delta B = -B_1$$
$$\overline{\mathrm{X}} \rightarrow \overline{\mathrm{C}} + \overline{\mathrm{D}} \text{ at rate } 1 - \bar{r},\ \Delta B = -B_2$$

When the Universe is at a temperature much greater than the mass of the X, these processes ensure that the initial baryon number is zero. When the expansion rate becomes more rapid than the decay rate, a baryon number excess is created (*3*). The resultant excess (calibrated to the number of photons, where all the energy from particle and antiparticle annihilations eventually winds up) is

$$\frac{n_b}{n_\gamma} = \begin{pmatrix} \text{fraction of all} \\ \text{particles that} \\ \text{are X's} \end{pmatrix} (r - \bar{r})\,(B_1 - B_2)$$

There are roughly 100 particles in the grand unified schemes, so the first number is $\sim 10^{-2}$. A reasonable value of $(B_1 - B_2)$ is 1. Since the observed value of n_b/n_γ is $\sim 10^{-9}$, the theory must place $(r - \bar{r})$ in the range of 10^{-7}. Numerous schemes have been devised to do so.

Baryosynthesis is quite different from nucleosynthesis in that laboratory determinations of the cross sections will not be rapidly forthcoming. There are other ways to corroborate the result. The first one is to find proton decay. This is currently a topic of intense experimental effort, with an upper limit of 10^{31} years to the decay into what is believed to be the principal mode (*19*). These results will help constrain the range of possible theories (the simplest model is already ruled out), but proton decay experiments will not directly yield the reaction rates needed in baryosynthesis.

The measurement of the neutron electric dipole moment is an experiment that has direct bearing on the problem of baryosynthesis. It is a cherished belief that the combination CPT (T is time) is an exact symmetry (there is currently no way to build a local field theory without it), so a violation of CP must be accompanied by a matching violation of T. The neutron electric dipole moment is a sensitive probe of T violation which complements the CP violation necessitated by baryosynthesis. The net baryon excess observed implies a lower bound on the neutron electric dipole moment of 3×10^{-28} e − cm, whereas the current limit is 1.6×10^{-24} e − cm. It will be a decade or two before the electric dipole moment experiments are in this range (*20*).

Looking Ahead

This survey has considered processes at energies of 10^{17} GeV, some 13 orders of magnitude above current accelerator experiments and only two orders of magnitude below the realm of quantum gravity. Many surprises are certain to lurk in unknown corridors. Where does one start to look? If the answers were not many and varied, we would be facing the end of a science; but of course they are. Plate IX shows some of the tools that are windows on cosmological events and processes.

There is a clear experimental route toward elucidating the processes of baryosynthesis, as was described in the last section. In the next decade, experiments on proton decay and the neutron electric dipole moment (see Plate IX, a and b) will either provide confirmation of the scheme sketched here or insights into new theories.

The horizon and flatness problems are certainly not yet solved. Examining the large-scale structure of the Universe by using large telescopes on the ground, the Space Telescope, and the Cosmic Background Explorer satellite (see Plate IX, c and d) will help to elucidate the origin and evolution of the deviations from a homogeneous, flat cosmology. It will also shed light on the current value of the expansion rate, help to determine whether the Universe is open or closed, and place better limits on the current energy density of the vacuum.

An important problem in need of solution by theorists is the current tiny value of the vacuum energy density (the cosmological constant) relative to that expected in our current low-temperature vacuum state. The solution may involve a new, fundamental principle that is certain to have a broader impact than the resolution of the cosmological problems discussed here.

References and Notes

1. A. A. Penzias, *Rev. Mod. Phys.* **51**, 422 (1979).
2. P. J. E. Peebles, *Physical Cosmology* (Princeton Univ. Press, Princeton, N.J., 1971).
3. S. Weinberg, *Rev. Mod. Phys.* **52**, 515 (1980).
4. S. L. Glashow, *ibid.*, p. 539.
5. E. C. G. Stuckelberg, *Helv. Phys. Acta* **11**, 299 (1983); E. P. Wigner, *Proc. Am. Philos. Soc.* **93**, 521 (1949).
6. P. W. Anderson, *Science* **177**, 393 (1972).
7. C. Callan, R. H. Dicke, P. J. E. Peebles, *Am. J. Phys.* **33**, 105 (1965).
8. P. J. E. Peebles and R. H. Dicke, in *General Relativity: An Einstein Centenary Survey*, S. W. Hawking and W. Israel, Eds. (Cambridge Univ. Press, Cambridge, 1979), pp. 504–517.
9. A. H. Guth, *Phys. Rev. D* **23**, 347 (1981).
10. R. Brout, F. Englert, E. Gunzig, *Ann. Phys.*

(N.Y.) **115**, 78 (1978); R. Brout, F. Englert, P. Spindel, *Phys. Rev. Lett.* **43**, 417 (1979).
11. J. R. Gott, *Nature (London)* **295**, 304 (1982).
12. A. D. Linde, *Phys. Lett. B* **108**, 389 (1982).
13. A. Albrecht and P. J. Steinhardt, *Phys. Rev. Lett.* **48**, 1220 (1982).
14. S. Coleman and E. J. Weinberg, *Phys. Rev. D* **7**, 1888 (1973).
15. G. 'tHooft, *Phys. Rev. Lett.* **37**, 8 (1976).
16. R. Omnès, *ibid.* **23**, 38 (1969); G. Steigman, *Annu. Rev. Astron. Astrophys.* **14**, 339 (1976).
17. A. D. Sakharov, *Pis'ma Zh. Eksp. Teor. Fiz.* **5**, 32 (1967).
18. C. W. Misner, K. S. Thorne, J. A. Wheeler, *Gravitation* (Freeman, San Francisco, 1973).
19. R. Bionta *et al.*, *Phys. Rev. Lett.* **51**, 27 (1983).
20. J. Ellis, M. K. Gaillard, D. V. Nanopoulos, S. Rudaz, *Phys. Lett. B* **99**, 101 (1981).

This material originally appeared in *Science* **224**, 18 May 1984.

19. The Origin of Galaxies and Clusters of Galaxies

P. J. E. Peebles

The large-scale distribution of matter is strikingly clumpy; we see stars in galaxies, galaxies in groups and clusters, and clusters in superclusters (*1*). Only in the average over scales beyond approximately 30 megaparsecs (*2*) does the homogeneity of the conventional cosmological model emerge (*3*). Debate on the origin of this arrangement goes back to the 1930's. Lemaître (*4*) proposed that density fluctuations develop as a result of the gravitational instability of the expanding universe. Because galaxies are denser than clusters of galaxies he found it natural to suppose that galaxies were formed before clusters (*4*). Hubble (*5*), however, noted that the fact that early-type galaxies (which contain relatively few young stars and little interstellar dust and gas) tend to be concentrated in the densest clusters might be evidence that galaxies formed within protoclusters that existed before galaxies. A definite form of this latter "top-down" scenario is Zel'dovich's pancake theory (*6*). Zel'dovich showed that under not unreasonable conditions the first generation of objects in the expanding universe could be sheets or pancakes of gas defined by the intersection of orbits of the primeval matter velocity field. Since the mass of a pancake could be comparable to that of a supercluster, pancakes could be the protoclusters that fragmented to make galaxies. This theory has attracted considerable attention because remnants of the pancakes might be seen in a filamentary or cellular character of the present galaxy distribution. However, there are problems, such as the puzzle of where galaxies like the one we are in came from; it is a member of a loose and young-looking group that is most naturally interpreted as forming by Lemaître's "bottom-up" scenario. The following is a list of the clues to be assessed in trying to decide which scenario (if either!) is closer to the truth.

Filaments

Two examples of filaments are seen in the map of bright galaxies in Fig. 1. A prominent band slopes down to the right on the right side of the map, and one also can discern a straight line running from 5^h, 70° to 10^h, 45°. Giovanelli and Haynes (*7*) noticed the latter and found that it appears also in the distribution of galaxies with known distances (from red shifts) in the range 40 to 70 Mpc, but not in the nearer and further galaxies. This is an example of discovery based on crude distance information and confirmed with a subset of accurate distances. Hence it is a strong candidate for a physical filament of galaxies.

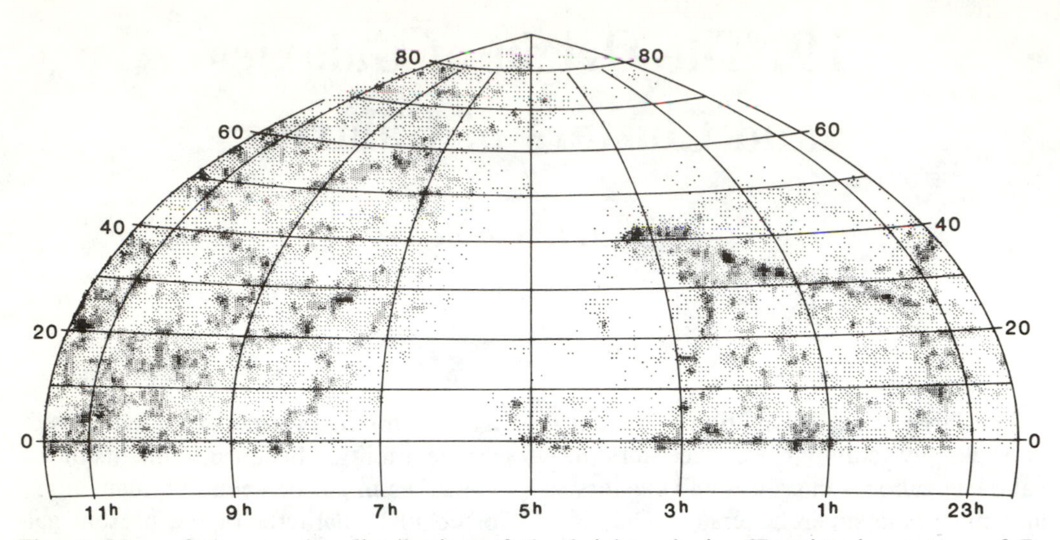

Fig. 1. Map of the angular distribution of the bright galaxies [Reprinted courtesy of R. Giovanelli and M. P. Haynes, *Astrophysical Journal* 87, 1355 (1982), published by the University of Chicago, © American Astronomical Society]. The shade density in each cell is proportional to the logarithm of the surface density of galaxies brighter than apparent magnitude $m \sim 15.7$. The blank zone running through the middle of the map is caused by obscuration by the dust in the disk of our galaxy.

The most natural interpretation of filaments is that they are fossils of preexisting linear structures, or of anisotropic stresses, or something equally interesting. Gravity could make large-scale filaments after galaxies had formed, just as it produced linear structures in the distribution of hydrogen in the pancake scenario, but that does require a special arrangement; the galaxies would have to approximate a "gas" with only large-scale density gradients. There is yet another possibility to consider: if galaxies were placed in a clustering pattern according to a random statistical process that had no preference for lines, the occasional filament would be produced by chance and, because the eye is so sensitive to patterns, we might attach undue significance to these accidents. To explain why some of us place so much emphasis on this point I will recall some cases where it has been a factor.

In the 1920's it was known that nebulae tend to be spirals, so it was natural to ask whether there are spiral patterns in globular star clusters, and it is not surprising that people were able to find some pretty good cases. In the example shown in Fig. 2, one can pick out a rudimentary four-arm spiral. Interest in

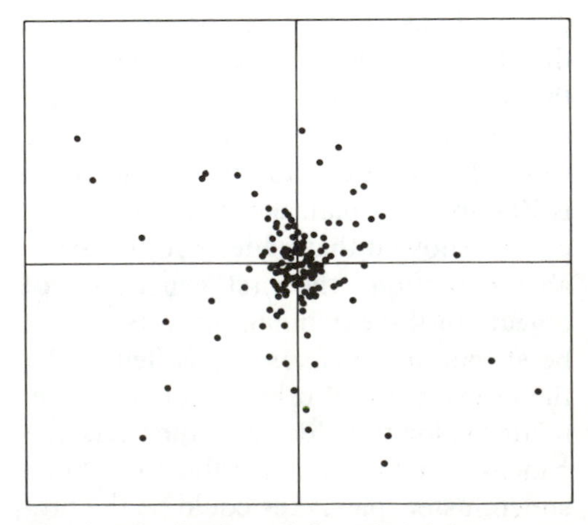

Fig. 2. Distribution of the 150 brightest stars in the globular star cluster M3 (*8*).

the idea soon faded because, as ten Bruggencate (*8*) emphasized, the pattern is not reproduced in the fainter cluster members, and it is now doubted that there is any reason why globular clusters should have spirals or that there is any evidence of spirals. The eye was similarly deceived in the case of canals on Mars. That was what Maunder and others argued at the time, and to test it he and Evans (*9*) made maps of the observed features of Mars, excluding the canals, and asked schoolboys to draw everything they could make out on the maps. Evans and Maunder found that when the maps were placed far enough away that the smaller features were just detectable, the students tended to add linear structures that look like the canals. [Details of this ''small boy theory'' are given by Hoyt (*10*).] Star chains are commonly seen, although not often reported because most people are convinced they really are accidents. The best example I could find in a brief survey of observers is the photograph in Fig. 3 that was kindly provided by Bob McClure (*11*). As the stars in this cluster are old, they surely have been thoroughly mixed since formation. Most striking are the two nearly horizontal chains of stars that define a branched path near the top of the figure, but one can pick out other chains, all of which are presumed to be accidents. It is true that this is the best example I could find, but then examples of galaxy filaments also are the results of searches for good cases. The moral I draw certainly is not that the filaments shown in Fig. 1 are only statistical accidents, but rather that we ought to bear that possibility in mind until we have tests that provide good evidence to the contrary.

It is more difficult to assess galaxy filaments because galaxies really are arranged in clusters. Hence one must judge the significance of chains of clumps of galaxies rather than chains of individual stars. We see a strikingly filamentary texture in the map in Fig. 4A of the angular distribution of the brightest one million galaxies (*12*, *13*). To test the significance of this texture, Ray Soneira and I (*14*) devised a prescription for a model galaxy distribution that matches the observed low-order galaxy correlation functions [the random process in this prescription is one of Mandelbrot's fractals (*15*)] and then made maps from the model. The result shown in Fig. 4B lacks some of the crispness of the real galaxy distribution, but one can pick out long sinuous filaments that we know are accidents because the model has no preference for lines. I am not aware of any test that reveals a significant difference in the abundance of large-scale linear arrangements in model and data (*16*).

Figure 5 shows a comparison of real and model galaxy distributions at about one-third the depth of Fig. 4, where we have distance information so we can look at three-dimensional structure. The radial coordinate in these maps is the galaxy red shift, which is presumed to be the sum of the cosmological term Hr and the Doppler shift from random motions, the latter introducing a scatter in the true relative radial distances of the galaxies. The two maps on the left are from the Cambridge Center for Astrophysics red-shift sample (*17*). The maps on the right are realizations of the random process mentioned above generalized to velocities as well as positions (*18*). One sees in model and data maps roughly similar clouds and knots and holes; no very prominent filaments as long (30 Mpc) as the Giovanelli-Haynes object (Fig. 1) are seen in the model maps or in the Center for Astrophysics red-shift maps (*17*).

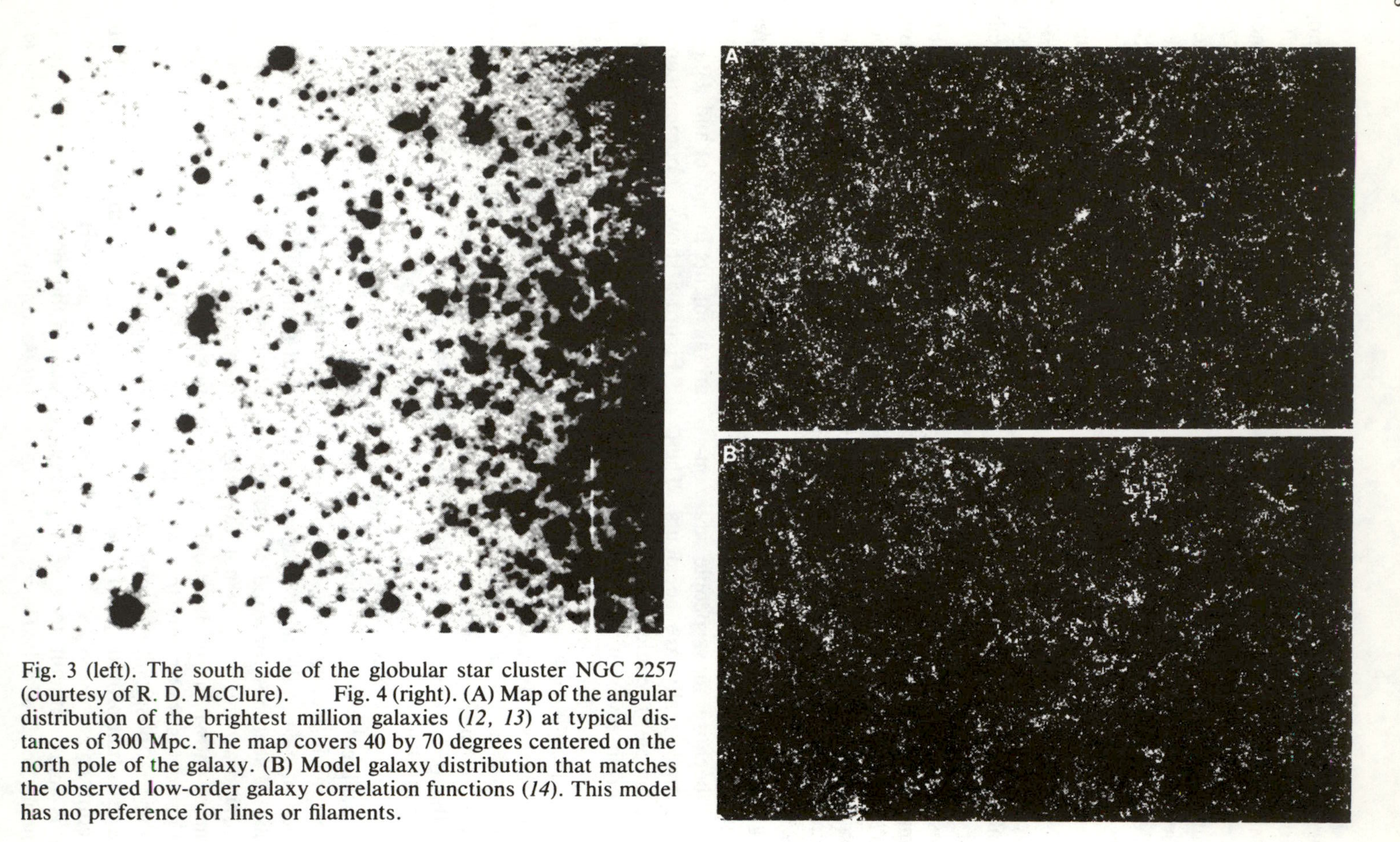

Fig. 3 (left). The south side of the globular star cluster NGC 2257 (courtesy of R. D. McClure). Fig. 4 (right). (A) Map of the angular distribution of the brightest million galaxies (*12, 13*) at typical distances of 300 Mpc. The map covers 40 by 70 degrees centered on the north pole of the galaxy. (B) Model galaxy distribution that matches the observed low-order galaxy correlation functions (*14*). This model has no preference for lines or filaments.

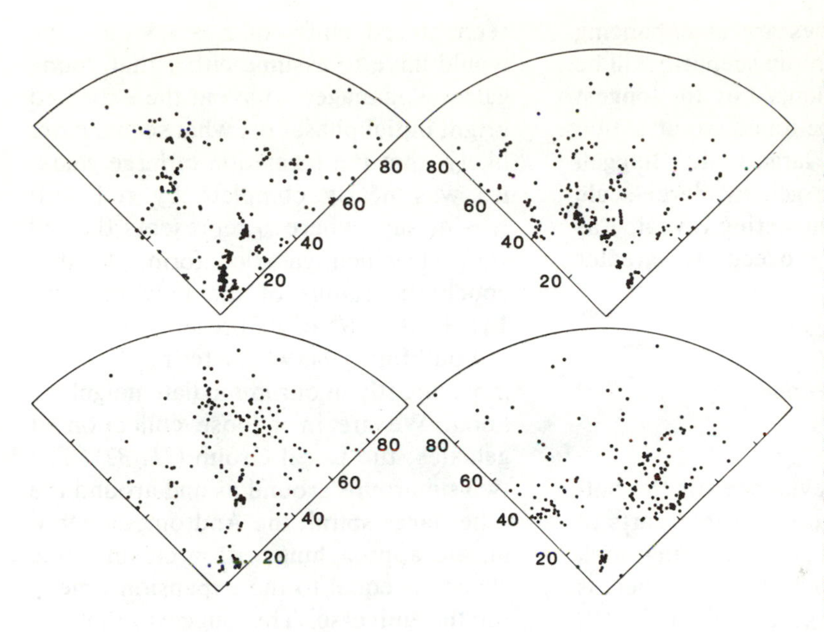

Fig. 5. Maps of galaxy red shifts. The radial coordinate is recession velocity in units of 100 km/sec, the angular coordinate is right ascension (azimuthal angle), and each map is a slice 10° thick in declination (polar angle θ at θ close to 90°). The two maps on the left are real distributions (*17*), and the two on the right are realizations of a random process originally designed for comparison with the deeper map of angular distributions in Fig. 4.

Remarkable indirect evidence of filaments emerged from Binggeli's study of the orientations of clusters of galaxies (*19*). He observed, as have others, that the long axis of the central brightest galaxy in a cluster tends to line up with the long axis of the cluster. That could be because both are remnants of a pancake, but arguing against this interpretation is that the other galaxies in the cluster generally show no marked alignment (*20*). Another interpretation is that we are seeing debris that tends to arrange itself symmetrically in the cluster potential well (*21*). Binggeli found that if a rich cluster happens to have a neighbor cluster at distance less than approximately 20 Mpc (compared to a mean separation of rich clusters of about 80 Mpc), the long axis of the cluster tends to point toward the neighbor. That could be a residuum of pancakes (*16*, *19*), but again other interpretations are possible; one could imagine the alignment was caused by tides operating at red shifts of $z \sim 10$ to 20 when these dense clusters might have been forming. Binggeli also found a tendency for the long axes of clusters to point at other clusters as far away as approximately 75 Mpc. This is a difficult measurement, and perhaps arguing against it is the fact that the clusters of rich clusters Bahcall and Soneira (*22*) identified do not look particularly elongated. Further tests of the Binggeli effect will be followed with great interest, for it is hard to see how alignment on scales of more than 20 Mpc or so could be anything but a new and very significant phenomenon.

To decide whether structures like those seen in Fig. 1 could arise in the simplest "bottom-up" scenario where the clustering pattern develops in a more or less continuous progression of increasing mass, we need to have a better theoretical understanding of how often such linear structures might appear by chance in a chaotic "frothy" (*17*) distribution and how effective such gravitational effects as tides are at enhancing linearity. The bottom-up scenario will be most seriously challenged by the longest filaments that can be found, so attention will focus on the large-scale Binggeli effect and on the general large-scale three-dimensional clustering revealed as red-shift surveys proceed to greater depths.

The Ages of Galaxies and Clusters of Galaxies

The most direct evidence for the bottom-up scenario is the fact that parts of the pattern of galaxy clustering look young, as if just now forming, whereas galaxies by and large are old (*23*, *24*). The ages of the oldest stars in our Milky Way galaxy are found to be 14 to 20 billion years (*25*), which may be consistent with the age of the universe derived from the observed rate of expansion if these stars formed when the universe was fairly young. We see what other galaxies were like in the past by observing distant ones. Galaxies observed at red shifts of $z \sim 1$ (*26*), when the universe was about half its present age, have luminosities and colors similar to nearby galaxies (*24*, *27*); this suggests that large galaxies had reached a fairly mature state by $z = 1$. If a galaxy formed out of a gas cloud, a young galaxy would have to have been very bright to make the heavy elements in the spheroid while the cloud was collapsing (*28*). To account for the fact that no such objects have been discovered, it is presumed that the luminous phase is hidden by a high red shift or else obscured by intergalactic dust (*29*). As quasars with appreciable heavy element abundances are seen at red shifts of $z \sim 3.5$ (*30*), we would have to assume either that young galaxies managed to avoid the expected bright initial phase, or, what seems more likely, that the formation of large galaxies was nearly complete by red shift $z_f \sim 4$, say, where z_f represents the red shift at which galaxies form. At that epoch the radius of the universe was $1/(1 + z_f) \lesssim 1/5$ of its present value.

Youthful galaxy clustering is seen most directly in our immediate neighborhood. We are in a loose collection of galaxies, the Local Group (*31*, *32*). The two subgroups around us and around the other large spiral, the Andromeda nebula, are approaching with a closing time r/v about equal to the expansion time t_e for the universe. This suggests that the group is only now forming (*33*), and, consistent with that, one finds that the crossing times for the outlying group members all are comparable to t_e (*31*, *32*). De Vaucouleurs (*31*) calls the Local Group a "typical loose group" of galaxies. For example, our nearest neighbor, the Sculptor group in the Southern Hemisphere, has very similar size and internal velocities and so also is presumed to be young. The Sculptor group is 2.5 Mpc away (compared to a radius of 1 Mpc for the Local Group), and the group is moving away from us at 230 kilometers per second, about what is expected from Hubble's law. The same is true of the other nearby groups. If we extrapolate these motions back in time to the latest epoch ($z_f \sim 4$) at which the above evidence suggests galaxies might have formed, we find that the four nearest groups listed by de Vaucouleurs (*31*) all were within the bounds of the Local Group when galaxies formed, which is hard to reconcile with the idea that these groups existed when galaxies formed. We see evidence of youth on a still larger scale in the Local Supercluster. This is a loose cloud of galaxies, of which we are an outlying member, with a radius of about 15 Mpc centered roughly on the Virgo cluster of galaxies (*34*). The galaxies in our immediate neighborhood are moving away from the Virgo cluster at 1000 km/sec, which is less than would be expected from pure Hubble flow by some 30 percent (*35*). The conventional interpretation is that the mass concentration in the Local Supercluster is slowing the general expansion and causing the cluster to grow; that is, that this system is in the process of forming now. It is thought that indications of immaturity are seen in some other clusters, whereas there are others that look well relaxed (*36*, *37*).

If protoclusters made galaxies, where did the Local Group come from? Could it and the whole Local Supercluster have been produced by one pancake? We would want this to have happened at red shift $z_f \gtrsim 4$, when the mean density of the universe was $(1 + z_f)^3 > 100$ times the present value, and the collapse of the pancake would have made the local density higher than the mean. But if the Local Supercluster formed at such high density, how do we account for the fact that the density of galaxies in our neighborhood averaged over spherical shells is only two or three times the present large-scale mean (*38*)? It is hard to imagine that gravity could have caused the pancake to collapse but then allowed its mean density to drop by such a large factor (*39*). Numerical simulations of the pancake theory point to the same problem: a pleasingly frothy structure is obtained, but the froth is young, tending to appear in regions that collapsed only recently. If at some point galaxy formation in the model is stopped, and the model expands by another factor of 5 or so to hide the young galaxies, much of the froth ends up in dense clumps (*40*, *41*).

Could cluster and field galaxies (found, respectively, in high- and low-density environments) form by different processes? Arguing for it is Hubble's point that there is a higher proportion of disk-shaped galaxies in the field, whereas elliptical galaxies tend to be concentrated in clusters (*5*, *36*). One could easily imagine that the difference in the collapse of "minipancakes" and "maxipancakes" that produced field and cluster galaxies would have produced the observed differences in morphologies. On the other hand, there is a striking similarity in the masses of large galaxies (*42*, *43*), which suggests that they all formed in much the same way (*44*). The circular velocity, v_c, in a spiral galaxy is a measure of the mass within a fixed radius, and v_c correlates well with the luminosity and morphology class of the galaxy. A similar measure for elliptical galaxies is the star velocity dispersion, σ, which correlates closely with luminosity. Spirals are found in a wide range of environments, from dense spots like the Virgo cluster to the edges of voids, and there is a like spread of environments for ellipticals; these correlations leave little room for a new parameter representing present ambient conditions. That is, whatever piled mass up into the structures we see as galaxies did so in quite a reproducible way, which suggests that galaxies formed under conditions a good deal more uniform than what we observe. As it is hard to find a parent protocluster for the Local Group, this argues against the idea that cluster galaxies formed in protoclusters.

Hybrid Scenarios and Continuity Between Galaxies and Clusters of Galaxies

It is consistent with the above arguments to assume that creation was a two-step process, groups and clusters forming by gravity and galaxies forming some other way. As Dekel (*41*) stressed, the advantage is that galaxies could form in the very early universe so that young galaxies are not seen, and clusters could form recently by pancaking of the ''gas'' of galaxies so that we get filaments. In a new version of this composite scenario, Ostriker (*44*) showed that galaxies might form at the surfaces of bubbles evacuated either by explosions or by isolated underdense spots in an otherwise initially homogeneous mass distribution. An isolated bubble tends to expand, which may eliminate overproduction of rich clusters, and remnants of the bubbles are a good source of filaments. Perhaps the major problem is that special initial conditions are needed to get bubbles to more or less fill space by red shift $z_f \gtrsim 4$ without strongly overlapping, since that would initiate the usual hierarchy of cluster formation. Another difficulty is that this really is a composite scenario, since objects like the Local Group and the Local Supercluster are not now growing by the bubble process. The problem with a composite scenario is that it does not fit very naturally with the following evidence of continuity of the mass-clustering hierarchy.

The distribution of galaxies approximates a scale-invariant clustering hierarchy, one of Mandelbrot's fractals (*15*), density (ρ) scaling with size as $\rho \propto r^{-\gamma}$, $\gamma = 1.77 \pm 0.04$, at $r \lesssim 15$ Mpc (*45*). When observed by starlight, galaxies stand out from the hierarchy as distinct islands, but, as Rubin (*43*) has described, the mass distribution in at least some galaxies is a good deal more spread out than the light (*43*, *46*, *47*). A measure of the mean mass around a galaxy is the root-mean-square relative velocity w of neighboring galaxies. One assumes that the mean relative acceleration w^2/r is balanced on the average by the gravitational acceleration due to the mass in the neighborhood, since otherwise the clustering pattern we see on small scales would dissolve well within the ages of the galaxies. Figure 6 shows estimates of w as a function of the projected separation r_p of the galaxy pair (*48*). At separations typical of the optical sizes of galaxies ($\sim$ 10 kpc), $w = 210 \pm 20$ km/sec, about what would be expected from observed star and gas motions within bright galaxies (*43*, *47*). The striking result is that $w(r_p)$ is close to flat from 5 kpc to 1 Mpc. Of course that just extends the by now familiar observation that the circular velocity of rotation in the disk of an isolated spiral is quite flat (*43*, *47*), and the interpretation is the same: if, as we almost always assume, Newtonian mechanics is an adequate approximation (*49*), the mean value of the mass (M) within distance r of a galaxy has to scale as $M(< r) \propto r^{\epsilon}$ with ϵ equal to or slightly larger than unity (*46*). This applies from within the galaxy ($r \lesssim 10$ kpc) to $r \sim 1$ Mpc, where we run out of data on $w(r_p)$, but since the galaxy clustering scales as the number of galaxies $N(< r) \sim \rho r^3 \propto r^{1.2}$, we see that if mass clusters like galaxies, the power law is a reasonable approximation to $r \sim 15$ Mpc, where $\rho(r)$ starts to flatten into the homogeneous

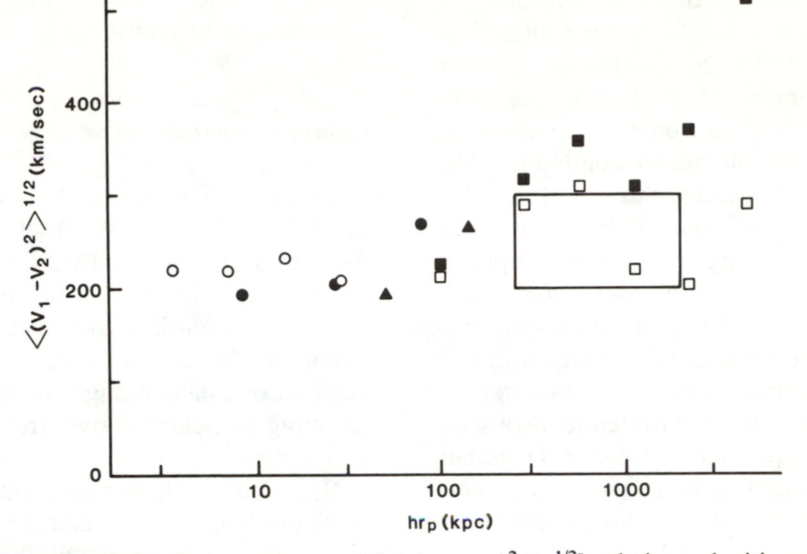

Fig. 6. Estimates of the root-mean-square [$< (v_1 - v_2)^2 >^{1/2}$] relative velocities of physical pairs of galaxies as a function of their projected separation r_p. The scale factor h reflects the uncertainty in the distance scale; $h = 0.75$ is adopted in this review.

mass distribution observed on very large scales.

Now we must assess the significance of these two clues. When measured by the distribution of starlight, galaxies stand out as islands in the clustering hierarchy. That means there are differences in the processes by which galaxies and clusters of galaxies form. But as we have seen, the mass distribution within galaxies and clusters is joined by a scaling law; the mean value of the mass within distance r of a galaxy scales as $M(< r) \sim r^{1.2}$ from within the galaxy out to $r \gtrsim 1$ Mpc. This scaling continuity suggests galaxies and clusters of galaxies formed by scaled versions of the same process. The formation of visible stars is poorly understood but surely is sensitive to physical conditions, so it seems to me easier to imagine that star formation broke the scaling symmetry, producing concentrations of starlight as frosting on the peaks of the mass distribution, than that the observed continuity in the mass distribution was the result of distinct processes of formation of galaxies and clusters of galaxies. If that is so, we are led to a bottom-up picture because galaxies are old and stable objects, whereas, at the top end of the hierarchy, superclusters are still forming. In a hybrid scenario we would need a special arrangement to hide the seams of the two-step process.

Non-Gaussian Large-Scale Density Fluctuations

If a sphere is placed at random, the root-mean-square fluctuation in the number of bright galaxies it contains is $\delta N/N = 1$ when the diameter (D) is 20 Mpc (*3*). This marks the transition from strongly nonlinear fluctuations on small scales to fluctuations that are small on the average at larger D. One might expect that at $D > 20$ Mpc the distribution of the number of bright galaxies contained rapidly approaches a Gaussian, reflecting simple initial conditions, but there are indications this is not so. The great void in Böotes (*50*) has a radius of about 40 Mpc, and if it is nearly empty, it is roughly a 3-standard-deviation downward fluctuation (*45*). Because the volume of the void is a few percent of the volume of space surveyed to its distance, we might not have expected to have seen such a large fluctuation in a Gaussian distribution. The largest known prominent concentration of galaxies is the Serpens Virgo cloud discovered by Shane and Wirtanen (*12*); it is number 14 in the Bahcall-Soneira catalog of clusters of rich clusters of galaxies (*22*). The galaxy density in the cloud averaged over a diameter of 30 Mpc is about ten times the large-scale mean, which is roughly a 10-standard-deviation upward fluctuation.

The great void could be due to a suppression of galaxy formation in the region (*50*), but because it is hard to explain the Serpens Virgo cloud that way, it does suggest that the mass distribution on scales greater than approximately 20 Mpc has non-Gaussian tails; that is, in the primeval mass distribution, regions of extremely high or low density occur more often than would be expected for a Gaussian probability distribution (*23*, *36*, *51*). This finds a natural interpretation in the top-down scenario as remnants of the network of protoclusters (*6*). It may be an embarrassment in the bottom-up scenario because the most popular candidate for the source of departures from homogeneity, quantum fluctuations in a nearly free field, leads to Gaussian noise. However, other sources, like the "vacuum strings" predicted in some gauge theories, could produce non-Gaussian perturbations (*52*).

Galaxy Formation with "Inos"

The preceding sections have dealt with general interpretations of clues to the formation of galaxies. I turn now to two more specific theories, both based on the idea that the dark matter needed to account for the dynamics of galaxies and clusters of galaxies might be weakly interacting particles left over from the very early universe.

Neutrinos with nonzero mass are a particularly attractive candidate because we know that neutrinos really do exist, and because there is a beautiful coincidence in the wanted neutrino mass. The abundance of neutrinos produced at high red shift in the Big Bang is known; if these neutrinos make an interesting but not excessive contribution to the mean mass density of the universe, their mass may be about 50 electron volts. That agrees with the lower bound on the mass if neutrinos are to be stuffed into galaxies to attain the wanted density without violating the exclusion principle (*53*). The mass also fixes the velocity distribution. Because these are weakly interacting particles, they move almost freely so that the velocity fixes the length smoothed by thermal motions; the result is comparable to that of superclusters. Thus we are led to a top-down pancake theory (*1*, *6*). There is, however, a problem. Analytic and numerical N-body model studies both suggest the predicted mass clustering length is too large (*54*). The analytic approach might be questionable because it is hard to be sure nonlinearities are properly handled, and the N-body approach is vexed by the limited dynamic range in space and time,

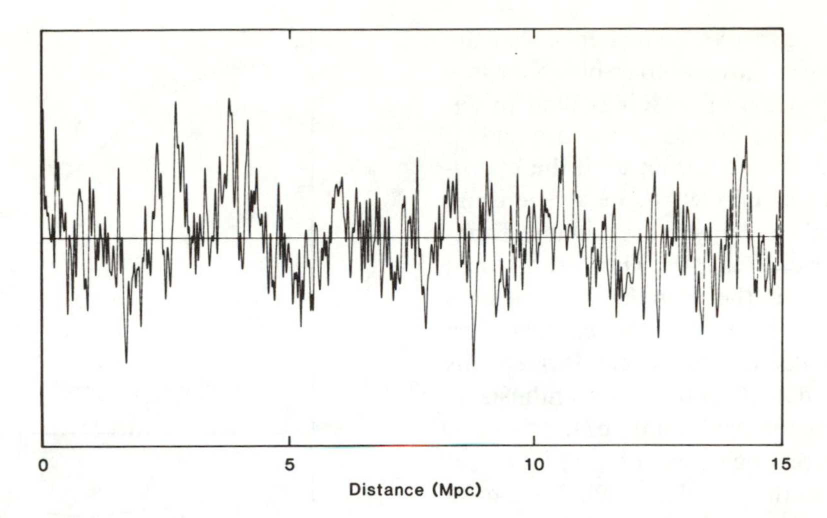

Fig. 7. Hydrogen distribution in linear perturbation theory in a cosmological model dominated by low-pressure weakly interacting matter.

which makes it hard to be sure the freedom of adjusting the initial spectrum of density fluctuations has been properly taken into account. But the fact that the two approaches yield such similar results suggests they are correct. Attempts to relieve the problem by adjusting the scenario of annihilation and decay of particles at high red shift have so far not been encouraging (*55*).

A new version of the bottom-up scenario has emerged from the realization that the dark matter could be weakly interacting particles with velocities much less than those in the massive neutrino model. These particles have names like "axions," "photinos," "selectrons," and "gravitinos" (*56*). It suffices for our purposes to notice that there is no empirical evidence that any of these particles exist; they are discussed in elementary particle physics because they appear in theories that are untested but attractive generalizations of successful theories, and they are considered in cosmology because they have some interesting and conceivably beneficial properties.

Density fluctuations in the very early universe are tightly constrained because at high density even a small mass excess promotes relativistic collapse to black holes. This must be avoided because it would produce an unacceptably high mean mass density. A currently fashionable assumption is that the initial density fluctuations on all scales are a fixed fraction of the threshold for black hole production (*57*). Figure 7 shows what the resulting hydrogen distribution would look like at fairly recent epochs ($z \sim 100$) if the universe were dominated by these low-velocity "inos" (*58*). The horizontal axis is evaluated at the present epoch under the assumption that the distribution followed the general expansion instead of breaking up into bound objects. The spikes are produced by gas pressure that suppresses density fluctuations on smaller scales. The spikes tend to appear in clumps (whose size is fixed by radiation pressure). It is an interesting and perhaps suggestive coincidence that the clump masses are comparable to masses of galaxies (*59*). The spikes tend to develop into gas clouds. Such a cloud, if left alone, would shrink until the hydrogen was hot enough to be ionized and then would collapse, presumably forming a star cluster (*60*). The result would be similar to the globular star clusters common in halos of galaxies. The idea that globular clusters were formed this way has not attracted wide enthusiasm but still seems viable (*59, 61*).

An interesting consequence of this picture is that the gas clouds would be born containing inos that would be left as dilute massive dark halos around the candidate globular star clusters. That could be observable, as is illustrated in the star cluster model shown in Fig. 8 (*62*). The upper curve is the surface density run of stars; it looks not unlike a real globular cluster. The bottom curve shows the star velocity dispersion. In standard models all three components decrease with increasing radius to produce the observed sharp drop in density at the cluster "surface." A dark massive halo can make the density drop another way, by the increase in gravity. In the model, the star orbits are isotropic in the core and close to radial near the surface, as might be expected if the stars in the envelope came from the core because of relaxation in the core. That makes the line-of-sight velocity dispersion, σ_l, decrease with increasing radius, about as observed from red-shift measurements (*63*). The dispersion, σ_r, in the plane of the sky and along the cluster radius vector is nearly constant; if there were no dark halo, σ_r would have to drop at the cluster surface. Cudworth (*64*) has shown that one can use proper motion measurements to estimate σ_r as a function of the angular distance θ from the cluster center, and the azimuthal component, $\sigma_a(\theta)$, so with some improvements in the data an interesting test may be possible.

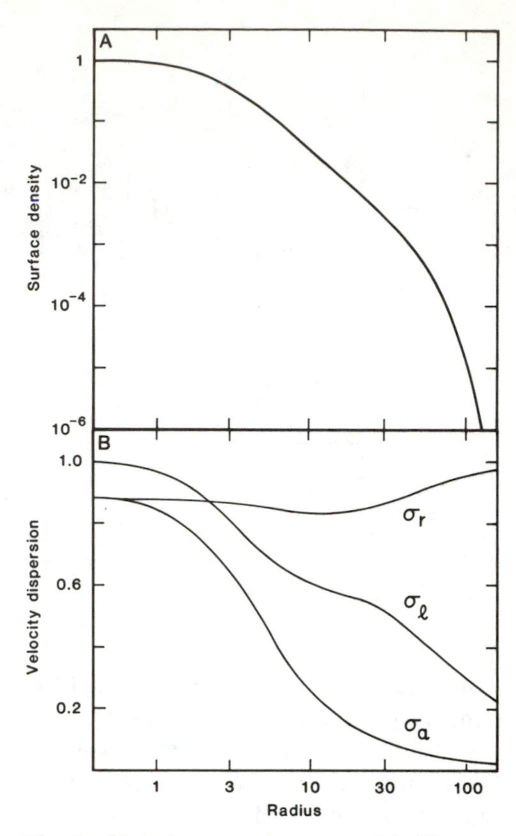

Fig. 8. Globular star cluster model. The top graph is the projected density of stars. The lower graph shows the star velocity dispersion σ_l along the line of sight and the dispersion σ_r and σ_a in the plane of the sky and along the radial and azimuthal directions.

A more immediate test for the bottom-up theory is whether it can account for the character of the galaxy distribution. Numerical *N*-body model simulations of the growth of galaxy clustering fail to produce quite as good an approximation to a scale-invariant (power law) clustering hierarchy as is observed, and in the models the clustering hierarchy is transient (*40, 41, 65*). The observed galaxy-clustering pattern is so close to scale-invariant (extending from 15 Mpc down to about 10 kpc, and, as discussed above, to the mass distribution within galaxies) that I am reluctant to believe it could be a transient effect. Perhaps these problems are only a reflection of the difficulties with *N*-body models, which have very limited dynamical range and deal with nongravitational forces crudely if at all. Perhaps the problems are due to errors in the details of the theory, like the assumption of initially Gaussian density fluctuations. Perhaps they are a reflection of something fundamentally wrong with the scenario.

Outlook

The strong divisions of opinion on how galaxies might have formed are a positive sign: it is only recently that the subject has advanced to the point that we can make out positions that seem defensible. It is proper that we should be guided by our assessment of what fundamental theory is telling us. The situation is not all that good: the "best" version of the top-down scenario, based on massive neutrinos, encounters problems with the galaxy-clustering length, the "best" bottom-up theory is based on matter whose existence is only conjectured; but neither problem need be permanent. The clustering length discrepancy arises in models that treat galaxy formation only crudely, and it is conceivable that as the treatment of nongravitational processes is sharpened the problem will go away (*1*). On the phenomenological side, perhaps the best motivation for the top-down scenario is that it offers a way to account for the large-scale character of the galaxy distribution. On the other hand it seems hard to avoid the conclusion that the bottom-up process has been operating because we see groups and clusters that surely are younger than the galaxies they contain, and the continuity arguments mentioned above support the idea that this has been the dominant process. The list of things to be done is promising; we have some fairly direct questions to address, like the significance of large-scale linearity in the galaxy distribution, and some interesting "long shots." I would count among the latter the point that in a universe dominated by low-pressure inos, one would expect to see the formation of objects like globular star clusters with dark halos. I also count as a long shot the possibility that discoveries in elementary particle physics will offer sharper constraints on the possible macroscopic properties of exotic dark matter, but this is a rapidly moving field so I will be watching the newspapers for developments. On the astronomical side there is the chance that galaxies in a bright initial phase might be detected at high red shift; if we saw what young galaxies are like it would settle a lot of arguments. Galaxies are detected at red shifts near unity, when the universe was about half its present age. The galaxy distribution at that depth is not well measured because of the confusion of clustering seen overlapping in projection, but statistical methods for dealing with that are known and I believe that in the next few years we will have a clean test of the prediction that the clustering pattern is growing because the universe is gravitationally unstable. The clustering pattern of relatively nearby galaxies can be studied in detail because it has become feasible to measure relative distances (through red shifts) wholesale (*17*). These data will strengthen the test of the proposed continuity of masses in galaxies and clusters of galaxies, but of course the main excitement will be the search for the largest possible things—holes or lines or sheets that might be fossils of something interesting.

References and Notes

1. For other reviews of the phenomena and their possible interpretation, see J. H. Oort, *Annu. Rev. Astron. Astrophys.* **21**, 373 (1983); P. R. Shapiro, in *Clusters and Groups of Galaxies*, F. Mardirossian, G. Giuricin, M. Mezzetti, Eds. (Reidel, Dordrecht, Netherlands 1984); M. M. Waldrop, *Science* **219**, 1050 (1983).
2. One megaparsec (Mpc) = 1000 kpc = 3.1×10^{24} cm ~ 3 million light years. For the most part the distances used here are calculated from red shifts by use of Hubble's law $c\Delta\lambda/\lambda = cz = v = Hr$, where c is the speed of light, $z = \lambda/\lambda$ is the ratio of the change in wavelength to the original wavelength, v is the velocity of recession, H is Hubble's constant, and r is the distance. I use $H = 75$ km sec^{-1} Mpc^{-1}.
3. P. J. E. Peebles, *The Large-Scale Structure of the Universe* (Princeton University Press, Princeton, N.J., 1980), section 2.
4. G. Lemaître, *Proc. Natl. Acad. Sci. U.S.A.* **20**, 12 (1934).
5. E. Hubble, *The Realm of the Nebulae* (Yale Univ. Press, New Haven, Conn., 1936), p. 81.
6. Ya. B. Zel'dovich, J. Einasto, S. F. Shandarin, *Nature (London)* **300**, 407 (1982); A. G. Doroshkevich *et al.*, *Comments Astrophys. Space Phys.* **9**, 265 (1982); J. Silk, A. S. Szalay, Ya. B. Zel'dovich, *Sci Am.* **249**, 72 (October 1983).
7. R. Giovanelli and M. P. Haynes, *Astron. J.* **87**, 1355 (1982).
8. P. ten Bruggencate, *Sternhaufen* (Springer, Berlin, 1927), p. 64.
9. J. E. Evans and E. W. Maunder, *Mon. Not. R. Astron. Soc.* **63**, 488 (1903).
10. W. G. Hoyt, *Lowell and Mars* (Univ. of Arizona Press, Tucson, 1976), chapter 10.
11. I should emphasize that McClure is not in the habit of collecting statistical freaks. He happened to notice this particularly striking case shortly before I asked him whether he knew of any good examples.
12. C. D. Shane and C. A. Wirtanen, *Publ. Lick Obs.* **22**, part 1 (1967).
13. M. Seldner, B. Siebers, E. J. Groth, P. J. E. Peebles, *Astron. J.* **82**, 249 (1977).
14. R. M. Soneira and P. J. E. Peebles, *ibid.* **83**, 845 (1978).
15. B. B. Mandelbrot, *The Fractal Geometry of Nature* (Freeman, San Francisco, 1982).
16. A positive effect is claimed in (*6*), but the test used there was questioned by A. Dekel, M. J. West, and S. J. Aarseth [*Astrophys. J.* **279**, 1 (1984)]. See also J. E. Moody, E. L. Turner, and J. R. Gott [*ibid.* **273**, 16 (1983)]. J. R. Kuhn and J. M. Uson [*Astrophys. J. Lett.* **263**, L47 (1982)] find a significant difference between model and data that could be due to filaments, but it is difficult to exclude the possibility that their statistic reveals, rather, a discrepancy in the character of the model on scales comparable to a cell size ~ 1 Mpc in the Lick catalog (*12*).
17. M. Davis, J. Huchra, D. W. Latham, J. Tonry, *Astrophys. J.* **253**, 423 (1982).
18. P. J. E. Peebles, *Astron. Astrophys.* **68**, 345 (1978); *Scripta Varia* **48**, 165 (1982). The motions of the galaxies within clusters in the model balance gravity on the average. Galaxy masses are adjusted to make a universe that is just closed (Einstein–de Sitter). That may make the mean velocity dispersion too high, but it does give reasonable velocities within rich clusters where the effect of the velocities is most noticeable. Three model maps were made with the consecutive string of random numbers starting from the first provided by the computer. The first and last maps are shown; the second looks messy because it happens to have a strong concentration of galaxies toward the origin.
19. B. Binggeli, *Astron. Astrophys.* **107**, 338 (1982). However, a more recent study by M. F. Struble and P. J. E. Peebles, *Astron. J.*, May 1985, using a larger sample fails to reproduce the large-scale alignment effect claimed by Binggeli.
20. A. Dressler, *Astrophys. J.* **226**, 55 (1978).
21. J. Binney, *Mon. Not. R. Astron. Soc.* **181**, 735 (1977); J. P. Ostriker and M. A. Hausman, *Astrophys. J. Lett.* **217**, L125 (1977).
22. N. A. Bahcall and R. M. Soneira, *Astrophys. J.* **277**, 27 (1984).
23. P. J. E. Peebles, *ibid.* **274**, 1 (1983); A. Dekel, *ibid.* **264**, 373 (1983).
24. J. E. Gunn, *Scripta Varia* **48**, 233 (1982).
25. A. Sandage, *Astron. J.* **88**, 1159 (1983); W. E. Harris, J. E. Hesser, B. Atwood, *Astrophys. J. Lett.* **268**, L111 (1983).
26. The red shift is $z = \lambda_o/\lambda_e - 1$, where the observed wavelength of a spectral feature is λ_o and the wavelength at the source is λ_e. In the conventional expanding cosmological model an object now at proper distance l was at distance $l/(1 + z)$ at red shift z.
27. M. J. Lebofsky, *Astrophys. J. Lett.* **245**, L59 (1981); J. B. Oke, in *Clusters and Groups of Galaxies*, F. Mardirossian, G. Giuricin, M. Mezzetti, Eds. (Reidel, Dordrecht, Netherlands, 1984).
28. R. B. Partridge and P. J. E. Peebles, *Astrophys. J.* **147**, 868 (1967).
29. J. P. Ostriker and J. Heisler, *ibid.* **278**, 1 (1984).
30. B. A. Peterson, A. Savage, D. L. Jauncey, A. E. Wright, *Astrophys. J. Lett.* **260**, L27 (1982).
31. G. de Vaucouleurs, in *Galaxies and the Universe*, A. Sandage, M. Sandage, J. Kristian, Eds. (Univ. of Chicago Press, Chicago, 1975), p. 557.
32. S. van den Bergh, *J. R. Astron. Soc. Canada* **62**, 1 (1968).
33. F. D. Kahn and L. Woltjer, *Astrophys. J.* **130**, 705 (1959); P. J. E. Peebles, *Physical Cosmology* (Princeton Univ. Press, Princeton, N.J., 1971), p. 81.

34. G. de Vaucouleurs, *Astron. J.* **63**, 253 (1958); R. B. Tully, *Astrophys. J.* **257**, 389 (1982).
35. M. Davis and P. J. E. Peebles, *Annu. Rev. Astron. Astrophys.* **21**, 109 (1983); A. Dressler, *Astrophys. J.*, in press.
36. A. Dressler, *Astrophys. J.* **236**, 351 (1980).
37. W. Forman and C. Jones, *Annu. Rev. Astron. Astrophys.* **20**, 547 (1982).
38. M. Davis, J. Tonry, J. Huchra, D. W. Latham, *Astrophys. J. Lett.* **238**, L113 (1980).
39. P. J. E. Peebles, in *Clusters and Groups of Galaxies*, F. Mardirossian, G. Giuricin, M. Mezzetti, Eds. (Reidel, Dordrecht, Netherlands, 1984).
40. A. L. Melott, *Mon. Not. R. Astron. Soc.* **202**, 595 (1983); C. S. Frenk, S. D. M. White, M. Davis, *Astrophys. J.* **271**, 417 (1983).
41. A. Dekel, in *Early Evolution of the Universe and its Present Structure*, G. O. Abell and G. Chincarini, Eds. (Reidel, Dordrecht, Netherlands, 1983), p. 249; A. Dekel and S. Aarseth, *Astrophys. J.*, in press.
42. S. M. Faber and R. E. Jackson, *Astrophys. J.* **204**, 668 (1976); J. Kormendy and G. Illingworth, *ibid.* **265**, 632 (1983).
43. V. C. Rubin, *Science* **220**, 1339 (1983).
44. J. P. Ostriker, *Scripta Varia* **48**, 473 (1982).
45. M. Davis and P. J. E. Peebles, *Astrophys. J.* **267**, 465 (1983).
46. J. P. Ostriker, P. J. E. Peebles, A. Yahil, *Astrophys. J. Lett.* **193**, L1 (1974); J. Einasto, A. Kaasik, E. Saar, *Nature (London)* **250**, 309 (1974).
47. A. Bosma, *Astron. J.* **86**, 1791 and 1825 (1981); D. Fabricant and P. Gorenstein, *Astrophys. J.* **267**, 535 (1983).
48. The circles refer to pairs of galaxies selected to be more or less isolated, as analyzed by P. J. E. Peebles (*Ann. N.Y. Acad. Sci.*, in press); the box to the Durham-Australia red-shift sample of A. J. Bean, G. Efstathiou, R. S. Ellis, B. A. Peterson, and T. Shanks [*Mon. Not. R. Astron. Soc.* **205**, 605 (1983)]; and the other symbols to the Center for Astrophysics red-shift sample (*17*). All symbols except the box are based on fits of the observed distribution of red-shift differences Δv to the exponential model $dP \propto \exp -2^{1/2}|\Delta v|/w$, which seems to give an adequate description of the data. The height of the box is the range of results obtained for a variety of assumed distributions of Δv, the width to the range of separations well sampled in the Durham-Australia data. For the relation between $w(r_p)$ and mass, see section 76 in (*3*).
49. A modification of Newtonian dynamics for the purpose of producing flat rotation curves without invoking dark mass is proposed by M. Milgrom [*Astrophys. J.* **270**, 365 (1983)].
50. R. P. Kirshner, A. Oemler, P. L. Schechter, S. A. Shectman, *Astrophys. J. Lett.* **248**, L57 (1981).
51. E. E. Salpeter, in *Early Evolution of the Universe and Its Present Structure*, G. O. Abell and G. Chincarini, Eds. (Reidel, Dordrecht, Netherlands, 1983), p. 211.
52. A. Vilenkin and Q. Shafi, *Phys. Rev. Lett.* **51**, 1716 (1983); N. Turok and D. N. Schramm, *Nature (London)*, in press.
53. R. Cowsik and J. McClelland, *Astrophys. J.* **180**, 7 (1973); S. Tremaine and J. E. Gunn, *Phys. Rev. Lett.* **42**, 407 (1979).
54. P. J. E. Peebles, *Astrophys. J.* **258**, 415 (1982); A. A. Klypin and S. F. Shandarin, *Mon. Not. R. Astron. Soc.* **204**, 891 (1983); S. D. M. White, C. S. Frenk, M. Davis, *Astrophys. J. Lett.* **274**, L1 (1983).
55. P. Hut and S. D. M. White, *Phys. Rev. Lett.*, in press.
56. H. Pagels, *Ann. N.Y. Acad. Sci.*, in press.
57. P. J. E. Peebles and J. T. Yu, *Astrophys. J.* **162**, 851 (1970); E. R. Harrison, *Phys. Rev.* **D1**, 2728 (1970); Ya. B. Zel'dovich, *Mon. Not. R. Astron. Soc.* **160**, 1P (1972). This spectrum of density fluctuations may be predicted in the inflationary cosmology, as discussed by A. H. Guth and S.-Y. Pi [*Phys. Rev. Lett.* **49**, 1110 (1982)] and S. W. Hawking [*Phys. Lett.* **115B**, 295 (1982)].
58. I am grateful to J. Uson for preparing Fig. 7. The initial power spectrum is $P \propto$ wave number. The final spectrum is computed in (*54*) and J. R. Bond and A. S. Szalay, *Astrophys. J.* **274**, 443 (1983). The figure assumes the density is a random Gaussian process. To resolve the spikes, we increased their width by a factor of 10.
59. P. J. E. Peebles, *Astrophys. J.* **277**, 470 (1984).
60. _____ and R. H. Dicke, *ibid.* **154**, 891 (1968).
61. For a discussion of globular clusters and another view of their possible significance see S. van den Bergh [in *Globular Clusters*, D. Hanes and B. Madore, Eds. (Cambridge Univ. Press, Cambridge, 1980), p. 175, and in *Clusters and Groups of Galaxies*, F. Mardirossian, G. Giuricin, M. Mezzetti, Eds. (Reidel, Dordrecht, Netherlands, 1984)].
62. The distribution function in single particle phase space in this model is $f \propto \exp - (\epsilon + h^2\beta^2/2\alpha^2)/v_o^2$, where ϵ and h are the energy and angular momentum per unit mass; $\alpha = v_o/(4\pi G\rho_o)^{1/2}$ is a length scale; v_o is the constant velocity dispersion in the radial direction; ρ_o is the central mass density in stars, and β is a dimensionless constant that characterizes the velocity anisotropy. This form for f was introduced by A. S. Eddington [*Mon. Not. R. Astron. Soc.* **75**, 366 (1915)]. The model assumes there is a homogeneous dark halo with mass density $\rho_x = 0.002\ \rho_o$, and uses $\beta = 0.3$, because these choices give reasonable-looking results.
63. J. E. Gunn and R. F. Griffin, *Astron. J.* **84**, 752 (1979); P. O. Seitzer, thesis, University of Virginia (1983).
64. K. M. Cudworth, *Astron. J.* **84**, 1312 (1979).
65. G. Efstathiou and J. W. Eastwood, *Mon. Not. R. Astron. Soc.* **194**, 503 (1981); R. H. Miller, *Astrophys. J.* **270**, 390 (1983); A. L. Melott, J. Einasto, E. Saar, I. Suisalu, A. A. Klypin, S. F. Shandarin, *Phys. Rev. Lett.* **51**, 935 (1983).

66. I have benefited from discussions with S. van den Bergh, A. Dekel, R. H. Dicke, A. Dressler, O. Gingerich, A. Melott, J. Oort, J. P. Ostriker, P. Shapiro, M. Struble, J. Uson, and D. T. Wilkinson. This research was supported in part by the National Science Foundation.

This material originally appeared in *Science* **224**, 29 June 1984.

20. Jets in Extragalactic Radio Sources

David S. De Young

Nearly three decades have passed since the discovery that many galaxies and quasi-stellar objects (QSO's) emit radiation at radio wavelengths. This radio emission is seen in two general locations. The first is a very small, bright region located in the very center of the galaxy or QSO, and the second is a double-structured region lying on either side of the associated galaxy or QSO. The size of this second class of radio-emitting volumes may range from dimensions comparable to a galaxy to regions of order a hundred times the size of the parent galaxy. Very often both kinds of structures are observed, and Fig. 1 shows a typical example of such a radio source as seen at radio wavelengths. The bright central cores or compact sources are quite small by astronomical standards, of order 1 parsec (3×10^{18} cm), whereas the large extended sources may be up to a million times this size. The total power radiated at radio wavelengths by the extended sources ranges from 10^{42} to 10^{46} erg sec^{-1}; the higher value is 100 times the energy radiated by all the stars in a typical galaxy. The enormous size and power of the extended radio sources make them unique: they are the largest single objects in the universe, and they contain very large amounts of energy in a very specialized form.

In all cases the radio emission from both compact and extended sources is almost certainly due to incoherent synchrotron radiation from an ensemble of relativistic electrons embedded in a magnetic field. Most of the radio frequency power lies in the range $\sim 10^2$ to 10^4 MHz, and the spectrum of the radiation is that due to an electron population with a power law distribution in energy (*1*).

The size and power of the extended sources lead immediately to speculation about how such large amounts of energy are produced, particularly in the specialized form of relativistic electrons and magnetic field, and how this energy can be transported over such great distances. Motivated by morphology similar to that shown in Fig. 1, investigators proposed in an early model (*2*) for the extended radio sources that two oppositely directed clouds of relativistic electrons, magnetic field, and cooler "thermal" plasma be ejected from the parent object by a gigantic explosion. The characteristic shape of the radio source would then arise from the ram pressure exerted by the surrounding medium on the cloud as it passes through at supersonic speeds.

However, subsequent observations have shown this kind of model to be inadequate in one important respect. For many extended radio sources, if the relativistic particles were ejected from the parent object in a single event, they would lose energy too rapidly to be ei-

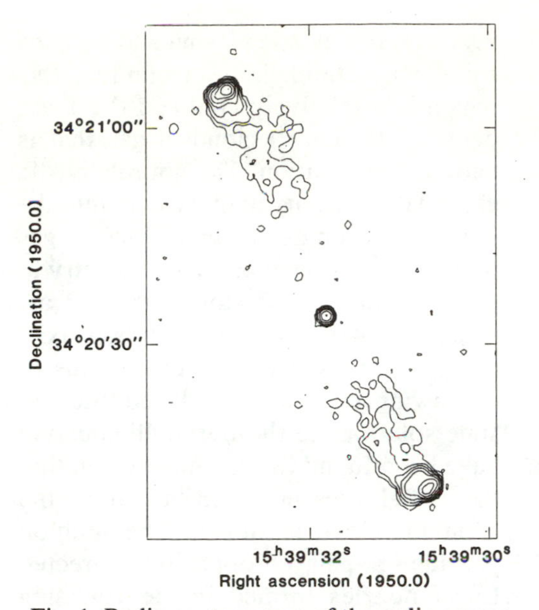

Fig. 1. Radio contour map of the radio source 1539 + 343 made at a wavelength of 6 cm with the Very Large Array. The radio source is associated with a distant galaxy with a red shift of 0.402. The overall extent of the source is 486 kiloparsecs, or about 20 times the size of the parent galaxy which is centered on the bright core of emission between the two lobes. The luminosity of this object at radio wavelengths is approximately 10^{43} erg sec^{-1}. [Reprinted courtesy of J. Burns, *Astrophysical Journal* 283, 515 (1984), published by the University of Chicago, © American Astronomical Society]

ther seen where they are seen or to be radiating at the highest frequencies observed. This means that all viable radio source models must provide for a continual replenishment of energy in some form. This "lifetime" problem can be seen to arise in a simple way. The frequency at which an electron radiates in this process is proportional to the square of the electron energy; thus, as the energy is radiated away, the frequency of the radiation decreases. Conversely, if the initial electron energy is known, the length of time it can radiate at a given frequency can be found. Although the initial electron energies are not known, minimum electron energies can be estimated, and these can be coupled with observations at very high frequencies to search for the frequency above which the radiation should abruptly decrease. Such cutoffs have not been found even at millimeter wavelengths (*3*), and this in turn implies net lifetimes of 5×10^9 years for nearby sources.

Given this limit on the lifetime, an overall radio source size of 1 megaparsec then requires that the energetic electrons must have traveled from the nucleus to the extreme edges of the source with average speeds of one quarter the speed of light, *c*. Sources up to 4 megaparsecs in size have been observed (*4*). Although only about 5 percent of all radio sources are 1 megaparsec or more in size (*5*), these exceptional sources present a severe lifetime problem.

More extreme cases are found in two nearby radio sources. The first is the giant elliptical galaxy NGC 4486, which lies at the center of the Virgo cluster of galaxies. This well-known radio galaxy exhibits a jetlike structure emanating from the nucleus, which contains "knots" or condensations of radio emission. The key feature is that these knots are also seen at optical wavelengths where they exhibit a blue, featureless continuum spectrum that is strongly polarized. The optical appearance strongly suggests that it is also due to synchrotron radiation; if so, the extremely high frequency of the radiation implies a very short radiative lifetime. This estimate can be made very quantitative if we use the best values for the magnetic field and particle energy recently obtained (*6*). These give a lifetime of $\approx$100 years. The radio and optical structure extends at least 1 kiloparsec from the nucleus (neglecting projection effects); hence the electrons would have to travel from the nucleus at speeds in excess of 30 times the speed of light in order to arrive at the end of the jet within their radiative lifetime.

This impossible situation is also found in the nearest radio galaxy, NGC 5128, where jetlike structure is seen in both radio and x-ray wavelengths (*7*). If the x-ray emission is electron synchrotron, then again lifetimes of ~100 years result, and these are less than one-tenth the travel time of light from the nucleus.

Thus these observations, together with other morphological properties of radio sources to be discussed below, have led to an inescapable conclusion: that the electron energies, if not the energetic electrons themselves, must be replenished throughout a considerable fraction of the radio source volume on a more or less continuous basis.

Jet Models

Motivated by the observations which demand in situ replenishment of the electron energy in radio sources, the basic idea underlying jet models is that there be a continuous, collimated outflow of energy from the nucleus to the radio source. An early version of this idea (*8*) proposed energy transport in the form of a highly collimated beam of electromagnetic radiation. This model failed to produce the observed polarization pattern of the radio emission and also suffered from difficulties in explaining how the beam was produced. A modification of this model was developed (*9*) and has become the "standard" beam or jet model. This model begins with the assumption that deep within the nucleus of the parent galaxy or QSO some (unspecified) process produces a steady amount of very hot, relativistic plasma and perhaps some magnetic field. Surrounding this region is a relatively cool ($\sim 10^8$ K) and dense ($\sim 10^3$ cm^{-3}) cloud of gas that is flattened by rotation. The hot relativistic gas inflates a cavity in this cloud until the pressure becomes sufficient for the gas to escape. Given rotational symmetry in the confining cloud, the path of least resistance lies along the rotation axis, and two oppositely directed streams of outflowing gas will be produced. Steady-state solutions to the Bernoulli equation have been found for this flow, given that the initial pressure distribution in the confining cloud is known. The solution provides a pair of oppositely directed Laval nozzles formed in the confining cloud, and, as the relativistic gas moves through these nozzles, its thermal energy is transformed into highly collimated, relativistic outflow with a very low internal temperature. Typical parameters for this "central engine" are a nozzle radius of ~10 parsecs and a nozzle length of ~200 parsecs, with a power output of $\sim 10^{45}$ erg sec^{-1}.

The subsequent propagation of the jets in this model has been developed in a more qualitative manner. It is assumed that the relativistic motion continues outward until it is stopped by interaction with the surrounding intergalactic or circumgalactic gas. This interaction occurs through a shock pair, the first being a bow shock propagating ahead of the jet termination and the second a shock internal to the jet through which the material in the jet is decelerated. This shock pair configuration is a well-known hydrodynamic phenomenon, and it is important to note that the end point of the jet propagates outward more slowly than the flow of material in the jet. The internal shock provides a means of converting the directed relativistic motion to near random relativistic motion of the particles. This is a necessary feature because the observations are not consistent with large-scale ordered motion of the relativistic electrons.

This basic jet model has several immediate advantages. It directly solves the lifetime problems described above. Moreover, a cold beam of particles moving at relativistic speeds is an extremely efficient way to transport energy. Thus the model places the minimum energy production requirements upon the central energy source in the nucleus. The model provides in a natural way for the high-intensity "hot spots" often observed near the leading edge of double radio sources; these would arise at the shock front where the flow becomes thermalized. After this thermalization, the relativistic electrons flow back around the jet as seen in a frame comoving with the jet terminus, and this "cocoon" of radiating particles can provide the bridge of emission that is often seen extending from the end of the source back toward the parent object. Last but far from least, the model provides in a very natural and fundamental way a means to produce the basic double structure seen in almost all extended radio sources.

Are Large-Scale Jets Relativistic?

Although the relativistic jet model has many virtues, recent observations provide evidence that it may require significant modification. Figure 2 shows the radio emission from a member of an important class of radio sources known as "head-tail" radio galaxies. These radio galaxies occur in clusters of galaxies, and it is believed that the radio morphology results from the continuous outflow of energetic particles being swept back by the ram pressure of the hot intracluster gas as the galaxy moves through this gas.

The radio map in Fig. 2 exhibits properties of great interest when one is considering relativistic flow models. The radio emission consists of several bright knots close to the nucleus that are linked by curved bridges of emission, and the problem that objects such as this pose for relativistic flow is fundamental. The model is based upon a beam of cold, collimated particles that emit synchrotron radiation when randomized by a shock at the end of the beam. Bright knots near the nucleus must imply, in the context of this model, disruption of the orderly flow long before it reaches the outer regions of the radio source. If this disruption occurs, how does the flow proceed beyond it in order to supply energy in the more removed portions of the radio source? Further, one must ask how a series of such knots can occur. Bright knots in linear arrays are not uncommon in extended radio sources (*5*), and it has been suggested that they arise from the sweeping up of interstellar clouds by the relativistic jet (*10*). In order for this process not to disrupt the jet, the cloud size must be much less than the cross section of the jet, and it is not clear whether the model parameters can be adjusted to accomplish this in all cases. Moreover, a continuous supply of clouds is required, yet very little interstellar gas is seen in the class of elliptical galaxies that produces extended radio sources.

Figure 3 shows another example of a very common type of radio source, a basic double with a bright jet on one side. Several elements are of importance, and these features are quite common (*5*); the brightest portion of the jet is

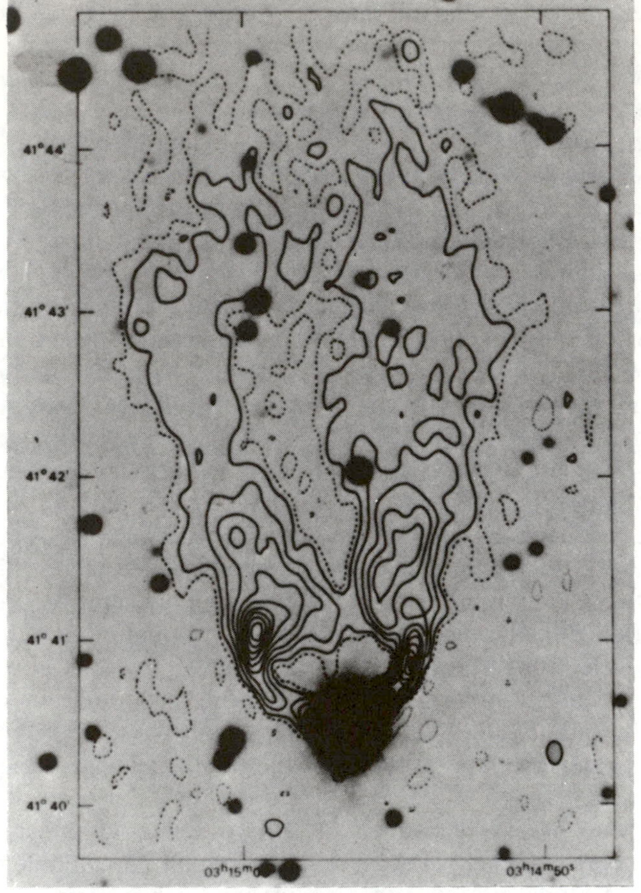

Fig. 2. Head-tail radio source associated with the galaxy NGC 1265 located in the Perseus cluster of galaxies. The radio contour map at 6-cm wavelength is superposed upon an optical photograph of the region. [Reproduced, with permission, from the *Annual Review of Astron. & Astrophys.*, Vol. 14 © 1976 by Annual Reviews Inc.]

nearest the nucleus, the jet is bent, and there is no bright spot at the end of the jet. None of these features would be expected in the standard relativistic flow picture. In order to bend a relativistic jet, some obstacle must be encountered, and such an encounter will inevitably produce shock waves. Thus it would be expected that a bright spot would occur whenever a bend is encountered, contrary to what is observed. There exist many radio sources with two oppositely directed jets similar to the one-sided jet shown in Fig. 3, and the general morphological characteristics of all these sources, which argue against relativistic flow, are high brightness features near the nucleus, generally strong radio emission all along the jet, lack of bright spots at the end of the jet, and bending or curvature of the jet. Indeed, the outermost features of the sources shown in

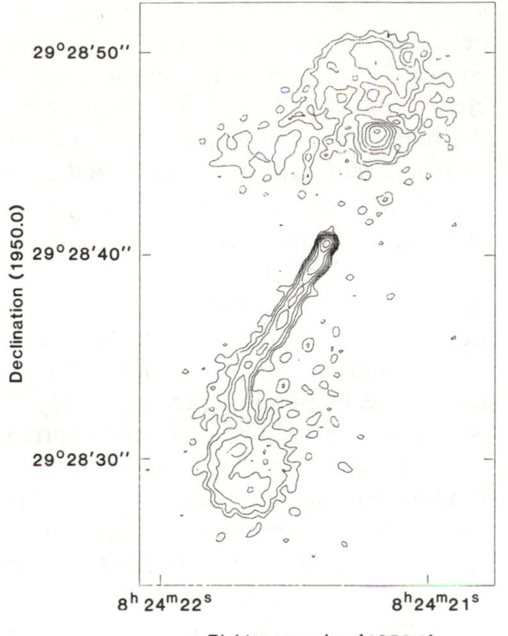

Fig. 3. Radio contour map of a double source with a large-scale jet at 6-cm wavelength. The parent object is a galaxy with a red shift of 0.458, and the galaxy is centered at the bright end of the jet. The power being emitted by the jet is 7×10^{42} erg sec^{-1}, and the overall radio luminosity of the source is 2×10^{43} erg sec^{-1}. [Reprinted courtesy of J. Burns, *Astrophysical Journal* 283, 515 (1984), published by the University of Chicago, © American Astronomical Society]

Figs. 2 and 3 and in several other sources, are suggestive of gentle, subsonic drift motion—far removed from what would be expected from the supersonically propagating termination point of a relativistic jet.

However, there do exist extended radio sources that are composed of two well-defined bright regions with sharp boundaries and no visible bridges of emission extending back to the nucleus. Such sources are still consistent with the relativistic jet model, and a question arises as to the existence of two distinct classes of radio source. That is, do there exist truly "quiet," efficient beams of energy propagating away from the central engine, or is there some "noise" or inefficiency in the energy propagation for all radio sources? The resolution of this question will require long observing periods with the most sensitive radio telescopes. Preliminary results (*11*) indicate that all sources may have observable radio jets. The principal conclusion is that for a great number of extended radio sources the jets are not cold, quiet, and relativistically propagating, and that the basic beam model does not apply to them. Current indications are that it may not apply to any extended radio source.

The Consequences of Large-Scale Slow Jets

The principal drawback to abandoning relativistic flow is the loss of efficiency and the consequent increase in the energy output required by the parent galaxy or QSO. This problem will be addressed subsequently, but it is important to first point out that slow jets can have significant virtues. The jet shown in Fig. 3 displays synchrotron losses all along its extent, and it is now clear that for the most lengthy examples of such jets there must be continuous reenergization of the electron population within the jet itself. This is a result of the lifetime arguments discussed above, and it shows that there cannot be efficient transport of energy to the end of the jet but that rather there must be a continual diverting of jet energy to the electron population all along its length.

For inefficient jets this energy input can arise in a natural way. The morphology of radio sources shows that there must exist a tenuous intergalactic or circumgalactic gas around the radio sources, otherwise the particles and field would be freely expanding into a vacuum and would not appear as they do (*12*). If the jet is moving supersonically but nonrelativistically, it is well known that for the high Reynolds number flow involved the jet interface with a surrounding gaseous medium is transacted via a turbulent boundary layer. (Although very little is known either experimentally or theoretically about relativistic jets, it would seem that some sort of turbulent boundary layer must exist for these jets as well. It may be that a quiet, efficient beam can exist only as a naïve theoretical construction.) Suppose the jet were to contain not only relativistic particles and magnetic field but significant amounts of cooler ($\sim 10^6$ to 10^8 K) gas as well. If most of the kinetic energy is carried by this cool gas, then in a turbulent boundary layer there exist numerous stochastic acceleration processes (*13*) which can serve to reenergize the electron population along the jet as required by observations.

Slow turbulent jets dominated by nonrelativistic gas may explain another recently observed feature associated with radio sources. Astronomers are finding an increasing number of radio sources which are coexistent with regions emitting spectral lines at optical wavelengths that indicate the presence of "heavy" elements such as sulfur and nitrogen (*14*). These lines are often seen well beyond the optical image of the parent galaxy, that is, outside the stellar population; yet such elements have clearly been through nucleosynthesis for their formation. An immediate question is how they came to be so far removed from their presumed site of production in the interstellar medium. A possible answer can be found in their association with radio sources, all of which fall in the "inefficient" class. The outflow that powers the radio source originates in the center of the parent galaxy or QSO, and, if a turbulent boundary layer develops, it must entrain gas from the interstellar medium or QSO envelope as it passes through. This entrained gas, which is rich in heavy elements, is shock-heated by the leading edge of the jet and may cool to the line-emitting temperatures of $\sim 10^4$ K only after it has been transported far beyond the stellar population that produced it. A more detailed consideration of this entrainment process (*15*) shows that it may also account for the presence of very young stars located along the edges of a radio jet in a nearby radio galaxy.

A final virtue of slow inefficient jets is that they may provide a means of supplying the required magnetic field for the extended radio sources. In the relativistic jet model and in radio source models in general, more attention has been paid to relativistic particle production than to magnetic field generation. Yet in the minimum energy configuration the energy density is comparable for both, and there is a need to understand how the required magnetic field comes about. If the energy transport from the nucleus is dominated by slow, relatively cool jets that are turbulent and that also produce turbulence throughout the radio source volume, then amplification of a dynamically unimportant seed field is possible. A fully nonlinear but geometrically restricted calculation of seed-field amplification by such turbulent processes shows that such a mechanism may produce the required field strengths (*16*).

The energy requirements for slow inefficient jets are simply calculated. Most stochastic particle-acceleration mechanisms proceed with rather low efficiency, of order 1 to 10 percent, although some shock-acceleration processes can proceed with efficiencies of order unity (*17*). Magnetic field amplification proceeds with efficiencies of 1 to 10 percent. If ϵ is the net efficiency of converting bulk kinetic energy into relativistic particles and field ($\epsilon \leq 1$) and if L is the radio luminosity of the source, then energy balance requires $L = \epsilon \rho v^3 A$, where ρ is the gas density in the jet, v is the jet velocity, and A is the jet cross-sectional area. A nominal luminosity could be 10^{44} erg sec^{-1}, and a nominal jet radius can be set at 1 kiloparsec. A truly slow beam could have $v \approx 10^8$ cm sec^{-1}, which implies the particle density in the jet would have to be 1 cm^{-3} or greater, depending on ϵ. This is the worst case for mass loss, but over a source life of 10^8 years (which produces a 100-kiloparsec jet for $v = 10^8$ cm sec^{-1}) the total mass lost is 10^9 to 10^{10} solar masses. This number is large but perhaps not impossible, given that the total masses of the giant elliptical galaxies that produce radio sources can be as high as 10^{12} to 10^{13} solar masses. Because of the v^3 dependence in the energy balance equation, raising v a small amount above 10^8 cm sec^{-1} clearly reduces the mass loss and yet keeps the jet velocity definitely nonrelativistic. For $\epsilon = 0.01$ the total energy required over 10^8 years is $\approx 10^{61}$ erg, or $\sim 10^7 M_\odot c^2$, where $M_\odot$ is the solar mass. This is a large but not totally intimidating number; given that nuclear burning proceeds at an efficiency of ~ 1 percent, this is the energy generated by only 10^9 solar type stars over their lifetimes.

Compact Radio Jets and the Nuclear Engine

In addition to the large-scale jets discussed above, very small jets have been observed emanating from deep in the central portions of radio galaxies and QSO's (*18*). Typical sizes are of the order of 1 parsec. In almost all cases the jet is one-sided and appears to emanate from a bright core at the center of the parent optical object. In those cases where there is a large-scale one-sided jet also present, the compact jet lies on the same side as the large-scale feature. With a few important exceptions, the compact jets appear linear, but this may be due in part to the lack of detail currently available in radio maps of these tiny objects. Compact jets are very much less common than large-scale jets, and, because of their small population and small size, we have much less detailed information about them than about the large-scale jets.

The general inference is that energetic outflow is being observed just as it emerges from the central engine. The fact that compact one-sided jets are seen in galaxies and QSO's that also have large-scale double structure has been interpreted as evidence that the compact jets are moving at relativistic velocities (*18*). This interpretation rests upon the assumption that there are really two compact jets, each of which powers one side of the large double structure, and that the compact jets are directed almost exactly along the line of sight between the radio source and the earthbound observer. If this is the case and if the flow is relativistic, then the Doppler effect will enhance the observed radiation from the oncoming jet and suppress the radiation from the receding jet to the point where only one jet is seen.

This interpretation does encounter some difficulties, however. The fact that the jets are visible implies that they are inefficient, although this per se does not argue against relativistic motion. Perhaps the most serious objection arises from the following argument. The compact jet is presumed to be the innermost portion of the outflow which powers the rest of the extended radio source, including those that possess large-scale jets on the same side as the compact jet. I have argued above that there is convincing evidence that the large-scale jets are not in relativistic motion. If this is the case and if the compact jet both is relativistic and contains enough energy to power the entire radio source, then in the region between the end of the compact jet and the onset of the large-scale jet the flow must be decelerated from relativistic to nonrelativistic speeds. This explanation requires that most of the kinetic energy used to power the entire radio source must be dissipated in this small region. Even if a mechanism could be found to do this, earlier arguments about shock-wave randomization of relativistic flow would lead one to expect vast amounts of radiation to be emitted from such a dissipation region. Observations do not confirm this, for the interval between the compact jet and the large-scale jet is often characterized by decreased radio emission and is never seen as the brightest part of the entire source, as might be expected. In addition, there is no evidence for enhanced emission from this region at x-ray, optical, or infrared wavelengths. A final difficulty with relativistic motion for the compact jets arises from very recent observations made by the Multi Element Radio Link Interferometer (MERLIN) (*19*). These reveal compact radio jets that are strongly curved or bent through 180° or more, and the configurations cannot be reasonably explained by the use of precessing jets plus projection effects. If these jets are relativisitic and fuel-extended sources, their momentum density and rigidity must be very high; thus it is difficult to see how they can be bent without being destroyed.

There remains one phenomenon associated with compact jets that is very strong evidence for relativistic flow, and it is illustrated in Fig. 4. What is observed is a change in the jet structure over a period of time. The number of cases where this is seen is very small (<10), but this may be due to selection effects as very few compact jets have been monitored for such structural changes. These jets are almost exclusively seen in association with QSO's, and, if the red shifts of the QSO's are taken as an indicator of their distance, the progress of the outwardly moving knots of radio emission implies velocities in excess of c. In the case of 3C 273 (Fig. 4), this apparent velocity is about $\sim 10c$. More recent observations of the radio source 3C 345 show apparent velocities that are even greater (*20*). Relativistic motion of a compact jet nearly along the line of sight can account in a simple way for this phenomenon, due to Doppler effects (*18*).

However, although relativistic beaming can account for this phenomenon, it is not a unique solution. Several other explanations have been put forward (*21*), although in their present form all of them have difficulties of varying severity. These alternative ideas involve a variety of mechanisms, such as phase effects from reflected photons instead of actual

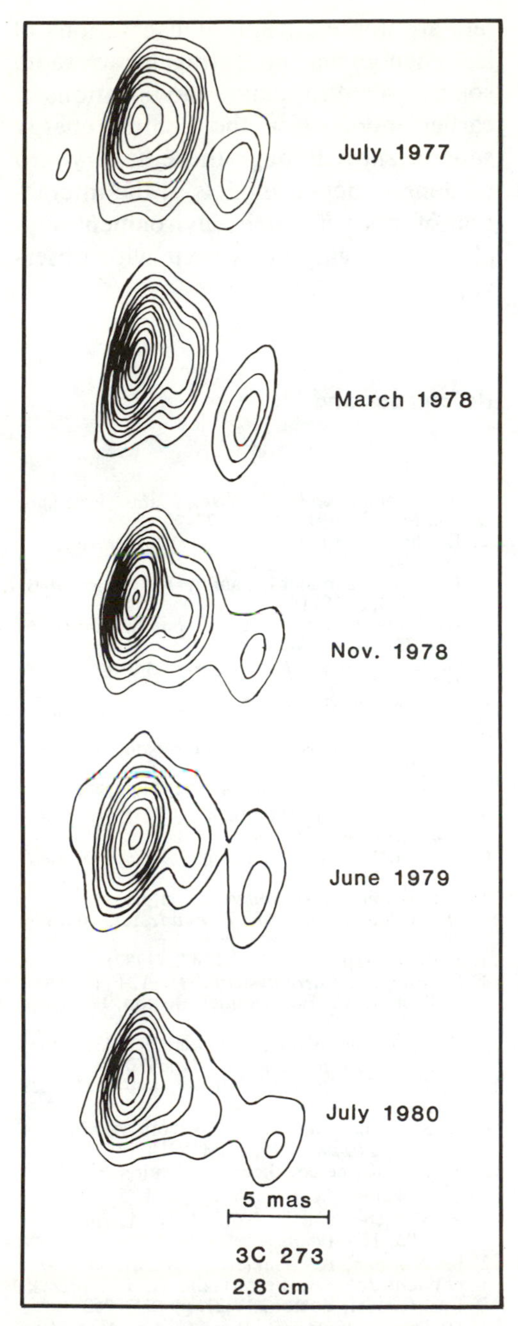

Fig. 4. Radio contour maps of the compact jet in the radio source 3C 273 made at different epochs; mas, milliarcseconds. [Reproduced, with permission, from the *Annual Review of Astron. & Astrophys.*, Vol. 19 © 1981 by Annual Reviews Inc.]

motion of material, gravitational lenses and screens, or noncosmological interpretations of the red shift. Although relativistic beaming may in some sense be the most "natural" way to explain the superluminal effect, the problems of energy deposition described above for compact jets in general still remains. There is in addition a serious consistency question. If relativistic motion explains the superluminal effect, there must be a large number of unseen relativistic compact jets that are not pointed nearly along the line of sight. The energy carried by these must be deposited at some point when the jet is stopped, yet this large population of objects is not clearly identified. In addition, because compact jets associated with large-scale jets clearly have an energy deposition problem if they are relativistic, one is placed in the awkward position of explaining why some compact jets are relativistic and others are not.

One of the questions of great interest about extended radio sources is the detailed nature of the central energy source. Although nonrelativistic motion of the jets will require modifications of the earlier models, no detailed calculations have yet been performed for a model that produces "slow" jets. It may well be that a very similar model based on nonrelativistic gas could be made to work, or a very different acceleration process may be required. Although compact jets are very close to this central energy source, most observations of them have provided little additional constraints as to its nature. Very high resolution maps such as the recent MERLIN results may change this in the near future.

One area where theoretical calculations have provided some extremely interesting results is in the modeling of the large-scale jets (*22*). Nonlinear numerical simulations of the hydrodynamics of supersonic but nonrelativistic jets have provided a wealth of detail to compare with observations. A key feature of these calculations is the role of the interaction of the jet with its environment. This interaction not only produces a bow shock but vortex sheets, surface instabilities, mixing layers, and internal shocks. These phenomena are precisely those required to drive the acceleration and entrainment processes needed to explain the observations.

Conclusion

The existence of jets or beams of outflowing material which provide a continuous supply of energy to the giant extragalactic radio sources is required by radiative lifetime arguments and is consistent with observational data. These same data also indicate that energy propagation in the jets is not highly efficient, and the morphology of large-scale jets argues strongly for nonrelativistic motion of the jets. If these jets are only mildly supersonic and dominated by nonrelativistic plasma, then the turbulent boundary layer that surrounds the jet may provide the required particle acceleration, magnetic-field amplification, and transport of metal-enriched gas far from the interstellar medium. Slow jets increase the total energy requirements for the source, but not beyond the limits of credibility.

Compact jets are also inefficient, and arguments can be made for their relativistic or nonrelativistic motion. At present there is no clear-cut resolution of this issue, but it is fair to say that a consistent picture could be constructed wherein all jets are nonrelativistic. Observations of jet phenomena in extragalactic radio sources seem to require modifications of earlier models for the central energy source, and, although these have yet to be done in detail, models of the interaction of jets with their environment supply encouraging agreement with observations.

References and Notes

1. A. Pacholczyk, *Radio Astrophysics* (Freeman, San Francisco, 1970), pp. 77–99.
2. D. De Young and W. I. Axford, *Nature (London)* **216**, 129 (1967).
3. K. Kellermann and I. Pauliny-Toth, *Astrophys. Lett.* **8**, 153 (1971).
4. A. Willis, R. Strom, A. Wilson, *Nature (London)* **250**, 625 (1974).
5. G. Miley, *Annu. Rev. Astron. Astrophys.* **18**, 165 (1980).
6. F. Owen, P. Hardee, R. Bignell, *Astrophys. J.* **239**, L11 (1980).
7. E. Schreier, J. Burns, E. Feigelson, *ibid.* **251**, 523 (1981).
8. M. Rees, *Nature (London)* **229**, 312 (1971).
9. R. Blandford and M. Rees, *Mon. Not. R. Astron. Soc.* **169**, 395 (1974).
10. R. Blandford and A. Konigl, *Astrophys. Lett.* **20**, 15 (1979).
11. J. Burns *et al.*, *Astrophys. J.*, in press.
12. D. De Young, *Ann. N.Y. Acad. Sci.* **302**, 669 (1977).
13. J. Eilek, *Astrophys. J.* **254**, 472 (1982).
14. G. Miley, in *Astrophysical Jets*, A. Ferrari and A. Pacholczyk, Eds. (Reidel, Boston, 1983), pp. 99–112.
15. D. De Young, *Nature (London)* **293**, 43 (1981).
16. ———, *Astrophys. J.* **241**, 81 (1980).
17. R. Blandford and J. Ostriker, *ibid.* **221**, L29 (1978).
18. K. Kellermann and I. Pauliny-Toth, *Annu. Rev. Astron. Astrophys.* **19**, 373 (1981).
19. T. Muxlow, personal communication.
20. S. Unwin *et al.*, *Astrophys. J.* **271**, 536 (1983).
21. A. Marscher and J. Scott, *Publ. Astron. Soc. Pac.* **92**, 127 (1980).
22. M. Norman, K. Winkler, L. Smarr, in *Astrophysical Jets*, A. Ferrari and A. Pacholczyk, Eds. (Reidel, Boston, 1983), pp. 227–252.
23. D. De Young, *Annu. Rev. Astron. Astrophys.* **14**, 447 (1976).
24. The National Optical Astronomy Observatories are operated by the Association of Universities for Research in Astronomy, Inc., under contract with the National Science Foundation.

This material originally appeared in *Science* **225**, 17 August 1984.

21. The Quest for the Origin of the Elements

William A. Fowler

All life on Earth depends on the energy in sunlight, which comes initially from the nuclear fusion of hydrogen into helium deep in the solar interior. But the sun did not produce the chemical elements which are found in the earth and in our bodies. The first two elements and their stable isotopes, hydrogen and helium, emerged from the first few minutes of the early high-temperature, high-density stage of the expanding Universe, the so-called big bang. A small amount of lithium, the third element in the periodic table, was also produced in the big bang, but the remainder of the lithium and all of beryllium, element 4, and boron, element 5, are thought to have been produced by the spallation of still heavier elements in the interstellar medium by the cosmic radiation. These elements are in general very rare, in keeping with this explanation of their origin (*1*).

Where did the heavier elements originate? The generally accepted answer is that all of the elements from carbon, element 6, up to long-lived radioactive uranium, element 92, were produced by nuclear processes in the interior of stars in our own Galaxy. The stars which synthesized the heavy elements in the solar system were born, aged, and eventually ejected the ashes of their nuclear fires into the interstellar medium over the lifetime of the Galaxy before the solar system itself formed 4½ billion years ago.

The lifetime of the Galaxy is thought to be more than 10 billion but less than 20 billion years. The ejection of the nuclear ashes or newly formed elements took place by slow mass loss during the old age of the stars, called the giant stage of stellar evolution, or during the relatively frequent outbursts which astronomers call novae, or during the final spectacular stellar explosions called supernovae.

In any case the sun, the earth, and all the other planets in the solar system condensed under gravitational and rotational forces from a gaseous solar nebula in the interstellar medium consisting of big-bang hydrogen and helium mixed with the heavier elements synthesized in earlier generations of galactic stars.

This idea can be generalized to successive generations of stars in the Galaxy, with the result that the heavy-element content of the interstellar medium and the stars which form from it increases with time. The oldest stars in the galactic

---

This chapter is based on the lecture delivered by William A. Fowler in Stockholm on 8 December 1983 when he received the Nobel Prize in Physics, which he shared with S. Chandrasekhar.

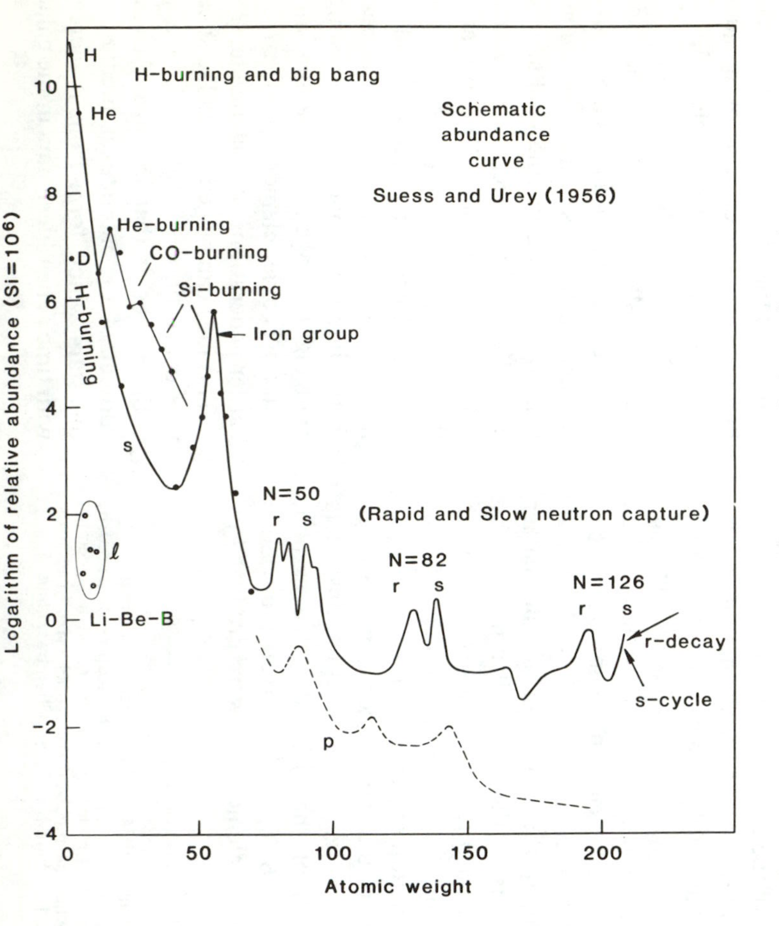

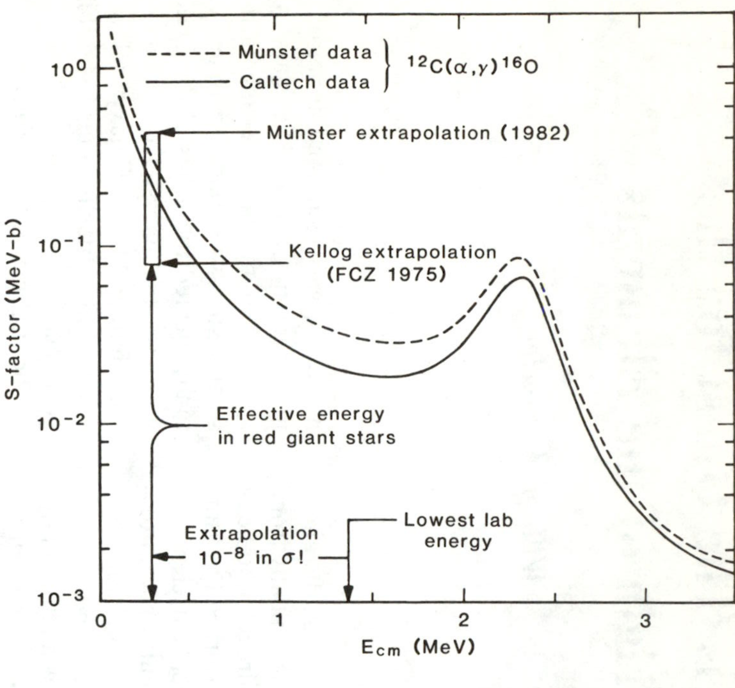

Fig. 1 (left). Schematic curve of atomic abundances relative to Si = 10^6 versus atomic weight for the sun and similar main-sequence stars. Fig. 2 (right). Cross-section factor *S* (in MeV-barns) versus center-of-momentum energy (in MeV) for $^{12}C(\alpha,\gamma)^{16}O$. The dashed and solid curves are the theoretical extrapolations of the Münster and Kellogg Caltech data by Langanke and Koonin (*36*). [Reprinted with modification and with permission of Kettner et al., *Z. Phys. A* 308, 73 (1982)]

halo—those we believe to have formed first—are found to have heavy-element abundances less than 1 percent of that of the solar system. The oldest stars in the galactic disk have approximately 10 percent. Only the less massive stars among those first formed can have survived to the present as so-called Population II stars. Their small concentration of heavy elements may have been produced in a still earlier but more massive generation of stars, Population III, which rapidly exhausted their nuclear fuels and survived for only a very short time. Stars formed in the disk of the Galaxy over its lifetime are referred to as Population I stars.

We speak of this element building as nucleosynthesis in stars. It can be generalized to other galaxies, such as our twin, the Andromeda Nebula, and so this mechanism can be said to be universal. We refer to the basic physics of energy generation and element synthesis in stars as nuclear astrophysics.

The field of nuclear astrophysics has two main goals. First, it attempts to understand energy generation in the sun and other stars at all stages of stellar evolution. Energy generation by nuclear processes requires the transmutation of nuclei into new nuclei with lower mass. The small decrease in mass is multiplied by the velocity of light squared, as Einstein taught us, and a relatively large amount of energy is released.

Thus the first goal is closely related to the second goal, that of understanding the nuclear processes which produced under various astrophysical circumstances the relative abundances of the elements and their isotopes in nature. Figure 1 shows a curve of atomic abundances as a function of atomic weight. The data for this curve were first systematized from a plethora of terrestrial, meteoritic, solar, and stellar data by Suess and Urey (*2*) and have been periodically updated by Cameron (*3*). Major contributions to the experimental measurement of atomic transition rates needed to determine solar and stellar abundances have been made by Whaling (*4*).

The curve in Fig. 2 is frequently referred to as "universal" or "cosmic," but it primarily represents relative atomic abundances in the solar system and in main-sequence stars similar in mass and age to the sun. In current usage the curve is described as "solar." How this curve serves as a goal can be simply put. Calculations of atomic abundances produced under astronomical circumstances at various postulated stellar sites are almost invariably reduced to ratios relative to "solar" abundances.

Early Research on Element Synthesis

George Gamow and his collaborators Alpher and Herman (*5*) attempted to synthesize all of the elements during the big bang by using a nonequilibrium theory involving neutron (n) capture with gamma-ray (γ) emission and electron (e) beta-decay by successively heavier nuclei. The synthesis proceeded in steps of one mass unit, since the neutron has approximately unit mass on the mass scale used in all the physical sciences. As they emphasized, this theory meets grave difficulties beyond mass 4 (^{4}He) because no stable nuclei exist at atomic mass 5 and 8. Enrico Fermi and Anthony Turkevich attempted valiantly to bridge these "mass gaps" without success. Seventeen years later Wagoner *et al.* (*6*), armed with nuclear reaction data accumulated over the intervening years, succeeded only in producing ^{7}Li at a mass fraction of at most 10^{-8} compared to hydrogen plus helium for acceptable model universes. All heavier elements totaled less than 10^{-11} by mass. Wagoner *et al.* (*6*) did succeed in producing ^{2}D, ^{3}He, ^{4}He, and ^{7}Li in amounts in reasonable agreement with observations at the time. More recent observations and calculations are frequently used to place constraints on models of the expanding universe, and in general favor open models in which the expansion continues indefinitely unless there exists an abundance of so-called "missing mass."

It was in connection with the gap at mass 5 that the W. K. Kellogg Radiation Laboratory first became involved, albeit unwittingly, in astrophysical and cosmological phenomena. At the laboratory, in 1939, Staub and Stephens (*7*) detected resonance scattering by ^{4}He of neutrons with orbital angular momentum equal to one in units of $\hbar$ (Planck's constant divided by 2π) and energy somewhat less than 1 MeV. This confirmed previous reaction studies by Williams *et al.* (*8*) and showed that the ground state of ^{5}He is unstable—as fast as ^{5}He is made it disintegrates. The same was later shown to be true for ^{5}Li, the other candidate nucleus at mass 5. The Pauli exclusion principle dictates for fermions that the third neutron in ^{5}He must have at least unit angular momentum, and not zero as permitted for the first two neutrons with antiparallel spins. In classical terminology, the attractive nuclear force cannot match the outward centrifugal force. Still later, at the laboratory, Tollestrup *et al.* (*9*) confirmed, with improved precision, the first quantitative proof by Hemmendinger (*10*) that the ground state of ^{8}Be is unstable. The Pauli exclusion principle is again at work in the instability of ^{8}Be. As fast as ^{8}Be is made it disintegrates into two ^{4}He nuclei. The latter may be bosons, but they consist of fermions. The mass gaps at 5 and 8 spelled the doom of Gamow's hopes that all nuclear species could be produced in the big bang one unit of mass at a time.

The eventual commitment of the Kellogg Radiation Laboratory to nuclear astrophysics came about in 1939, when Bethe (*11*) brought forward the operation of the CN cycle as one mode of the fusion of hydrogen into helium in stars (since oxygen has been found to be involved the cycle is now known as the CNO cycle). Charles Lauritsen, his son Thomas Lauritsen, and I were measuring the cross sections of the proton bombardment of the isotopes of carbon and nitrogen which constitute the CN cycle. Bethe's paper (*11*) told us that we were studying in the laboratory processes which are occurring in the sun and other stars. It made a lasting impression on us. World War II intervened, but in 1946, on returning the laboratory to nuclear experimental research, Lauritsen decided to continue in low-energy, classical nuclear physics with emphasis on the study of nuclear reactions thought to take place in stars. In this he was strongly supported by Ira Bowen, a Caltech professor of physics who had just been appointed director at the Mount Wilson Observatory, by Lee DuBridge, the new president of Caltech, by Carl Anderson, Nobel Prize winner in 1936, and by Jesse Greenstein, newly appointed to establish research in astronomy at Caltech.

Although Bethe (*11*) and others still earlier had previously discussed energy generation by nuclear processes in stars, the grand concept of nucleosynthesis in stars came from Fred Hoyle (*12*). In two classic papers the basic ideas of the concept were presented within the framework of stellar structure and evolution with the use of the then known nuclear data.

Again the Kellogg Laboratory played a role. Before his second paper Hoyle was puzzled by the slow rate of formation of ^{12}C nuclei from the fusion of three alpha particles (α's) of ^{4}He nuclei in red giant stars. Hoyle was puzzled because work with Schwarzschild (*13, 14*) had convinced him that helium burning through $3\alpha \rightarrow {}^{12}C$ should commence in red giants just above 10^8 K rather than at 2×10^8 K as required by the reaction rate calculation of Salpeter (*15*). Salpeter made his calculation while a visitor at Kellogg in 1951 and used the Kellogg value (*9*) for the energy of ^{8}Be in excess of two ^{4}He to determine the resonant rate for the process ($2\alpha \leftrightarrow {}^{8}Be$) which takes into account both the formation and decay of ^{8}Be. However, in calculating the next step, $^{8}Be + \alpha \rightarrow {}^{12}C + \gamma$, Salpeter had treated the radiative fusion as nonresonant.

Hoyle realized that this step would be speeded up by many orders of magnitude, thus reducing the temperatures for its onset, if there existed an excited state of ^{12}C with energy 0.3 MeV in excess of $^{8}Be + \alpha$ at rest and with the angular momentum and parity (0^+, 1^-, 2^+, 3^-, . . .) dictated by the selection rules for these quantities. He came to Kellogg early in 1953 and questioned the staff about the possible existence of his proposed excited state. Whaling and his visiting associates and graduate students (*16*) decided to go into the laboratory and search for the state, using the $^{14}N(d,\alpha)^{12}C$ reaction (d = deuteron). They found it almost exactly where Hoyle had predicted. It is now known to be at 7.654 MeV excitation in ^{12}C, or 0.2875 MeV above $^{8}Be + \alpha$ and 0.3794 MeV above 3α. Cook *et al.* (*17*) then produced the state in the decay of radioactive ^{12}B and showed that it could break up into 3α and thus by reciprocity could be formed from 3α. They argued that the spin and parity of the state must be 0^+, as is now known to be the case.

The $3\alpha \rightarrow {}^{12}C$ fusion in red giants jumps the mass gaps at 5 and 8. This process could never occur under big-bang conditions. By the time ^{4}He was produced in the early expanding Universe the density and temperature were too low for helium fusion to carbon. In contrast, in red giants, after hydrogen conversion to helium during the main-sequence stage, gravitational contraction of the helium core raises the density and temperature to values where helium fusion is ignited. Hoyle and Whaling showed that conditions in red giant stars are just right.

Fusion processes can be referred to as nuclear burning in the same way we speak of chemical burning. Helium burning in red giants succeeds hydrogen burning in main-sequence stars and is in turn succeeded by carbon, neon, oxygen, and silicon burning to reach the elements near iron and somewhat beyond in the periodic table. With these nuclei of intermediate mass as seeds, subsequent processes similar to Gamow's involving neutron capture at a slow rate (s-process) or at a rapid rate (r-process) continued the synthesis beyond ^{209}Bi, the last stable nucleus, up through short-lived radioactive nuclei to long-lived ^{232}Th, ^{235}U, and ^{238}U, the parents of the natural radioactive series. This last requires the r-process, which actually builds beyond mass 238 to radioactive nuclei which decay back to ^{232}Th, ^{235}U, and ^{238}U rapidly at the cessation of the process.

The need for two neutron-capture processes was explained by Suess and Urey (*2*). With the adroit use of relative isotopic abundances for elements with several isotopes, they demonstrated the existence of the double peaks (r and s) in Fig. 1. It was immediately clear that these peaks were associated with neutron shell filling at the magic neutron numbers $N = 50$, 82, and 126 in the nuclear shell model of Hans Jensen and Maria Goeppert-Mayer.

In the s-process the nuclei involved have low capture cross sections at shell closure and thus large abundances to maintain the s-process flow. In the r-process it is the proton-deficient radioactive progenitors of the stable nuclei which are involved. Low capture cross sections and small beta-decay rates at shell closure lead to large abundances, but after subsequent radioactive decay these large abundances appear at lower values of the mass number A than for the s-process, since the atomic (proton) number Z is less and thus $A = N + Z$ is less. In Hoyle's classic papers (*12*) stellar nucleosynthesis up to the iron-group elements was attained by charged-particle reactions. Rapidly rising Coulomb barriers for charged particles curtailed further synthesis. Suess and Urey (*2*) made the breakthrough which led to the extension of nucleosynthesis in stars by neutrons unhindered by Coulomb barriers all the way to ^{238}U.

The complete run of the synthesis of the elements in stars was incorporated into a paper by Burbidge, Burbidge, Fowler, and Hoyle (*18*), commonly referred to as B^2FH, and was independently developed by Cameron (*19*). Notable contributions to the astronomical aspects of the problem were made by Greenstein (*20*) and many other observational astronomers. Since that time nuclear astrophysics has developed into a full-fledged scientific activity including the exciting discoveries of isotopic anomalies in meteorites. The following account will highlight some of the experimental and theoretical research under way at present or carried out in the past few years. It cannot include details of the nucleosynthesis of all the elements and their isotopes, which would involve discussing all the reactions producing a given nuclear species and all those which destroy it. The reader will find some of these details for ^{12}C, ^{16}O, and ^{55}Mn.

It is noted that the measured cross sections for the reactions are customarily very small at the lowest energies of measurement, for $^{12}C(\alpha,\gamma)^{16}O$ even less than 1 nanobarn (10^{-33} cm^2) near 1.4 MeV. This means that experimental nuclear astrophysics requires accelerators with large currents of well-focused, monoenergic ion beams, thin targets of high purity and stability, detectors of high sensitivity and energy resolution, and experimentalists with great tolerance for long running times and with patience in accumulating data of statistical significance. Classical Rutherfordian measurements of nuclear cross sections are required, and the results are essential to our understanding of the physics of nuclei.

A comment on nuclear reaction notation is necessary at this point. In the reaction $^{12}C(\alpha,\gamma)^{16}O$ discussed above, ^{12}C is the laboratory target nucleus, α is the incident nucleus (^4He) accelerated in the laboratory, γ is the photon produced and detected in the laboratory, and ^{16}O is the residual nucleus, which can also be detected if it is desirable to do so. If ^{12}C is accelerated against a gas target of ^4He and the ^{16}O products are detected but not the gamma rays, the laboratory notation is $^4He(^{12}C,^{16}O)\gamma$. The stars could not care less. In stars all the particles are moving and only the center-of-momentum system is important for the determination of stellar reaction rates. In $^{12}C(\alpha,n)^{15}O(e^+\nu)^{15}N$, n is the neutron promptly produced and detected and e^+ is the beta-decay positron, which can also be detected.

Stellar Reaction Rates from Laboratory Cross Sections

Thermonuclear reaction rates in stars are customarily expressed as $N_A<\sigma\sigma>$ reactions per second per mole per cubic centimeter, where $N_A = 6.022 \times 10^{23}$ mole^{-1} is Avogadro's number and $<\sigma v>$ is the Maxwell-Boltzmann average as a function of temperature for the product of the reaction cross section σ (in square centimeters) and the relative velocity of the reactants v (in centimeters per second). Multiplication of $<\sigma v>$ by the product of the number densities per cubic centimeter for the two reactants is necessary to obtain rates in reactions per second per cubic centimeter. The N_A is incorporated so that mass fractions can be used (*21*).

Early work on the evaluation of stellar reaction rates from experimental laboratory cross sections was reviewed in Bethe's Nobel lecture (*11*). Fowler *et al.* (*21*) have provided detailed numerical and analytical procedures for converting laboratory cross sections into stellar reaction rates. It is first necessry to accommodate the rapid variation of the nuclear cross sections at the low energies which are relevant in astrophysical circumstances. For neutron-induced reactions this is accomplished by defining a cross-section *S*-factor equal to the cross section (σ) multiplied by the interaction velocity (v) in order to eliminate the usual v^{-1} singularity in the cross section at low velocities and low energies.

For reactions induced by charged particles such as protons, alpha particles, or ^{12}C, ^{16}O, . . . nuclei it is necessary to accommodate the decrease by many orders of magnitude from the lowest laboratory measurements to the energies of astrophysical relevance. This is done in the way first suggested by Salpeter (*22*) and emphasized in Bethe's Nobel lecture (*11*). A relatively slowly varying *S*-factor can be defined by eliminating the rapidly varying term in the Gamow penetration factor governing transmission through the Coulomb barrier. Stellar reaction rates can be calculated as an average over the Maxwell-Boltzmann distribution for both nonresonant and resonant cross sections. Expressions for reaction rates derived from theoretical statistical model calculations are given by Woosley *et al.* (*23*).

Although the extrapolation from cross sections measured at the lowest laboratory energies to cross sections at the effective stellar energy can often involve a decrease by many orders of magnitude, the elimination of the Gamow penetration factor, which causes this decrease, is based on the solution of the Schroedinger equation for the Coulomb wave functions, in which one can have considerable confidence. The main uncertainty lies in the variation of the *S*-factor with energy, which depends primarily on the value chosen for the radius at which formation of a compound nucleus between two interacting nuclei or nucleons occurs (*18*). The radii used by my colleages and me in recent work are given in (*23*). There is, in addition, an uncertainty in the ''intrinsic nuclear factor'' used in the definition of σ, and this can be eliminated only by recourse to laboratory experiments. The effect of a resonance in the compound nucleus just below or just above the threshold for a given reaction can often be ascertained by determining the properties of the resonance in other reactions which are easier to study.

Hydrogen Burning in Main-Sequence Stars and the Solar Neutrino Problem

Hydrogen burning in main-sequence stars has contributed only about 20 percent more helium than resulted from the big bang. However, hydrogen burning in the sun has posed a problem for many years. In 1938 Bethe and Critchfield (*24*) proposed the proton-proton or pp chain as one mechanism for hydrogen burning in stars. From many cross-section measurements at Kellogg and elsewhere it is now known that the pp chain, rather than the CNO cycle, is the mechanism which operates in the sun.

Our knowledge of the weak nuclear interaction tells us that two neutrinos are emitted when four hydrogen nuclei are converted into helium nuclei. Detailed elaboration of the pp chain (*25, 26*) showed that a small fraction of these neutrinos, those from the decay of ^{7}Be and ^{8}B, should be energetic enough to be detectable through interaction with the nucleus ^{37}Cl to form radioactive ^{37}Ar (*27, 28*). Davis (*29*) and his collaborators have attempted for more than 25 years to detect these energetic neutrinos by employing a 380,000-liter tank of perchloroethylene ($C_2{}^{35}Cl_3{}^{37}Cl_1$) located 1 mile deep in the Homestake Gold Mine in Lead, South Dakota. They find only about one-third of the number expected on the basis of the model-dependent calculations of Bahcall *et al.* (*30*).

Something is wrong—either the standard solar models or the relevant nuclear cross sections are in error, or the electron-type neutrinos produced in the sun are converted in part into undetectable muon neutrinos or taon neutrinos on the way from the sun to the earth. There indeed have been controversies about the nuclear cross sections which have been for the most part resolved (*31, 32*).

It is generally agreed that the next step is to build a detector which will detect the much larger flux of low-energy neutrinos from the sun through neutrino absorption by the nucleus ^{71}Ga to form radioactive ^{71}Ge. This will require 20 to 50 tons of gallium at a cost (for 20 tons) of approximately $10 million. An international effort is being made to obtain the necessary amount of gallium. We are back at square one in nuclear astrophysics. Until the solar neutrino problem is resolved, the basic principles underlying nuclear processes in stars are in question.

The CNO cycle operates at the higher temperatures which occur during hydrogen burning in main-sequence stars somewhat more massive than the sun, because the CNO cycle reaction rates rise more rapidly with temperature than do those of the pp chain. The cycle is important because ^{13}C, ^{14}N, ^{15}N, ^{17}O, and ^{18}O are produced from ^{12}C and ^{16}O as seeds. The role of these nuclei as sources of neutrons during helium burning is discussed next.

Synthesis of ^{13}C and ^{16}O and Neutron Production in Helium Burning

The human body is 65 percent oxygen and 18 percent carbon by mass; with the remainder mostly hydrogen. Oxygen (0.85 percent) and carbon (0.39 percent) are the most abundant elements heavier than helium in the sun and similar main-sequence stars. It is little wonder that the determination of the ratio $^{12}C/^{16}O$ produced in helium burning is a problem of paramount importance in nuclear astrophysics. This ratio depends in a fairly complicated manner on the density, temperature, and duration of helium burning, but it depends directly on the relative rates of the $3\alpha \rightarrow {}^{12}C$ process and the ${}^{12}C(\alpha,\gamma){}^{16}O$ process. If $3\alpha \rightarrow {}^{12}C$ is much faster than ${}^{12}C(\alpha,\gamma){}^{16}O$ then no ${}^{16}O$ is produced in helium burning. If the reverse is true then no ${}^{12}C$ is produced. For the most part the subsequent reaction ${}^{16}O(\alpha,\gamma){}^{20}Ne$ is slow enough to be neglected.

There is general agreement about the rate of the $3\alpha \rightarrow {}^{12}C$ process (*33*). However there is a lively controversy about the laboratory cross section for ${}^{12}C(\alpha,\gamma){}^{16}O$ and about its theoretical extrapolation to the low energies at which the reaction effectively operates. Data obtained at Caltech in the Kellogg Laboratory (*34*) and data obtained at Münster (*35*) have been compared with theoretical calculations, and the theoretical curves which yield the best fit to the two sets of data are from Langanke and Koonin (*36*). The crux of the situation is made evident in Fig. 2, which shows the extrapolations of the Caltech and Münster cross-section factors from the lowest measured laboratory energies ($\sim$1.4 MeV) to the effective energy $\sim$0.3 MeV, at $T = 1.8 \times 10^8$ K, a representative temperature for helium burning in red giant stars. The extrapolation in cross sections covers a range of 10^{-8}. The rise in the cross-section factor is due to the contributions of two bound states in the ${}^{16}O$ nucleus just below the ${}^{12}C(\alpha,\gamma){}^{16}O$ threshold. These contributions plus differences in the laboratory data produce the current uncertainty in the extrapolated S-factor.

With so much riding on the outcome it will come as no surprise that both laboratories are engaged in extending their measurements to lower energies with higher precision. In the discussion of quasi-static silicon burning that follows it will be found that the abundances produced in that stage of nucleosynthesis depend in part on the ratio of ${}^{12}C$ to ${}^{16}O$ produced in helium burning and that the different extrapolations shown in Fig. 2 are in the range crucial to the ultimate outcome of silicon burning. These remarks do not apply to explosive nucleosynthesis.

Recently, the ratio of ${}^{12}C$ to ${}^{16}O$ produced under the special conditions of helium flashes during the asymptotic giant phase of stellar evolution has become of great interest. The hot blue star PG 1159-035 has been found to undergo nonradial pulsations with periods of 460 and 540 seconds and others not yet accurately determined. The star is obviously highly evolved, having lost its hydrogen atmosphere, leaving only a hot dwarf of about 0.6 solar mass (0.6 $M_\odot$) behind. Theoretical analysis of the pulsations (*37*) requires substantial amounts of oxygen in the pulsation-driving regions where the oxygen is alternately ionized and deionized. Carbon is completely ionized in these regions and only diminishes the pulsation amplitude. It is not yet clear that sufficient oxygen is produced in helium flashes, which certainly involve $3\alpha \rightarrow {}^{12}C$ but may not last long enough for ${}^{12}C(\alpha,\gamma){}^{16}O$ to be involved. The problem may not lie in the nuclear reaction rates, according to (*37*). We shall see.

In what follows in this chapter β^+-decay is designated by $(e^+\nu)$, since both a positron (e^+) and a neutrino (ν) are emitted. Similarly, β^--decay will be designated by $(e^-\bar{\nu})$, since both an electron (e^-) and an antineutrino $(\bar{\nu})$ are emitted. Electron capture (often indicated by ϵ) will be designated by $e^-,\nu)$, the comma indicating that an electron is captured and a neutrino emitted. The notations $(e^+,\bar{\nu})$,(ν,e^-), and $(\bar{\nu},e^+)$ should now be obvious.

Neutrons are produced when helium burning occurs under circumstances in

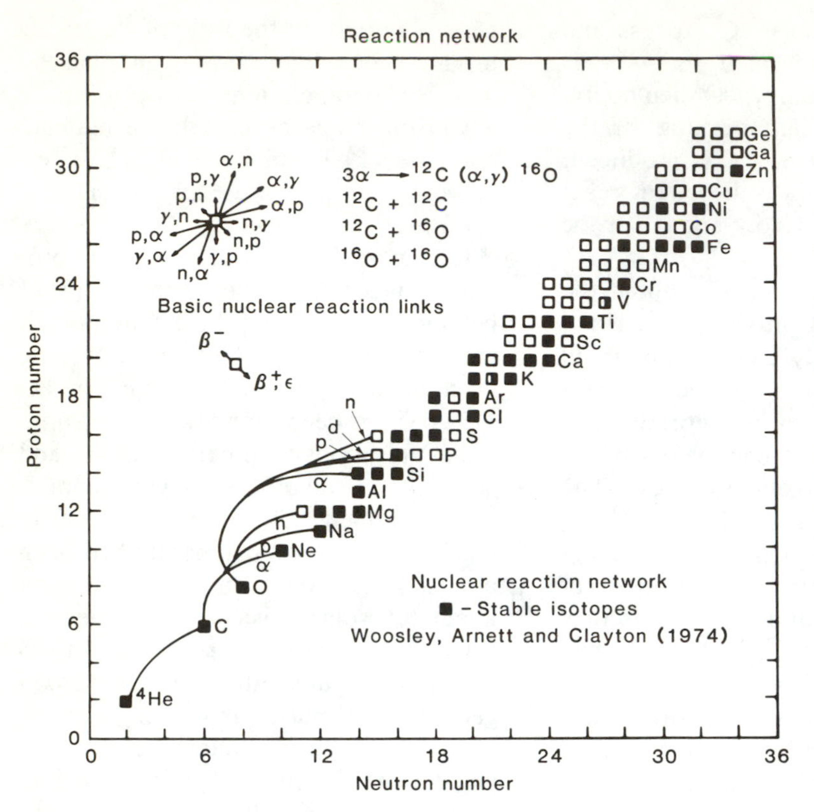

Fig. 3. Reaction network for nucleosynthesis involving the most important stable and radioactive nuclei with $N = 2$ to 34 and $Z = 2$ to 32. Stable nuclei are indicated by solid squares. Radioactive nuclei are indicated by open squares.

which the CNO cycle has been operative in the previous hydrogen burning. When the cycle does not go to completion, copious quantities of ^{13}C are produced in the sequence of reactions $^{12}C(p,\gamma)^{13}Ne(e^{+}\nu)^{13}C$. In subsequent helium burning, neutrons are produced by $^{13}C(\alpha,n)^{16}O$. When the cycle goes to completion the main product (>95 percent) is ^{14}N. In subsequent helium burning, ^{18}O and ^{22}Ne are produced in the sequence of reactions $^{14}N(\alpha,\gamma)^{18}F(e^{+}\nu)^{18}O(\alpha,\gamma)^{22}Ne$, and these nuclei in turn produce neutrons through $^{18}O(\alpha,n)$ $^{21}Ne(\alpha,n)^{24}Mg$ and $^{22}Ne(\alpha,n)^{25}Mg$. However, the astrophysical circumstances and sites under which the neutrons produce heavy elements through the s-process and the r-process are, even today, matters of some controversy and much study.

Carbon, Neon, Oxygen, and Silicon Burning

The advanced burning processes discussed in this section involve the network of reactions shown in Fig. 3. Because of the high temperature at which this network can operate, radioactive nuclei can live long enough to serve as live reaction targets. Excited states of even the stable nuclei are populated and also serve as targets. Determination of the nuclear cross sections and stellar rates of the approximately 1000 reactions in the network has involved and will continue to involve extensive experimental and theoretical effort.

The following discussion applies to stars massive enough that electron degeneracy does not set in as nuclear evo-

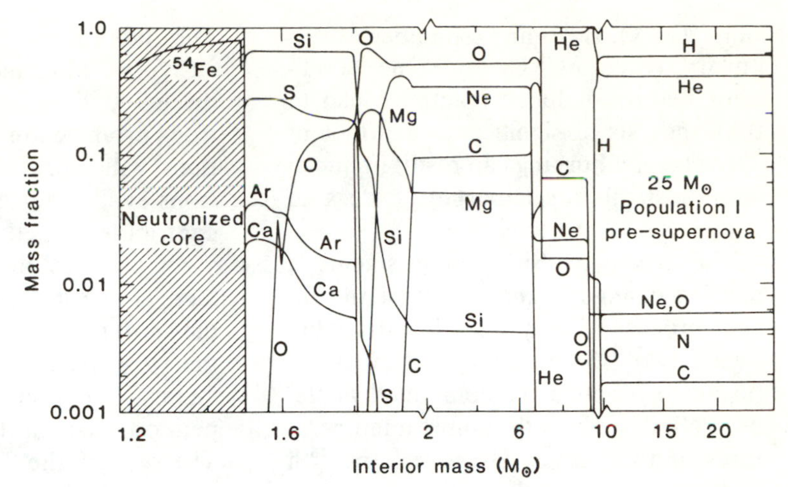

Fig. 4. Pre-supernova abundances by mass fraction versus increasing interior mass in solar masses, $M_\odot$, measured from zero at the stellar center to 25 $M_\odot$, the total stellar mass from Woosley and Weaver (*38*) for a Population I star.

lution proceeds through these various burning stages. In less massive stars electron degeneracy can terminate further nuclear evolution at certain stages, with catastrophic results leading to the disruption of the stellar system.

Figure 4, taken from Woosley and Weaver (*38*), applies to the pre-supernova stage of a young (Population I) star of 25 $M_\odot$ and shows the result of various nuclear burnings in the following mass zones: (i) $>10\ M_\odot$, convective envelope with the results of some CNO burning; (ii) 7 to 10 $M_\odot$, products mainly of H burning; (iii) 6.5 to 7 $M_\odot$, products of He burning; (iv) 1.9 to 6.5 $M_\odot$, products of C burning; (v) 1.8 to 1.9 $M_\odot$, products of Ne burning; (vi) 1.5 to 1.8 $M_\odot$, products of O burning; and (vii) $<1.5\ M_\odot$, products of Si burning in the partially neutronized core (these are not shown in detail but consist mainly of ^{54}Fe and other neutron-rich nuclei such as ^{48}Ca, ^{50}Ti, ^{54}Cr, and ^{58}Fe). Figure 4 has been evaluated shortly after photodisintegration has initiated core collapse, which will then be sustained by the reduction of the outward pressure through electron capture and the resulting almost complete neutronization of the core.

It must be realized that the various burning stages took place originally over the central region of the star and finally in a shell surrounding that region. Subsequent stages modify the innner part of the previous burning stage. For example, in the star of Fig. 4, C burning took place in the central 6.5 $M_\odot$ of the star but the inner 1.9 $M_\odot$ were modified by subsequent Ne, O, and Si burning.

Helium burning produces a stellar core consisting mainly of ^{12}C and ^{16}O. After core contraction the temperature and density rise until carbon burning through ^{12}C + ^{12}C fusion is ignited. The main product of carbon burning is ^{20}Ne, produced primarily in the ^{12}C(^{12}C,α)^{20}Ne reaction. When the ^{12}C is exhausted, ^{20}Ne and ^{16}O are the major remaining constituents. As the temperature rises from further gravitational contraction, the ^{20}Ne is destroyed by photodisintegration, ^{20}Ne(γ,α)^{16}O. This occurs because the alpha particle in ^{20}Ne is bound to its closed-shell partner, ^{16}O, by only 4.731 MeV (for comparison, the binding of an alpha particle in ^{16}O is 7.162 MeV).

The next stage is oxygen burning through ^{16}O + ^{16}O fusion. The main product is ^{28}Si through the primary reaction $^{16}O(^{16}O,\alpha)^{28}Si$ and a number of secondary reactions. Under some conditions neutron-induced reactions lead to the synthesis of significant quantities of ^{30}Si. Oxygen burning can result in nuclei with a small but important excess of neutrons over protons.

The onset of Si burning signals a marked change in the nature of the fusion process. The Coulomb barrier between two ^{28}Si nuclei is too great for fusion to produce the compound nucleus, ^{56}Ni, directly at the ambient temperatures and densities. However, the ^{28}Si and subsequent products are easily photodisintegrated by (γ,α), (γ,n), and (γ,p) reactions. As Si burning proceeds, more and more ^{28}Si is reduced to nucleons and alpha particles, which can be captured by the remaining ^{28}Si nuclei to build through the network in Fig. 3 up to the iron-group nuclei. The main product in explosive Si burning is ^{56}Ni, which transforms eventually through two beta-decays to ^{56}Fe.

In quasi-static Si burning the weak interactions are fast enough that ^{54}Fe, with two more neutrons than protons, is the main product. Because of the important role played by alpha particles (α) and the inexorable trend to equilibrium (e) involving nuclei near mass 56, which have the largest binding energies per nucleon of all nuclear species, B^2FH (*18*) broke down what is now called Si burning into their α-process and e-process. Quasi-equilibrium calculations for Si burning were made by Bodansky *et al.* (*39*), who cite the original papers in which the basic ideas of Si burning were developed. Modern computers permit detailed network flow calculations to be made (*38*, *40*).

The extensive laboratory studies of Si burning reactions are reviewed in (*33*). The laboratory excitation curves for $^{54}Cr(p,n)^{54}Mn$ and $^{54}Cr(p,\gamma)^{55}Mn$ are discussed here as examples. The neutrons produced in the first of these reactions will increase the number of neutrons available in Si burning but will not contribute directly to the synthesis of ^{55}Mn as does the second reaction. In fact, above its threshold at 2.158 MeV the (p,n) reaction competes strongly with the (p,γ) reaction, which is of primary interest, and produces a pronounced competition cusp in the excitation curve. The rate of the $^{54}Cr(p,\gamma)^{55}Mn$ reaction at very high temperatures will be an order of magnitude lower because of the cusp than would otherwise be the case.

The element manganese has only one isotope, ^{55}Mn, and in nature is produced in quasi-static Si burning, probably through the $^{54}C(p,\gamma)^{55}Mn$ reaction. The reactions $^{51}V(\alpha,\gamma)^{55}Mn$ and $^{52}V(\alpha,n)^{55}Mn$ may also contribute, especially in explosive Si burning. The overall synthesis of ^{55}Mn involves a balance in its production and destruction. In quasi-static Si burning the reactions which destroy ^{55}Mn are probably $^{55}Mn(p,\gamma)^{56}Fe$ and $^{55}Mn(p,n)^{55}Fe$, which are discussed and illustrated in (*41*). $^{55}Mn(\alpha,\gamma)^{59}Co$, $^{55}Mn\ (\alpha,p)^{58}Fe$, and $^{55}Mn(\alpha,n)^{58}Co$ may also destroy some ^{55}Mn in explosive Si burning. Calculations of the overall synthesis of ^{55}Mn yield values that are in fairly close agreement with the abundance of this nucleus in the solar system. Unfortunately, the same cannot be said about many other nuclei.

Laboratory measurements on Si burning reactions have covered only about 20 percent of the reactions in the network of Fig. 3 involving stable nuclei as targets. Direct measurements on short-lived radioactive nuclei and the excited states of all nuclei are impossible at present, although the production of radioactive ion beams, pioneered by Richard Boyd and Haight *et al.* (*42*), hold great promise for the future.

In any case, it has been clear for some time that experimental results on Si burning reactions must be systematized and supplemented by comprehensive theory. Fortunately, theoretical average cross sections will suffice in many cases, because the stellar reaction rates integrate the cross sections over the Maxwell-Boltzmann distribution. For most Si burning reactions resonances in the cross section are closely spaced and even overlapping, and the integration covers a wide enough range of energies that the detailed structure in the cross sections is averaged out. The statistical model of nuclear reactions developed by Hauser and Feshbach (*43*), which yields average cross sections, is ideal for the purpose. Accordingly, Holmes *et al.* (*44*) undertook the task of developing a global, parametrized Hauser-Feshbach theory and computer program for use in nuclear astrophysics (*23*). The free parameters are the radius, depth, and compensating reflection factor of the blackbody, square-well equivalent of the Woods-Saxon potential characteristic of the interaction between n, p, and α with nuclei having $Z \geq 8$. Two free parameters must also be incorporated to adjust the intensity of electric and magnetic dipole transitions for gamma radiation. Weak interaction rates must also be specified, and ways of doing this will be discussed later.

It is well known that the free parameters can always be adjusted to fit the cross sections and reaction rates of any one particular nuclear reaction. This is not done in a global program. The parameters are, in principle, determined by the best least-squares fit to all reactions for which experimental results are available. This lends some confidence in predictions for cases where experimental results are unavailable.

The original program (*23*, *44*), has produced reaction rates in numerical or analytical form as a function of temperature. Ready comparison with integrations of laboratory cross sections for target ground states are possible. Using the same global parameters which apply to reactions involving the ground states of stable nuclei, the theoretical program calculates rates for the ground states of radioactive nuclei and the excited states of both stable and radioactive nuclei. Summing over the statistically weighted contributions of the ground and known excited states or theoretical level density functions yields the stellar reaction rate for the equilibrated statistical population of the nuclear states. After summing, division by the partition function of the target nucleus is necessary.

Sargood (*45*) compared experimental results from a number of laboratories for protons and alpha particles reacting with 80 target nuclei (which are, of course, in their ground states) with the theoretical predictions of (*23*). Ratios of statistical model calculations to laboratory measurements for 12 cases are shown in Table 1 for temperatures in the range 1×10^9 to 5×10^9 K. The double entry for $^{27}Al(p,n)^{27}Si$ signifies ratios of theory to measurements made in two different laboratories. The theoretical calculations match the experimental results within 50 percent with a few marked exceptions. For the rather light targets in Table 1, especially at low temperature, the global mean rates can be in error whenever more and stronger resonances or fewer

Table 1. Statistical model calculations versus measurements. Ratio of reaction rate (ground state of target) from Woosley *et al.* (*110*) to reaction rates from experimental yield measurements (1970–1982) at Bombay, Caltech, Colorado, Kentucky, Melbourne, and Toronto.

Reaction	$T_9 = T/10^9$ K				
	1	2	3	4	5
^{23}Na(p,n)^{23}Mg	1.4	1.2	1.1	1.1	1.0
^{25}Mg(p,γ)^{26}Al	1.2	1.1	1.0	0.9	0.8
^{25}Mg(p,n)^{25}Al	1.1	1.0	0.9	0.8	0.8
^{27}Al(p,γ)^{28}Si	3.7	2.1	1.5	1.3	1.1
^{27}Al(p,n)^{27}Si	1.8	1.4	1.3	1.3	1.2
	0.9	0.9	0.9	1.0	1.0
^{28}Si(p,γ)^{29}P		1.2	1.3	1.2	0.9
^{29}Si(p,γ)^{30}P		1.0	1.6	1.6	1.5
^{39}K(p,γ)^{40}Ca	15	4.5	3.0	2.6	2.5
^{41}K(p,γ)^{42}Ca	0.5	0.5	0.5	0.4	0.4
^{41}K(p,n)^{41}Ca	0.8	1.0	1.1	1.2	1.3
^{40}Ca(p,γ)^{41}Sc				0.1	0.2
^{42}Ca(p,γ)^{43}Sc	1.3	1.4	1.4	1.4	1.3
	0.8	1.1	1.3	1.4	1.4

and weaker resonances than expected on average occur in the excitation curve of the reaction at low energies.

Sargood (*45*) also compared the ratio of the stellar rate of a reaction with target nuclei in a thermal distribution of ground and excited states with the rate for all target nuclei in their ground state. The latter is determined from laboratory measurements. In many cases, notably for reactions producing gamma rays, the ratio of stellar to laboratory rates is close to unity. In other cases the ratio can be high by several orders of magnitude. This frequently occurs when the ground state can interact only through partial waves of high angular momentum, resulting in small penetration factors and thus small cross sections and rates. This makes clear that it is frequently not valid to assume that a statistical theory which does well predicting ground state results will do equally well in predicting excited state results. Bahcall and Fowler (*46*) have shown that in a few cases laboratory measurements on inelastic scattering involving excited states can be used indirectly to determine reaction cross sections for those states.

Ward and Fowler (*47*) have investigated in detail the circumstances under which long-lived isomeric states do not come into equilibrium with ground states. When this occurs it is necessary to incorporate into network calculations the stellar rates for both the isomeric and ground states. An example of great interest is the nucleus ^{26}Al. The ground state has spin and parity $J^\pi = 5^+$ and isospin $T = 0$, and has a mean lifetime for positron emission to ^{26}Mg of 10^6 years. The isomeric state at 0.228 MeV has $J^\pi = 0^+$, $T = 1$, and mean lifetime of 9.2 seconds. Ward and Fowler (*47*) showed that the isomeric state effectively does not come into equilibrium with the ground state for temperatures $< 4 \times 10^8$ K. At these low temperatures both the isomeric state and the ground state of ^{26}Al must be included in the network of Fig. 3.

Astrophysical Weak Interaction Rates

Weak nuclear interactions play an important role in astrophysical processes, as indicated in Fig. 3. Only through the weak interaction can the overall proton number and neutron number of nuclear matter change during stellar evolution, collapse, and explosion. The formation of a neutron star requires that protons in ordinary stellar matter capture electrons. Gravitational collapse of a type II supernova core is retarded as long as electrons remain to exert outward pressure.

Many years of theoretical and experimental work on weak interaction rates in the Kellogg Laboratory and elsewhere have culminated in the calculation and tabulation by Fuller *et al.* (*48*) of electron and positron emission rates and continuum electron and positron capture rates, as well as the associated neutrino energy loss rates, for free nucleons and 226 nuclei with mass numbers between 21 and 60. Extension to higher and lower values of A is under way.

The detailed nature and the difficulty of the theoretical aspects of the combined atomic, nuclear, plasma, and hydrodynamic physics problems in type II supernova implosion and explosion were brought home to us by Hans Bethe during his stay in our laboratory early in 1982. His visit resulted in the preparation of two seminal papers (*49, 50*).

Current ideas on the nuclear equation of state predict that early in the collapse of the iron core of a massive star the nuclei present will become so neutron-rich that allowed electron capture on protons in the nuclei is blocked. Allowed electron capture, for which $\Delta l = 0$, is not permitted when neutrons have filled the subshells having orbital angular momentum, l, equal to that of the subshells occupied by the protons. This neutron shell blocking phenomenon, and several unblocking mechanisms operative at high temperature and density, including forbidden electron capture, have been studied in terms of the simple shell model by Fuller (*51*). Typical conditions result in a considerable reduction of the electron capture rates on heavy nuclei, leading to significant dependence on electron capture by the small number of free protons and a decrease in the overall neutronization rate.

Recent work on the weak interaction has concentrated on making the previously calculated reaction rates as efficient as possible for users of the published tables and computer tapes. The stellar weak interaction rates of nuclei are in general very sensitive functions of temperature and density. For electron and positron emission, most of the temperature-dependence is due to thermal population of parent excited states at all but the lowest temperatures and highest densities. In general, only a few transitions will contribute to these decay rates and hence the variation of the rates with temperature is usually not so large that rates cannot be accurately interpolated in temperature and density with standard grids (*48*). The density-dependence of these decay rates is minimal. In electron emission, however, there may be considerable density-dependence; but this does not present much of a problem for practical interpolation since the electron emission rate is usually very small under these conditions.

The temperature- and density-dependence of continuum electron and positron capture is a much more serious problem. In addition to temperature sensitivity introduced through thermal population of parent excited states, there are considerable effects from the lepton distribution functions in the integrands of the continuum-capture phase-space factors. This means that interpolation in temperature and density on the standard grid to obtain a rate can be difficult, especially for electron capture processes with threshold above zero energy.

We have found that the interpolation problem can be greatly eased by defining a simple continuum-capture phase-space integral, based on the value for the transition from the parent ground state to the daughter ground state, and then dividing this by the tabulated rate (*48*) at each temperature and density grid point to obtain values that are much less dependent on temperature and density. This procedure requires a formulation of the capture phase-space factors which is simple enough to use many times in the inner loop of stellar evolution nucleosynthesis computer programs. Such a formulation in terms of standard Fermi integrals has been found, along with approximations for the requisite Fermi integrals. When the chemical potential (Fermi energy) which appears in the Fermi integrals goes through zero these approximations have continuous values and continuous derivatives.

We have recently found expressions for the reverse reactions to e^-, e^+-capture, (that is, $\nu,\bar{\nu}$-capture) and for $\nu,\bar{\nu}$-blocking of the direct reactions when $\nu,\bar{\nu}$-states are partially or completely filled. These reverse reactions and the blocking are important during supernova core collapse when neutrinos and antineutrinos eventually become trapped, leading to equilibrium between the two directions of capture. General analytic expressions have been derived and approximated with computer-usable equations. All of these new results will be published in Fuller *et al.* (*52*) and new tapes including $\nu,\bar{\nu}$-capture will be made available to users on request.

Calculated Abundances for $A \lesssim 60$ and Comments on Explosive Nucleosynthesis

Armed with the available strong and weak nuclear reaction rates which apply to the advanced stages of stellar evolution, theoretical astrophysicists have attempted to derive the elemental and isotopic abundances produced in quasi-static pre-supernova nucleosynthesis and in explosive nucleosynthesis occurring during supernova outbursts. Although there is reasonably general agreement on nucleosynthesis during the various preexplosive stages, explosive nucleosynthesis is still an unsettled matter, subject to intensive study (*53*).

The abundances produced in explosive nucleosynthesis must depend on the detailed nature of supernova explosions. Ideas concerning the nature of type I and type II supernova explosions were published many years ago (*54, 55*). It was suggested that type I supernovae of small mass were precipitated by the onset of explosive carbon burning under conditions of electron degeneracy, where pressure is approximately independent of temperature. Carbon burning raises the temperature to the point where the electrons are no longer degenerate and explosive disruption of the star results. For type II supernovae of larger mass it was suggested that Si burning produced iron-group nuclei, which have

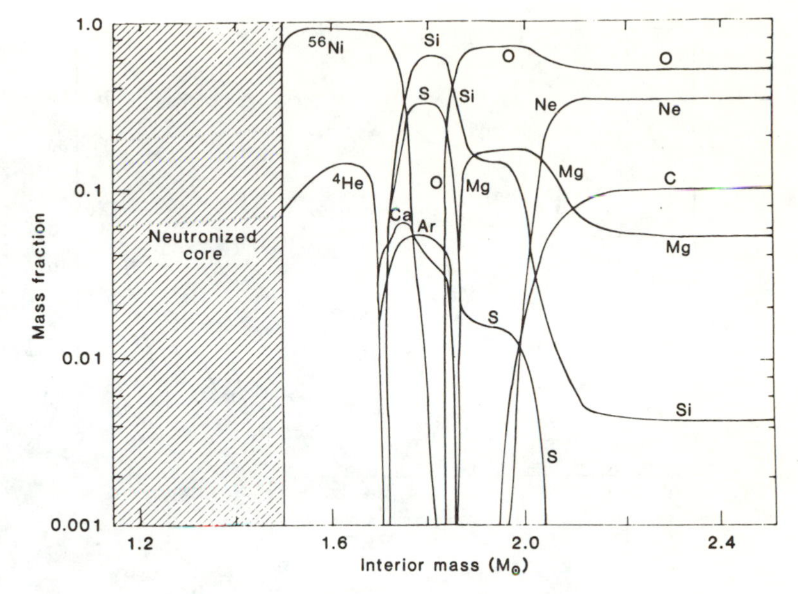

Fig. 5. Final abundances by mass fraction versus increasing interior mass in solar masses, $M_{\odot}$, in Type II supernova ejecta from a Population I star with total mass equal to 25 $M_{\odot}$, from Woosley and Weaver (*38*).

the maximum binding energies of all nuclei, so that nuclear energy is no longer available. Subsequent photodisintegration and electron capture in the stellar core lead to core implosion and ignition of explosive nucleosynthesis in the infalling inner mantle, which still contains nuclear fuel. These ideas have "survived" but, to say the least, with considerable modification over the years (*56*). Modern views on type II supernovae are given in (*40, 49, 50, 57*), and on type I supernovae in (*58*).

We can return to the nuclear abundance problem by reference to Fig. 5, which shows the distribution of final abundances by mass fraction in the supernova ejecta of a 25-$M_{\odot}$ Population I star. The pre-supernova distribution is that shown in Fig. 4. The modification of the abundances for mass zones interior to 2.2 $M_{\odot}$ is very apparent. Mass exterior to 2.2 $M_{\odot}$ is ejected with little or no modification in nuclear abundances. The supernova explosion was simulated by arbitrarily assuming that the order of 10^{51} ergs was delivered to the ejected material by the shock generated in the bounce or rebound of the collapsing and hardening core.

Integration over the mass zones of Fig. 5 for 1.5 $M_{\odot} < M < 2.2$ $M_{\odot}$ and of Fig. 4 for $M > 2.2$ $M_{\odot}$ yielded the isotopic abundances ejected into the interstellar medium by the simulated supernova (*38*). The results relative to solar abundances are shown in Fig. 6. The relative ratios are normalized to unity for ^{16}O, for which the overproduction ratio was 14; that is, for each gram of ^{16}O originally in the star, 14 grams were ejected. This overproduction in a single supernova can be expected to have produced the heavy-element abundances in the interstellar medium just prior to formation of the solar system, given the fact that supernovae occur approximately every 100 years in the Galaxy. The ultimate theoretical calculations will yield a constant overproduction factor of the order of 10.

The results shown in Fig. 6 are disappointing if one expects the ejecta of 20-

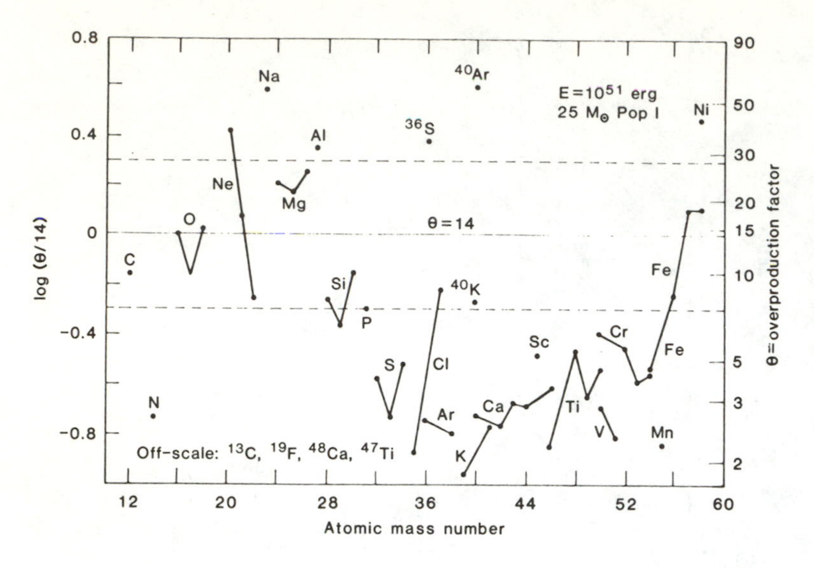

Fig. 6. Overabundance (ϑ) relative to 14 times solar abundances versus atomic mass number for nucleosynthesis resulting from a Type II, Population I supernova with total mass equal to 25 $M_\odot$, from Woosley and Weaver (*38*).

$M_\odot$ Population I supernovae to match solar system abundances with a relatively constant overproduction factor. The dip in abundances from sulfur to chromium is readily apparent. However, calculations must be made for other stellar masses and properly integrated over the mass distribution for stellar formation, which is roughly inversely proportional to mass (*38*). Woosley *et al.* (*53*) discuss their expectations of the abundances produced in stellar explosions for stars in the mass range 10 to 10^6 $M_\odot$. They show that a 200-$M_\odot$ Population III star produces abundant quantities of sulfur, argon, and calcium, which may compensate for the dip in Fig. 6. Population III stars are massive stars in the range 100 $M_\odot < M < 300$ $M_\odot$ which are thought to have formed from hydrogen and helium early in the history of the Galaxy and evolved very rapidly. Since their heavy-element abundance was zero they have no counterparts in presently forming Population I stars or among old, low-mass Population II stars.

Other authors have suggested a number of solutions to the problem depicted in Fig. 6. Nomoto *et al.* (*59*) calculated the abundances produced in carbon deflagration models of type I supernovae and, by adding equal contributions from type I and type II supernovae, obtained a result which is somewhat more satisfactory than Fig. 6. Arnett and Thielemann (*60*) recalculated quasi-static pre-supernova nucleosynthesis for $M \approx 20$ $M_\odot$, using a value for the $^{12}C(\alpha,\gamma)^{16}O$ rate equal to three times that given in (*21*); this would seem to be justified by the recent analysis of $^{12}C(\alpha,\gamma)^{16}O$ data, as discussed earlier. They then assumed that explosive nucleosynthesis would not substantially modify their quasi-static abundances and obtained results in which the average overproduction ratio is roughly 14. However, their assumption of minor modification during explosion and ejection is questionable.

I feel that the results discussed in this section and those obtained by numerous other authors show promise of an eventual satisfactory answer to the question where and how the elements from carbon to nickel originated.

Isotopic Anomalies in Meteorites and Evidence for Ongoing Nucleosynthesis

Almost a decade ago it became clear that nucleosynthesis occurred in the Galaxy up to the time of formation of the solar system or at least up to several million years before the formation. For slightly over a year it has been clear that nucleosynthesis has continued up to the present time or at least within several million years of the present. The decay of radioactive ^{26}Al ($\bar{\tau} = 1.04 \times 10^6$ years) is the key to these statements, which bring great satisfaction to most experimentalists, theorists, and observers in nuclear astrophysics.

Isotopic anomalies in meterorites produced by the decay of short-lived radioactive nuclei were first demonstrated in 1960 by Reynolds (*61*), who found large enrichments of ^{129}Xe in the Richardson meterorite. Jeffery and Reynolds (*62*) demonstrated that the excess ^{129}Xe was correlated with ^{127}I in the meteorite and that it resulted from the decay in situ of ^{129}I ($\bar{\tau} = 23 \times 10^6$ years). Quantitative results indicated that ^{129}I/^{127}I $\approx 10^{-4}$ at the time of meteorite formation. On the assumption that ^{129}I and ^{127}I were produced in roughly equal abundances in nucleosynthesis (most probably in the r-process) over a period of $\sim 10^{10}$ years in the Galaxy prior to formation of the solar system, and taking into account that only the ^{129}I produced over a period of the order of its lifetime survives, Wasserburg *et al.* (*63*) suggested that a period of free decay of the order of 10^8 years or more occurred between the last nucleosynthetic event which produced ^{129}I and its incorporation in meteorites in the solar system. There remains evidence for such a period in some cases, notably ^{244}Pu, but probably not in the history of the nucleosynthetic events which produced ^{129}I and other "short-lived" radioactive nuclei such as ^{26}Al and ^{107}Pd ($\bar{\tau} = 9.4 \times 10^6$ years).

The substantial meteoritic anomalies in ^{26}Mg from ^{26}Al, in ^{107}Ag from ^{107}Pd, in ^{129}Xe from ^{129}I, and in the heavy isotopes of Xe from the fission of ^{244}Pu ($\bar{\tau} = 117 \times 10^6$ years; fission tracks also observed) as well as searches in the future for anomalies in ^{41}K from ^{41}Ca ($\bar{\tau} = 0.14 \times 10^6$ years), in ^{60}Ni from ^{60}Fe ($\bar{\tau} = 0.43 \times 10^6$ years), in ^{53}Cr from ^{53}Mn ($\bar{\tau} = 5.3 \times 10^6$ years), and in ^{142}Nd from ^{146}Sm ($\bar{\tau} = 149 \times 10^6$ years; α-decay) are discussed exhaustively by Wasserburg and Papanastassiou (*64*). They espouse in situ decay for the observations to date, but Clayton (*65*) argues that the anomalies occur in interstellar grains preserved in the meteorites and originally produced by condensation in the expanding and cooling envelopes of supernovae and novae. Wasserburg and Papanastassiou write (*64*, p. 90), "There is, as yet, no compelling evidence for the presence of preserved presolar grains in the solar system. All of the samples so far investigated appear to have melted or condensed from a gas, and to have chemically reacted to form new phases." With mixed emotions I accept this.

Before turning to some elaboration of the ^{26}Al/^{26}Mg case it is appropriate to return to a discussion of the free decay interval mentioned above. It is the lack of detectable anomalies in ^{235}U from the decay of ^{247}Cm ($\bar{\tau} = 23 \times 10^6$ years) in

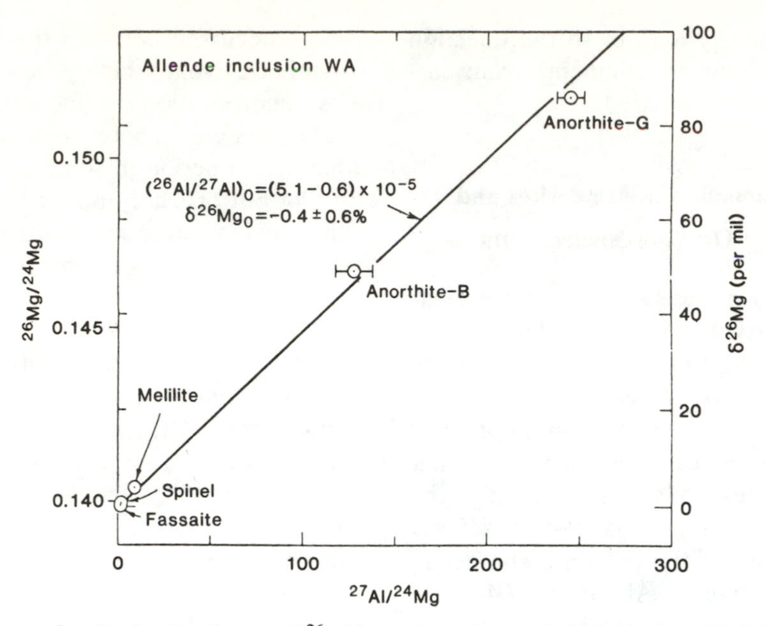

Fig. 7. Evidence for the in situ decay of ^{26}Al in various minerals in inclusion WA of the Allende meteorite, from Lee *et al.* (*68*). The linear relation between $^{26}Mg/^{24}Mg$ and $^{27}Al/^{24}Mg$ implies that $^{26}Al/^{27}Al = (5.1 \pm 0.6) \times 10^{-5}$ at the time of formation of the inclusion with ^{26}Al considered to react chemically in the same manner as ^{27}Al.

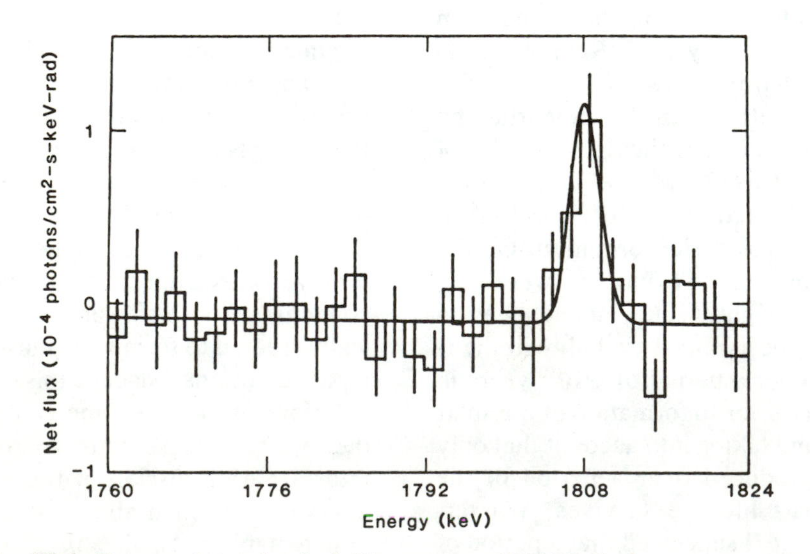

Fig. 8. The High Energy Astrophysical Observatory (HEAO 3) data on gamma rays in the energy range 1760 to 1824 keV emitted from the galactic equatorial plane, from Mahoney *et al.* (*69*). The line at 1809 keV is attributed to the decay of radioactive ^{26}Al ($\tau^{-} = 1.04 \times 10^{6}$ years) to the excited state of ^{26}Mg at this energy.

meteorites (*66*) coupled with the demonstrated occurrence of heavy Xe anomalies from the fission of ^{244}Pu ($\bar{\tau}$ = 117 × 10^6 years) (*67*) which demands a free decay interval of the order of several times 10^8 years. This interval is measured from the "last" r-process nucleosynthesis event (supernova?) which produced the actinides, Th, U, Pu, Cm, and beyond, up to the "last" nucleosynthesis events (novae?, supernovae with short-run r-processes?) which produced the short-lived nuclei ^{26}Al, ^{107}Pd, and ^{129}I before the formation of the solar system. The fact that the anomalies produced by these short-lived nuclei relative to normal abundances all are of the order of 10^{-4} despite the wide range in their mean lifetimes (1.04 × 10^6 years to 23 × 10^6 years) indicates that this anomaly range must be the result of inhomogeneous mixing of exotic materials with much larger quantities of normal solar system materials over a short time rather than the result of free decay. The challenges presented by this conclusion are manifold. Figure 14 of (*64*) shows the time scale for the formation of dust, rain, and hailstones in the early solar system and for the aggregation into chunks and eventually terrestrial planets. The solar nebula was almost but not completely mixed when it collapsed to form the solar system. From ^{26}Al it becomes clear that the mixing time down to an inhomogeneity of only one part in 10^3 was the order of 10^6 years.

Evidence that ^{26}Al was alive in interstellar material in the solar nebula which condensed and aggregated to form the parent body (planet in the asteroid belt?) of the Allende meteorite is shown in Fig. 7, taken with some modification from Lee *et al.* (*68*). The Allende meteorite fell near Pueblito de Allende in Mexico on 8 February 1969 and is a carbonaceous chondrite, a type of meteorite thought to contain the most primitive material in the solar system, unaltered since its original solidification.

Figure 7 depicts the results for ^{26}Mg/^{24}Mg versus ^{27}Al/^{24}Mg in different mineral phases from a Ca-Al-rich inclusion called WA obtained from a chondrule found in Allende. The excess ^{26}Mg is linearly correlated with the amount of ^{27}Al in the mineral phases. Since ^{26}Al is chemically identical with ^{27}Al, it can be inferred that phases rich in ^{27}Al were initially rich in ^{26}Al, which subsequently decayed in situ to produce excess ^{26}Mg. Aluminum-26 existed with abundance 5 × 10^{-5} that of ^{27}Al in one part of the solar nebula when the WA inclusion aggregated during the earliest stages of the formation of the solar system. The unaltered inclusion survived for 4.5 billion years to tell its story. Other inclusions in Allende and other meteorites yield ^{26}Al/^{27}Al from zero up to $\sim 10^{-3}$, with 10^{-4} a representative value. The reader is referred to (*68*) for the details of the story and the significance of non–accelerator-based contributions to nuclear astrophysics.

Evidence that ^{26}Al exists in the interstellar medium today appears in Fig. 8 from Mahoney *et al.* (*69*), which shows the gamma-ray spectrum observed in the range 1760 to 1824 keV by instruments aboard the High Energy Astronomical Observatory HEAO 3, which searched for diffuse gamma-ray emission from the galactic equatorial plane. The discrete line in the spectrum at 1809 keV, detected with a significance of nearly 5 standard deviations, is without doubt due to the transition from the first excited state at 1809 keV in ^{26}Mg to its ground state. Radioactive ^{26}Al decays by $^{26}Al(e^+\nu)^{26}Mg(\gamma)^{26}Mg$ to this state and thence to the ground state of ^{26}Mg. Given the mean lifetime (1.04×10^6 years) of ^{26}Al, this shows that ^{26}Al was produced no more than several million years ago and is probably being produced continuously. It is no great extrapolation to argue that nucleosynthesis in general continues in the Galaxy. Quantitatively, the observations indicate that $^{26}\text{Al}/^{27}\text{Al} \sim 10^{-5}$ in the interstellar medium. This average value was probably much the same when the solar system formed, but the variations in $^{26}\text{Al}/^{27}\text{Al}$ in meteoritic inclusions show that there were wide variations in the solar nebula about this value ranging from zero to 10^{-3}.

The question immediately arises, what is the site of synthesis of the ^{26}Al? Since the preparation of (*47*) I have been convinced that ^{26}Al could not be synthesized in supernovae at high temperatures where neutrons are copiously produced because of the expectation of a large cross section for $^{26}\text{Al}(n,p)^{26}\text{Mg}$. This expectation has been borne out by measurements on the reverse reaction $^{26}\text{Mg}(p,n)^{26}\text{Al}$ in the Kellogg Laboratory by Skelton *et al.* (*70*). There is little doubt that the stellar rate for $^{26}\text{Al}(n,p)^{26}\text{Mg}$ is very large indeed.

It has been suggested that ^{26}Al is produced in novae (*65, 70, 71*). This is quite reasonable on the basis of nucleosynthesis in novae (*72*). In current models for novae, hydrogen from a binary companion is accreted by a white dwarf until a thermal runaway involving the fast CN cycle occurs. Similarly, a fast MgAl cycle may occur with production of $^{26}\text{Al}/^{27}\text{Al} \gtrsim 1$, as shown in figure 9 of (*47*) and substantiated by the recent experiental measurements cited in (*47*). Clayton (*65*) argues that the estimated 40 novae occurring annually in the galactic disk can produce the observed $^{26}\text{Al}/^{27}\text{Al}$ ratio of 10^{-5} on average. He assumes that each nova ejects $10^{-4}\ M_\odot$ of material containing an ^{26}Al mass fraction of 3×10^{-4}.

Another possible source of ^{26}Al is spallation induced by irradiation of protoplanetary material by high-energy protons from the young sun as it settled on the main sequence. This possibility was discussed very early by Fowler *et al.* (*73*), who also attempted to produce D, Li, Be, and B in this way, requiring such large primary proton and secondary neutron fluxes that many features of the abundance curve in the solar system would have been changed substantially. A more reasonable version of the scenario was presented by Lee (*74*) but without notable success. I find it difficult to believe that an early irradiation produced the anomalies in meteorites. The ^{26}Al in the interstellar medium today certainly cannot have been produced in this way.

Anomalies have been found in meteorites in the abundances compared with normal solar system material of the stable isotopes of many elements: O, Ne, Mg, Ca, Ti, Kr, Sr, Xe, Ba, Nd, and Sm. The possibility that the oxygen anomalies are nonnuclear in origin has been raised by Thiemens and Heidenreich (*75*), but the anomalies in the remaining elements are generally attributed to nuclear processes.

One example is a neutron-capture/beta-decay (nβ) process studied by Sandler *et al.* (*76*). The seed nuclei consist of all of the elements from Si to Cr with normal solar system abundances. With this process at neutron densities of $\sim 10^7$ mol cm^{-3} and exposure times of $\sim 10^3$ seconds, small admixtures ($\lesssim 10^{-4}$) of the exotic material produced are sufficient to account for most of the Ca and Ti isotopic anomalies found in the Allende meteorite inclusion EK-1-4-1 by Niederer *et al.* (*77*). The anomalies in stable isotope abundances are of the same order as those for short-lived radioactive nuclei and strongly support the view that the solar nebula was inhomogeneous, with regions containing exotic materials up to 10^{-4} or more of normal material.

Agreement for the ^{46}Ca and ^{49}Ti anomalies in EK-1-4-1 was obtained only by increasing the theoretical Hauser-Feshbach cross sections for $^{46}K(n,\gamma)$, and $^{49}Ca(n,\gamma)$ by a factor of 10 on the basis of probable thermal resonances just above threshold in the compound nuclei ^{47}K and ^{50}Ca, respectively. Huck *et al.* (*78*) reported an excited state in ^{50}Ca just 0.16 MeV above the $^{49}Ca(n,\gamma)$ threshold which can be produced by s-wave capture and fulfills the requirements of (*76*).

Sandler *et al.* (*76*) suggest that the exposure time scale of $\sim 10^3$ seconds is determined by the mean lifetime of ^{13}N (862 seconds), produced through $^{12}C(p,\gamma)^{13}N$ by a jet of hydrogen suddenly introduced into the helium-burning shell of a red giant star where a substantial amount of ^{12}C has been produced by the $3\alpha \rightarrow {}^{12}C$. The beta-decay $^{13}N(e^{+}\nu)^{13}C$ is followed by $^{13}C(\alpha,n)^{16}O$ as the source of the neutrons. All of this is very interesting, if true. More to the point, Sandler *et al.* (*76*) predict the anomalies to be expected in the isotopes of chromium, and attempts to measure these anomalies are under way by Wasserburg and his colleagues.

Observational Evidence for Nucleosynthesis in Supernovae

Over the years there has been considerable controversy concerning elemental abundance observations at optical wavelengths on galactic supernova remnants. To my mind the most convincing evidence for nucleosynthesis in supernovae has been provided by Chevalier and Kirshner (*79*), who obtained quantitative spectral information for several of the fast-moving knots in the supernova remnant Cassiopeia A (approximately dated 1659, but a supernova event was not observed at that time). The knots are considered to be material ejected from various layers of the original star in a highly asymmetric, nonspherical explosion. In one knot, KB33, the following ratios relative to solar were observed: S/O = 61, Ar/O = 55, Ca/O = 59. It is abundantly clear that oxygen burning to the silicon-group elements in the layer in which KB33 originated has depleted oxygen and enhanced the silicon-group elements. Other knots and other features designated as filaments show different abundance patterns, albeit not so easily interpreted. The moral for supernova modelers is that spherically symmetric supernova explosions may be the easiest to calculate but are not to be taken as realistic.

Most striking of all has been the payoff from the NASA investment in the High Energy Astronomy Observatory HEAO 2, now called the Einstein Observatory. With instruments aboard this satellite Becker *et al.* (*80*) observed the x-ray spectrum in the range 1 to 4 keV of Tycho Brahe's supernova remnant (1572), showing the *K*-level x-rays from Si, S, Ar, and Ca. Shull (*81*) has used a single-velocity, non–ionization-equilibrium model of a supernova blast wave to calculate abundances in Tycho's remnant relative to solar and finds Si = 7.6, S = 6.5, Ar = 3.2, and Ca = 2.6. With considerably greater uncertainty he gives Mg = 2.0 and Fe = 2.1. He finds different enhancements in Kepler's remnant (1604) and in Cassiopeia A. One more lesson for the modelers: no two supernovae are alike. Nucleosynthesis in supernovae depends on their initial mass, rotation, mass loss during the red giant stage, degree of symmetry during explosion, initial heavy-element content, and probably other factors. These details aside, it seems clear that supernovae produce enhancements in elemental abundances up to iron and probably beyond. Detection of the much rarer elements beyond iron will require more sensitive x-ray detectors operating at higher energies. The nuclear debris of supernovae eventually enriches the interstellar medium, from which succeeding generations of stars are formed. It becomes increasingly clear that novae also enrich the interstellar medium. Sorting out these two contributions poses interesting problems for research in all aspects of nuclear astrophysics.

Explosive Si burning in the shell just outside a collapsing supernova core primarily produces ^{56}Ni, as shown in Fig. 5. It is generally believed that the initial energy source for the light curves of type I supernovae is electron capture by ^{56}Ni ($\bar{\tau}$ = 8.80 days) to the excited state of ^{56}Co at 1.720 MeV with subsequent gamma-ray cascades to the ground state. The subsequent source of energy is electron capture and positron emission by ^{56}Co ($\bar{\tau}$ = 114 days) to a number of excited states of ^{56}Fe with gamma-ray cascades to the stable ground state of ^{56}Fe. Both the positrons and gamma rays heat the ejected material. If ^{56}Co is an energy source there should be spectral evidence for cobalt in newly discovered type I supernovae, since its lifetime is long enough for detailed observations to be possible after the initial discovery.

The cobalt has been observed. Axelrod (*82*) studied the optical spectra of SN1972e obtained by Kirshner and Oke (*83*) and assigned the two emission lines near 6000 Å to Co III. The lines are clearly evident in spectra obtained at 233 and 264 days after Julian day 2441420, assigned as the initial day of the explosive event, but are only marginally evident at 376 days ($\sim \bar{\tau}$ later). The lines decay in reasonable agreement with the mean lifetime of ^{56}Co.

Branch *et al.* (*84*) studied absorption spectra during the first hundred days of SN1981b. Deep absorption lines of Co II are clearly evident near 3300 and 4000 Å.

It is my conclusion that there is substantial evidence for nucleosynthesis of elements produced in oxygen and silicon burning in supernovae. The role of neutron capture processes in supernovae will be discussed next.

Neutron-Capture Processes in Nucleosynthesis

In an earlier section I discussed the need for two neutron-capture processes for nucleosynthesis beyond $A \geqslant 60$: the s-process and the r-process. For a given element the heavier isotopes are frequently bypassed in the s-process and produced only in the r-process; thus the designation r-only. Lighter isotopes are frequently shielded by more neutron-rich stable isobars in the r-process and are produced only in the s-process; thus the designation s-only. The lightest isotopes are frequently very rare because they are not produced in either the s- or the r-process and are thought to be produced in what is called the p-process, involving positron production and capture, proton capture, neutron photoproduction, and/or (p,n)-reactions (*85*).

The s-process has the clearest phenomenological basis of all processes of nucleosynthesis, primarily as a result of the correlation of s-process abundances

(N) (*86*) with a beautiful series of measurements on neutron capture cross sections (σ) in the range 1 to 100 keV (*87*). In first-order approximation the product σN should be constant in s-process synthesis: a nucleus with a small (large) neutron capture cross section must have a large (small) abundance to maintain continuity in the s-capture path. When σN is plotted against atomic mass, this is demonstrated in plateaus found from $A = 90$ to 140 and from $A = 140$ to 206. Nuclear shell structure introduces the precipices shown in such a plot at $A \sim 84$, ~ 138, and ~ 208, which correspond to the s-process abundance peaks in Fig. 1. At these values of A the neutron numbers are "magic," $N = 50$, 82, and 126; the cross sections for neutron capture into new neutron shells are very small, and with a finite supply of neutrons the σN product must drop to a new plateau as observed. Iben has argued convincingly that the site of the s-process is the He-burning shell of a pulsating red giant (*88*) and the neutron source is the $^{22}Ne(\alpha,n)^{25}Mg$ reaction. Critical discussions have been given in (*89*) and (*90*).

The r-process has been customarily treated by the waiting point method of B^2FH (*16*). Under explosive conditions a large flux of neutrons drives nuclear seeds to the neutron-rich side of the valley of stability where the (n,γ)-reaction and the (γ,n)-reaction reach equality. The nuclei wait at this point until electron beta-decay transforms neutrons in the nuclei into protons and further neutron capture can occur. At the cessation of the r-process the neutron-rich nuclei decay to their stable isobars. In first order, this means that the abundance of an r-process nucleus multiplied by the electron beta-decay rate of its neutron-rich r-process isobar progenitor will be roughly constant. At magic neutron numbers in the neutron-rich progenitors, beta-decay must open the closed neutron shell in transforming a neutron into a proton and there the rate will be relatively small. Accordingly, the abundance of progenitors with $N = 50$, 82, and 126 will be large. The associated number of protons will be less than in the corresponding s-process nuclei with a magic number of neutrons. It follows that the stable daughter isobars will have smaller mass numbers, and this is indeed the case, the r-process abundance peaks occurring at $A \sim 80$, ~ 130, and ~ 195, all below the corresponding s-process peaks as illustrated in Fig. 1.

A phenomenological correlation of r-process abundances with beta-decay rates by Becker and Fowler (*91*) is too phenomenological to satisfy critical nuclear astrophysicists, who wish to know the site of the high neutron fluxes demanded for r-process nucleosynthesis and the details of the r-process path through nuclei far from the line of beta-stability. There is also a general belief at present that the waiting point approximation is a poor one and must be replaced by dynamical r-process flow calculations taking into account explicit (n,γ),(γ,n), and beta-decay rates with time-varying temperature and neutron flux.

Many suggestions have been made for possible sites of the r-process, almost all in supernova explosions where the basic requirement of a large neutron flux of short duration is met. These suggestions are reviewed in Schramm (*92*) and Truran (*90*). To my mind the helium core thermal runaway r-process of Cameron *et al.* (*93*) is the most promising. These authors do not rule out $^{22}Ne(\alpha,n)^{25}Mg$ as the source of the neutrons, but their detailed results are based on $^{13}C(\alpha,n)^{13}O$

as the source. They start with a star formed from material with the same heavy-element abundance distribution as in the solar system but with smaller total amount. They assume that the helium core of the star after hydrogen burning contains a significant amount of ^{13}C, which was produced by the introduction of hydrogen into the core which had already burned half of its helium into ^{12}C. The electrons in the core are initially degenerate, but the rise in temperature with ^{13}C burning lifts the degeneracy, producing a thermal runaway with expansion and subsequent cooling of the core. This event is the second helium-flash episode in the history of the core, and if it occurs only a small amount of the r-process material produced need escape into the interstellar medium to contribute the r-process abundance in solar system material. It is my belief that a realistic astrophysical site for the thermal runaway, perhaps with different initial conditions, will be found.

Nucleocosmochronology

Armed with r-process calculations of the abundances of the long-lived parents of the natural radioactive series, ^{232}Th, ^{235}U, and ^{238}U, and with the then-current solar system abundances of these nuclei, B^2FH (*18*) determined the duration of r-process nucleosynthesis from its beginning in the first stars in the Galaxy up to the last events before the formation of the solar system. The abundances used were those observed in meteorites, assumed to be closed systems since their formation, taken to have occurred 4.55 billion years ago. It was necessary to correct for free decay during this period in order to obtain abundances for comparison with calculations based on r-process production plus decay before the meteorites became closed systems. The calculations required only the elemental ratio Th/U in meteorites, since the isotopic ratio $^{235}U/^{236}U$ was assumed to be the same for meteoritic and terrestrial samples. The Apollo Program added lunar data to the meteoritic and terrestrial data.

B^2FH considered a number of possible models, including r-process nucleosynthesis uniform in time and an arbitrary time interval between the last r-process contribution to the solar nebula and the closure of the meteorite systems. A zero value for this time interval indicated that uranium production started 18 billion years ago. When this time interval was taken to be 0.5 billion years, the production started 11.5 billion years ago. These results are in remarkable, if coincidental, concordance with current values.

It is appropriate to point out that nucleocosmochronology yields, with additional assumptions, an estimate for the age of the expanding Universe independent of astronomical redshift-distance observations of distant galaxies. These assumptions are that the r-process started soon (<1 billion years) after the formation of the Galaxy and that the Galaxy formed soon (<1 billion years) after the big-bang origin of the Universe. Adding a billion years or so to the start of r-process nucleosynthesis yields an independent value, based on radioactivity, for the age or time back to the origin of the expanding Universe.

Much has transpired over recent years in the field of nucleocosmochronology. I have kept my hand in most recently in (*94*). Sophisticated models of galactic evolution were introduced by Tinsley (*95*). A method for model-independent determinations of the mean age of nuclear chronometers at the time of solar system formation was developed by Schramm and Wasserburg (*96*). The most recent results are those of Thielemann *et al.* (*97*), who calculated that r-process nucleosynthesis in the Galaxy started 17.9 billion years ago, with uncertainties of +2 billion and −4 billion years. This is to be compared with my value of 10.5 ± 2.3 billion years (*94*). Thielemann and I are now recomputing the new value for the duration, using an initial spike in galactic synthesis plus uniform synthesis thereafter.

The results of Thielemann *et al.* (*97*) indicate that the age of the expanding Universe is 19 billion years, give or take several billion years. This is to be compared to the Hubble time or reciprocal of Hubble's constant, given by Sandage and Tammann (*98*) as 19.5 ± 3 billion years. However, the Hubble time is equal to the age of the expanding Universe only for a completely open Universe with mean matter density much less than the critical density for closure, which can be calculated from the value for the Hubble time just given to be 5×10^{-30} g cm^{-3}. The observed visible matter in galaxies is estimated to be 10 percent of this, which reduces the age of the Universe to 16.5 billion years. Invisible matter, neutrinos, black holes, and so on may add to the gravitational forces which decrease the velocity of expansion and may thus decrease the age to that corresponding to critical density, which is 11.1 billion years. If the expansion velocity was greater in the past, the time to the present radius of the Universe is correspondingly less. Moreover, others have obtained results for the Hubble time equal to about one-half that of Sandage and Tammann (*98*), as reviewed in van den Bergh (*99*).

A completely independent nuclear chronology involving radiogenic ^{187}Os produced during galactic nucleosynthesis by the decay of ^{187}Re ($\bar{\tau} = 65 \times 10^9$ years) was suggested by Clayton (*100*). Schramm (*92*) discusses still other chronometric pairs. Clayton's suggestion involves the s-process even though ^{187}Re is produced in the r-process, as it requires that the abundance of ^{187}Re be compared to that of its daughter, ^{187}Os, when the s-only production of this daughter nucleus is subtracted from its total solar system abundance. This was to be done by comparing the neutron capture cross section of ^{187}Os with that of its neighboring s-only isotope ^{186}Os, which does not have a long-lived radioactive parent, and using the $N\sigma$ = constant rule for the s-process. However, Fowler (*101*) pointed out that ^{187}Os has a low-lying excited state at 9.75 keV which is practically fully populated at the temperature (3.5×10^8 K) at which the s-process is customarily assumed to occur. Moreover, with spin $J = 3/2$ this state has twice the statistical weight of the ground state with spin $J = 1/2$, so that measurements of the ground state neutron capture cross section yield only one-third of what one needs to know.

All of this led to a series of beautiful and difficult measurements for neutron-induced reactions on the isotopes of osmium, yielding values for the cross-section ratio of ^{186}Os(n,γ) relative to ^{187}Os(n,γ). This ratio must be multiplied by a theoretical factor to correct the ^{187}Os cross section for that of its excited state. Woosley and Fowler (*102*) obtained estimates for this factor in the range 0.8 to 1.10, which translate into a time for the beginning of the r-process in the Galaxy in the range 14 to 19 billion years. Measurements of the cross sections for neutron scattering off the ground state of ^{187}Os to its excited state at 9.75 keV (*103, 104*) supported the lower value of the Woosley and Fowler (*102*) factor and thus a value for the time back to the beginning of r-process nucleosynthesis in the range 18 to 20 billion years. This is concordant with the latest value from Th/U nucleocosmochronology. Measurements of the neutron capture cross section on the ground state of ^{189}Os might be helpful, since ^{189}Os has a ground state with the same spin and Nilsson numbers as the excited state of ^{187}Os and an excited state corresponding to the ground state of ^{187}Os. Such measurements have been made by Browne and Berman (*105*) but are now being checked.

It will be clear that the lifetime of ^{187}Re comes directly into the calculations under discussion, and there has been some discrepancy in the past between lifetimes measured geochemically and those measured directly by counting the electrons emitted in the 2.6-keV decay $^{187}\mathrm{Re}(e^- \nu)^{187}\mathrm{Os}$. This is treated in considerable theoretical detail by Williams *et al.* (*106*), who found that the direct measurements by Payne and Drever (*107*), which agree with the geochemical measurements of Hirt *et al.* (*108*), are correct. There is also the vexing problem of a possible decrease in the effective lifetime of ^{187}Re in the galactic environment, where ^{187}Re is subject to destruction by the s-process as well as being produced by the r-process. This decreases all times based on the Re/Os chronology (*109*). The time back to the beginning of r-process nucleosynthesis could be as low as 12 billion years. It is appropriate to end this section with the considerable uncertainity in nucleocosmochronology, indicating that, as in all nuclear astrophysics, there is much exciting experimental and theoretical work to be done for many years to come.

Conclusion

In spite of the past and current research in experimental and theoretical nuclear astrophysics, the ultimate goal of the field has not been attained. Hoyle's grand concept of element synthesis in the stars will not be truly established until we attain a deeper and more precise understanding of many nuclear processes operating in astrophysical environments. Hard work must continue on all aspects of the cycle: experiment, theory, observation. It is not just a matter of filling in the details. There are puzzles and problems in each part of the cycle which challenge the basic ideas underlying nucleosynthesis in stars. Not to worry—that is what makes the field active, exciting, and fun. It is a great source of satisfaction to me that the Kellogg Laboratory continues to play a leading role in experimental and theoretical nuclear astrophysics.

And now permit me to pass along one final thought. My major theme has been that all of the heavy elements from carbon to uranium have been synthesized in stars. Our bodies consist for the most part of these heavy elements. Apart from hydrogen, we are 65 percent oxygen and 18 percent carbon, with smaller percentages of nitrogen, sodium, magnesium, phosphorus, sulfur, chlorine, potassium, and traces of still heavier elements. Thus it is possible to say that each one of us and all of us are truly and literally a little bit of stardust.

References and Notes

1. J. Audouze and H. Reeves, in *Essays in Nuclear Astrophysics*, C. A. Barnes, D. D. Clayton, D. N. Schramm, Eds. (Cambridge Univ. Press, Cambridge, 1982), p. 355.

2. H. E. Suess and H. C. Urey, *Rev. Mod. Phys.* **28**, 53 (1956).
3. A. G. W. Cameron, in *Essays in Nuclear Astrophysics*, C. A. Barnes, D. D. Clayton, D. N. Schramm, Eds. (Cambridge Univ. Press, Cambridge, 1982), p. 23.
4. W. Whaling, in *ibid.*, p. 65.
5. R. A. Alpher and R. C. Herman, *Rev. Mod. Phys.* **22**, 153 (1950).
6. R. V. Wagoner, W. A. Fowler, F. Hoyle, *Astrophys. J.* **148**, 3 (1967).
7. H. Staub and W. E. Stephens, *Phys. Rev.* **55**, 131 (1939).
8. J. H. Williams, W. G. Shepherd, R. O. Haxby, *ibid.* **52**, 390 (1937).
9. A. V. Tollestrup, W. A. Fowler, C. C. Lauritsen, *ibid.* **76**, 428 (1949).
10. A. Hemmindinger, *ibid.* **73**, 806 (1948); *ibid.* **74**, 1267 (1949).
11. H. A. Bethe, *ibid.* **55**, 434 (1939); in *Les Prix Nobel 1967* (Almquist & Wiksell, Stockholm, 1968).
12. F. Hoyle, *Mon. Not. R. Astron. Soc.* **106**, 343 (1946); *Astrophys. J. Suppl.* **1**, 121 (1954).
13. ______ and M. Schwarzschild, *Astrophys. J. Suppl.* **2**, 1 (1955).
14. A. R. Sandage and M. Schwarzschild, *Astrophys. J.* **116**, 463 (1952). In particular, see last paragraph on p. 475.
15. E. E. Salpeter, *ibid.* **115**, 326 (1952).
16. D. N. F. Dunbar, R. E. Pixley, W. A. Wenzel, W. Whaling, *Phys. Rev.* **92**, 649 (1953).
17. C. W. Cook, W. A. Fowler, C. C. Lauritsen, T. Lauritsen, *ibid.* **107**, 508 (1957).
18. E. M. Burbidge, G. R. Burbidge, W. A. Fowler, F. Hoyle, *Rev. Mod. Phys.* **29**, 547 (1957); hereafter referred to as B^2FH (*18*). Also see F. Hoyle, W. A. Fowler, E. M. Burbidge, G. R. Burbidge, *Science* **124**, 611 (1956).
19. A. G. W. Cameron, *Publ. Astron. Soc. Pac.* **69**, 201 (1957).
20. J. L. Greenstein, in *Modern Physics for the Engineer*, L. N. Ridenour, Ed. (McGraw-Hill, New York, 1954), chapter 10; in *Essays in Nuclear Astrophysics*, C. A. Barnes, D. D. Clayton, D. N. Schramm, Eds. (Cambridge Univ. Press, Cambridge, 1982), p. 45.
21. W. A. Fowler, G. R. Caughlan, B. A. Zimmerman, *Annu. Rev. Astron. Astrophys.* **5**, 525 (1967); *ibid.* **13**, 69 (1975). Also see M. J. Harris *et al.*, *ibid.* **21**, 165 (1983); G. R. Caughlan *et al.*, *At. Data Nucl. Data Tables*, in press.
22. E. E. Salpeter, *Phys. Rev.* **88**, 547 (1957); *ibid.* **97**, 1237 (1955).
23. S. E. Woosley, W. A. Fowler, J. A. Holmes, B. A. Zimmerman, *At. Data Nucl. Data Tables* **22**, 371 (1978).
24. H. A. Bethe and C. L. Critchfield, *Phys. Rev.* **54**, 248 (1938).
25. W. A. Fowler, *Astrophys. J.* **127**, 551 (1958).
26. A. G. W. Cameron, *Annu. Rev. Nucl. Sci.* **8**, 249 (1958).
27. B. Pontecorvo, *Chalk River Lab. Rep. PD-205* (1946).
28. L. W. Alvarez, *Univ. Calif. Radiat. Lab. Rep. UCRL-328* (1949).
29. R. Davis, Jr., in *Science Underground*, M. M. Nieto *et al.*, Eds. (American Institute of Physics, New York, 1983), p. 2.
30. J. N. Bahcall, W. F. Huebner, S. H. Lubow, P. D. Parker, R. K. Ulrich, *Rev. Mod. Phys.* **54**, 767 (1982).
31. R. G. H. Robertson, P. Dyer, T. J. Bowles, R. E. Brown, N. Jarmie, C. J. Maggiore, S. M. Austin, *Phys. Rev. C* **27**, 11 (1983); J. L. Osborne, C. A. Barnes, R. W. Kavanagh, R. M. Kremer, G. J. Mathews, J. L. Zyskind, *Phys. Rev. Lett.* **48**, 1664 (1982)
32. R. T. Skelton and R. W. Kavanagh, *Nucl. Phys. A* **414**, 141 (1984).
33. C. A. Barnes, in *Essays in Nuclear Astrophysics*, C. A. Barnes, D. P. Clayton, D. N. Schramm, Eds. (Cambridge Univ. Press, Cambridge, 1982), p. 193.
34. P. Dyer and C. A. Barnes, *Nucl. Phys. A* **233**, 495 (1974); S. E. Koonin, T. A. Tombrello, G. Fox, *ibid.* **220**, 221 (1974).
35. K. U. Kettner, H. W. Becker, L. Buchmann, J. Gorres, H. Kräwinkel, C. Rolfs, P. Schmalbrock, H. P. Trautvetter, A. Vlieks, *Z. Phys. A* **308**, 73 (1982).
36. K. Langanke and S. E. Koonin, *Nucl. Phys. A* **410**, 334 (1983); private communication (1983).
37. S. G. Starrfield, A. N. Cox, S. W. Hodson, W. D. Pesnell, *Astrophys. J.* **268**, L27 (1983); S. A. Becker, private communication (1983).
38. S. E. Woosley and T. A. Weaver, in *Essays in Nuclear Astrophysics*, C. A. Barnes, D. D. Clayton, D. N. Schramm, Eds. (Cambridge Univ. Press, Cambridge, 1982), p. 381.
39. D. Bodansky, D. D. Clayton, W. A. Fowler, *Astrophys. J. Suppl.* **16**, 299 (1968).
40. T. A. Weaver, S. E. Woosley, G. M. Fuller, in *Proceedings of the Conference on Numerical Astrophysics*, R. Bowers, J. Centrella, J. LeBlanc, M. LeBlanc, Eds. (Science Books International, Boston, 1983).
41. L. W. Mitchell and D. G. Sargood, *Aust. J. Phys.* **36**, 1 (1983).
42. R. N. Boyd, in *Proceedings of the Workshop on Radioactive Ion Beams and Small Cross Section Measurements* (Ohio State Univ. Press, Columbus, 1981); R. C. Haight, G. J. Mathews, R. M. White, L. A. Avilés, S. E. Woodward, *Nucl. Instrum. Methods* **212**, 245 (1983).
43. W. Hauser and H. Feshbach, *Phys. Rev.* **78**, 366 (1952).
44. J. A. Holmes, S. E. Woosley, W. A. Fowler, B. A. Zimmerman, *At. Data Nucl. Data Tables* **18**, 305 (1976).
45. D. G. Sargood, *Phys. Rep.* **93**, 61 (1982); *Aust. J. Phys.* **36**, 583 (1983).
46. N. A. Bahcall and W. A. Fowler, *Astrophys. J.* **157**, 645 (1969).
47. R. A. Ward and W. A. Fowler, *ibid.* **238**, 266 (1980). For recent experimental data on the production of ^{26}Al through $^{25}Mg(p,\gamma)^{26}Al$, see A. E. Champagne, A. J. Howard, P. D. Parker, *ibid.* **269**, 686 (1983). For recent experimental data on the destruction of ^{26}Al through $^{26}Al(p,\gamma)^{27}Si$, see L. Buchmann, M. Hilgemaier, A. Krauss, A. Redder, C. Rolfs, H. P. Trautvetter, in press.
48. G. M. Fuller, W. A. Fowler, M. J. Newman, *Astrophys. J. Suppl.* **42**, 447 (1980); *ibid.* **48**, 279 (1982); *Astrophys. J.* **252**, 715 (1982).
49. H. A. Bethe, A. Yahil, G. E. Brown, *Astrophys. J. Lett.* **262**, L7 (1982).

50. H. A. Bethe, G. E. Brown, J. Cooperstein, J. R. Wilson, *Nucl. Phys. A* **403**, 625 (1983).
51. G. M. Fuller, *Astrophys. J.* **252**, 741 (1982).
52. ———, W. A. Fowler, M. J. Newman, in preparation.
53. S. E. Woosley, T. S. Axelrod, T. A. Weaver, in *Stellar Nucleosynthesis*, C. Chiosi and A. Renzini, Eds. (Reidel, Dordrecht, 1984).
54. F. Hoyle and W. A. Fowler, *Astrophys. J.* **132**, 565 (1960).
55. W. A. Fowler and F. Hoyle, *Astrophys. J. Suppl.* **9**, 201 (1964).
56. J. C. Wheeler, *Rep. Prog. Phys.* **44**, 85 (1981).
57. G. E. Brown, H. A. Bethe, G. Baym, *Nucl. Phys. A.* **375**, 481 (1982).
58. K. Nomoto, *Astrophys. J.* **253**, 798 (1982); *ibid.* **257**, 780 (1982); in *Stellar Nucleosynthesis*, C. Chiosi and A. Renzini, Eds. (Reidel, Dordrecht, 1984).
59. ———, F.-K. Thielemann, J. C. Wheeler, *Astrophys. J.* **279**, L23 (1984); erratum **283**, L25 (1984).
60. W. D. Arnett and F.-K. Thielemann, in *Stellar Nucleosynthesis*, C. Chiosi and A. Renzini, Eds. (Reidel, Dordrecht, 1984).
61. J. H. Reynolds, *Phys. Rev. Lett.* **4**, 8 (1960).
62. P. M. Jeffery and J. H. Reynolds, *J. Geophys. Res.* **66**, 3582 (1961).
63. G. J. Wasserburg, W. A. Fowler, F. Hoyle, *Phys. Rev. Lett.* **4**, 112 (1960).
64. G. J. Wasserburg and D. A. Papanastassiou, in *Essays in Nuclear Astrophysics*, C. A. Barnes, D. D. Clayton, D. N. Schramm, Eds. (Cambridge Univ. Press, Cambridge, 1982), p. 77.
65. D. D. Clayton, *Astrophys. J.* **199**, 765 (1975); *Space Sci. Rev.* **24**, 147 (1979); *Astrophys. J.* **268**, 381 (1983). Also see D. D. Clayton and F. Hoyle, *Astrophys. J. Lett.* **187**, L101 (1974); *Astrophys. J.* **203**, 490 (1976).
66. J. H. Chen and G. J. Wasserburg, *Earth Planet. Sci. Lett.* **52**, 1 (1981).
67. D. S. Burnett, M. I. Stapanian, J. H. Jones, in *Essays in Nuclear Astrophysics*, C. A. Barnes, D. D. Clayton, D. N. Schramm, Eds. (Cambridge Univ. Press, Cambridge, 1982), p. 144.
68. T. Lee, D. A. Papanastassiou, G. J. Wasserburg, *Astrophys. J. Lett.* **211**, L107 (1977).
69. W. A. Mahoney, J. C. Ling, W. A. Wheaton, A. S. Jacobson, *Astrophys. J.* **278**, 784 (1984); see also W. A. Mahoney, J. C. Ling, A. S. Jacobson, R. E. Lingenfelter, *ibid.* **262**, 742 (1982).
70. R. T. Skelton, R. W. Kavanagh, D. G. Sargood, *ibid.* **271**, 404 (1983).
71. M. Arnould, H. Nørgaard, F.-K. Thielemann, W. Hillebrandt, *ibid.* **237**, 931 (1980).
72. J. W. Truran, in *Essays in Nuclear Astrophysics*, C. A. Barnes, D. D. Clayton, D. N. Schramm, Eds. (Cambridge Univ. Press, Cambridge, 1982), p. 467.
73. W. A. Fowler, J. L. Greenstein, F. Hoyle, *Geophys. J.* **6**, 148 (1962).
74. T. Lee, *Astrophys. J.* **224**, 217 (1978).
75. M. H. Thiemens and J. E. Heidenreich, *Science* **219**, 1073 (1983).
76. D. G. Sandler, S. E. Koonin, W. A. Fowler, *Astrophys. J.* **259**, 908 (1982).
77. F. R. Niederer, D. A. Papanastassiou, G. J. Wasserburg, *Astrophys. J. Lett.* **228**, L93 (1979).
78. A. Huck, G. Klotz, A. Knipper, C. Miéhé, C. Richard-Serre, G. Walter, *CERN 81-09* (1981), p. 378.
79. R. A. Chevalier and R. P. Kirshner, *Astrophys. J.* **233**, 154 (1979).
80. R. H. Becker *et al.*, *Astrophys. J. Lett.* **234**, L73 (1979).
81. J. M. Shull, *Astrophys. J.* **262**, 308 (1982); private communication (1983).
82. T. S. Axelrod, thesis, University of California, Berkeley (1980).
83. R. P. Kirshner and J. B. Oke, *Astrophys. J.* **200**, 574 (1975).
84. D. Branch, C. H. Lacy, M. L. McCall, P. G. Sutherland, A. Uomoto, J. C. Wheeler, B. J. Wills, *ibid.* **270**, 123 (1983).
85. J. Audouze and S. Vauclair, *An Introduction to Nuclear Astrophysics* (Reidel, Dordrecht, 1980), p. 92.
86. P. A. Seeger, W. A. Fowler, D. D. Clayton, *Astrophys. J. Suppl.* **11**, 121 (1965).
87. R. L. Macklin and J. H. Gibbons, *Rev. Mod. Phys.* **37**, 166 (1965). Also see B. J. Allen, R. L. Macklin, J. H. Gibbons, *Adv. Nucl. Phys.* **4**, 205 (1971).
88. I. Iben, Jr., *Astrophys. J.* **196**, 525 (1975).
89. J. Almeida and F. Käppeler, *ibid.* **265**, 417 (1983).
90. J. W. Truran, *International Physics Conference Series No. 64* (Institute of Physics, London, 1983), p. 95.
91. R. A. Becker and W. A. Fowler, *Phys. Rev.* **115**, 1410 (1959).
92. D. N. Schramm, in *Essays in Nuclear Astrophysics*, C. A. Barnes, D. D. Clayton, D. N. Schramm, Eds. (Cambridge Univ. Press, Cambridge, 1982), p. 325.
93. A. G. W. Cameron, J. J. Cowan, J. W. Truran, in *Proceedings of the Yerkes Observatory Conference on "Challenges and New Developments in Nucleosynthesis,"* W. D. Arnett, Ed. (Univ. of Chicago Press, Chicago, 1984).
94. W. A. Fowler, in *Proceedings of the Welch Foundation Conferences on Chemical Research, XXI, Cosmochemistry*, W. D. Milligan, Ed. (Robert A. Welch Foundation, Houston, 1977), p. 61; also see W. A. Fowler, in *Cosmology, Fusion and Other Matters*, F. Reines, Ed. (Colorado Associated Universities Press, Boulder, 1972), p. 67.
95. B. M. Tinsley, *Astrophys. J.* **198**, 145 (1975).
96. D. N. Schramm and G. J. Wasserburg, *ibid.* **163**, 75 (1970).
97. F.-K. Thielemann, J. Metzinger, H. V. Klapdor, *Z. Phys. A* **309**, 301 (1983); private communication.
98. A. Sandage and G. A. Tammann, *Astrophys. J.* **256**, 339 (1982).
99. S. van den Bergh, *Nature (London)* **229**, 297 (1982).
100. D. D. Clayton, *Astrophys. J.* **139**, 637 (1964).
101. W. A. Fowler, *Bull. Am. Astron. Soc.* **4**, 412 (1972).
102. S. E. Woosley and W. A. Fowler, *Astrophys. J.* **233**, 411 (1979).
103. R. L. Macklin, R. R. Winters, N. W. Hill, J. A. Harvey, *ibid.* **274**, 408 (1983).
104. R. L. Hershberger, R. L. Macklin, M. Balakrishnan, N. W. Hill, M. T. McEllistrem, *Phys. Rev. C* **28**, 2249 (1983).

105. J. C. Browne and B. L. Berman, *ibid.* **23**, 1434 (1981).
106. R. D. Williams, W. A. Fowler, S. E. Koonin, *Astrophys. J.* **281**, 363 (1984).
107. J. A. Payne, thesis, University of Glasgow (1965); R. W. P. Drever, private communication (1983).
108. B. Hirt, G. R. Tilton, W. Herr, W. Hoffmeister, in *Earth Sciences and Meteorites*, J. Geiss and E. D. Goldberg, Eds. (North-Holland, Amsterdam, 1963).
109. K. Yokoi, K. Takahashi, M. Arnould, *Astron. Astrophys.* **117**, 65 (1983).
110. S. E. Woosley, W. A. Fowler, J. A. Holmes, B. A. Zimmerman, *At. Data Nucl. Data Tables* **22**, 371 (1978).
111. My work in nuclear astrophysics has involved collaborative team work with many people, and I am especially grateful to F. Ajzenberg-Selove, J. Audouze, C. A. Barnes, E. M. Burbidge, G. R. Burbidge, G. R. Caughlan, R. F. Christy, D. D. Clayton, G. M. Fuller, J. L. Greenstein, F. Hoyle, J. Humblet, R. W. Kavanagh, S. E. Koonin, C. C. Lauritsen, T. Lauritsen, D. N. Schramm, T. A. Tombrello, R. V. Wagoner, G. J. Wasserburg, W. Whaling, S. E. Woosley, and B. A. Zimmerman. For aid and helpful cooperation in all aspects of my scientific work, especially in the preparation of publications, I am grateful to E. Gibbs, J. Rasmussen, K. Stapp, M. Watson, and E. Wood. I acknowledge support for my research over the years by the Office of Naval Research (1946 to 1970) and by the National Science Foundation (1968 to present).

This material originally appeared in *Science* **226**, 23 November 1984.

22. The Dark Night-Sky Riddle: A "Paradox" That Resisted Solution

E. R. Harrison

In recent decades the dark night-sky riddle has become widely known as "Olbers's paradox." This popular title, introduced by Hermann Bondi (*1*) in 1952, persists even though many writers (*2–8*) have shown that Wilhelm Olbers did not originate the riddle. In this chapter I show that the word "paradox" is also an unfortunate choice; it causes us to interpret too narrowly the scientific meaning of the riddle and to misjudge the historical evidence.

Olbers in 1823 discussed the darkness of the night sky in a universe uniformly sown with luminous stars and proposed that the most distant stars remain invisible owing to interstellar absorption of starlight (*9*). In 1744 Jean-Philippe Loys de Chéseaux had said much the same (*10*), and Jaki has discussed the circumstances relating to the remarkable similarity of the proposals made by Chéseaux and Olbers (*11*). The Chéseaux-Olbers solution of the dark night-sky riddle fails because the interstellar medium heats up, as shown by John Herschel (*12*), and emits as much radiation as it absorbs. Olbers made no reference to the work by Chéseaux; he referred to Edmund Halley's (*13*) papers, which Chéseaux, though undoubtedly influenced by them (*5, 14*), had failed to acknowledge. In two short but important papers Halley discussed in 1721 the riddle of a dark night sky in an infinite universe; in the second paper he referred to the riddle as a "Metaphysical Paradox," and in the first he attributed the riddle to an undisclosed source by stating:

> Another argument I have heard urged, that if the number of Fixt stars were more than finite, the whole superficies of their apparent Sphere would be luminous.

I have argued elsewhere (*15*) that Johannes Kepler was probably the first to realize that a dark night sky is in direct conflict with the idea of an infinite universe filled with luminous stars. In *Conversation with the Sidereal Messenger* (*16*), Kepler wrote in 1610,

> If this is true, and if they are suns having the same nature as our Sun, why do not these suns collectively outdistance our Sun in brilliance? Why do they all together transmit so dim a light to the most accessible places?

The historical evidence indicates that "Kepler's paradox" rather than "Halley's paradox" might be a more fitting title. Kepler, however, saw nothing paradoxical in a dark night sky; he believed in a finite bounded universe, and darkness at night confirmed his belief.

Alternative Interpretations

Unquestionably the riddle of darkness in an unbounded homogeneous universe of luminous stars raises cosmological issues of extraordinary subtlety. We observe the heavens studded with a finite number of visible stars, and we notice how they are separated by empty gaps of darkness. Why—in a universe that stretches away apparently without limit and contains possibly an unlimited number of luminous stars—do we observe dark gaps? When observing these dark gaps, what do we look at? From the outset we have a choice of alternative interpretations.

The first interpretation takes for granted the idea that the dark gaps are actually filled with distant stars. This idea is supported by the argument that a line of sight in any direction must always intercept the surface of a star, no matter how distant the star. If we suppose that most stars resemble the Sun, then the sky at every point should blaze as bright as the Sun's disk. This startling contradiction between theory and observation justifies the term "paradox."

This interpretation, frequently accepted without question by astronomers and historians who pay attention to the subject, assumes that in an infinite star-populated universe the observed dark gaps are completely filled with a continuous background of invisible stars. Rays of light emitted by the background of stars hurry toward us, and yet for some puzzling reason never reach Earth. Guided by this interpretation, historians have critically examined the various resolutions of the paradox proposed by astronomers in the preceding four centuries. Halley receives recognition as a pioneer, but earlier investigators such as Thomas Digges, William Gilbert, Johannes Kepler, and Otto von Guericke are begrudged recognition because they failed to stress the paradoxical aspect of the riddle of darkness. "Paradox" becomes the operative word, and it is the paradox of a dark night sky that must be resolved.

The second interpretation adopts the apparently simpleminded view that the dark gaps are mostly empty and not filled with a background of invisible stars. Of course, larger and better telescopes reveal more and fainter stars and also numerous extragalactic systems of stars, but however far we look out into the depths of space we always see the most distant stars immersed in pools of darkness. According to this second interpretation, the stars do not cover the sky, and the problem is reduced to explaining why the observed gaps remain unfilled in a star-populated universe of unlimited extent. Because no startling contradiction now exists between observation and expectation, "paradox" is no longer the operative word; we have a puzzle but not a paradox.

The first interpretation assumes that the stars actually cover the entire sky in contradiction of the evidence; hence the riddle ranks as a paradox, and we must explain why most stars remain unseen. The second interpretation assumes that the stars do not cover the entire sky in agreement with the evidence; hence the riddle falls short of being a paradox, and we must explain why stars are insufficient to fill the dark gaps and form an intensely luminous background. Constant use of the word "paradox" encourages us to overlook the possibility of the unparadoxical second interpretation.

All paradoxes are riddles, but not all riddles are paradoxes. Life abounds with paradox, yet "What is life?" is a riddle, not a paradox. Time flows, we say, and when asked, "How can time flow?", we are beset by a riddle that amounts to a paradox. Riddles are puzzles, problems, or paradoxes, which we attempt to unriddle. Paradoxes generally contain contradictory elements and consist of propositions contrary to known facts or received opinions. Calling a thing a riddle (or, as some would say, a mystery) does not exclude it from being also a paradox. Thus the title "dark night-sky riddle" allows for either of the aforementioned interpretations. Admittedly we lose the sensational appeal of "paradox," but this is a small price to pay for an uncommitted mind. A title such as "Olbers's paradox" or "Halley's paradox" or "dark night-sky paradox" (all of which I have used myself) has the disadvantage of stressing an aspect of the riddle that may be unwarranted.

Of those who adopted the paradoxical first interpretation, Halley gave a geometrical argument (*6, 17*) and concluded that the beams from distant stars "are not sufficient to move our Sense"; Chéseaux and Olbers appealed to interstellar absorption of starlight; and Bondi (*1*), working within the framework of the expanding universe, proposed that the remote stars are invisible because of their large red shifts. These and many other writers assumed that stars cover the sky at every point and tried to explain why the light emitted by most stars remains unseen.

Of those who adopted the unparadoxical second interpretation, Kepler assumed that we look out between the stars and in effect see a dark enclosing wall; John Herschel (*18*), Richard Proctor (*19*), Fournier d'Albe (*20*), and Carl Charlier (*21*) considered a hierarchical universe arranged in larger and larger systems in such a manner that distant stars remain insufficient to cover the sky; and Edgar Allan Poe (*22*) suggested that we look out in space and back in time and see the nothingness that existed before the birth of stars. These and other authors took the contrary view and assumed that the entire sky is not covered by stars.

Thomas Digges

Calling the puzzle a "paradox" tends to distort the historical picture. Halley referred to the paradoxical aspect of the riddle, and we may justly say that he originated the paradox of a dark night sky. Kepler, and other astronomers before Halley, failed to see anything paradoxical in the riddle. With our attention fixed on the first interpretation, we feel strongly tempted to criticize these earlier astronomers for failing to contribute anything significant to the cause of "Olbers's paradox." Notice how our attitude changes when we abandon the paradox template and regard the problem as an unqualified riddle. We must ask, Who was the first person to realize that the gaps of darkness between visible stars require an explanation? Without doubt, from this more general viewpoint, Kepler preceded Halley, and Digges preceded Kepler.

Thomas Digges, the foremost astronomer in England, revised in 1576 his father's book *Prognostication Everlastinge*. In an appended work (*23*) entitled

"A Perfit Description of the Caelestiall Orbes," Digges explained the Copernican system to a wide audience and introduced a major modification. He wrote,

> Especially of that fixed Orbe garnished with lightes innumerable and reachinge up in *Sphaericall altitude* without ende. Of which lightes Celestiall it is to bee thoughte that we onely behoulde sutch as are in the inferioure partes of the same Orbe, and as they are hygher, so seeme they of lesse and lesser quantity, even tyll our sighte beinge not able farder to reache or conceyve, the greatest part rest by reason of their wonderfull distance invisible unto us.

By grafting endless space on to the Copernican system and dispersing the sphere of affixed stars of the Ptolemaic system, Digges pioneered in 16th-century astronomy the idea of an infinite universe filled with countless stars, a universe "fixed infinitely up" and "garnished with perpetuall shininge glorious lightes innumerable" (Figs. 1 and 2). Copernicus revived the heliocentric theory of Aristarchus, and Digges revived the infinite universe of Democritus. It would be unfair to deny Digges the distinction of taking this extremely important step in astronomy on the grounds that he referred to the star-filled depths of space as the "pallace of foelicitye" and the "court of coelestiall angelles devoyd of greefe and replenished with perfite endlesse joye." Possibly he felt it imperative to make this concession to the theological convictions of his audience. If we deny Digges the innovation of an infinite universe, we must with equal injustice deny the more mystical Kepler his discovery of the three laws of planetary motion.

In Digges's *Perfit Description*, published 33 years after the death of Copernicus, we see the beginning of the dark night-sky riddle. Conceived in the Copernican Revolution and born when the infinite universe entered Western European astronomy, the riddle had yet to mature into the burning question, Why is the sky dark at night? This step was taken 34 years later by Kepler (*16*) in response to Galileo's discoveries with the recently developed telescope. The riddle has emerged as a realization that the invisibility of distant stars demands an explanation. The response given by Digges and accepted by many astronomers who followed was that most stars could not be seen because "the greatest part rest by reason of their wonderfull distance invisible unto us."

What could be more natural, given the rudimentary state of optical science in the 16th century, than to suppose that the most distant stars were too faint to be seen? At its birth the riddle received what seemed a perfectly sensible solution. Yet in the spirit of the first interpretation, Digges contributed nothing of significance to "Olbers's paradox" because he found nothing paradoxical in the darkness of the night sky. But in the spirit of the second interpretation, unencumbered by paradox, Digges originated the riddle of a dark night sky because he was the first person to realize that the dark gaps between visible stars need an explanation.

Did Halley Mislead Chéseaux?

Using the word "paradox" also tends to distort the cosmological picture. When we suppose the riddle to be paradoxical, we take for granted that stars cover the entire sky. The problem then consists of explaining why this immense multitude of stars remains unseen. Chéseaux avoided the fault in Halley's argument; he understood that stars increase

Folio.43

A perfit description of the Cælestiall Orbes,

according to the most auncient doctrine of the Pythagoreans. &c.

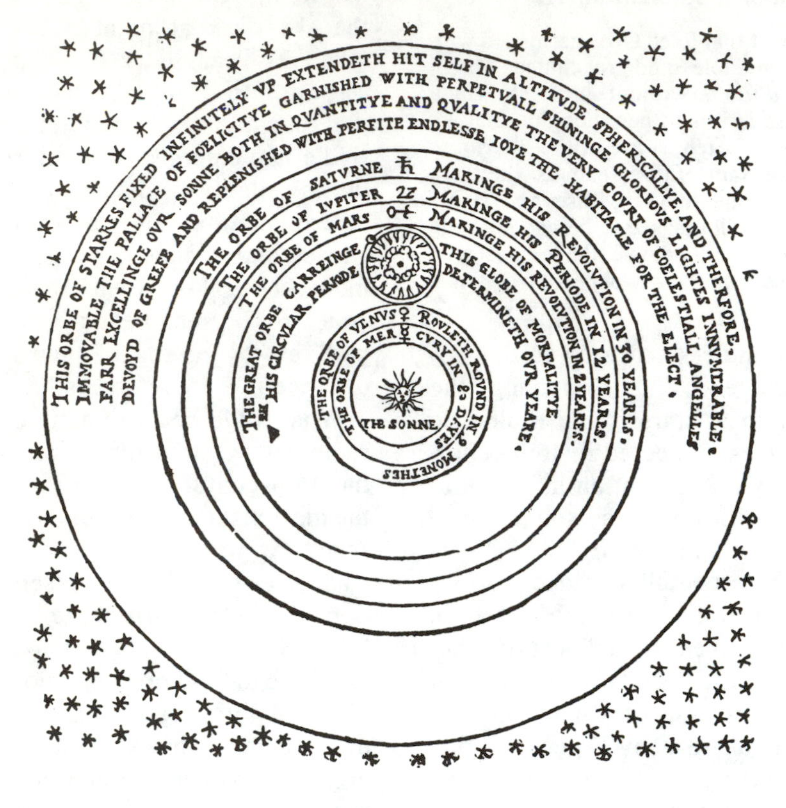

M. i. A PER-

Fig. 1. Thomas Digges's representation of the infinite universe in his *Perfit Description* (*23*). Because of *The Nature of Things* by Lucretius (first century B.C.), discovered in 1417, and because of *Learned Ignorance*, written by Nicholas of Cusa in 1440, the idea of an infinite universe was in the air in the 16th century. Digges, who influenced Giordano Bruno and William Gilbert (see Fig. 2), was the first astronomer to champion openly the idea of an infinite universe. He dismantled the sphere of affixed stars in the Ptolemaic and Copernican systems and dispersed the stars throughout infinite space. He was the first to realize that in an infinite universe the darkness of the night sky needs explanation. The legend in the diagram reads, "This orbe of starres fixed infinitely up extendeth hit self in altitude sphericallye, and therefore * immovable the pallace of foelicitye garnished with perpetuall shininge glorious lightes innumerable * farr excellinge our sonne both in quantitye and qualitye the very court of coelestiall angelles * devoyd of greefe and replenished with perfite endlesse joye the habitacle for the elect."

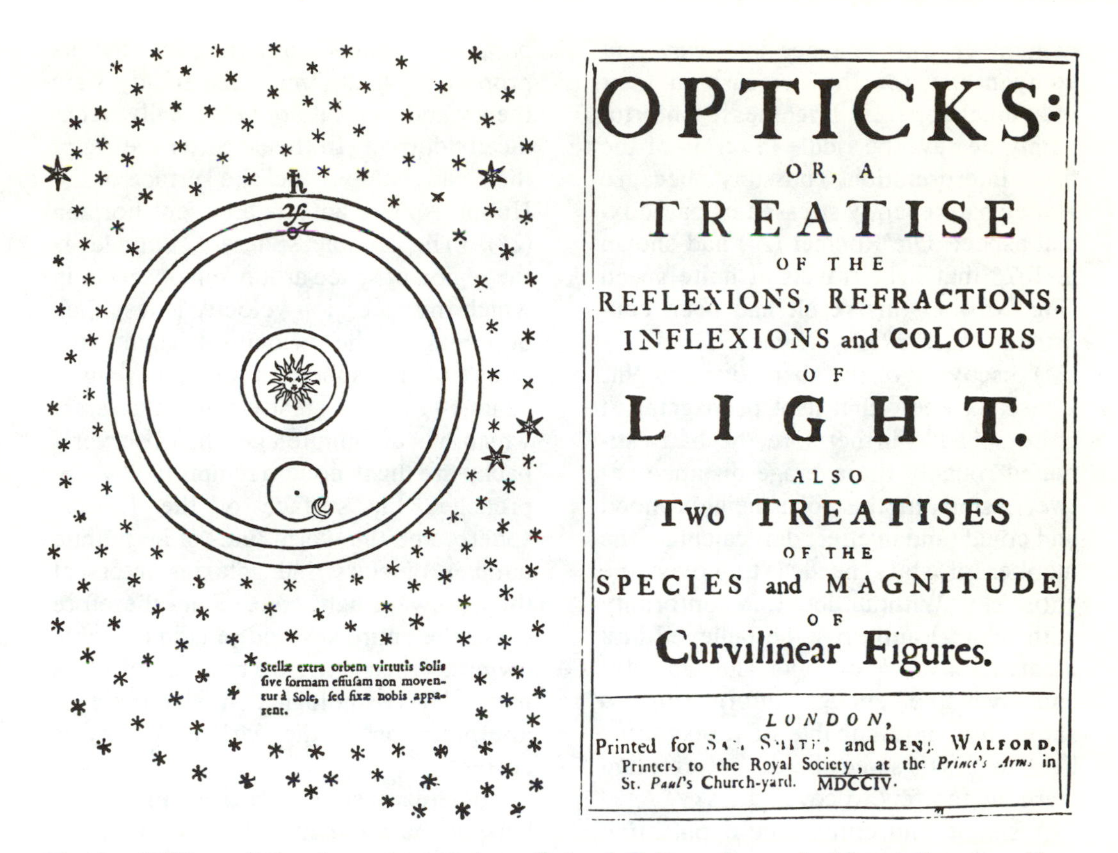

OPTICKS:
OR, A
TREATISE
OF THE
REFLEXIONS, REFRACTIONS, INFLEXIONS and COLOURS
OF
LIGHT.
ALSO
Two TREATISES
OF THE
SPECIES and MAGNITUDE
OF
Curvilinear Figures.

LONDON,
Printed for Sam. Smith, and Benj. Walford, Printers to the Royal Society, at the *Prince's Arms* in St. *Paul's* Church-yard. MDCCIV.

Fig. 2. William Gilbert's representation of the infinite universe in his posthumous *New Philosophy* (Amsterdam, 1651). His cosmological ideas were similar to those of Digges and Bruno; he rejected the geocentric system and championed the infinite universe introduced by Digges. In his great work *On the Magnet*, published in 1600, Gilbert wrote, "It is evident that all the heavenly bodies, set as if in destined places, are there formed unto spheres, that they tend to their own centres, and that round them there is a confluence of all their parts" (*34*). Fig. 3 (right). Title page of Newton's *Opticks*, published in 1704. The heading of proposition XI, book 2, part III, reads, "Light is propagated from luminous Bodies in time, and spends about seven or eight Minutes of an Hour in passing from the Sun to the Earth." The first sentence of the proposition states, "This was observed by Roemer, and then by others, by means of the Eclipses of the Satellites of Jupiter." The only thing needed to solve Digges's riddle within the context of Judaic-Christian-Islamic cosmology of a created universe of finite age was the discovery of the finite speed of light by Roemer in 1676. When Halley read his two papers to the Royal Society in 1721, Newton as president was in the chair (*6*). Why did Newton remain silent when the answer to the dark night-sky riddle lay at hand? Why did Halley not realize that he had the information and ideas needed to solve the riddle? Why, among the hundreds of astronomers in the last three centuries who have commented on the riddle and who believed in a universe of finite age, did no one realize how simple the solution was? This is now the only remaining puzzling aspect of the dark night-sky riddle.

in number with distance in a way that compensates for the decrease in their individual apparent brightness. Unfortunately he saw the riddle in terms of the first interpretation, possibly because Halley had recently stressed its paradoxical aspect. Ole Roemer (*24*) had shown in 1676 that light travels at finite speed (Fig. 3), and this result had been confirmed in 1729 by James Bradley's (*25*) discovery of the aberration of light. Chéseaux knew that light propagates at finite speed. Furthermore, he had estimated roughly the average distance between stars in the solar neighborhood and could (and in effect did) calculate the number of stars needed to cover the entire sky. Without doubt, in conformity with biblical authority, he believed in a created universe of finite age (*26*). He had available enough information to show that stars capable of transmitting light to Earth since the day of creation were far too few to cover the sky.

A simple and convincing explanation of the darkness of the night sky lay at hand. Instead, Chéseaux explored the idea of absorption and embarked on a complicated calculation in radiative transfer. Perhaps if Halley had asked the simple question, Why do dark gaps exist between the stars? instead of the paradoxical question, Why cannot we see the stars that fill the dark gaps? (*27*), Chéseaux might have solved the dark night-sky riddle.

The Expanding Universe

In an expanding universe the radiation received from receding extragalactic sources is enfeebled by red shift. This is the solution of the riddle first proposed by Bondi (*1*) that immediately springs to mind. Much of the subtlety of "Olbers's paradox" comes from the fact that its proponents had in mind the steady-state theory and were, if not historically, technically correct. In the de Sitter metric of the steady-state model the surface of the Hubble sphere acts as an event horizon (*28*). (The Hubble sphere is defined as the region of space around an observer in which the recession velocity is less than or equal to the velocity of light; the radius of this sphere is about 10^{10} light-years.) In the steady-state universe, which has an infinite age, the observer's backward light cone asymptotically approaches the surface of the Hubble sphere, and the world lines of an infinite number of stars and galaxies intersect the backward light cone. Stars therefore cover the entire sky and remain invisible owing to their extreme red shift. In this particular cosmological model the first interpretation of the dark night sky is correct.

But steady-state theorists and members of their audience believed that the red-shift solution had general validity and could be applied to all expanding cosmological models, including those with big bangs having particle horizons instead of event horizons. We may say, roughly speaking, that particle horizons exist in cosmological models of finite age (*28*). For several years, after Bondi had drawn attention to the subject, many investigators stated that the darkness of the night sky provides proof that the universe is expanding; some even claimed that expansion is the necessary and sufficient condition for darkness. Most writers who discussed the red-shift resolution of Olbers's paradox deemed mathematical confirmation quite unnecessary. Calculations in any case were awkward for evolving models and required integrations over the backward light cone in curved expanding space, and the results were generally difficult to interpret (*29*).

To find the right answer we must first ask the right question. When we ask a paradoxical question, we must not be surprised if we receive a paradoxical answer. Undoubtedly something about the red-shift solution is paradoxically odd. It implies that in a static universe the night sky is intensely bright. But a static universe of the kind imagined by Olbers and earlier workers, as we now realize, does not contain enough energy to create a bright night sky (*30*). If all matter in the universe were suddenly annihilated and the released energy converted into thermal radiation, the night sky would still be dark. The energy density of the radiation would be too low by a factor 10^{-13}. Why then must we appeal to the red shift of an expanding universe to resolve the so-called paradox when in fact a bright night sky is already impossible in a static universe? Come to that, why must we appeal to hierarchy, or any of the many variants of a static homogeneous universe (*5*), when already a bright night sky is impossible?

Something else about the red-shift solution, and about all other solutions in the category of the first interpretation, strikes one as rather odd. Simple estimates (*7*) show that the number of stars required to cover the entire sky has the enormous value of 10^{60}. But the number of stars in the observable region of an evolving universe (a region roughly equal in size to the Hubble sphere) is of order 10^{20}. When we observe the night sky, we look out in space 10^{10} light-years and back in time 10^{10} years to the early universe. The stars in the observable region of the universe, independently of whether they are huddled together in galaxies or uniformly scattered, are insufficient to fill more than 10^{-13} of the solid angle of the sky. Hence the dark gaps cannot be entirely filled with invisible stars. Interestingly enough, much the same situation occurs in a static model of the universe: The stars have typical luminous lifetimes of 10^{10} years, and, on looking out beyond a distance of 10^{10} light-years (or a few times this distance for several stellar generations), we look back to a dark era before the origin of stars. Again luminous stars cover only 10^{-13} of the sky.

Calculations made with powerful thermodynamic methods (*30, 31*) show that the extragalactic red shift of stellar light in most cosmological models is quite unimportant for the solution of the riddle. Far more important is the fact that stars cannot shine long enough to fill the universe with radiation equal in intensity to that at the surface of stars (*7, 15, 28*). Even in the standard model of an infinite homogeneous static universe of finite age, of the kind in which the riddle was conceived, the light emitted by stars falls a long way short of creating a bright sky, simply because the stars exhaust their energy reserves long before the universe fills with starlight. The time needed to fill the universe with starlight in equilibrium with the stars is roughly 10^{23} years, which greatly exceeds the luminous lifetime of stars. With the new methods (*32*) it is easy to design theoretical bright-sky as well as dark-sky static, nonstatic, and even steady-state cosmological models.

Conclusions

The sky happens to be dark in an evolving universe simply because stellar disks fail to cover the entire sky. The first interpretation, which treats the riddle as a paradox, is therefore wrong and the second interpretation correct. Since the time of Halley the riddle has usually been stated in the form of a paradox, and this may explain why earlier astronomers, who were unable to regard it as a paradox, have been denied credit for its discovery and development. I have shown that probably Thomas Digges was the first to realize that the dark gaps between the visible stars call for an explanation, and therefore he should receive credit as the originator of the riddle.

The habit of stating the riddle in paradoxical form may have greatly delayed the discovery of the explanation of why we live in a dark-sky universe. Both Halley and Chéseaux had sufficient knowledge to give a finite-velocity-of-light solution that would have been obvious and acceptable to everyone.

Of the various assumptions attributed to Olbers by authors who have discussed "Olbers's paradox," the one of crucial importance, and never mentioned, is his assumption that we can afford to ignore the speed of light. Olbers, repeating Chéseaux's argument, estimated with reasonable accuracy the average distance between stars in the neighborhood of the Sun. All he had to do was multiply the speed of light by the age of the universe—using either the Mosaic chronology or one of many estimates by earlier scientists, such as the 100,000 years from Immanuel Kant's (*33*) cosmogony of 1755—and in a sphere of this radius he would have found insufficient stars to cover the whole sky.

The region accessible to observation in a universe of finite age is always of finite size. The failure to realize this important truth by many scientists in the 18th, 19th, and 20th centuries constitutes the only puzzling feature that survives in the dark night-sky riddle.

Telescopes reveal numerous stars in the Galaxy that are invisible to the unaided eye; telescopes also reveal numerous galaxies stretching away to the horizon of the observable universe. The stars accessible to observation are found to be insufficient to cover all points of the sky. Through the gaps between stars we look back to the beginning of the universe. Poe (*22*) in 1848 correctly solved the riddle in a static model of the universe by supposing that "the distance of the invisible background so immense that no ray has yet been able to reach us at all." This finite-velocity-of-light solution needs only slight modification when adapted to an expanding universe originating from a singular state. A similar remark applies to the suggestion made by the British scientist Fournier d'Albe (*20*), who stated in 1907:

> If the world were created 100,000 years ago, then no light from bodies more than 100,000 light-years away from us could possibly have reached us up to the present; but light from stars further and further away would be continually arriving at the earth's surface, and thus our vision into space, confined at present by the Milky Way, would be expanding at the rate of 186,000 miles per second.

In the modern expanding and evolving universe we look through the gaps and "see" the big bang. The red-shift of starlight, which is of only minor importance in solving the riddle, now assumes paramount importance: We see the high-temperature big bang red-shifted into the feeble 3-degree afterglow that fills the universe.

The riddle of a dark night sky unfolds as an extraordinary story in the annals of science. One of the most remarkable features of the riddle is the mischief played by the seductive word "paradox," which has misled the astronomer into an unwarranted interpretation of the phenomena and the historian into too narrow an appreciation of the issues involved.

References and Notes

1. H. Bondi, *Cosmology* (Cambridge Univ. Press, Cambridge, 1952); "Theories of cosmology," *Adv. Sci.* **12**, 33 (1955). See also F. Hoyle, *Frontiers in Astronomy* (Heinemann, London, 1955); D. W. Sciama, *The Unity of the Universe* (Faber & Faber, London, 1959).
2. O. Struve, "The constitution of diffuse matter in interstellar space," *J. Wash. Acad. Sci.* **31**, 217 (1941); "Some thoughts on Olbers' Paradox," *Sky Telescope* **25**, 140 (1963).
3. G. A. Tammann, "Jean-Philippe Loys de Chéseaux and his discovery of the so-called Olbers' paradox," *Scientia (Milan)* **60**, 22 (1966).
4. S. L. Jaki, "Olbers', Halley's, or whose paradox?" *Am. J. Phys.* **35**, 200 (1967).
5. ———, *The Paradox of Olbers' Paradox* (Herder & Herder, New York, 1969). Contains much bibliographical material.
6. M. Hoskins ["Dark skies and fixed stars," *J. Br. Astron. Assoc.* **83**, 4 (1973)] criticizes the use of the word "paradox."
7. E. R. Harrison, "Why is the sky dark at night?" *Phys. Today* **28**, 69 (February 1974).
8. D. D. Clayton, *The Dark Night Sky* (Quadrangle, New York, 1975).
9. H. W. M. Olbers, "Ueber die Durchtigkeit des Weltraumes," *Astronomische Jarhbuch für das Jahr 1826*, J. E. Bode, Ed. (C. F. E. Späthen, Berlin, 1823), p. 10; reproduced in (*5*) as appendix 3; translated: "On the transparency of space," *Edinburgh New Philos. J.* **1**, 141 (1826).
10. J.-P. Loys de Chéseaux, *Traité de la Comète* (M. M. Bousequet, Lausanne, 1744), p. 223; reproduced in (*5*) as appendix 2. Chéseaux elaborated on Halley's idea of spherical shells of equal thickness and found, unlike Halley, that all shells give equal increments of light. Olbers followed this procedure and came to the same conclusion; also he used the novel argument that any line of sight must ultimately intercept the surface of a distant star and thus demonstrated that clustering of a finite number of levels in a multilevel universe cannot avert a bright sky.
11. S. L. Jaki, "New light on Olbers's dependence on Chéseaux," *J. Hist. Astron.* **1**, 53 (1970).
12. J. F. W. Herschel, "Humboldt's *Kosmos*," *Edinburgh Rev.* **87**, 170 (1848); reproduced in *Essays* (Longman, Brown, Green, Longmans and Roberts, London, 1857), p. 257. Strictly speaking, the statement that absorption fails to prevent a bright sky is inadequate; we must show that the absorbing matter heats up in less than the lifetime of the universe or the lifetime of the luminous stars, whichever is the smaller. For example, Herschel's criticism fails in the case of Fournier d'Albe's (*20*) proposal that for every luminous star there are 10^{12} nonluminous stars.
13. E. Halley, "Of the infinity of the fix'd stars," *Philos. Trans.* **31**, 22 (1720–1721); "Of the number, order, and light of the fix'd stars," *ibid.*, p. 24; both papers are reproduced in (*5*) as appendix 1. According to the *Journal Book* of the Royal Society, these papers were read in March 1721, and this is the year to which they should be assigned (*6*).
14. Possibly Chéseaux assumed that the educated reader was familiar with the literature on the subject, particularly with the papers by Halley, who had only recently died in 1742. Halley's two papers (*13*) on the infinity of the universe and the riddle of darkness were undoubtedly well known in astronomical circles; they had been reproduced, with other selected papers by different authors from 1719 to 1733, in a special edition of the *Philosophical Transactions* in six volumes; see vol. 6, part I (Brotherton, London, 1734), p. 147.
15. E. R. Harrison, "The dark night sky paradox," *Am. J. Phys.* **45**, 119 (1977).
16. E. Rosen, *Kepler's Conversation with the Sidereal Messenger* (Johnson, New York, 1965). See also A. Koyré, *From the Closed World to the Infinite Universe* (Harper, New York, 1958).
17. G. J. Whitrow, "Why is the sky dark at night?" *Hist. Sci.* **10**, 128 (1971).
18. J. F. W. Herschel, "Humboldt's *Kosmos*," *Edinburgh Rev.* **87**, 170 (1848); reproduced in *Essays* (Longman, Brown, Green, Longmans and Roberts, London, 1857), p. 285.
19. R. A. Proctor, *Other Worlds than Ours* (Appleton, New York, 1871), p. 286.
20. E. E. Fournier d'Albe, *Two New Worlds* (Longmans, Green, New York, 1907). This author makes several proposals, including hierarchy in a "multi-universe," absorption of light by nonluminous stars, and a finite-velocity-of-light solution similar to that made by Poe (*22*).
21. C. V. L. Charlier, "Ist die Welt endlich oder unendlich in Raum and Zeit?" *Arch. syst. Philos. (Berlin)* **2**, 477 (1896); "Wie eine unendliche Welt aufgebaut sein kann," *Ark. Mat. Astron. Fys.* **4**, No. 24 (1908). Charlier showed that the night sky remains uncovered by stars if, at each level, $R_i/R_{i-1} > \sqrt{N_i}$, where R_i is the radius of clusters of the ith level, containing N_i subclusters of radius R_{i-1}. He showed that, when hierarchy satisfies this condition, it resolves also the "gravity paradox," which concerned Hugo Seeliger ["Ueber das Newton'sche Gravitationsgesetz," *Astron. Nachr.* **137**, No. 3273 (1895)]. According to this so-called paradox, the gravity potential is everywhere infinite in a universe of unlimited extent and finite density. Advocates of this paradox omit to mention that without proper boundary conditions the Newtonian gravity potential is undefined. General relativity gives the correct theoretical treatment, and this paradox can be decently buried.
22. E. A. Poe, *Eureka: A Prose Poem* (G. Putnam, New York, 1848). Reprinted in *The Science Fiction of Edgar Allan Poe*, H. Beaver, Ed. (Penguin, Harmondsworth, Middlesex, England, 1976). Perhaps other solutions similar to Poe's treatment, in addition to that by Fournier d'Albe (*20*), exist in the literature, and, though

interesting, have been viewed as irrelevent because their authors failed to conform to the paradox convention.

23. T. Digges, "A perfit description of the caelestiall orbes" in *Prognostication Everlastinge* (London, 1576); reproduced by F. R. Johnson and S. V. Larkey, "Thomas Digges, the Copernican system, and the idea of the infinity of the universe in 1576," *Huntington Libr. Bul.* (Harvard Univ. Press, Cambridge, Mass., 1934), No. 5, p. 69. See also F. R. Johnson, "Thomas Digges and the infinity of the universe," *Astronomical Thought in Renaissance England* (Johns Hopkins Press, Baltimore, 1937), chap. 4; reproduced in M. K. Munitz, *Theories of the Universe* (Free Press, New York, 1957).
24. I. B. Cohen, "Roemer and the first determination of the velocity of light (1676)," *Isis* **31**, 327 (1940). Roemer studied the phase shift in the period of Io's Jovian orbit owing to the Doppler shift produced by Earth's motion. In effect, he measured the astronomical unit (the Sun-Earth distance) in light-travel time and found it to be 11 minutes. Halley [in *Philos. Trans.* **18** (No. 214), 237 (1694)] reviewed the eclipse observations of Io and obtained a value of 8.5 minutes—only a few seconds more than the modern value. Halley and Newton knew from the investigations by James Gregory that the separating distance between neighboring stars is of order 10^6 astronomical units—roughly 10 light-years. [See M. Hoskin, "The English background to the cosmology of Wright and Herschel," in *Cosmology, History, and Theology*, W. Yourgrau and A. D. Breck, Eds. (Plenum, New York, 1977.)] Neither Halley nor Newton seemed to realize that the age of the universe, when divided by the light-travel time from the Sun to Earth, would give the maximum possible visible distance measured in astronomical units. By using the 6000 years from biblical records for the age of the universe, derived by James Ussher, in *The Annals of the World Deduced from the Origin of Time* (1658), or by Newton, in *The Chronology of Ancient Kingdoms Ammended* (1710), or by any of the numerous reputable scholars such as Dante and Kepler, or by using from whatever source any cosmic age less than 10^{17} years, Halley in 1721 could have shown quite easily that the visible stars were too few to cover the sky. The riddle is why he thought the subject was a paradox.
25. J. Bradley, "A new discovered motion of the fix'd stars," *Philos. Trans.* (No. 406) (1729), p. 637. A reluctance (by Cartesians but not Newtonians) to accept Roemer's discovery of the finite speed of light vanished in 1729 when James Bradley discovered the aberration of light [see G. Sarton, "Discovery of the aberration of light," *Isis* **16**, 233 (1931)].
26. F. C. Haber, *The Age of the World: From Moses to Darwin* (Johns Hopkins Press, Baltimore, 1959).
27. Halley in his first paper (*13*) states, "so that, tho' it were true, that some such Stars are in such a place, yet their Beams, aided by any help yet known, are not sufficient to move our Sense."
28. W. Rindler, "Visual horizons in world-models," *Mon. Not. R. Astron. Soc.* **116**, 662 (1956). See also E. R. Harrison, *Cosmology: The Science of the Universe* (Cambridge Univ. Press, New York, 1981); "Cosmological horizons," *Phys. Today*, in press.
29. G. J. Whitrow and B. D. Yallop, "The background radiation in homogeneous isotropic world-models I," *Mon. Not. R. Astron. Soc.* **127**, 301 (1964); "The background radiation in homogeneous isotropic world-models II," *ibid.* **130**, 31 (1965).
30. E. R. Harrison, "Olbers' paradox," *Nature (London)* **204**, 271 (1964); "Olbers' paradox and the background radiation density in an isotropic homogeneous universe," *Mon. Not. R. Astron. Soc.* **132**, 1 (1965).
31. W. Davidson, "The cosmological implications of the recent counts of radio sources. II an evolutionary model," *Mon. Not. R. Astron. Soc.* **124**, 79 (1962); "Local thermodynamics and the universe," *Nature (London)* **206**, 249 (1965).
32. E. R. Harrison, "Radiation in isotropic and homogeneous models of the universe," *Vistas Astron.* **20**, 341 (1977). Contains a moderately complete bibliography on radiation in various cosmological models.
33. G. J. Whitrow, Ed., *Kant's Cosmogony*, translated by W. Hastie (Johnson, New York, 1970).
34. W. Gilbert, *On the Loadstone*, translated by S. P. Thompson (Chiswick, London, 1900).
35. I am indebted to S. Kleinmann and V. Trimble for their comments and corrections.

This material originally appeared in *Science* **226**, 23 November 1984.

Part IV
Instrumentation

23. Radio Astronomy with the Very Large Array

R. M. Hjellming and R. C. Bignell

The largest ground-based astronomical instrument ever built has now been completed on a 2100-meter plain in New Mexico. The instrument is called the Very Large Array (VLA) because it consists of 27 radio antennas spread in various sizes of Y-shaped configurations up to 35 kilometers in maximum extent. It is the most advanced instrument in radio astronomy. Funded by the National Science Foundation and built and operated by the National Radio Astronomy Observatory, the VLA is primarily an instrument for making two-dimensional images of astronomical sources of radio emission (*1*). It can produce images of radio sources with a resolution comparable to or better than that of the best ground-based optical telescopes.

Figure 1 shows several VLA antennas, a small section of the railroad track system for moving antennas, and the antenna assembly and maintenance building. The inner and outer portions of the array are schematically shown in Fig. 2, where one can see the relationship of the antenna stations, the twin railroad track antenna transportation system, and a waveguide communication system that allows operational control and data acquisition by computers in the control building.

Why the Very Large Array?

All sciences are fundamentally rooted in observation and experimentation. Astronomy outside the solar system is founded solely on observations of astronomical sources of radiation (electromagnetic) and matter (cosmic rays) reaching the earth. Because the earth's atmosphere is transparent to radiation mainly at optical (29 to 120 micrometers) and radio (1 centimeter to 100 meters) wavelengths, ground-based astronomy—and indeed all of astronomy—is dominated by observation in these wavelength ranges. The angular resolution (θ) of any telescope is fundamentally determined by the ratio of the wavelength L of the radiation and the size of the measuring instrument D, namely $\theta = L/D$.

Historically, optical observations have been the foundation and main support of astronomy. If one wishes to do radio astronomy with a resolution comparable to that at optical wavelengths, the radio instrument must be larger than the optical instrument by the ratio of their wavelengths, roughly 10,000. Large optical telescopes have optics with dimensions up to meters in size, so a radio instrument with comparable resolution must have dimensions of tens of kilometers. Radio antennas on this size scale cannot be constructed on the surface of the earth. Fortunately, electronically linked antennas with tens of kilometers of separation can be used as a single instrument to accomplish radio observations with a resolution which sometimes exceeds that possible with optical telescopes. The VLA is an array of 27 antennas, each with a surface 25 m in diameter, linked by a waveguide system along each of the three arms of a Y-shaped configuration. The four possible configurations for the antennas correspond to the range of sizes 1, 3.5, 10, and 35 km, so one can obtain comparable resolution for the four primary wavelengths, 1.3, 2, 6, and 21 cm. All antennas are movable and can be operated from any of the 72 three-piered concrete stations. They are moved by the antenna transporters along the twin railroad track system every few months to form one of the four standard antenna configurations. The antennas were assembled from prefabricated sections in the maintenance building (Fig. 1) between 1975 and November 1979.

Principle of Aperture Synthesis

The VLA parameters were established by the principal design goal: to build an instrument that would have variable, high-resolution capability for all accessible regions of the sky. In particular, this requirement set the number of antennas, their location along the arms of a Y-shaped railroad system, and the typical mode of operation by which radio sources to be mapped are tracked across the sky as the earth rotates. The basic principle of operation, called aperture synthesis, was pioneered by Sir Martin Ryle. Aperture synthesis is based on the fact that the cross-correlation of signals between any two antennas i and j is a measurement of a complex visibility that can be expressed, in a two-dimensional approximation valid under most circumstances, as

$$V_{ij}(u,v) = \int\int I(x,y)\, f(x - x_0,\, y - y_0) \times \exp[-2\pi i(u_{ij}x + v_{ij}y)]\, dx\, dy \qquad (1)$$

where $I(x,y)$ is the intensity distribution of radio emission on the sky as a function of sky coordinates x and y, $f(x-x_0, y-y_0)$ is the antenna sensitivity function when pointing at and tracking a sky position (x_0,y_0), and u_{ij} and v_{ij} are two coordinates describing the separation of the ith and jth antennas as seen by an observer located in the sky at (x_0,y_0). The measured complex visibilities are basically a two-dimensional Fourier transform of the radiation distribution on the sky weighted by the antenna sensitivity pattern, which localizes the observed radiation to the area of the sky covered by the antenna beam. Furthermore, since we know the sensitivity pattern for each antenna, Eq. 1 means that with a sufficient number of measurements of $V(u, v)$, one can reconstruct an image of the radio emission in the antenna beam by using the properties of Fourier transformation, namely

$$I(x,y) = \frac{1}{f(x - x_0,\, y - y_0)} \int\int V(u,v) \times \exp[2\pi i(ux + vy)]\, du\, dv \qquad (2)$$

The quality of images reconstructed from the numerical equivalents of Eq. 2 is determined by a combination of the quality of individual visibility measurements and the completeness of the sampling of (u,v) points inside the area of maximum radius D. Figure 3 shows plots

Fig. 1. Photograph of the inner portions of the array which gives close-ups of the antenna station piers, the railroad track system, and some of the antennas. The rectangular building in the background is the antenna assembly building and the view is to the southwest. [Courtesy of the National Radio Astronomy Observatory, operated by Associated Universities, Inc. under contract with the National Science Foundation]

of sampled (u,v) points as seen by an observer located in the sky at the position of a source at different declinations (angular distances from the projection of the earth's equator on the sky northward to the location of the radio source), assuming all 27 antennas of the VLA observe the source continuously from horizon to horizon. These are pictures of the "aperture" being synthesized by the VLA. The trade-offs between cost and sampling with as many antennas as possible resulted in the design compromise of 27 antennas in a Y-shaped configuration with sampling capabilities as shown in Fig. 3. An ideal instrument would have completely uniform sampling within a specific area.

Antennas and Electronics

During the years between first conception and final construction, the modes of operation of the VLA were made more complex by adding capabilities to accomplish most of the conceivable types of observations at centimeter wavelengths. This has resulted in an instrument that can switch rapidly (in tens of seconds) between the principal wavelength bands and tune to any frequency

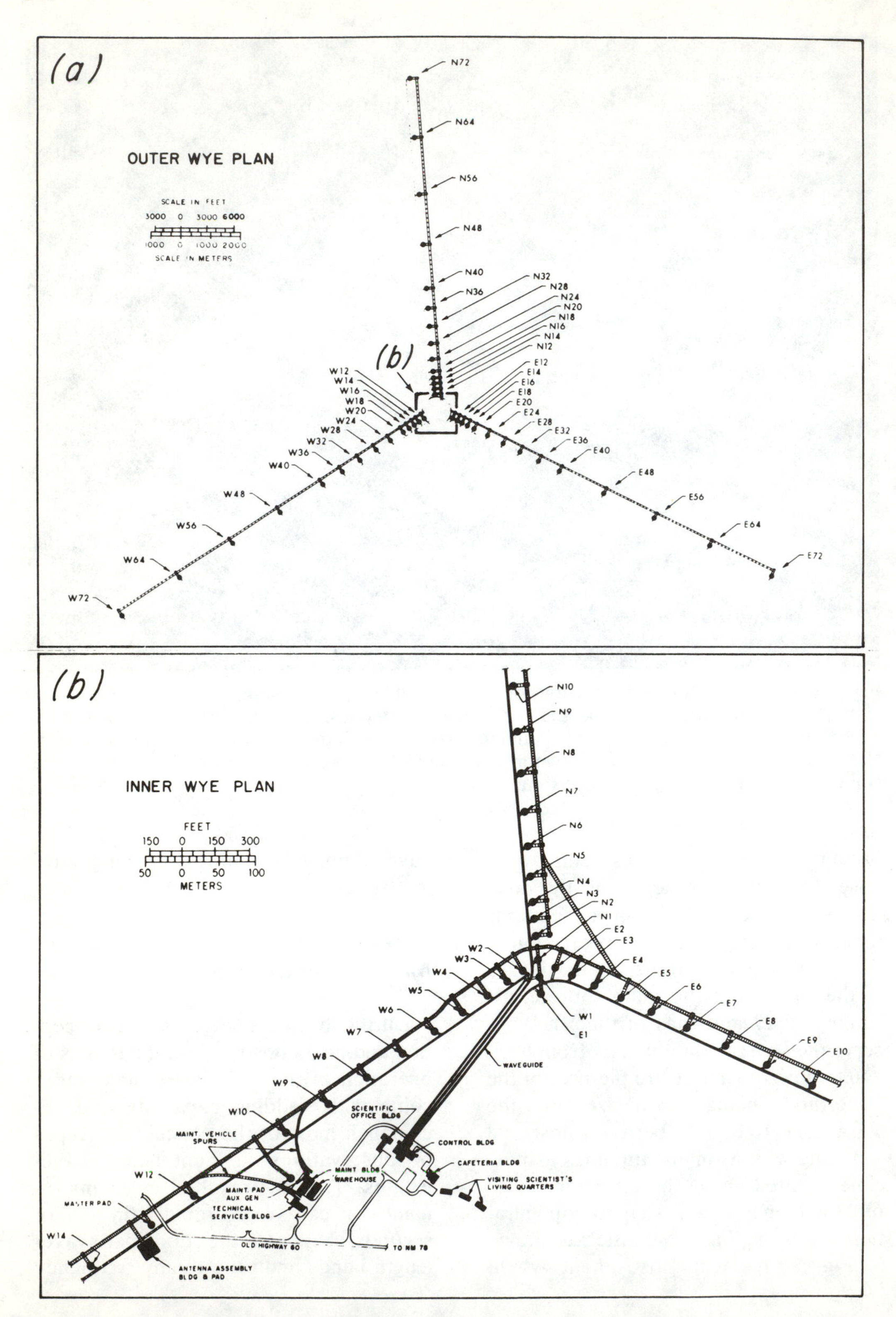
(a)
OUTER WYE PLAN
SCALE IN FEET
3000 0 3000 6000
1000 0 1000 2000
SCALE IN METERS
N72
N64
N56
N48
N40
N36
N32
N28
N24
N20
N18
N16
N14
N12
E12
E14
E16
E18
E20
E24
E28
E32
E36
E40
E48
E56
E64
E72
W12
W14
W16
W18
W20
W24
W28
W32
W36
W40
W48
W56
W64
W72
(b)
INNER WYE PLAN
FEET
150 0 150 300
50 0 50 100
METERS
N10
N9
N8
N7
N6
N5
N4
N3
N2
N1
E2
E3
E4
E5
E6
E7
E8
E9
E10
W1
E1
W2
W3
W4
W5
W6
W7
W8
W9
W10
W12
W14
WAVEGUIDE
SCIENTIFIC OFFICE BLDG
CONTROL BLDG
CAFETERIA BLDG
VISITING SCIENTIST'S LIVING QUARTERS
MAINT VEHICLE SPURS
MAINT BLDG
WAREHOUSE
MAINT. PAD AUX GEN
TECHNICAL SERVICES BLDG
MASTER PAD
OLD HIGHWAY 60
TO NM 78
ANTENNA ASSEMBLY BLDG & PAD

for observation of spectral lines, such as those listed in Table 1. As seen in Fig. 3, each antenna consists of a collecting surface with a nearly parabolic cross section, which focuses radiation on a rotatable, hyperbolic surface subreflector that is held in position by four support structures. Rotation of this subreflector under computer control results in focusing the radiation to one of the (currently) four feeds that collect at the 20-, 6-, 2-, and 1.3-cm wavelengths. The feeds transmit the observed radio signals to electronics in a room built under the surface of each antenna, where the cryogenically cooled electronics amplify, carry out various frequency selection processes, and convert the signals to antenna-dependent frequency regimes. These signals are sent up and down the waveguide of each arm in the array for nine antennas. During one out of every 52 milliseconds the signals in the waveguide are antenna and electronics control signals sent from the control computers in the control building, and during the other 51 msec they are signals containing astronomical information sent from each antenna to the electronics systems in the control building. The signals from each of n (=27) antennas are converted to the same frequency range and, after insertion of delays that compensate for the different times of arrival at different antennas, the signals for all $n(n-1)/2$ (=351) antenna pairs are multiplied (cross-correlated) and averaged over 10 seconds to produce measurements of complex visibility functions. Once all the observations of a particular radio source are stored in the off-line computer system (and on magnetic tape), the astronomer uses an extensive set of computer programs to apply known corrections, carry out an empirical calibration, compute the radio images by using the properties of Eq. 2, and display the results.

Extragalactic Radio Sources

It is not possible here to fairly summarize the hundred or so VLA observing programs that are carried out each year; however, the real reasons for building and using an instrument like the VLA are found only in the reasons for carrying out these observing programs. In this and the following sections we will therefore briefly discuss a sample of VLA observing programs.

The primary motivation for many of the U.S. astronomers who conceived the VLA was the mapping of radio sources associated with distant galaxies. The primary advantages of the VLA over present arrays in England and the Netherlands are a wide range of resolutions, including fine resolution hitherto unavailable, a greater ability to detect weak radio emission, greater flexibility, and a greatly improved imaging capability over the sky north of −48° declination. Maps of extragalactic radio sources have been one of the main products of the VLA ever since it became possible in 1977 to map sources with more than several antennas.

The radio source 3C388, mapped by Burns and Christiansen (*2*) (Fig. 4),

Fig. 2. (facing page) (a) Schematic diagram of the outer stations of the Y-shaped (*wye*) VLA array. The circles show antenna stations, the dashed line shows waveguide locations, and the solid lines correspond to the railroad track system. The designations *Xnn* are station identifiers; $X =$ *N, W*, and *E* for the north, southwest, and southeast arms and *nn* = station numbers, (b) Inner portions of the VLA showing buildings, roads, stations, the rail system, and the waveguide runs for each arm connected to the control building.

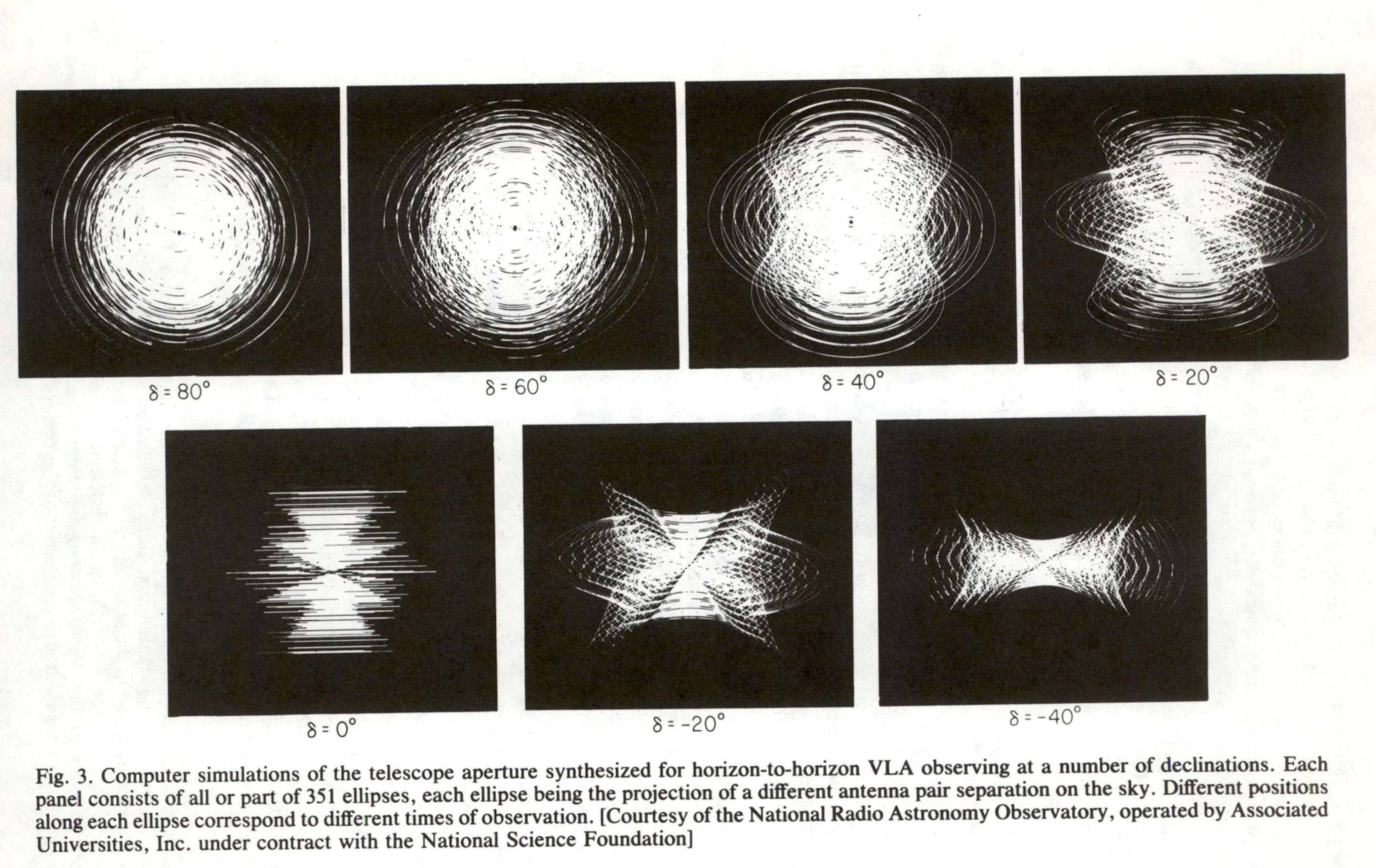

Fig. 3. Computer simulations of the telescope aperture synthesized for horizon-to-horizon VLA observing at a number of declinations. Each panel consists of all or part of 351 ellipses, each ellipse being the projection of a different antenna pair separation on the sky. Different positions along each ellipse correspond to different times of observation. [Courtesy of the National Radio Astronomy Observatory, operated by Associated Universities, Inc. under contract with the National Science Foundation]

Table 1. VLA observing frequencies and associated spectral lines.

Possible frequencies (GHz)	Protected* frequencies (GHz)	Atomic and molecular lines
1.34 to 1.73	1.40 to 1.427	Neutral H: 1420.4 MHz H, He, and so on: recombination lines $HCONH_2$ (formamide): 1538 to 1542 MHz OH: 1612, 1665, 1667, and 1721 MHz HCOOH (formic acid): 1639 MHz
4.5 to 5.0	4.99 to 5.0	$HCONH_2$: 4617 to 4620 MHz OH: 4660, 4751, and 4766 MHz H_2CO (formaldehyde): 4830 MHz H, He, and so on: recombination lines
14.4 to 15.4	15.35 to 15.40	H_2CO: 14.489 GHz H, He, and so on: recombination lines
22.0 to 24.0	23.6 to 24.0	H_2O: 22.235 GHz NH_3: 22.834 to 23.870 GHz

*Frequencies specifically allocated for radio astronomy by international treaties.

shows some of the principal features of large extragalactic radio sources. At the center of a large ellipical galaxy there is a core radio source whose dimensions are about one-fourth those of the surrounding double or lobed radio source in Fig. 4. The bright features in the lobes that curve back toward the central source are rudimentary versions of the "jets" that are found in many radio sources. These jets are a visible manifestation of the channeled regions through which the central portions of the galaxy supply energy for the magnetic fields and relativistic electrons whose interactions produce the observed radio emission by synchrotron radiation processes. Detailed mapping of the intensity and polarization structure in extragalactic jets has been and is one of the main types of VLA observing programs.

The imaging capability of the VLA at low declinations is greatly improved over that of previous instruments because of the Y-shaped distribution of antennas. The 20-cm radio image of a radio galaxy at a declination of −42° has been mapped by Ewald (*3*) and is shown in Fig. 5. The observations were taken with the 3.5-km configuration of the VLA in August 1980. The x in Fig. 5 corresponds to the location of a large ellipical galaxy that is the brightest member of a rich cluster of galaxies. The large radio structures to the east and north of the galaxy are typical of the so-called head-tail radio sources. Such radio tails are commonly interpreted as due to radio-emitting material that is ejected from the center of the galaxy and swept back by the dynamical interaction with the gas through which the galaxy is moving. Polarization maps of such radio sources reveal details about their magnetic field structures. Illustrating the need for the various sizes of the VLA, the extended structure in Fig. 5 is properly mapped only in the smaller configurations, whereas the small structures existing in the "head" of such head-tail radio sources can only be mapped with the larger configurations of the VLA.

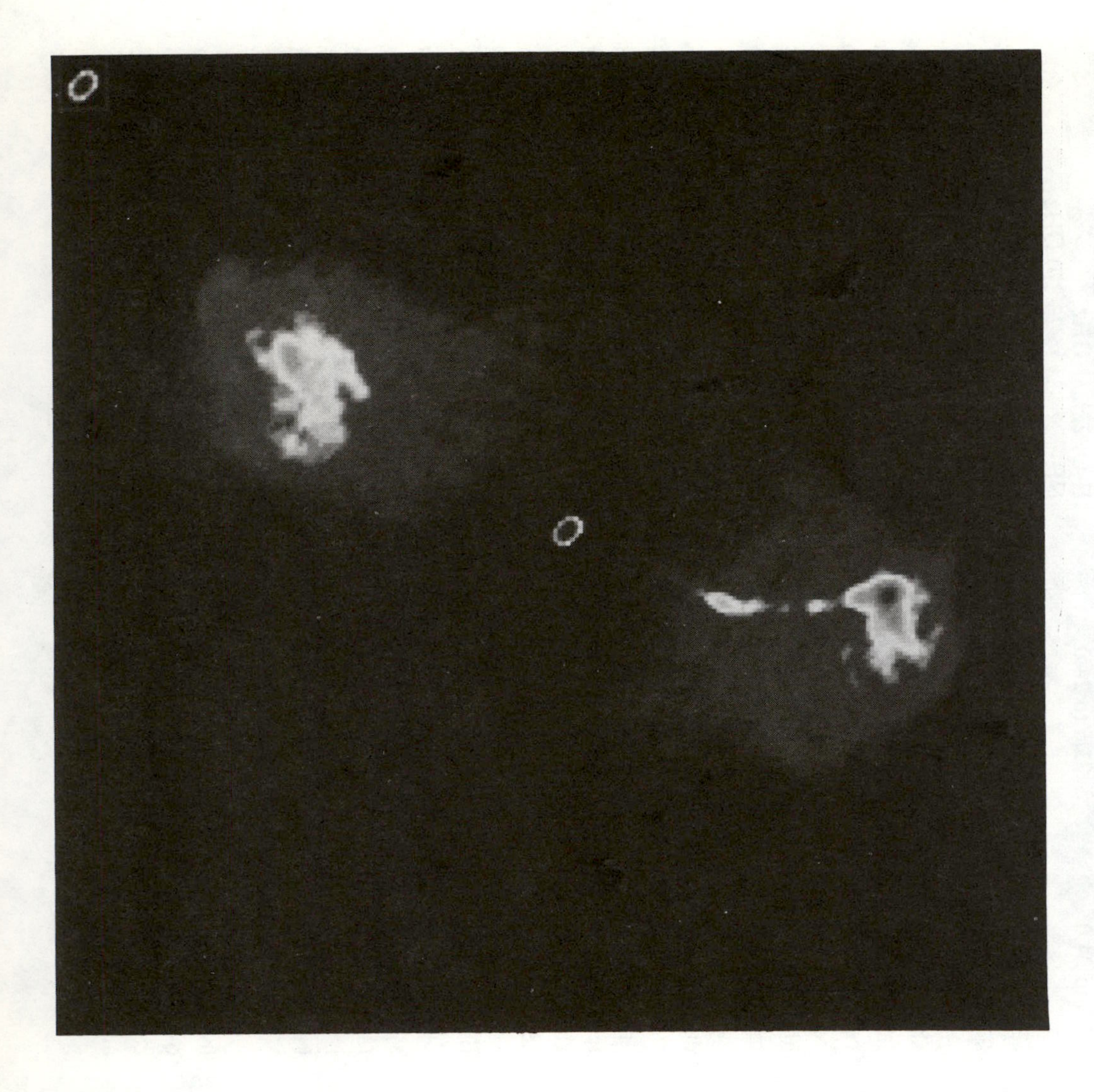

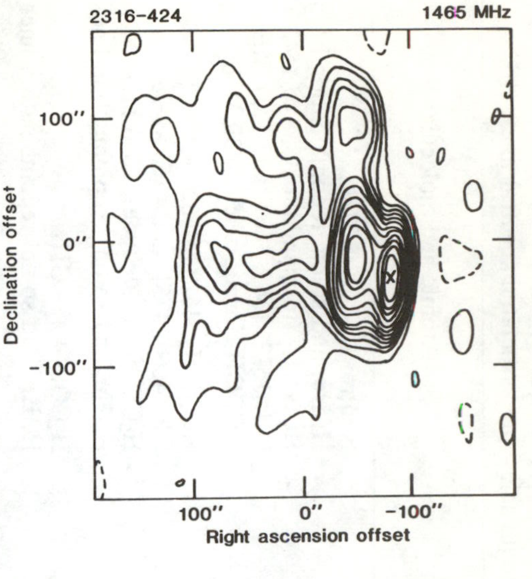

Fig. 4 (left). Radio image of the extragalactic radio source 3C388 at 20 cm as obtained by Burns and Christiansen (*2*). The appearance of a point source is indicated inside the box in the upper left corner. [Courtesy of the National Radio Astronomy Observatory, operated by Associated Universities, Inc. under contract with the National Science Foundation] Fig. 5 (above). The "head-tail" radio source 2316-424 as mapped by Ewald (*3*) at 20 cm. With a declination of −42°, it is an example of the good imaging characteristics of the VLA at very low declinations.

The high resolution of the VLA is being utilized by many observing programs searching for and mapping the small structures of radio sources associated with quasars. Many of these observations indicate that the only differences between quasar radio sources and radio sources associated with galaxies are their size scale. Jets seem to be a very common component of quasars, and one of the principal puzzles they introduce is the common occurrence of one-sided jets.

Radio Emission Associated with Stars

There are several types of stars in our galaxy which have observable radio emission. In some of these objects the high-resolution capabilities of the VLA now give astronomers the opportunity to map radio structures produced by stars and stellar systems.

The ultimate resolution possible with the VLA is obtained by mapping objects with the 35-km array at 1.3-cm wavelength. Such a map of the stellar winds surrounding the star V1016 Cygni, obtained in December 1980 by Newell (*4*), is shown in Fig. 6. This source was previously interpreted as one where thermal radio emission is produced by an ionized and massive stellar wind, and the VLA radio map shows that such stellar winds can have complex structure. With levels of detail of this type, one discovers that either there must be two interacting winds or the observed radio emission is due to a nova-like ejection of material into a previously existing stellar wind. The principal structures in V1016 Cygni are qualitatively similar to the equatorial ring and polar "blob" structures of nova shells.

A more exotic example of a radio source associated with a stellar system is SS 433. Figure 7 shows a VLA map of SS 433 made at 6-cm wavelength on 4 December 1980 by Hjellming and Johnston (*5*). The remarkable structures associated with this object change on time scales of a week or two. The radio emission is caused by synchrotron radiation from relativistic electrons and magnetic fields generated by a central star system. SS 433 has become famous in recent years (*6*) for the optical jets seen in the form of intense emission lines which indicate matter moving both toward and away from the earth with speeds that are projections of a total speed of one-fourth the speed of light. Changes in the apparent speeds of motion have been used to derive parameters of a twin-jet model in which the jets rotate with a period of 164 days about the jet axis.

Detailed structures in the radio jets of SS 433 are found to move outward from the star with an angular speed of 3 arc seconds per year. Hjellming and Johnston (*5*) showed that all the structures in maps of SS 433 at different times lie on a "corkscrew" pattern on the surface of a cone with an axis at a position angle of 100° oriented 80° from the line of sight. The conical surface is 20° from the cone axis, and the corkscrew pattern is achieved because the oppositely directed ejection vectors rotate with a period of 164 days. The corkscrew pattern for 4 December 1980 is superimposed on the radio map in Fig. 7. Because of time travel effects across the source, the apparent corkscrews of SS 433 are perceptibly "distorted" in a way that allows absolute determination of the velocity of motion. This velocity turns out to be one-fourth the speed of light, and all the geometric parameters of the rotating corkscrew match the parameters of the optical jets of SS 433. Thus the radio jets

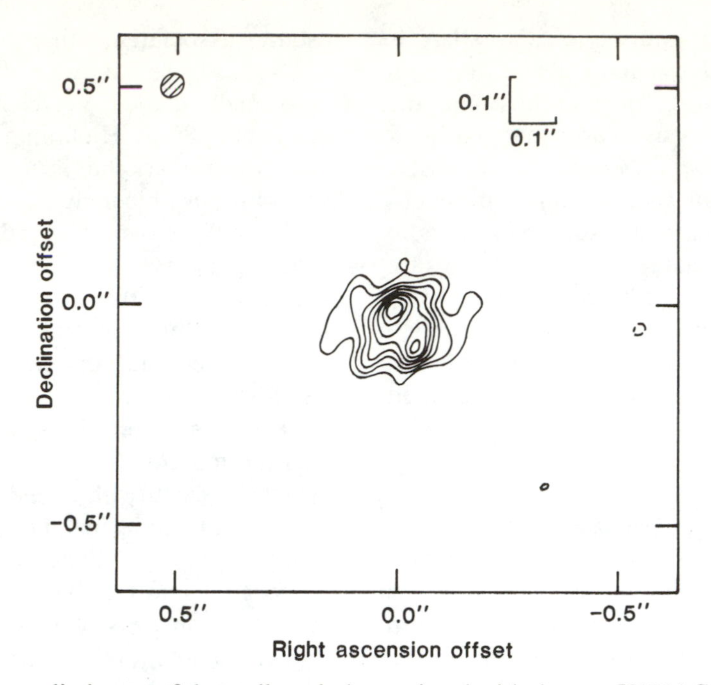

Fig. 6. A 1.3-cm radio image of the stellar wind associated with the star V1016 Cygni as mapped by Newell (*4*) with the largest (35 km) configuration of the VLA. This is an example of the high resolution in a VLA radio map, which cannot be obtained with ground-based optical telescopes because of the effects of the earth's atmosphere.

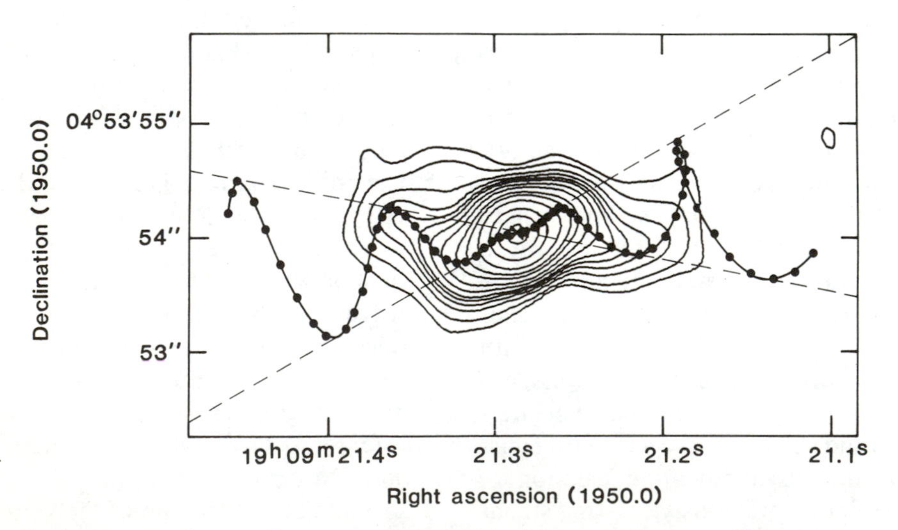

Fig. 7. Contour map of the radio source associated with the star system SS 433. The map was obtained in 5 minutes of observation at 6-cm wavelength with the largest configuration of the VLA on 4 December 1980 by Hjellming and Johnston (*5*). The corkscrew indicates the observed paths of motion due to the 164-day period of rotation of the twin-jet ejection vector about the central axis. The filled circles are located at intervals of 10 days in time of ejection from the center.

are additional observable manifestations of the moving material emitting optical lines very close to the star system, probably due to flows perpendicular to a precessing accretion disk.

One of the capabilities of the VLA which has become fully believable only with the completion of the instrument is the power of a single short observation or "snapshot." The map of SS 433 in Fig. 7 was made from data taken during a single 5-minute period. More extensive observations produce only small degrees of improvement, although they are essential for detecting the weakest levels of radio emission. The snapshot mode of observing makes it possible to map hundreds of strong sources in a day. Many observing programs can therefore be carried out in only an hour or two of scheduled observing. Because of this, many more scientists will be able to use the VLA in a given period of time than would have been possible without such a large number of antennas.

Observations of Radio Spectral Lines

Many atomic and molecular spectral lines can be observed with the VLA, some of which are listed in Table 1. In Fig. 8 a portion of the spectrum due to the 1612-megahertz line of the OH radical is shown, in addition to two high-resolution VLA maps made at different frequencies from data obtained by Bowers *et al.* (*7*). The 1612-MHz line is spread out in frequency due to line-of-sight Doppler shifts of the emitting material. In this case the OH emission is maser radiation coming from an expanding cloud of gas around a red giant star that is undergoing extensive mass loss. For a saturated maser like this, the amplification at a given velocity depends on the path length over which the velocity variation along the line of sight is very small. The peak feature in the spectrum in Fig. 8 is material approaching us on the near side of the star, and the second strongest feature is material on the other side of the star moving away from us. The weakness in the emission between these two features is due to the large velocity spread of matter moving perpendicular to the line of sight. The result is that the VLA map of the approaching material (velocity of −60 km/sec) shows the more compact emission with high maser gain along the line of sight, whereas the map at −50 km/sec is mainly a map of the extended shell structure of OH-emitting material perpendicular to the line of sight. The results shown in Fig. 8 are only a small fraction of the data from which the OH maser shell can be constructed in great detail. This provides specific information about the expanding atmosphere around the star.

Although spectral line work played only a small role in the original arguments for building the VLA, observations of most of the spectral lines listed in Table 1 have shown that detailed mapping of spectral line emission will be a major task of the VLA in the coming years.

Solar System Observations

The high resolution and sensitivity of the VLA make possible new types of observations of solar system objects. A spectacular example of this is the mapping of the radiation belts surrounding Jupiter. Shown in Fig. 9 is an early map of Jupiter made by Roberts *et al.* (*8*). The thermal emission from the central disk of the planet is clearly perceptible, and surrounding it is the nonthermal radio

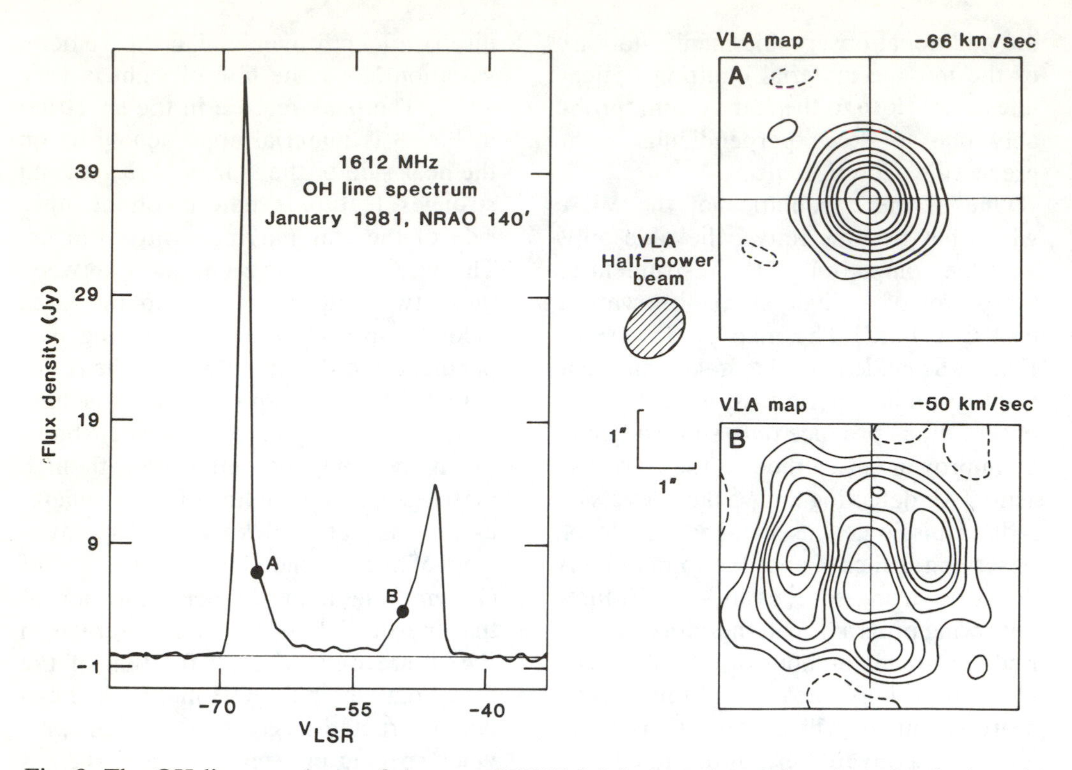

Fig. 8. The OH line spectrum of the star OH127.8-0.0 plotted as a function of velocity with respect to the local standard of rest (LSR), together with VLA maps made at specific frequencies corresponding to −66 and −50 km/sec by Bowers *et al.* (*7*).

emission from Jupiter's radiation belts. These structures are analogous to the Van Allen radiation belts around the earth.

Observations of radio events on the sun with the VLA have been carried out by groups from the California Institute of Technology, the University of Maryland, Tufts University, and others. With the high resolution and high frequencies of the VLA solar radio astronomers can observe deep into the solar atmosphere to see radio emission associated with the sites of origin of major flares. A VLA radio map of a solar active region obtained by Velusamy and Kundu (*9*) is shown in Fig. 10. The radio contours are superimposed on an optical photograph in the hydrogen alpha line taken by R. Robinson at Sacramento Peak Observatory.

Among the many complex problems being investigated by solar radio astronomers with the VLA, there is one specific theme that occurs with great frequency. With the VLA one can make very good, high-resolution maps of potential flare sites. This makes it possible to locate and study radio emission from material participating in the motions and acceleration processes involved in the conversions of energy between magnetic fields and plasma which are basic to the physics of active regions and flare sites.

A final example of the use of the VLA in solar system studies is the observation of asteroids. C. M. Wade, K. J. Johnston, and P. K. Seidelmann are using the

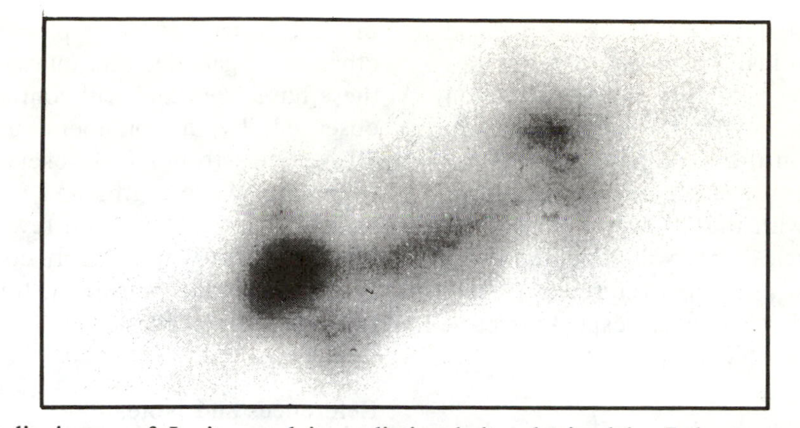

Fig. 9. Radio image of Jupiter and its radiation belts obtained by Roberts *et al.* (*8*) at a wavelength of 20 cm. [Courtesy of the National Radio Astronomy Observatory, operated by Associated Universities, Inc. under contract with the National Science Foundation]

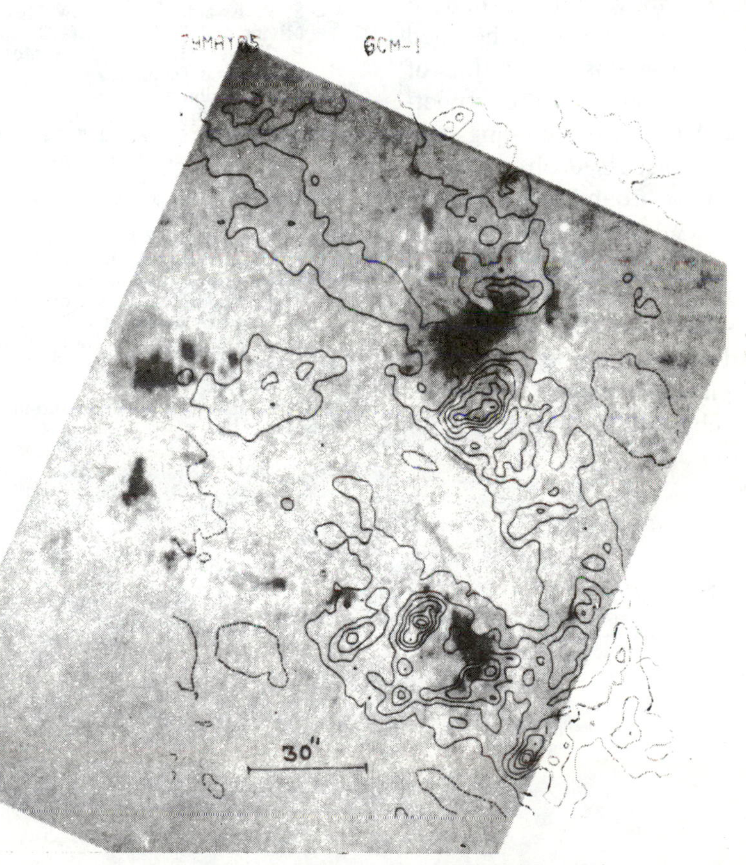

Fig. 10. Contour map of active regions on the sun made on 5 May 1978 at 6 cm by Velusamy and Kundu (*9*), superimposed on an optical picture taken in the Hα line by R. Robinson of Sacramento Peak Observatory.

VLA to observe and track Ceres and other asteroids. This is one of few cases where both radio and optical emission are due to exactly the same (thermal) processes in the same physical regions. Thus successful simultaneous tracking of asteroids with the VLA and optical astrometric telescopes will allow the radio and optical observing reference frames to be established with respect to each other to high accuracy.

Future of Astronomy with the Very Large Array

In the survey above I have had to neglect the vast majority of scheduled VLA observing programs. An outline of these programs summarizes the expected role of the VLA in the coming decades. I have not discussed observations of comets, moons around solar system planets, ordinary stars, double stars, flare stars, pulsars, gaseous nebulas, novas, supernovas, supernova remnants, x-ray sources, interstellar molecules, interstellar neutral hydrogen, the structure of nearby spiral galaxies, supernovas and gaseous nebulas in other galaxies, or the full variety of radio phenomena in other radio galaxies and quasars. All of these have been and will continue to be observed by astronomers using the VLA. For astronomical observations at centimeter wavelengths and resolutions from 0.05 arc second to a few arc minutes, the VLA will probably continue to be the dominant instrument for at least the next two decades.

References and Notes

1. A. R. Thompson, B. G. Clark, C. M. Wade, P. J. Napier, *Astrophys. J. Suppl.* **44**, 151 (1980).
2. J. Burns and W. Christiansen, *Nature (London)* **287**, 20 (1980).
3. S. P. Ewald, thesis, New Mexico Institute of Mining and Technology (1981).
4. R. T. Newell, thesis, New Mexico Institute of Mining and Technology (1981).
5. R. M. Hjellming and K. J. Johnston, *Astrophys. J. Lett.* **246**, L141 (1981).
6. B. Mangon, *Science* **215**, 247 (1982).
7. P. Bowers, K. J. Johnston, J. Spencer, *Astrophys. J.*, in press.
8. J. Roberts, G. Berge, R. C. Bignell, in preparation.
9. T. Velusamy and M. R. Kundu, in *Radio Physics of the Sun*, M. R. Kundu and T. Gergely, Eds. (Reidel, Boston, 1980), pp. 105–108.
10. The National Radio Astronomy Observatory is operated by Associated Universities, Inc., under contract with the National Science Foundation.

This material originally appeared in *Science* **216**, 18 June 1982.

24. Space Research in the Era of the Space Station

Kenneth J. Frost and Frank B. McDonald

As the first 25 years of the space age come to an end, we can reflect on a series of remarkable advancements and achievements in space science. In the planetary disciplines, there have been detailed analyses of the Apollo moon-rock samples, the Viking landings on Mars, the Pioneer entry and orbiter probes for Venus, the Mariner 10 close-up look at Mercury, and the Voyager measurements of the magnetospheres, atmospheres, rings, and satellite systems of Jupiter and Saturn. Observations of the earth, its atmosphere, land surface, and oceans have provided us with a global view of our own planet. These observations have advanced the scientific basis for weather forecasting and helped characterize the surface of the planet, from the correlation of phytoplankton concentrations with ocean circulation patterns, to the existence of ancient, now subsurface land forms in the Egyptian desert. These pioneering efforts also established the basis for the current operational satellites that daily provide data for geologists, agricultural scientists, meteorologists, and others.

In the solar-terrestrial sciences, studies of the solar wind, the corona, and the release of energy through solar flares have extended our knowledge of our own star and its surrounding heliosphere. The sun interacts in a complex way with the earth's magnetosphere and a large number of plasma processes have been identified and investigated by in situ measurements.

Probes such as Pioneer 10 are providing the first information on the distant regions of the outer heliosphere as they extend the distance space vehicles have traveled. The complete electromagnetic spectrum from radio waves to gamma rays can now be observed, and telescopes sensitive in the near and far ultraviolet regions of light have changed our concepts of the interstellar medium by revealing great expanses of clumped, hot plasmas that are produced by long-lived shocks from supernova explosions. X-ray astronomy provides an unexpected window on astrophysical phenomena, ranging from neutron stars, white dwarfs, and black holes to the hot gaseous medium in superclusters of galaxies. Investigations now concentrate on quasars and other active galactic nuclei, which release great amounts of energy via processes that are only dimly understood at best. The first steps in gamma-ray astronomy offer a new way to study the structure of our galaxy, and infrared astronomy has illuminated the "cool universe,'' from the birth of solar systems to celestial cirrus clouds of thin dust that may pervade the galaxy. Comparable progress has been made in many other fields ranging from the physics of the upper atmosphere to the nature of cosmic rays.

Many factors contributed to the rapid evolution of these scientific disciplines in space. A necessary condition was the continued development of rocket technology that began in the early part of this century with the pioneering work of Tsiolkovsky, Goddard, and Oberth and underwent a rapid development during World War II and the postwar era (*1*). In the late 1940's and early 1950's, sounding rockets and large skyhook balloons became valuable tools for studying the upper atmosphere, ionosphere, and cosmic rays and opened the field of ultraviolet astronomy. In order to take full advantage of these tools scientists learned how to conduct experiments on remote platforms and, aided by new silicon devices such as solar cells and transistors, how to design light-weight, low-power experiments. It was the coming together of these and other technological developments, as well as the maturity of individual scientific disciplines, that provided the foundation for rapid advances in space research.

This year President Reagan announced that the design and development of a space station will proceed. The program includes a manned facility with co-orbiting platforms, an orbital maneuvering system, and a polar platform (Fig. 1). This space station embodies many national goals, of which science is one.

The station should become operational in the early 1990's. The initial cost estimate is $8 billion (in 1984 dollars). This is on the same order as the development cost of the shuttle and less than 20 percent of the development cost of the Apollo program when compared in 1984 dollars. The space station is not as formidable an engineering challenge as either of those programs. The program is seen as an international venture, and discussions about it are already underway with Canada, Japan, and the countries of Western Europe.

The combination of a shuttle and space station along with continued technological advances in such areas as data transmission and microprocessors plus the maturity of space-related science disciplines offer significant increases in our research capabilities during the 1990's. The transition to the era of the space station will be complex and challenging. It will require bringing closer together NASA's manned and unmanned programs, which had been only loosely coupled until the advent of the shuttle.

In this chapter, we will illustrate how these new capabilities could be used. Since we are in the design phase of the space station project, it is crucial that scientists in different disciplines consider how they might best use the planned research facilities. To put this future in perspective, we will begin with a survey of the current U.S. program for space science.

The Current Program

Voyager II will explore Uranus and Neptune. The Galileo mission will send probes into the Jovian atmosphere, which will provide long-term synoptic observations of the Jovian cloud system, its moons, and its extended magnetosphere. The Venus Radar Mapper, with its synthetic aperture radar, will look

Fig. 1. Artist's concept of space station design showing modular construction. The actual architecture of the system is still in the preliminary planning stage.

through the thick clouds of Venus and map the topology of that planet, and the Mars Geochemistry and Climatology Orbiter will survey the global distribution of the elements on the Martian surface and record the climatic changes over a Martian year. Subsequent planetary exploration will focus on extended detailed studies of cometary nuclei and representative asteroids, and further study of the Saturian system, including Titan, is also under consideration.

In earth science, research satellites such as Nimbus 7, the Solar Mesosphere Explorer, and the International Sun Earth Explorers will continue to provide data along with meteorological and land satellites. The Upper Atmosphere Research Satellite will provide data on the stratosphere to determine how the chemical, dynamic, and radiative processes of this region determine the structure of the ozone layer. The Ocean Topography experiment together with a new research scatterometer will provide observations of the large-scale circulation of the oceans and their response to the atmospheric winds.

Our understanding of solar and terrestrial physics will be advanced by the three dimensional exploration of the heliosphere by the International Solar Polar mission, a joint effort with the European Space Agency. In the planning phase is an International Solar Terrestrial Physics program to better understand the sun and its coupling to the earth's magnetosphere and upper atmosphere. In the astrophysics area truly dramatic advances are expected. The combination of missions now planned includes the Space Telescope, the Cosmic Background Explorer, the Extreme Ultraviolet Explorer, and the Gamma Ray Observatory, along with the development of a new generation of observing instruments on shuttle Spacelab flights and major new observatories such as the Advanced X-Ray Facility and the Space Infrared Telescope Facility. These missions should provide an unprecedented increase in astrophysical knowledge.

In the near term, however, the most valuable elements of the current NASA program are the 17 active scientific satellites returning data to investigators. These missions are the sources of the results presented at meetings and published in the journals, and they maintain the vitality and productivity of our space science program.

On the one hand the U.S. space science program is in a well-balanced state with a level of financial support well above that of Western Europe, Japan, or the U.S.S.R. However, there has been a long-term change toward sustained observations from larger, more complex, longer-lived observatories and planetary orbiters. This evolution has occurred as the exploratory phase of space research has been completed. These programmatic changes, as well as a drop in the level of financial support (Fig. 2), have led to a dramatic decrease in the number of launches of science missions from an average of six per year in the late 1960's to 1.5 per year in the 1980's. The changes in funding for space science are complex with the large peaks from 1964 to 1966 and from 1972 to 1975 caused primarily by transient increases in the planetary program. Decreases in flight programs and funding since 1965 have forced dramatic reductions in many space research groups.

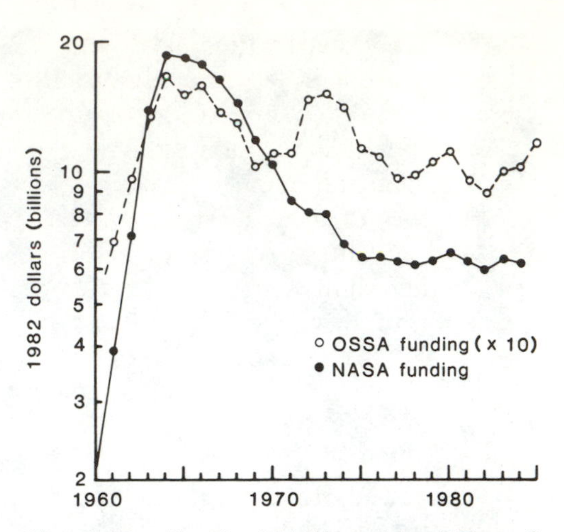

Fig. 2. Funding for NASA and its Office of Space Science and Applications (OSSA) from 1960 to 1985 (estimated) expressed in 1982 dollars. The OSSA share of the total NASA funding has doubled from approximately 9 percent in the peak years (1964–1966) to approximately 18 percent in 1985.

The space shuttle was expected to provide the means for quick, low-cost experiments, which would maintain the vitality of the experimental groups. With the shuttle having just reached operational status, its promise for science has not yet been realized. The fulfillment of the shuttle's capability is a major concern of the space science community. There are several promising developments, such as the Hitchhiker, Get-Away Special, and Spartan concepts, which would allow the use of the shuttle's cargo bay for moderate-size experiments on an as-available basis. Spacelab offers great capabilities, and with increasing use it is becoming a more flexible and cost-effective tool. It is of utmost importance that both the scientific community and the manned space program learn how to use the space shuttle to greater advantage for science. For it is with the shuttle that the systematic merging of the unmanned science program and the manned program will begin. This experience will provide the foundation for the era of the space station.

As viewed from our present experience, the scientific use of the space station can be divided into three broad categories: (i) for in-orbit assembly, refurbishment, and repair of spacecraft and experiments; (ii) as a laboratory for conducting experiments; and (iii) as a base for missions to and from synchronous orbit, the moon, planets, and other distant locations.

In-Orbit Assembly, Refurbishment, and Repair

Repair and refurbishment are key elements in current NASA planning. The repair of the Solar Maximum mission spacecraft in April 1984 by the crew of the space shuttle mission 41C was impressive. The initial effort by the mission specialist to capture the satellite using an untethered manned-maneuvering unit was not successful due to unexpected material that was mounted adjacent to the spacecraft's trunion pin. However, engineers from the Goddard Space Flight Center were able to stabilize the satellite, and the shuttle crew was able to capture it using the remote manipulator system (RMS). The RMS was then used to bring the satellite into the cargo bay where it was secured to a service table and electrically connected to the Orbiter via umbilical cables.

First, the attitude control system was replaced. The satellite, which was the first multimission, modular spacecraft, was designed to promote the use of common spacecraft subsystems. With slight modification of the modular design, the satellite could also be made repairable in orbit using the space shuttle. The attitude control system was removed by undoing two large jackscrew-type bolts. The control system package was then pulled back, which automatically disconnected the electrical circuits. The replacement system was then brought up and positioned, and the two jackscrew-type bolts were torqued down, pulling the module into place and mating the electrical connectors.

The second repair involved one of the scientific instruments in the payload—the coronagraph/polarimeter. Here it was necessary to replace an electronics box located inside the instrument module. Contrary to the current design philosophy for spacecraft, none of the satellite's scientific instruments had been designed with repair in mind. In an operation that required demating 12 small electrical connectors, the astronauts removed the electronics box and inserted a replacement box. After a complete check-out, the Solar Maximum mission satellite was found to be fully operational and was returned to free flight by the RMS. The repairs took less than half of the time estimated for them.

The RMS, which played such a crucial role in all phases of the repair activity, is a sophisticated and powerful device. Although not a robotics system in the rigorous sense, it is a stepping stone to the partnership between man and robotics that is planned for the space station. The system was supplied by the Canadian government and was designed and built by Canadian industry. When manipulating the Solar Maximum Observatory, it extended to a length of almost 12 meters and guided the 2200-kilogram spacecraft into and out of the cargo bay with impressive precision. Yet on the ground, the RMS cannot support its own weight in an extended position.

The repair of the Solar Maximum mission satellite is an example of the maintenance and refurbishment that will maintain semipermanent observatories in space, such as the space telescope and the Advanced X-ray Facility. The space station can greatly enhance this capability and enable more complex repair and refurbishment operations and delicate procedures such as the optical alignment of experiments.

Permanent Observatories in Space

In the future, remote-sensing satellites will observe the universe, the planets, the activity of our sun, and the global functioning of the earth. For astrophysics, there will be a family of observatories that will concentrate on various portions of the electromagnetic spectrum. There is a close analogy with large ground-based observatories, such as the Mount Palomar Telescope. When this telescope was first put into operation in the late 1940's, its huge area for collecting light (a 200-inch primary mirror) allowed us to peer deeper into space than ever before. Today its capability to do fundamental astrophysical research has not been exhausted. Progress in the technology used to view the images formed by this telescope and to analyze the data from them allows this telescope to be used in ways not envisaged by its designers.

There is an enormous advantage to placing a large optical telescope and other astrophysical observatories in space. The atmosphere allows only a small portion of the electromagnetic spectrum to reach the surface of the earth. The optical character of our atmosphere and its turbulence and temperature variation along the direction of observation permanently blurs our vision and limits how well, how far, and what we can see. The space telescope, in principle, will provide benefits similar to those that would accrue if the Mount Palomar Telescope were moved into space (Fig. 3).

The capability of the Space Transportation System to maintain the space telescope, once it is in orbit, is integral to ensuring the telescope's success. Along with being serviced, the observing instruments could also be upgraded to keep pace with advances in science and detector technology.

The space telescope is the prototype for long-lived observatories in all fields of remote sensing. For astrophysics, this concept will be extended to other spectral regions by the Gamma-Ray Observatory along with the proposed Advanced X-ray Astronomy Facility and the Space Infrared Telescope Facility.

For earth sciences, the first adaptation of this idea will be the Earth Observing System (EOS) that is under consideration for the space station's polar platforms. To understand the dynamic physical, chemical, and biogeochemical processes that comprise the global earth system, simultaneous observations are required from the full set of remote-sensing instruments in the EOS. To understand the changes at work in our

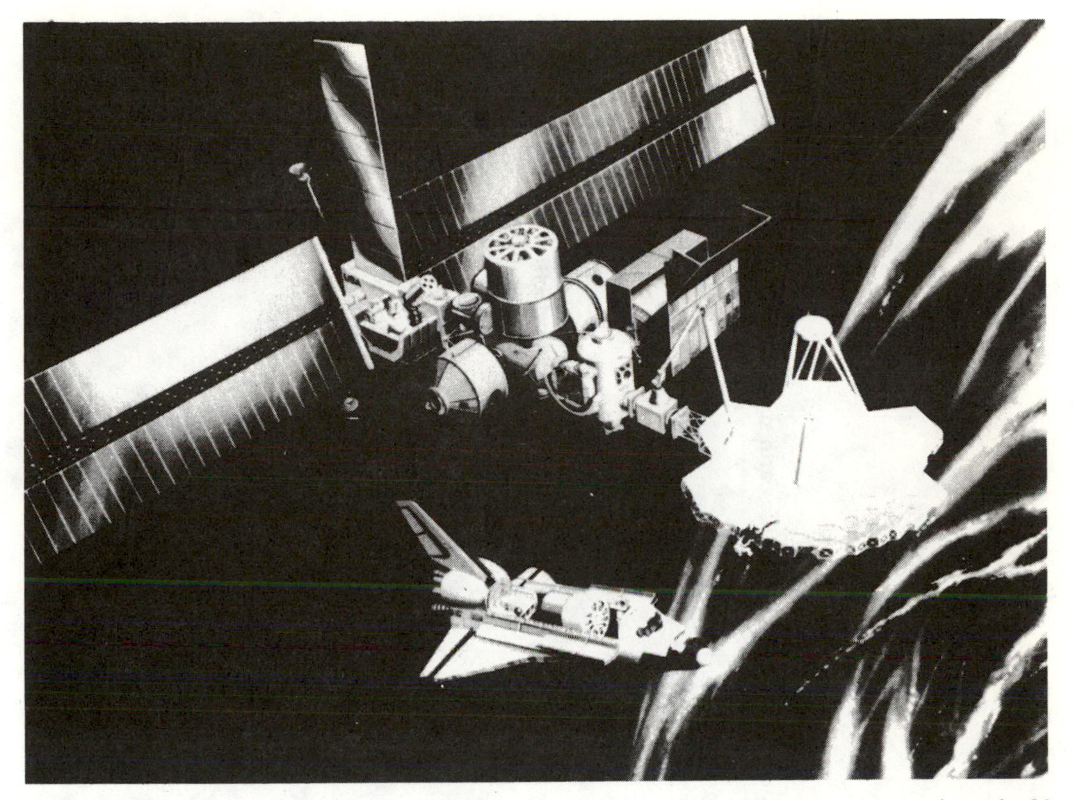

Fig. 3. Artist's concept of the construction in space of a large (diameter, approximately 20 meters) reflector array for astronomical studies in the far-infrared region.

environment, such as the seasonal cycles and the influence of man on the environment, sustained observations are required for a decade or longer. The ability to accommodate and service an integrated set of instruments is the key to accomplishing both tasks.

Long-lived observatories, made possible by the ability to repair and update both the satellites and the experiments they carry, would introduce fundamental changes in the manner in which the U.S. space science program is conducted. For example, there would be fewer new missions. A long-term commitment from NASA and from the scientific community will be required to ensure that the scientific vitality of these observatories will be maintained, but in return, however, there would be a continuous flow of data and opportunities for new experiments. The ability to replace experiments should stimulate the development of detector technology and information processing.

For planetary exploration, the space station can play a vital role as a staging point for the assembly, fueling, and inspection of large outbound spacecraft. In addition, it could act as a receiving laboratory or quarantine facility for samples from the surfaces of planets, comets, or asteroids. Missions to bring back such samples from Mars and from a comet could take advantage of the new space station, leading to simpler and less costly

Table 1. Major space observatories planned for astronomy and the earth sciences.

Observatory	Mass (kilograms)	Length (meters)	Diameter (meters)	Objectives
Hubble Space Telescope	9500	13	4.3	High resolution (0.01 second); for visible and ultraviolet regions
Advanced X-ray Astrophysics Facility	~9000	~13	~4.5	High resolution (0.1 second); for imaging in the x-ray region, 0.1 to 8 kiloelectron volts
Advanced Solar Observatory	~11,000	~8*	~4*	Space platform with high-resolution cluster (0.1 second); for visible through x-ray regions
Large deployable reflector	20,000 to 40,000	20 to 30	10 to 20	High resolution (1 second); for submillimeter region
COSMIC optical interferometer	10,000 to 15,000	15 to 30	4.5	Very high resolution (0.001 second); for imaging in the visible and ultraviolet regions

*For the high-resolution cluster only.

implementation than might have been possible otherwise. NASA is in the process of establishing a program of missions to explore the solar system based on technology developed during the last decade. With the space station acting as a receiving laboratory, the program could include the goal of systematically acquiring samples from throughout the solar system.

Finally, the space station will serve as a microgravity laboratory. The ability to manually conduct extended experiments in this environment will make possible a broad range of fundamental experiments in the life sciences, material sciences, physics, and chemistry.

Requirements for a Space Station

The space station will allow us to go beyond how we presently do science in space. To meet this challenge, however, we must learn to effectively combine our technologies, our structures, data systems, detectors, and so forth, into permanent observatories in space. A number of such observatories are in the early stages of planning and we should reexamine them. In the present program, the space telescope, the Advanced X-Ray Facility and the Space Infrared Telescope Facility each will completely fill the shuttle payload bay at launch. The larger observatories of the future must be assembled in orbit and will require special facilities at the space station.

Examples of large space observatories for astronomy and the earth sciences are listed in Table 1. To provide a specific example, it is useful to consider the large deployable reflector (Fig. 3), for which early definition studies have already begun. This instrument will study the far-infrared (30 to 200 micrometers) and submillimeter (200 micrometers to 1 millimeter) radiation from newly forming stars and planetary systems as well as the ancient, red-shifted signals from galaxies and pregalactic gas located at the edge of the universe. A 1–arc-second resolution, sensitive observations of spectral lines, and a large collecting area are necessary to undertake these investigations. As currently envisaged, the reflector is approximately 20 meters in diameter and composed of 60 mirror segments. It is expected that the mirror segments will have to be continuously adjusted in order to maintain acceptable image quality because of the thermal and mechanical disturbances during a given orbit.

This instrument is but a sample of what the 1990's hold for space science. Similar missions are planned for earth observations, communications, and experiments to detect gravity waves and to verify aspects of the general and special theories of relativity. Many anticipated payloads are beyond the launch capability of any available booster, and all press the capability of the shuttle itself. The shuttle would be able to carry into orbit the materials needed to assemble one of these observatories. However, the time, the crew, and the facilities required to both assemble and check these observatories are beyond what the shuttle can handle. The space station, however, will make the assembly and implementation of such observatories feasible.

From the Shuttle to the Space Station

Concurrent with developing the space station architecture, NASA should enhance the level of science done with the space shuttle during the remainder of the 1980's. Such an approach would move space science into the era of the space station with optimal advantage to both the scientific community and the space station itself. Such an approach would also avoid the difficulties that have been associated with the transition to the space shuttle. Contrary to widespread belief, the shuttle program appears to have had little impact on the funding of space science (Fig. 2). However, the decision to eliminate some types of expendable launch vehicles before the shuttle became operational did delay programs such as the Galileo mission to Jupiter and the International Solar Polar mission, which resulted in significantly higher costs for these programs.

The scientific features and capabilities that have been broadly sketched in this article are not necessarily an automatic, predetermined part of a space station program. For example, the Soviet Union has had a very active space station program starting in 1971 and has launched six Salyut space stations into orbit (*2*). With Salyut, the Soyuz spacecraft for ferrying crews, and Progress vehicles for the resupply and refueling operations, the Soviet Union has demonstrated the ability to maintain man in space for periods as long as 6 months. Salyut 7 and COSMOS 1443 were linked together to form an orbital complex capable of housing as many as six crew members. These missions have made important advances in studying the long-term effects of microgravity on human physiology, and the Russian cosmonauts have carried out a number of on-board experiments including material science studies and astronomical and earth science experiments. In fact, the long-term objective could well be toward the colonization of space. Repair, maintenance, and refurbishment of the Salyut station itself appears to have been a major activity. However, the Soviet Union has apparently not attempted to maintain scientific observatories in space nor begun the on-orbit assembly of large structures for scientific experiments.

The Future

The limits we face in putting experiments in space are not fundamental, like those posed by the earth's atmosphere for a ground-based telescope. Our present limits are technological and managerial. The space station is an opportunity to move those limits with creativity and imagination. It should be approached in that manner and not viewed as an alternate opportunity to implement our current program.

The experience, expertise, and creativity that resides in our space research communities are unique resources that must be used to assure that the space station will be a step forward for space research. How to effectively utilize these resources is a crucially important challenge in managing the development of the station.

Under the auspices of its Advisory Council, NASA has established a study group, with Peter Banks of Stanford University as chairman, to define the scientific requirements of the space station. The Space Science Board of the National Academy of Sciences, under the leadership of Thomas M. Donahue of the University of Michigan, is undertaking a 2-year study to define the nation's long-range objectives in space science for the period 1992 to 2015. These studies will be closely linked and will form a crucial part of NASA's long-range planning for effective scientific utilization of the space station.

The space station will be judged on the magnitude of the advance it affords in the utilization of space. Effective utilization is as important as that of engineering the hardware itself. The responsibility for designing and building a scientifically useful space station is a mutual one. It is shared by NASA and the scientific community jointly, and jointly they are responsible to the public for our progress in space.

References and Notes

1. H. Newell, *Beyond the Atmosphere* (NASA SP4211, National Aeronautics and Space Administration, Washington, D.C., 1980).
2. Office of Technology Assessment, *Salyut, Soviet Steps Toward Permanent Human Presence in Space* (GPO Stock No. 052-003-00937-4, Government Printing Office, Washington, D.C., 1983).

This material originally appeared in *Science* **226**, 21 December 1984.

About the Authors

James R. Arnold is professor in the Department of Chemistry, University of California at San Diego, La Jolla, and director of the California Space Institute.

R.C. Bignell is an astronomer working for the National Radio Astronomy Observatory in Socorro, New Mexico.

David C. Black is an astrophysicist in the Space Science Division at NASA Ames Research Center, Moffett Field, California. He was a member of the Galactic Astronomy Working Group of the National Academy of Sciences Astronomy Survey Committee.

Leo Blitz is assistant professor in the Astronomy Program at the University of Maryland, College Park.

Erika Böhm-Vitense is professor in the Department of Astronomy at the University of Washington, Seattle.

S. Chandrasekhar is Morton D. Hull Distinguished Service Professor at the University of Chicago, Chicago, Illinois.

Kris Davidson is associate professor in the Astronomy Department at the University of Minnesota, Minneapolis.

David S. De Young is an astronomer at Kitt Peak National Observatory, National Optical Astronomy Observatories, Tucson, Arizona.

J.V. Evans is chairman of the Committee on Solar-Terrestrial Research at the National Academy of Sciences, Washington, D.C., and assistant director of Lincoln Laboratory at the Massachusetts Institute of Technology, Lexington.

Paul D. Feldman is professor of physics at Johns Hopkins University, Baltimore, Maryland.

Michel Fich is a research assistant in the Astronomy Department at the University of California, Berkeley.

William A. Fowler is Institute Professor of Physics Emeritus at the California Institute of Technology, Pasadena.

Kenneth J. Frost is a senior scientist at the Goddard Space Flight Center, Greenbelt, Maryland. He has also served as science adviser to the space station task team at NASA Headquarters.

Robert D. Gehrz is professor of physics and astronomy at the University of Wyoming, Laramie. He chaired the Galactic Astronomy Working Group of the National Academy of Sciences Astronomy Survey Committee.

E.R. Harrison is professor of astronomy in the Department of Physics and Astronomy at the University of Massachusetts, Amherst, and a member of the Five College Astronomy Department of Amherst, Hampshire, Mount Holyoke, and Smith colleges, and the University of Massachusetts.

R.M. Hjellming is an astronomer working for the National Radio Astronomy Observatory in Socorro, New Mexico.

Roberta M. Humphreys is associate professor in the Astronomy Department at the University of Minnesota, Minneapolis.

Richard G. Kron is associate professor at the University of Chicago, Yerkes Observatory, Williams Bay, Wisconsin.

Shrinivas Kulkarni is a research assistant in the Astronomy Department at the University of California, Berkeley.

George Lake is with the Physical Research Laboratory, AT&T Bell Laboratories, Murray Hill, New Jersey.

Devendra Lal is director of the Physical Research Laboratory, Navrangpura, Ahmedabab, India, and visiting professor at Scripps Institute of Oceanography, La Jolla, California.

Frank B. McDonald is NASA's chief scientist at NASA Headquarters in Washington, D.C.

Bohdan Paczyński is professor of astrophysical sciences at Princeton University, Princeton, New Jersey. He is on leave from the N. Copernicus Astronomical Center, Polish Academy of Sciences, Warsaw, and is a long-term member of the Institute for Advanced Study at Princeton.

P.J.E. Peebles is Albert Einstein University Professor of Science, Princeton University, Princeton, New Jersey.

M. Peimbert is an astronomer at Instituto de Astronomía, Universidad Nacional Autónoma de México, México, México.

Robert C. Reedy is a staff member in the Nuclear Chemistry Group at Los Alamos National Laboratory, Los Alamos, New Mexico.

Morton S. Roberts is senior scientist and former director at the National Radio Astronomy Observatory in Charlottesville, Virginia.

Vera C. Rubin is a staff member of the Department of Terrestrial Magnetism at the Carnegie Institution of Washington, Washington, D.C., and an adjunct staff member of the Mount Wilson and Las Campanas observatories.

David M. Rust is senior staff scientist at American Science and Engineering, Inc., Cambridge, Massachusetts.

A. Serrano is an astronomer at Instituto de Astronomía, Universidad Nacional Autónoma de México, México, México.

Philip M. Solomon is professor of astronomy at the State University of New York, Stony Brook. He was a member of the Galactic Astronomy Working Group of the National Academy of Sciences Astronomy Survey Committee.

Lyman Spitzer, Jr. is Charles A. Young Professor of Astronomy Emeritus at the Princeton University Observatory, Princeton, New Jersey.

K.M. Strom is a visiting astronomer at Kitt Peak National Observatory, National Optical Astronomy Observatories, Tucson, Arizona.

S.E. Strom is an astronomer at Kitt Peak National Observatory, National Optical Astronomy Observatories, Tucson, Arizona.

S. Torres-Peimbert is an astronomer at Instituto de Astronomía, Universidad Nacional Autónoma de México, México, México.

Edwin L. Turner is associate professor of astrophysical sciences and an Alfred P. Sloan Research Fellow at the Princeton University Observatory, Princeton, New Jersey.

Arthur H. Vaughan is a scientific staff member of the Applied Optics Division, Perkin-Elmer Corporation, Garden Grove, California, and also a staff associate of the Mount Wilson and Las Campanas observatories, Carnegie Institution of Washington, Pasadena, California.

Index

3C 13, high redshift for, 214
3C 65, redshifts measured from colors, 213
3C 133, as strong emission line galaxy, 213
3C 273, radio contour maps of compact jet in, 298–299
3C 295, redshift of emission line of, 213, 215
3C 388, features of, 349–352
3C 405, as strong emission line galaxy, 213
3C 427, high redshift for, 214
30 Doradus (also Tarantula nebula), 97–99
0014.4+1551, extreme blueness of, 216
0957+561, as gravitational lens, 261
1305.4+2941, redshift of, 216
1305+2952, redshift of, 214–215
1539+343, radio contour map of, 291
2316–424, radio map of, 351–352

Accretion disk model of quasars, 258–259
Accretion disks in binary systems, 102–103, 148–150
Acoustic waves in the photosphere, 114, 116–118
Active galactic nuclei, 256, 257
Advanced Solar Observatory, 366
Advanced X-Ray Astronomy Facility, 362–364, 366
Alfvén waves, 118–119
Allende meteorite
  Ca and Ti isotopic anomalies in, 322–323
  in situ decay of aluminum-26 in, evidence for, 322–323
Alpha Centauri AB, 143
Alpha Centauri ABC, 143
Aluminum-26
  cosmic ray bombardment, production by, 47–48, 54–55
  decay of, 319–322
  synthesis of, 314, 322
AM Canum Venaticorum, 148
Ammonia emission in Orion Molecular Cloud, 135
Andromeda Nebula
  antimatter hypothesis of, 271
  binary relation to the Milky Way, 250
  brightest stars, lack of surveys for, 94
  brightness of, 211
  Hubble-Sandage variables in, 97
  line spectrum of, 241
  nucleosynthesis in, 303
  rotation and distance of, 242, 251–252
Antarctic ice sheet, meteorites in, 50, 56
Antimatter. *See* Matter-antimatter asymmetry
Aperture synthesis, 346–347
Apollo 17 orange glass, 48
^{39}Ar/^{38}Ar ratio in meteorites, 54
Aroos iron meteorite, 53
Asteroid belt origin of meteorites, 48
Asteroids, radio observations of, 356–358
Astro orbiting observatory, 72–73
Atmosphere, terrestrial, 7, 23n
  fair-weather electric field in, 22
  properties of, 10, 19–22
  solar radiation, interaction with, 9–10
  solar wind, interaction with, 12–13, 17
  tides in, 20
  transparency of, 345
Auroral heating, 19–20
Auroral precipitation, 22
Auroras, 7, 18–19
Axions, 285

Background (2.7 K) radiation
  big bang cosmology, evidence for, 123, 340
  early Universe, probe of, 263
Baryon number
  conservation of, 263–264
  nonconservation of, 272–273
Becklin-Neugebauer (BN) star, 135
Becklin-Neugebauer-Kleinmann-Low (BNKL) object, 136
Beryllium-10, 47, 54–56
Big bang
  and causal structure, 267–268
  evidence for, 123
  neutrinos produced in, 284
  nucleosynthesis in, 131–132, 301, 303, 305
  standard models of, 124–125
Binary galaxies, 82, 250
Binary stars
  cataclysmic, 146, 148–149
  components of, separation between, 143
  enhanced line emission in, 116

formation and evolution of, 138, 143–146, 150–153, 164–169
mass measurements for, 144–147
soft and hard, 165, 168
x-ray emission from, 118
BL Lacerta objects (BL Lacs), 256
Black holes
candidates for, 103, 147–148, 227
evolution of, 151
formation of, 200
galactic nuclei, in model of, 149
in globular cluster dynamics, 164, 168
mathematical theory of, 204–205
nonluminous matter, as form of, 250
production of, threshold for, 285
quasars, in model of, 258–259
Blue compact galaxies, chemical abundances in, 128
Bow shock, 19
Butterfly diagram, 173–174, 185

Calcium *H* and *K* lines
in 1899 spectrum of the Andromeda Nebula, 244
flux records of, 178–186
in stellar spectra, 176–181
in the sun's spectrum, 115–116, 172, 176, 179
Carbon burning in stars, 311
Carbon-14
nuclear weapons, produced by, 55
solar activity, related to, 46–47, 52, 55, 175
Carbon monoxide
comets, observed in, 64, 71–72
emission along the inner galactic plane, Color Plate IV
emission from the interstellar medium, 133–134
emission from interstellar molecular clouds, Color Plate V
emission from Orion Molecular Cloud, Color Plate VI
Milky Way, observed in, 79
outflows in stellar evolution, 136
Carina arm of the Milky Way, 85–86
Carina nebula (also NGC 3372), very massive stars in, 96–97, 102
Cassiopeia A, evidence for nucleosynthesis in, 323
Cataclysmic binaries
accretion disks in, 148–149
model of formation of, 152–153
schematic picture of, 146
Catastrophe models, 156–160
Cepheid variables, discoveries of, 242
Ceres, radio tracking of, 358
Chandrasekhar limit of mass, 146
Chromosphere, solar
emission lines from, 106, 115–116, 176–180
energy balance in, 109–116
energy emanating from, 10, 12
flash spectrum of, 105
magnetic cycle, related to, 171
outer convection zone, association with, 176
temperature stratification in, 8, 106–108
See also Chromospheres, stellar
Chromospheres, stellar
classes of stars having, 171–173
in cool stars, 115–116
emission lines from, 115–116, 118, 176–180
HK fluxes from, 181–182
outer convection zones, association with, 176
rotation rate and age, activity related to, 184–185
search for, 172–173
See also Chromosphere, solar
^{36}Cl/^{36}Ar ratio in meteorites, 54
Climate
correlations with solar indices and ^{14}C activity, 46–47
possible correlation with solar constant fluctuations, 9
Clusters of galaxies
ages of, 280–282
brightest (first-ranked), 210–212
hot gas in, distribution of, 250
Oemler types of, 224–225
origin of, 275–288
CN cycle, 304, 322
CNO cycle, 97, 104n, 304, 308, 310
Color-magnitude diagram. *See* Hertzsprung-Russell diagram
Coma cluster
disk sizes in, 229, 238
galaxies in, distribution of, 228
star formation in, 232
Comet Austin, 67, 69
Comet Bennett, 60, 71
Comet Borrelley, 67
Comet Bowell, 67
Comet Bradfield, 64–65, 69–72
Comet Encke, 62, 69
Comet Grigg-Skjellerup, 67
Comet Kohoutek, 60–61, 65, 68
Comet Meier, 67
Comet Mrkos, 71
Comet Panther, 67
Comet Seargent, 67
Comet Stephan-Oterma, 67, 71

Comet Tago-Sato-Kosaka, 60, 71
Comet Tuttle, 67
Comet West, 60, 65–66, 70–72
Comets
  characteristics of, 59
  far-ultraviolet observations of, 64–72
  models for, 62–64
Convection zone
  solar, 8
  in stars, 176
Corona
  emission lines from, 176–180
  energy balance in, 109–111, 113
  energy emanating from, 10, 12
  heating of, 116–119
  magnetic field lines in, 14
  spectra from, types of, 105, 107
  temperature stratification of, 8, 13, 105–106, 109
  x-ray pictures of, 106–107, 117
Coronal holes, 13–15, 106, 117
Cosmic Background Explorer satellite, 273, 362, 366, Color Plate IX
Cosmic rays
  history of, 50–55
  matter, interaction with, 44–46
  solar events, variation with, 13, 22, 42–43, 45, 55, 175
  See also Galactic cosmic rays; Solar cosmic rays
Cosmic spherules in deep-sea sediments, 50, 566
Cosmogenic nuclides, 44–47, 50–53
Cosmological constant, 266–267, 270, 273
Cosmological parameters, 261
Current sheet, 15–16
Cygnus arm of the Milky Way, 85–86
Cygnus X-1, candidate black hole in, 147–148

Dark matter. *See* Nonluminous matter
Dark night-sky riddle, 332–340
de Vaucouleur's law, 220–221
Deep Springs iron meteorite, 49
Density fluctuations in the Universe, 283–285
Density wave theory, 78
Density waves, galactic, 135, 237
Detectors, spectroscopic, 213–214
Dhajala chondrite, 53
Differential rotation
  in massive stars, 101–102
  of the sun, 8–9, 173–176
Disk, galactic
  single to multiple star systems in, ratio of, 143
  surface brightness in, variation of, 245
Disk galaxies
  chemical composition, 233
  evolution of, factors affecting, 235
  evolution of, models for, 236–239
  frequency distribution of, 229
  gas content and star formation, 230–233
  infrared properties of, 233–234
  as laboratories for understanding star formation, 234
  mass/luminosity ratio of, 234–235
  morphology of, 220–224
  proportion in the field, 281
  size and shape, 229–230
  See also Irregular galaxies; Spiral galaxies; S0 galaxies
Dwarf galaxies, 82, 250
Dwarf novae, accretion disks in, 149–150
Dwarf stars
  chromospheres in, 171
  HK flux records of, 185
  rotation rates of, 180
Dynamo theory, 180–182

Eclipse, total solar, 105–106, 108
Eddington limit, 93, 101–102
Eddington's paradox, 192–193
Einstein Observatory
  clusters, observation of, 166, 212
  data on gamma rays from the galactic equatorial plane, 320
  quasars, surveys of, 256
  stars, observation of, 117
  supernova remnant, observation of, 323–324
Electron degeneracy and stellar evolution, 310–311
Elemental abundances
  in binary stars, 144–145
  estimates, theoretical, 127–129
  in HII regions, 120–123
  of hydrogen, helium, and lithium, 263
  pregalactic, 122–124
  for the sun and similar stars, 302–303
Elements
  formation of, 131–132, 301–328
  primary and secondary, 127–129
  See also Nucleosynthesis
Elliptical galaxies
  clusters, concentration in, 281
  evolution of, 228–229
  frequency distribution of, 224–225
  properties of, 220–221, 225–228
  as radio sources, 293, 297
  star formation in, 135, 226–227

Energy, sun's
 and the earth, 10
 source of, 7, 175, 301, 303
 transport of, 8
Eta Carinae, 93, Color Plate III
 on Hertzsprung-Russell diagram, 95, 103
 mass ejection from, 100
 observation of, history of, 96
 properties of, 96, 99, 101, 102
Event horizon, 204, 338
Expansion of the Universe
 big bang cosmology, evidence for, 123
 deceleration of, 210
 dynamical equations for, 264–267
 models of, constraints on, 304
 nature of, 250
 rate of, 267
 redshifts ascribed to, 209–213
 See also Big bang; Redshifts; Universe
Extreme Ultraviolet Explorer, 362
Extreme ultraviolet (EUV) radiation
 absorption in the earth's atmosphere, 10
 flux of, variability of, 12
 heating caused by, 20
 solar processes, formed in, 8

Faculae, 9
Farmington chondrite, 49
Ferromagnetism, model of, 270
Field galaxies, formation of, 281
Filaments
 in Cassiopeia A, 323
 interpretation of, 276–277, 282
 in maps of bright galaxies, 275–280
Flares, solar
 chronology of, 27–30
 cosmic rays from, 41–43
 earth's magnetic field, effect on, 23n
 energy source for, 29
 forecasting of, 26–27, 30–31
 giant, evidence for, 52
 loops in, 34–35
 magnetic fields in, 32
 mechanism of, 9
 ozone abundance, effect on, 22
 proton energies in, 25
 risks to astronauts from, 25–27
 streams produced by, 15
 x-ray flux during, 12
Flares, terrestrial analogs of solar, 19
Flatness problem, 269–270, 273
Formaldehyde, interstellar, 79
Fraunhofer corona, 105–106, 115–116
Fraunhofer lines. *See* Calcium *H* and *K* lines
FU Orionis stars, accretion disks in, 149–150

G-type stars, change in luminosity in, 9
Galactic cosmic rays
 carbon-14 activity, correlation with, 46–47
 matter, interactions with, 44–46, 52
 properties of, 42–44
 solar effects on, 13, 22
 studied through exposure histories of targets, 53–55
 See also Cosmic rays
Galactic nuclei
 binary star systems, compared with, 148
 jets ejected from, 150
 model for, 149
Galactic shock model of star formation, 236–237
Galaxies
 ages of, 280–282
 angular distribution of the brightest million, 277–278
 chemical evolution of, 120, 126–129
 distribution of, Color Plate VIII
 distributions of, real and model, 277, 279
 era of formation of, 132, 219
 evolution of, 209–211, 213, 219–239
 formation of, models of, 250
 integrated colors of, 230–231
 masses of, determination of, 242
 morphology of, 219–224
 origin of, 275–288
 as radio sources, 212
 surface brightness of, 211
 very distant, 209–217
 See also Disk galaxies; Elliptical galaxies
Galaxy. *See* Milky Way
Galileo mission, 360, 368
Gamma Ray Observatory, 362, 364
Gaseous nebulae. *See* Planetary nebulae
Gauge (local) symmetries, 264
General relativity
 black holes, prediction of, 147
 expansion of the Universe, in treatment of, 265
 instability for rotating configurations, prediction of, 203–204
Giant molecular clouds
 HII regions, associated with, 83, 86
 in the interstellar medium, 132–135
 in the Milky Way, 79–80, 86
 around newly formed stars, 233
 properties of, 133–135
 protostellar collapse in, 137
 the Rosette Nebula, associated with, 80
 See also Orion Molecular Cloud
Giants and supergiants
 chromospheres in, 171
 circumstellar dust grains of, 132

detection of coronas in, lack of, 117
evolution of, 151
helium fusion in, 305
on the Hertzsprung-Russell diagram, 95, 115
See also Stars, very luminous
Giotto spacecraft, 63
Glaciations, 9, 23n
Global symmetries, 263–264, 272–273
Globular clusters
binaries in, 164–167
around elliptical galaxies, 227
iron abundance in, 123
mass distribution in, 250
around the Milky Way, 77, 79, 83–84, 242
models for, 154–164, 286–287
properties of, 169
star formation in, 135
stellar motions in, 155
Grand unified theories, 270, 272
Granulation, solar, 8, 114
Gravitational collapse
in giant molecular clouds, 134–135
onset of, 199–200
in protostellar collapse, 138
Gravitational lenses
major uses for, 260–262
quasars viewed through, 254, 257
Gravitational radiation
evidence for, 148
general relativity, predicted by, 203–204
Gravitinos, 250, 285
Gravity waves, 20
Gravothermal instability, 158, 162–164, 166, 168

Halley's comet, 59–60, 63–64, 69–73
Halos
chemical abundances in, 123
of elliptical galaxies, 227–228
evidence for, 82, 248–251
globular clusters in, 160–162, 164
of the Milky Way, 88–90
properties of, 250
single stars in, predominance of, 143
Hard x-ray imaging spectrometer (HXIS), 34–35
HD 1835, possible sunlike activity cycle in, 184
HD 18256, possible activity cycle in, 184
HD 25998, possible sunlike activity cycle in, 184
HD 29645, HK flux records of, 178–179
HD 32147, sunlike activity cycle in, 183
HD 38268. *See* Radcliffe 136a
HD 76151, activity variations in, 184
HD 81809, HK flux record of, 172, 179–180
HD 82885, activity variations in, 184
HD 93129A, luminosity of, 96–97
HD 149661, activity variations in, 184–185
HD 152391, sunlike activity cycle in, 183–184
HD 156026, possible activity cycle in, 183
HD 160346, activity cycle in, 183
HD 161239, activity cycle in, 183
HD 190007, activity cycle in, 183
HD 190406, activity variations in, 184
HD 201092, activity cycle in, 183
HD 207978, HK flux records of, 178–179
He^{++} shell, 15
Head-tail radio sources, 351–352
Heavy-element abundance
galaxy age, related to, 209
in intergalactic clouds, 259–260
in interstellar medium and stars, 301–303, 317–319
pre-supernova, 311, 316–317
Helium
burning in stars, 308–311
content in stars, 144–146
in HII regions, 120, 123
in nebulae, 122, 124
origin in the big bang, 131, 301, 305
pregalactic abundance of, 123–124
Hercules cluster, size-luminosity relation for galaxies in, 229
Hertzsprung-Russell (H-R) diagram
for main sequence, giant, and supergiant branches, 115
for the most luminous stars, 94–95
stellar properties related to, 144
High Energy Astrophysical Observatory (HEAO 3). *See* Einstein Observatory
HK photometer, 176–177, 185
Homunculus, 96
Hopkins Ultraviolet Telescope (HUT), 73
Horizon problem, 269–270, 273
HII regions, 83, 120
chemical abundances in, 120–125, 128
giant molecular clouds, associated with, 83, 86, 133
the Rosette Nebula, associated with, 80
in stellar evolution, 136
Hubble constant, 262n, 266
Hubble sequence of galaxies
categories in, 220–221
as a luminosity sequence, 230–231
mass and luminosity related to, 249
rotational properties related to, 243–245, 249
Hubble-Sandage variables, 95, 97, 100, 103
Hubble sphere, 338
Hydra cluster, redshift of, 213

Hydrogen
burning, 305, 308
envelope around comets, 60–61
mass/luminosity ratio, 231–232, 234
origin in the big bang, 131, 301
Hydrogen, atomic
deviations from the galactic equator, Color Plate II
in disk galaxies, 231
in edge-on galaxies, 87
in elliptical galaxies, 226
emission in the interstellar medium, 132–133
extension beyond optical galaxy, 250
in the Milky Way, 77–79, 81, 84–88
in S0 galaxies, 232
surface density of, Color Plate I
21-cm emission from, 243
See also HII regions
Hydrogen, molecular, observation in the Milky Way, 79

IC 724, mass distribution in, 248
IC 1613, Hertzsprung-Russell diagram for, 94–95
Infall rate in galaxies, 127, 129
Inflationary universe, 264, 269–271
Infrared Astronomical Satellite (IRAS), observations of Vega from, 139
Instantaneous recycling approximation, 126
Intergalactic gas clouds
populations of, 259–260
quasars, observation of, 254
See also Interstellar medium
International Magnetosphere Study, 15
International Solar Polar mission, 361, 368
International Solar Terrestrial Physics program, 362
International Sun-Earth Explorer, 33, 37, 361
International Ultraviolet Explorer (IUE) satellite
chromospheres, observations of, 115
comets, observations of, 64–65, 67–72
elemental abundances from, 122
Eta Carinae, data on, 96–97
spectra of Hubble-Sandage variables from, 97
Interplanetary magnetic field, 17–18, 22, 42
Interstellar medium
continuing star formation in, 132
dust and gas in, 78–79
elemental abundances in, 120, 122–123, 125–126, 301
molecular clouds in, 132–135
supernovae, enrichment by, 324
See also Intergalactic gas clouds
Interstellar molecules, 61, 72, 132–134

Ionization fronts, 135
Ionosphere, 7, 12, 17
Irradiance, solar, 10–12
Irregular galaxies
chemical abundance ratios in, 128
gas content and star formation in, 230–232
Large Magellanic Cloud as an example of, 224
morphology of, 220
star formation rate in, 129

Jet stream, 21
Jets
in binary systems, 150
compact radio jets, 297–300
large-scale slow jets, consequences of, 295–297
models for, 292–295
quasars, associated with, 353
radio galaxies, observed in, 291–292
of SS433, 353–354
Jupiter, radio image of, 355–357
Jupiter-like planets, 250

$^{40}K/^{41}K$ ratio in meteorites, 49, 54
King models, 164
$^{81}Kr/^{83}Kr$ ratio in meteorites, 49, 55

Large Magellanic Cloud
candidate black hole in, 147
chemical abundances in, 122
irregular star-forming pattern in, 224
See also Magellanic Clouds
Ledoux-Schwarzschild-Härm limit, 93
Little ice age, 21
LMC X-3, candidate black hole in, 147–148
Local Group, 280–282
Local Supercluster, 281–282
Loop structures in the corona, 9, 33–35, 106–107, 117–118
Luminosity
of disk galaxies, 229–233
and elliptical galaxy size, 225
increase with time, 144
of quasars, 257–258
of radio sources, 212
and star-forming ability, 237
of the sun, 9
upper envelope for normal stars, 94–95, 100, 102
See also Mass/luminosity ratio
Lunar samples, records of cosmic-ray nuclei in, 41, 44–46, 48, 51–53, 55–56

M3, distribution of brightest stars in, 276
M8 (also Lagoon nebula), photograph of, 121
M19 (also NGC 6273)
  photograph of, 155
  properties of, 154
M33
  chemical abundances in, 128
  Hertzsprung-Russell diagram for, 94–95
  Hubble-Sandage variables in, 97
  rotational velocities in, 251
  Wolf-Rayet stars in, 95, 99
M83, chemical abundance ratios in, 128
M87
  elliptical galaxy, example of, 221
  globular clusters surrounding, 227
M101, chemical abundance ratios in, 128
Magellanic Clouds
  luminosity of massive stars in, 93–95, 97–99
  Magellanic stream from, 82
  oldest stars in, 231
  See also Large Magellanic Cloud; Small Magellanic Cloud
Magic neutron numbers, 325
Magnetic field, solar, 15, 171–176
Magnetic field, terrestrial
  carbon-14, variation correlated with, 46–47
  disturbance in, types of, 23n
  solar wind, interaction with, 13, 15–17
Magnetic fields, stellar
  decrease with age in stars, 116
  evidence for, 173, 176–180, 186
  sources of, 115–118
Magnetic storms, 22, 29
Magnetohydrodynamic waves. *See* Alfvén waves
Magnetopause, 17, 19
Magnetosphere, terrestrial, 7, 15–19
Main sequence stars, 95
Malakal chondrite, 53
Manganese-53, production by cosmic-ray bombardment, 47–48, 54–55
Manganese-rich nodules, 47
Mars Geochemistry and Climatology Orbiter, 361
Maser emission
  from expanding atmosphere around a star, 355
  from molecular cloud cores, 136
Mass/luminosity ratio
  in disk galaxies, 234–235
  in elliptical galaxies, 227–228
  Hubble type, indicator of, 245
  nonluminous mass from, evidence of, 248–250
  in spiral galaxies, 129, 245
  stellar population, measure of, 249
  in very massive stars, 93
  See also Luminosity
Matter-antimatter asymmetry, 271–273
Maunder minimum, 21, 174-175
  galactic cosmic-ray spectrum during, 42–43, 53–54, 56
  stars, possible occurrence in, 183–184
Messier 8. *See* M8
Messier 19. *See* M19
Messier 31. *See* Andromeda Nebula
Messier 87. *See* M87
Metallicity
  in disk galaxies, 230, 233–234
  in elliptical galaxies, 226–230
  and gas fraction in galactic evolution, 126, 129
  low-mass stars, and relative number of, 129
Meteorites
  cosmic-ray bombardment in, records of, 41, 44–46, 48–50, 56
  cosmic-ray exposure ages of, 49
  cosmogenic nuclides in, activities of, 53
  gas-rich, 52
  isotopic anomalies in, 138, 319–323
  with known orbits, 53
MgII *h* and *k* lines, 115–116
Milky Way
  ages of oldest stars in, 280
  chemical abundance ratios in, 128
  Hertzsprung-Russell diagram for, 94–95
  lifetime of, 301
  luminosities of massive stars in, 93
  rotation of, 80–84
  spiral structure of, 85–86
  stellar population in spiral arms of, 94–95
  undetected matter in, 82
"Mini" black holes, 199
Missing mass, 304
Monopoles, 250, 270
Moon, effects of cosmic rays on, 47–48
Multi Element Radio Link Interferometer (MERLIN), 298–299

N galaxies
  quasars, in classification of, 256
  radio identification of, 212
N97, chemical abundance ratios and predicted mass of, 124–125
$^{22}Ne/^{21}Ne$ ratios in meteorites, 49
Nebulae, 262n
Neutral sheet, 32
Neutrinos
  in big bang models, 124–125

massive galactic halo, possible constituents of, 90
nonluminous matter, form of, 250, 284
Neutron capture processes
astrophysical circumstances and sites of, 310
in nucleosynthesis, 324–328
products of, 321, 324
two processes, need for, 305–306
Neutron electric dipole moment, 273
measuring, apparatus for, Color Plate IX
Neutron stars
discovery of, 146
evolution of, 151
formation of, 200, 315
masses of, 146–147
as x-ray sources in globular clusters, 166–168
NGC 205, star formation in, 226
NGC 801
mass distribution in, 248
rotation curve for, 247
NGC 891, dark matter in, 89
NGC 1265, head-tail radio source associated with, 294
NGC 1365, chemical abundance ratios in, 128
NGC 1421
mass distribution in, 248
rotation curve for, 246–247
NGC 2257, star chains in photograph of, 277–278
NGC 2403, rotational velocity of, 252
NGC 2742, rotation curve for, 246–247
NGC 2998, rotation curve for, 246–247
NGC 3115, structure and dynamics of, 251
NGC 3372, *See* Carina nebula
NGC 3603, analogy to Radcliffe 136a, 98
NGC 4406, 221
NGC 4486, radio emission structure of, 291
NGC 4594 (also Sombrero galaxy), spectrum of, 242, 251
NGC 4698, mass distribution in, 248
NGC 5128, jetlike structure in radio and x-ray, 292
NGC 5471, chemical abundance ratios in, 124
NGC 6302
chemical abundance ratios in, 124–125
mass of, predicted, 125
NGC 6822, Hertzsprung-Russell diagram for, 94–95
Nimbus 7
in current space research program, 361
total solar irradiance by, measurements of, 10–11
Nonluminous matter (or dark matter)
in disk galaxies, 235
evidence for, 248–251
expansion of the Universe, relation to, 250
as low-mass stars in halos, 129
measure of, 282
in the Milky Way, 82, 88–90
models for, 261
in spiral galaxies, 82, 88
studied with gravitational lenses, 260–261
Novae, 301
suggested production of aluminum-26 in, 322
Nuclear astrophysics, 303, 306
Nuclear cross sections, stellar reaction rates from, 307–308, 310, 313–314
Nucleocosmochronology, 326–328
Nucleosynthesis
ongoing, evidence for, 319–323
primordial, 131–132
in protostellar collapse, 139
reaction network for, 310–315
in stars, 301–315
in supernovae, 316–318, 323–324
See also Elements

O-type stars, 95, 103n
O3 stars, 93, 96
OAO-2 observatory, 60, 71
OB complexes, 232
OB stars, evolution of, 136
Ocean cores, 23n
Ocean Topography experiment, 361
Oemler clusters (also Butcher-Oemler clusters), 224–225, 238–239
OGO-5 observatory, 60
OH emission
in comets, 61–62, 67–69
radio observation of, 355–356
Olber's paradox, 332, 334–335, 340
Orion arm of the Milky Way, 85–86
Orion Molecular Cloud, 133–136
CO emission from, Color Plate VI
Orion Nebula, 122–123, 125, 133–134
Oxygen, burning in stars, 311–312
Ozone, in the earth's atmosphere, 10, 22

P Cygni, 97, 99, 100
Pancake theory of galaxy formation, 275–276, 279, 281–282, 284, 287
Particle horizon, 338
Perseus arm of the Milky Way, 85–86
PG 1159-035, nonradial pulsations of, 309
Photinos, 285
Photosphere, solar, 8–9, 106, 109, 114
Plages, 12

Planetary nebulae, 120–125, 153
Planetary systems, 138–139
Plasma sheet, 18–19
Population I, II, and III stars, 303, 318
Post-Newtonian approximation, 147–148
Potassium-40, record in iron meteorites, 54, 56
Procyon B, 146
Proton decay
  experiments to detect, 272–273
  measuring, apparatus for, Color Plate IX
Proton showers
  frequency of, 30
  observations of, 33, 36–38
  risks to astronauts from, 25–27
Protons, solar, average fluxes of, 42–43, 51
Protostar, 137
Protostellar collapse, 131–132, 135, 137–139
Proxima Centauri (also Alpha Centauri C), 143
PSR 1913 + 16, masses of components of, 146–148
Pulsars
  masses of, 200
  as neutron stars, 203
  radio and x-ray, 146–147
  rotational periods of, 203

QSS, luminosity function of, 258
Quasars
  classification of, 256
  empirical properties of, 255–256
  epoch of formation of, 257–258
  jets from, 150
  luminosity of, 211
  model of, 149, 258–259
  population statistics of, 256–258
  as radio sources, 212, 290, 292, 295–298, 353
  redshifts of, 211–212, 254, 256–258
Quasi-stellar objects (QSO's). *See* Quasars

r-process. *See* Neutron capture processes
Radcliffe 136a (R136a)
  black hole, possible existence of, 103
  on Hertzsprung-Russell diagram for most luminous stars, 95
  properties of, 97–99
Radiation pressure
  circumstellar dust, effect on, 132
  in massive stars, 92–93, 100
  in stellar evolution, 136, 189–192, 198
Radio sources
  compact and extended, 290
  galaxies as, 212, 256, 293–294
  head-tail, 293–294
  mapping of, 349–353
  models for, 290–297
  quasars as, 255–256, 353
  solar flare sites as, 356–357
  stars as, 353–355
Radionuclide-stable product pairs, 48–49, 54
Redshifts
  colors, derived from, 212–213
  distance, relation to, 210
  of galaxies, 214, 277, 287
  highest values of, 214
  of intergalactic clouds, 259–260
  of quasars, 211–212, 254, 256–258
  of radio galaxies, 212
  surface brightness on, dependence, 211
Relativistic instabilities, 200–204
Richardson meteorite, 319
Roche lobe, 146, 151–152
Rosette Nebula, 180
Rossby relation, 180–182, 185–186
Rotation curves
  of disk galaxies, 234–235
  of the Milky Way, 81–84
  of the solar system, 80
  of spiral galaxies, 82, 244–252

S Doradus, 97
s-process. *See* Neutron capture processes
S0 galaxies
  bulge-to-disk ratios of, 238
  colors of, 232
  frequency distribution of, 224–225
  gas content and star formation in, 232–233
  hydrogen-poor, 232
  morphology of, 220, 222–223
  size-luminosity relation for, 229
Sagittarius, 242
Salyut space stations, 368
Satellite orbits, most frequently used, 30
Scalloping, 87–88
Schwarzschild radius, 201, 268
Scorpio X-1, 150
Sculptor group, 280
Sectors, solar magnetic, 15, 21
Selectrons, 285
Serpens Virgo cloud, 284
Seyfert galaxies, quasar-like objects in, 256
Sharpless 106
  false-color image of, Color Plate VII
  stage of star formation of, 137
Shock transition, 13
Shortwave blackout, 12
σ^2 Eridani B, 146

Silicon burning, 312–313, 316
Sirius B, 146
Sky radiation, 213
Skylab, 32–34, 60
  observations of comet Kohoutek from, 60
Small Magellanic Cloud, 122
  See also Magellanic Clouds
SN1972e, observation of cobalt in, 324
SN1981b, observation of cobalt in, 324
Solar abundances, 303, 317–318
Solar constant, 9–10
Solar cosmic rays, 41
  exposure histories of targets, studied through, 50–53, 56
  matter, interactions with, 44–46
  the moon, effects observed on, 47–48
  properties of, 42–44
  See also Cosmic rays
Solar Maximum Mission satellite, 10–11, 33–34, 36, 363–364
Solar Maximum Year, 25, 33–36
Solar Mesosphere Explorer, 361
Solar neutrino problem, 175, 308
Solar system, 132, 138–139
Solar wind
  Alfvén waves in, 118
  coronal energy loss due to, 113
  cosmic-ray fluxes, influence on, 175
  energy, input of to the atmosphere by, 12–13, 17
  implanted ions from, 44
  processes leading to, 113
  properties of, 13
  streams in, 13–15
  sun's angular momentum, influence on, 185
Solar-terrestrial research, 7, 22
South Ray crater at Apollo 16 site, 48
Space Infrared Telescope Facility, 362, 364, 367
Space radiation risks, 25–27
Space science
  current U.S. program, 360–363
  funding, 362
Space shuttle, 360, 362–363, 367–368
Space station, 360, 361, 363–369
Space Telescope
  model of, Color Plate IX
  properties of, 73, 366
  uses of, 98, 217, 232, 362–364
Spacelab, 362–363
Spectral indices (also color indices, colors), 212–213
Spherical stars, instability of, 200–203
Spiral galaxies
  bright knots in, 223
  chemical abundances in, 125–126
  disk light in, 81
  examples of, 222–223
  frequency distribution of, 224–225
  giant molecular clouds in, location of, 134
  in the Hubble sequence, 220–221
  infall rate in, 129
  mass distribution in, 244–248
  protogalaxies, formation from, 132
  rotation of, 242–252
  star formation in, 230–232, 236–237
  undetected matter in, 82
Spiral structure of the Milky Way, 78
SS433
  as a binary system, 150
  standard model for quasars, comparison to, 258–259
  the Very Large Array, observed with, 353–355
St. Mesmin chondrite, 50
Star chains, 277–278
Stars
  evolution of, 150–151, 188, 194, 197–200
  formation of, 79, 86, 131–139, 220
  number and lifetime of, 128, 339
  ratios of single to multiple, 143
  rotation of, 116
  very luminous, 94–95, 99
  very massive, 92, 100–103
  See also Binary stars
Starspots, 186
Stellar activity variations, 183–186
Stellar structure theory, 9
Stellar winds
  in cool luminous stars, 117
  deceleration of stars, related to, 185
  in evolving young stars, 136
  the Very Large Array, mapped with, 353–354
  in very massive stars, 100
Streamers, coronal, 117–118
Substorms, 19–20
Suess wiggles, 47, 55
Sun
  differential rotation of, 173–176
  elemental abundances of, 123
  revolution of the galaxy by, 248–250
  rotation period of, 12, 22
Sunspot cycle
  CaII *K* emission related to, 116
  carbon-14 activity, relation to, 46–47
  coronal holes during, 14
  cosmic-ray flux, relation to, 42–43, 45, 53
  magnetic fields, relation to, 115–116, 171
  mechanism of, 8–9

solar proton fluxes during, 51
thermosphere, effect on, 12
weather, possible correlation with, 21–22
Sunspots, 8–10
butterfly diagram of, 173–174, 185
enhanced x-ray emission from, 117
fluence curve for, 43
magnetic polarity of, 171
mean annual number of, 173–174
Supernova shocks, 135
Supernovae
cataclysmic binaries, possible evolution from, 148
cosmic rays, as sources of, 42, 54
events leading to, 199–200
frequency in the Galaxy, 317
interstellar medium, enrichment of, 120, 125, 129, 132, 139
nucleosynthesis during, 316–318, 323–324
type I, 200, 316, 324
type II, 200, 315–318
SV Centauri, 143
Synchrotron radiation, 290–291

Tamarugal iron meteorite, 50
Tarantula nebula. *See* 30 Doradus
Telescopes, resolution of, 345–346
Thermoluminescence, 44
Thermonuclear reactions
in stars, 144, 307–308
in the sun, 7, 8, 10, 175
Thermosphere, 12, 17
Tidal capture, 167–168
Tidal forces, 210
Transition region, solar, 108–109, 112–113
Trapezium, 136
Tree ring dates, 47, 54–55
Troposphere, 7, 22n
Tycho Brahe's supernova remnant, 323

UGC 2885, rotation curve for, 247
Ultraviolet Imaging Telescope (UIT), 73
Universe
age from nucleocosmochronology, 326–327
expanding, 338–339
infinite, 332–333, 335–337
models of, 264–269
See also Expansion of the Universe
Upper Atmosphere Research Satellite, 361

V1016 Cygni, radio map of stellar winds of, 353–354
Vacuum energy, 266–268, 273
Van Allen radiation particles, 19, 25
Var 83, physical properties of, 99
Var A, variability of, 97
Vega, 139
Venus Radar Mapper, 360–361
Very Large Array, 35, 345–358
Virgo cluster, 229, 281
Voyager II, 360

W Ursae Majoris, 143
Warp of galactic planes, 86–88
Weak interactions, 315–316
Weather, 20–22
White dwarfs
electron degeneracy in, 192–193
energy paradox in, 192
evolution of, 151
in globular clusters, 167–168
with known masses, 146
mass loss from, 120, 125
theory of, 193–197, 199
Wisconsin Ultraviolet Photo-Polarimeter Experiment (WUPPE), 73
Wolf-Rayet stars
characteristics of, 100–102
on the Hertzsprung-Russell diagram, 95
possible identification in M33, 99

X-ray sources
brightest, 150
bursters, 147
discovery of, 146–147
in globular clusters, 154, 166, 168
ratio of Hubble types in, 238
star formation in, 229

Zeta Phoenicis (also HD 6882 or HR 338), properties of, 144–146